Nanoantennas and Plasmonics

The ACES Series on Computational and Numerical Modelling in Electrical Engineering

Andrew F. Peterson, PhD – Series Editor

The volumes in this series encompass the development and application of numerical techniques to electrical and electronic systems, including the modelling of electromagnetic phenomena over all frequency ranges and closely related techniques for acoustic and optical analysis. The scope includes the use of computation for engineering design and optimization, as well as the application of commercial modelling tools to practical problems. The series will include titles for senior undergraduate and postgraduate education, research monographs for reference, and practitioner guides and handbooks.

Published Books in the ACES Series:

W. Yu, X. Yang, and W. Li, "VALU, AVX and GPU Acceleration Techniques for Parallel FDTD Methods," 2014.

A.Z. Elsherbeni, P. Nayeri, and C.J. Reddy, "Antenna Analysis and Design Using FEKO Electromagnetic Simulation Software," 2014.

A.Z. Elsherbeni and V. Demir, "The Finite-Difference Time-Domain Method in Electromagnetics with MATLAB® Simulations, 2nd Edition," 2015.

M. Bakr, A.Z. Elsherbeni, and V. Demir, "Adjoint Sensitivity Analysis of High Frequency Structures with MATLAB®," 2017.

O. Ergul, "New Trends in Computational Electromagnetics," 2019.

K. F. Warnick, "Numerical Methods for Engineering: An introduction using MATLAB® and computational electromagnetics examples, 2nd Edition", 2020.

Nanoantennas and Plasmonics

Modelling, Design and Fabrication

Edited by
Douglas H. Werner, Sawyer D. Campbell and Lei Kang

The Institution of Engineering and Technology

Published by SciTech Publishing, an imprint of The Institution of Engineering and Technology, London, United Kingdom

The Institution of Engineering and Technology is registered as a Charity in England & Wales (no. 211014) and Scotland (no. SC038698).

First published 2020

The Institution of Engineering and Technology
Michael Faraday House
Six Hills Way, Stevenage
Herts, SG1 2AY, United Kingdom

www.theiet.org

British Library Cataloguing in Publication Data
A catalogue record for this product is available from the British Library

ISBN 978-1-78561-837-6 (hardback)
ISBN 978-1-78561-838-3 (PDF)

Typeset in India by MPS Limited

Contents

About the editors

Douglas Werner is a professor of electrical engineering and the director of the Computational Electromagnetics and Antennas Research Lab (CEARL), as well as a member of the Materials Research Institute (MRI), at Penn State University. Prof. Werner has received numerous awards and recognitions for his work in the areas of electromagnetics and optics. He is a Fellow of the IEEE, the IET, the OSA, the ACES, and the PIERS Electromagnetics Academy. He is also a senior member of the National Academy of Inventors (NAI) and the International Society for Optics and Photonics (SPIE).

Sawyer Campbell is an assistant research professor in the Computational Electromagnetics and Antennas Research Lab (CEARL) in the Department of Electrical Engineering at Penn State University. He has published over 80 technical papers and proceedings articles and is the author/co-author of three book chapters. He is a member of the IEEE, the APS, the OSA, and SPIE and is the past Chair and current Vice-Chair/Treasurer of the IEEE Central Pennsylvania Section.

Lei Kang is an assistant research professor in the Computational Electromagnetics and Antennas Research Lab (CEARL) in the Department of Electrical Engineering at Penn State University. He has published over 60 peer-reviewed papers and is the co-author of one book chapter. His research interests include metamaterials at both RF and optical frequencies and nanophotonics.

Preface

The subject of this book is the rapidly growing area of nanophotonics which is enabled by devices composed of plasmonic and dielectric optical nanoantennas. Owing to their resonant nature and subwavelength geometries, optical nanoantennas can dramatically enhance light-matter interaction on the nanometer scale, which not only have offered unprecedented flexibility in the control of light, but also have provided a unique opportunity to investigate the optical, electronic, thermal, mechanical, etc. properties of materials in both linear and nonlinear regimes. In addition, based upon their scattering properties, nanoantennas have been arranged in intelligent ways to produce a wide range of optical functionalities, including the tailoring of intrinsic characteristics (e.g., intensity, polarization, and phase) of light waves and the enhancement of nonlinearities such as harmonic generations, extremely large nonlinear coefficients, ultrafast optical processes, etc. Moreover, nanoantenna arrays have been used to demonstrate physical phenomena such as reconfigurability, strong coupling, and non-reciprocity, accomplished through integration with optically active materials. Theoretically, the underlying mechanism of many of these phenomena has not been fully understood and, from an application point of view, the design of nanophotonics devices is expected to be challenging owing to the complicated and strong interaction between light and nanoantennas of complex geometrical and material composition. This book has compiled a comprehensive treatment from a group of leading scholars and researchers in the field on subjects ranging from fundamental theoretical principles and new technological developments, to state-of-the-art device design and optimization and fabrication, as well as cutting-edge examples encompassing a wide range of related sub-areas.

This volume seeks to address questions such as "What fabrication techniques exist to realize nanophotonic devices?", "How does one design and/or optimize plasmonic and dielectric optical nanoantennas?", "How does one take advantage of nanophotonic devices to realize active manipulation of light waves?", and "How to theoretically describe the nonlinear processes in nanophotonics systems and how to achieve optimized nonlinearities?" The theoretical discussions are complemented by a broad range of design examples and measurement results. The applications encompass highly-directive nanoantennas, flat optics, reconfigurable/tunable optical components, field-localizing nanoantennas for enhanced nonlinearities and sensing applications, and non-reciprocal optical devices. This edited volume is intended to serve as a reference to the fast-evolving and exciting research area of nanophotonics and its application to the design of revolutionary and disruptive new devices. We hope that this book will be an indispensable resource for graduate students, researchers, and professionals in the greater optical and electromagnetics communities who have interest in applied nanophotonic device design, optimization, and fabrication.

This book comprises a total of 11 invited chapters contributed from leading experts in the fields of nanophotonics, plasmonics, and metamaterials. All artwork and illustrations are provided in color to supply the reader with important visual aids to understand the complex physical phenomena and devices presented in this book. A brief summary of each chapter is provided as follows. Chapter 1 presents an overview of the analytical theory for circular and elliptical impedance-loaded plasmonic nanoloop antennas. The theory is validated by commercial solvers and is used to aid in the design of broadband super-directive nanoloops. Multi-objective optimization studies show the inherent tradeoffs between electrical size, directivity, and gain for single- and multi-nanoloop configurations. Chapter 2 discusses how to design passive and active multi-layered nano-cylinders to achieve enhanced- and highly-directive scattering behaviors. Both symmetric and asymmetric configurations of these cylindrical nanoantennas are explored in pursuit of controlling both near- and far-field directivity. Chapter 3 discusses the coherent control of light scattering to achieve coherent perfect absorption and coherently enhanced wireless power transfer. Based on a scattering matrix method, theoretical analyses of the various scattering effects in photonics are addressed. Chapter 4 presents a generalized dispersion model (GDM) for time-domain modeling of arbitrary materials which is crucial for nanophotonic meta-device simulation where the underlying materials may possess complex dispersion behaviors. Moreover, most commercial time domain electromagnetics modeling tools have a limited set of time domain material models that they support. To this end, classical descriptions of the electric permittivity such as the Debye, Lorentz, Drude, Sellmeier, and critical point models can all be represented by the GDM. Chapter 5 presents an introduction to applied optimization of plasmonic and dielectric optical nanoantennas. Following a lengthy discussion of local, global, and multi-objective optimization paradigms, key examples are presented, which showcase the power and necessity of optimization algorithms in the nanophotonic meta-device generation process. Both planar (*i.e.*, 2D) and 3D nanoantenna-based meta-devices are optimized to achieve static and reconfigurable beam-steering behaviors. Chapter 6 presents multi-level time domain models for non-linear light-matter interactions. Such models are critical for accurate time domain modeling of phenomena such as saturable- and reverse-saturable absorption and lasing. Multilevel rate equations and Jablonski diagrams are provided to describe each physical phenomenon. Simulation and measurement results are provided, which validate the underlying models. Chapter 7 presents a strategy for the nonlinear interference of electric and magnetic multipoles to achieve non-reciprocal behaviors and realize devices such as a one-way mirror. Chapter 8 discusses the principle of using plasmonic metasurfaces to control the phase, intensity, and polarization states in harmonic generation processes. This control can be exploited to realize devices for nonlinear imaging, holography, and chiral-exploiting applications. Chapter 9 presents an overview of photoconductive nanoantennas as promising sources for enhanced THz emission, in which both a comprehensive comparison between a variety of implementation approaches and development perspectives of the field are provided. Chapter 10 discusses two employed strategies to achieve active photonics metadevices: by hybridizing metallic nanostructures and phase change materials and by varying the arrangement of nanoparticle assemblies. Measurement results of the fabricated devices validate the multifunctional meta-device modeling setup. Chapter 11 provides an overview of the techniques and materials used for state-of-the-art nanofabrication of sub-wavelength scale optics. Plasmonic, silicon, III–V semiconductor, chalcogenide, phase change materials, and 2-D van der Waals materials are discussed along with their associated fabrication implementations and examples of realized devices are presented in the literature.

Chapter 1

Optical properties of plasmonic nanoloop antennas

Jogender Nagar[1], Ryan J. Chaky[1], Arnold F. McKinley[2], Mario F. Pantoja[3] and Douglas H. Werner[1]

Wireless communication systems are expected to play a crucial role in a variety of important technologies including tablets, smartphones, energy-harvesting devices, and medical devices [1]. The need for compact systems with a high data rate means that nanotechnology-enabled devices will likely be the backbone of next-generation wireless communication systems [1]. In particular, these devices require high-performance nanoantennas which are capable of operating anywhere from the optical to terahertz regimes [2]. Metallic devices at these frequencies can no longer be treated as perfect electric conductors (PECs). Instead, they exhibit significant dispersion and loss [3] which results in a drastic impact on the antenna parameters, including directivity and gain [4]. While closed-form analytical expressions are widely available for antennas operating in the RF regime [5], optical antennas are mostly designed using numerical simulation tools [6,7]. Though these tools are very useful for high-fidelity simulations of complex structures, they typically require a large amount of computational resources and time. The availability of exact analytical expressions lead to a deeper intuitive understanding of the underlying physics and also allow for much faster, less computationally intensive design iterations. While there is a large amount of literature devoted to the analysis and design of nanodipole antennas [8,9], there has been less analytical study of the nanoloop antenna despite its potential applications in sensing [10], spectroscopy [11], and light harvesting in solar cells [12]. This is due to the extremely complex form of the integrals which must be solved [13,14].

This book chapter will present a summary of recent work on the theory of nanoloop antennas. First, Section 1.1 presents a theoretical formulation for the antenna parameters of nanoloops with an arbitrary number of impedance loads placed around the periphery. In particular, useful exact analytical expressions will be derived for the current, input impedance, far-zone electric field, radiated power,

[1]Department of Electrical Engineering, Pennsylvania State University, Pennsylvania, USA
[2]Department of Electronic and Electrical Engineering, University College London, London, UK
[3]Department of Electromagnetics and Physics Matter, University of Granada, Granada, Spain

directivity, and gain. These expressions will be specialized for the case of a simple closed loop with no additional impedance loads. Next, Section 1.2 presents analytical expressions for evaluating the coupling between two loops at arbitrary locations. Array factor theory can be employed to compute the resulting far-field properties of arrays of nanoloops. After the theoretical derivations have been presented, a variety of interesting results will be showcased. Section 1.3 presents a surprising discovery, the fact that superdirectivity over a broad bandwidth occurs for a nanoloop comprised of the appropriate material and size. A physical explanation of this phenomenon will be presented, along with comparisons to dielectric nanoantennas. Then, Section 1.4 shows a variety of trade-off curves between electrical size, directivity and gain. The geometries under consideration will include closed and impedance-loaded nanoloops, along with arrays of nanoloops. Finally, Section 1.5 introduces an exciting new research topic, elliptical nanoloop antennas. The analytical theory and some interesting results are presented.

1.1 Analytical theory of impedance-loaded nanoloops

Closed circular loops of wire, intended to act as antennas, were first studied analytically by Hallen [15] in the first half of the last century. These were loops driven from one point in the periphery with a "delta-function" voltage source under a perfectly conducting, thin-wire approximation. After the Second World War, Storer [16] and Wu [17] solved several issues relating to divergences in the Fourier series, allowing them to find the circulating currents in terms of the Fourier modes. As an extra benefit, they managed to characterize quite accurately the input impedance of the loop up to 2.5 times the driven wavelength. Not long afterward, Iizuka [18] characterized loops containing load impedances placed along arbitrary points in the periphery.

Loops as plane wave scatterers were studied by Pocklington himself in 1897 [19]. Harrington returned to those studies in 1966 [20], deriving radar cross sections of closed and loaded circular thin loops using an illuminated plane wave. Rao [21] developed expressions for the far-field based on the Storer/Wu currents which were accurate up to 2.5λ. In 1996, Werner [22] developed an exact integration procedure for the near zone.

The past decade has seen the analytical theory of loops and nano-scaled rings extended from the radio frequency (RF) region to the high microwave, far infra-red, and low optical regimes. These advances are particularly important with the current interest in finding antenna structures for the high GHz and low THz regions of the spectrum. The fabrication processes in the THz region lead to thick loops, for which the thin-wire approximations begin to break down and a new analytical theory needs to be developed. In this section, it will be shown that the thin-wire theory works well for loops of moderate thicknesses. However, of more importance is the theory related to the material characteristics of loops. All previous works had focused on perfectly conducting metals, a characteristic of wires that limits the theory to regions below the 150–200 GHz range. Understanding how loss in the material influences the current, and in turn the radiation pattern, enables the theory to be extended to the THz regions. This occurred in a series of papers from 2012 [23–26].

In 1999, Pendry's key paper [27] on the phenomenon of metamaterials appeared, featuring two, nested and twisted, flat copper rings, each with a gap. Those gaps introduced capacitance to the rings, which, together with the natural inductance, gave rise to a very prominent, high Q, resonance in the region below 0.5λ. That region came to be known as the "sub-wavelength" regime. Many variations on these rings have been used to generate the negative index of refraction phenomena associated with metamaterials, but up until then the use of capacitance in loops to generate the sub-wavelength resonance for communication purposes had only been explored in the experimental world, never in the analytical world. The past few years have seen this lack rectified for the low THz region [28], opening up new avenues in communication at these frequencies and perhaps some new understandings of how metamaterials behave.

The aim of this section is to summarize these new developments and provide the reader with an overview that allows calculational use of the results.* A brief overview of the lossy and dispersive character of materials at high frequencies introduces the section. It uses an expanded Drude model for the region from 150 GHz to the low THz. The literature uses additionally, a critical point, expanded Drude model for the optical region to handle inter-band transitions, which are covered in detail in [28]. The general theory of closed and impedance loaded loops follow, taking into account these material characteristics. The section concludes with the far-zone fields and radiation properties of these loops, supported by an example using a split ring resonator.

1.1.1 Material characteristics

Loop antennas constructed of copper, steel or aluminum for the RF region, as they ordinarily are, may be considered perfectly conducting (PECs: perfect electrical conductors). As the operating frequency is raised into the high GHz, however, electrons in the metal cannot keep up with the rapidly oscillating electric field and hence lag behind. This causes some delay in the phase and, as a consequence, an imaginary part creeps into the complex conductivity. This is particularly true of gold and silver, metals used in the low- and high-THz regions. This delay and associated loss of power needs to be taken into account in the representative equations for the material. The main consequences of these phenomena are an effective lengthening of the current wavelength in the metal and increased absorption.

Since the skin depth in metals does not affect behavior until about 25 THz, an effective method of handling the problem is to assume a surface impedance for the

*Specifics of the historical development of the analytical theory are available elsewhere [29].

metal wire. A number of authors [30–32] have noted that it is dependent only on the index of refraction, η. Assuming a round wire, the Bessel functions of order zero, $\mathbf{J}_0(z)$, and one, $\mathbf{J}_1(z)$, play an essential role in the surface impedance:

$$Z_s = -j\xi_0\left(\frac{\eta}{\eta^2-1}\right)\frac{\mathbf{J}_0(\gamma a)}{\mathbf{J}_1(\gamma a)} \tag{1.1}$$

where $\gamma = \frac{\omega}{c}\eta$, and

$$c = \text{the speed of light and } \xi_0 \text{ is the impedance, both of free space.} \tag{1.2}$$

The index of refraction can be adequately represented using an extended Drude model which ties the high GHz to the DC conductivity in the following way.

$$\eta^2 = \varepsilon_r = 1 - f_0\frac{\omega_p^2}{\omega}\left(\frac{1}{\omega - j2\Gamma_0} + \frac{\alpha}{\omega - j2\beta\Gamma_0}\right) \tag{1.3}$$

where ω_p is the "plasma" frequency and Γ_0 is the damping frequency due to the loss. f_0 is a coefficient used to match the measured data in the high GHz region whereas α and β are used to tie in the DC value of the conductivity. The conductivity is:

$$\sigma(\omega) = -j\omega_p^2\varepsilon_0\left(\frac{1}{\omega - j2\Gamma_0} + \frac{\alpha}{\omega - j2\beta\Gamma_0}\right) \tag{1.4}$$

The DC value of the conductivity is found by taking $\omega \to 0$. Then,

$$\sigma_0 = \frac{\omega_p^2\varepsilon_0}{2\Gamma_0}\left[1 + \frac{\alpha}{\beta}\right] \tag{1.5}$$

The values for all of the coefficients are given in Table 1.1 for the metals gold, silver, and copper

Table 1.1 Coefficients for the extended Drude model useful in representing metals in the GHz and low THz regions.

Material	σ_0 $\times 10^6$	α	β	ω_p (eV)	f_0	Γ_0 (eV)
Au	45	1.540	13.18	9.0	0.37	0.005
Ag	63	0.100	0.433†	9.0	0.94	0.010
Cu	60	0.035	0.005	8.4	1.00	0.064

†The value of Silver's β parameter was changed from that in [23] to rid its associated conductivity of an imaginary part around 4 eV.

Table 1.2 The first two resonances of the closed PEC circular loop. Note that the thicker $\Omega = 10$ loop has a definite fundamental, but not a definite second harmonic. Note that the resonances do not correspond to the integers. The modal resonances are reported in Table 1.3.

Resonances and circuit element values							
Thickness factor Ω	**k_{br} at $X = 0$**	**R Ohms**	**L µH**	**X_L Ohms**	**C pF**	**X_C Ohms**	**Q**
12	1.088	148	1.26	857	1.7	863	5.8
	2.155	206	1.16	1,572	0.5	1,577	7.6
10	1.150	165	0.98	710	1.93	719	4.3

1.1.2 The closed thin-wire loop

The resonances of closed circular PEC loops operating below the 150 GHz region depend entirely on the circumference (see Table 1.2). Indeed, the resonances occur very close to $2\pi b = n\lambda$, where b is the radius of the loop and n are integers; $n > 0$. The resonances do not exactly correspond to the integers as expected, because no wire is perfectly thin, and charge can be found to flow around the wire on the inside circumference, a distance of $2\pi(b - a)$, where a is the radius of the wire. Hence the actual resonances are a bit higher.

The current at any driven frequency can be decomposed into its Fourier components, the "modal" currents:

$$I_\phi(\phi) = \sum_{n=-\infty}^{\infty} I_n e^{jn\phi}, \quad \text{where } n \text{ are integers} \tag{1.6a}$$

Since the closed circular loop is perfectly symmetric, the modal currents are symmetric around $n = 0$ and therefore the current may also be written in the form:

$$I_\phi(\phi) = I_0 + 2\sum_{n=1}^{\infty} I_n \cos(n\phi) = 2\sum_{n=0}^{\infty} \varepsilon_n I_n \cos(n\phi)$$

where

$$\varepsilon_n = \begin{cases} 1/2, & \text{for } n = 0. \\ 1, & \text{otherwise.} \end{cases} \tag{1.6b}$$

The coefficients themselves depend upon the form of the driving source. A Dirac delta function, $\delta(\phi)$, is usually used across an infinitesimal gap at $\phi = 0$ as the voltage source, V_s, driving the loop, where

$$E_\delta(\phi) = -\frac{V_s \delta(\phi)}{b} \tag{1.7}$$

Integrating over the circumference gives

$$\int_{-\pi}^{\pi} E_\delta(\phi) b d\phi = -V_s \tag{1.8}$$

The Fourier coefficients of E_δ are

$$E_{\delta n} = -\frac{1}{2\pi}\int_{-\pi}^{\pi} \frac{V_s}{b}\delta(\phi)e^{-jn\phi}d\phi = -\frac{V_s}{2\pi b} \tag{1.9}$$

The governing equations come from Maxwell's wave equations and the solution determines the final form of the current and of the Fourier modal coefficients. The solution allows a modal admittance that accumulates ε_n into its definition, so that:

$$\begin{aligned} Y_0' &\equiv \frac{1}{j\pi\xi_0 a_0 + (b/a)Z_s}, \quad Y_n' \equiv \frac{1}{j\pi\xi_0 a_n/2 + (b/a)Z_s/2} \quad \text{and} \\ I_n &= Y_n' V_s \\ I(\phi) &= V_s \sum_{n=-\infty}^{\infty} Y_n' e^{jn\phi} \end{aligned} \tag{1.10}$$

where

$$a_n = k_b\left(\frac{K_{n+1} + K_{n-1}}{2}\right) - \frac{n^2 K_n}{k_b} = a_{(-n)} \tag{1.11}$$

The term k_b represents the electrical size of the loop and is defined as $k_b = \frac{2\pi b}{\lambda}$. Wu's solution for the K_n coefficients, and the one most commonly used, is

$$\begin{aligned} K_n &= \frac{1}{\pi}\left[\mathbf{K}_0\left(\frac{|n|a}{b}\right)\mathbf{I}_0\left(\frac{|n|a}{b}\right) + \gamma - 2\sum_{k=0}^{|n|-1}\frac{1}{2k+1} + \ln(4|n|)\right] \\ &\quad + \frac{1}{2}\int_0^{2k_b}(\mathbf{\Omega}_{2n}(x) - j\mathbf{J}_{2n}(x))dx \end{aligned} \tag{1.12}$$

$$K_0 = \frac{1}{\pi}\ln\frac{8b}{a} + \frac{1}{2}\int_0^{2kb}[\mathbf{\Omega}_0(x) - j\mathbf{J}_0(x)]dx \tag{1.13}$$

$\mathbf{I}_0(z)$ and $\mathbf{K}_0(z)$ are modified Bessels of the first and second kind, respectively. $\mathbf{\Omega}_n(x)$ and $\mathbf{J}_n(x)$ are the Lommel–Weber function and the Bessel function of the first kind, respectively.‡

$$\begin{aligned} \mathbf{\Omega}_n(x) &= \frac{1}{\pi}\left[\int_0^{\pi} \sin(n\theta - x\sin\theta)d\theta\right] \\ \mathbf{J}_n(x) &= \frac{1}{\pi}\left[\int_0^{\pi} \cos(n\theta - x\sin\theta)d\theta\right] \end{aligned} \tag{1.14}$$

‡Storer and Wu both used the Jahnke–Emde definition, where $(x\sin\theta - m\theta)$ is assumed. This gives rise to a sign difference for $\mathbf{\Omega}_n(z)$ with the standard definition.

Note that: $\boldsymbol{\Omega}_{-n}(x) = (-1)^n \boldsymbol{\Omega}_n(x)$ and $\mathbf{J}_{-n}(x) = (-1)^n \mathbf{J}_n(x)$

The closed loop impedance and admittance is found from the ratio of the current to voltage at the feed point:

$$Y_{cl} \equiv \frac{I_\phi(0)}{V_s} = \sum_{n=-\infty}^{\infty} Y'_n \tag{1.15}$$

The resonances occur where the imaginary part of the input impedance crosses 0 (Figure 1.1). These are marked by high values of the admittance. These are distinguishable from the anti-resonances, which occur near the half integers ($\approx 0.5, \approx 1.5$, etc.) and are marked by very high resistance, making current flow in the loop impossible.

Table 1.2 gives the fundamental and second harmonic resonances for two thin-wire, PEC loops, one slightly thicker than the other. The second harmonic for the thicker loop is not definite, because the reactance never crosses zero above the fundamental. The values given are approximate, depending upon the implementation of the summations and the integrals. The table also shows the input resistance and reactance, the loop's inductance and capacitance and finally the Q.

Table 1.3 provides the resonances associated with each of the modes. At each modal resonance, the loop behaves like a series resonant circuit, and these RLC elements can be calculated [33].

1.1.3 The loaded loop

Impedances may be placed anywhere in the periphery of the loop, as shown in Figure 1.2. The corresponding angle is called ϕ_q. Any load impedance can also be

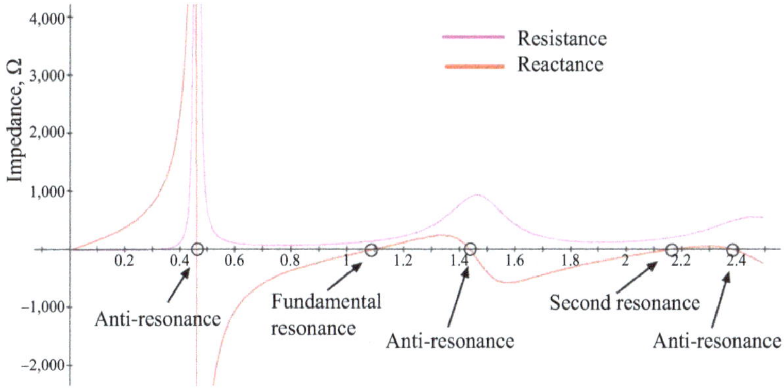

Figure 1.1 The impedance of a closed, thin-wire, $\Omega = 12$ PEC loop exhibiting the natural resonances and anti-resonances.

Table 1.3 The modal resonances for the fundamental resonances of the closed PEC loop with corresponding circuit element values.

Ω k_{br}	n	$\tilde{k}_{brn}$	$\tilde{R}_n$ Ohms	$\tilde{L}_n$ μH	$\tilde{X}_{Ln}$ Ohms	$\tilde{C}_n$ pF	$\tilde{X}_{Cn}$ Ohms
	0	0	217	2.8	1,931	∞	0
12	1	1.07	292	2.5	1,680	0.90	1,632
1.088	2	1.95	62	2.4	1,632	0.28	5,228
	3	2.96	4	2.0	1,375	0.14	10,150
	4	3.97	0	1.8	1,221	0.09	16,277
	0	0	264	2.2	1,618	∞	0
10	1	1.11	310	1.8	1,316	1.1	1,223
1.150	2	1.93	75	1.8	1,304	0.4	3,678
	3	2.93	5	1.4	1,035	0.2	6,729
	4	3.96	0	1.2	873	0.1	10,319

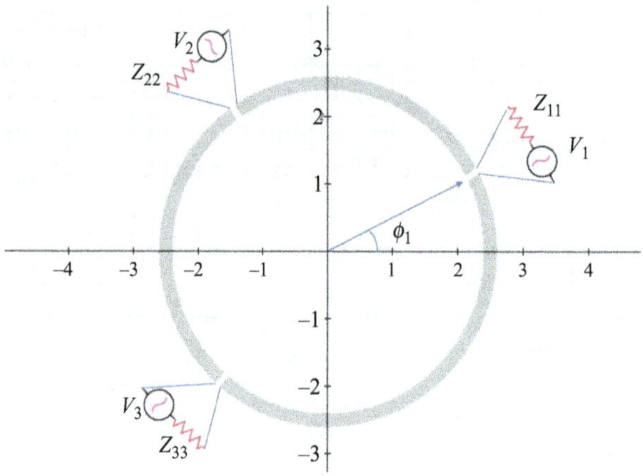

Figure 1.2 A loop with three sources and associated loads.

accompanied by a delta voltage generator, as used in the closed loop, if desired. The load acts as a voltage sink and is placed in series with the generator. Any given element is then added to the voltage source as

$$V_q - \sum_{k=1}^{M} Z_{qk} I_k \tag{1.16}$$

Each of the sources influence the current at the load, q, and therefore a sum is required. The current anywhere in the loop is the sum of the currents due to each of these generators. Assuming for the moment that all loads are 0, the current in the loop is due to all the generators. The phase difference between generators should be taken into account:

$$I(\phi) = \sum_{q=1}^{M} I_q(\phi) = \sum_{q=1}^{M}\left(\sum_{n=-\infty}^{\infty} Y_n' e^{jn(\phi-\phi_q)}\right)V_q \tag{1.17}$$

$$\equiv \sum_{q=1}^{M} Y(\phi,q)V_q \equiv \sum_{q=1}^{M} Y_q(\phi)V_q \tag{1.18}$$

where M is the number of generators. $Y_q(\phi)$ and V_q can be thought of as column vectors. In matrix notation, this is

$$I(\phi) = \mathbf{Y}^T(\phi)\mathbf{V} \tag{1.19}$$

In order to include a load at ϕ_q, V_q is replaced with

$$V_q - \sum_{k=1}^{M} Z_{qk}I_k \tag{1.20}$$

where Z_q is the load placed at q and I_k is the current due to the source V_k. The summation is required because each current affects the voltage across the Z_k. Z_{qk} is a square diagonal matrix with the complex impedances along the diagonal. I_k is a column vector of the currents at the various source generators at ϕ_k and needs to be derived before we can find the total current. Note that these are not the modal currents. Substituting, we have

$$I(\phi) = \sum_{q=1}^{M} Y_q(\phi)\left(V_q - \sum_{k=1}^{M} Z_{qk}I_k\right) \tag{1.21}$$

The currents at any given load p are found by assigning $\phi = \phi_p$ and solving for $I(\phi_p)$. Note that I_p and I_k are the same vector. Defining the matrix, Y_{pq}, and substituting, we have

$$\begin{aligned} Y_{pq} &\equiv Y_q(\phi_p) \\ I_p &= \sum_{q=1}^{M}\left[Y_{pq}\left(V_q - \sum_{k=1}^{M} Z_{qk}I_k\right)\right] \end{aligned} \tag{1.22}$$

Next, we define the square matrices

$$\mathbb{Y} \equiv Y_{pq} \quad \text{and} \quad \mathbb{Z} \equiv Z_{qk} \tag{1.23}$$

where $\mathbb{Z}$ is, in particular, square diagonal with the loads at q on the diagonal, and the column vectors $\mathbf{I}$ and $\mathbf{V}$. Then, (1.22) becomes

$$\mathbf{I} = \mathbb{Y}(\mathbf{V} - \mathbb{Z}\mathbf{I}) \tag{1.24}$$

which may be arranged to yield

$$\mathbf{I} = [\mathbb{I} + \mathbb{Y}\mathbb{Z}]^{-1}\mathbb{Y}\mathbf{V} \equiv \mathbb{F}^{-1}\mathbb{Y}\mathbf{V} \tag{1.25}$$

with the identity matrix $\mathbb{I}$. With the currents, **I**, due to the various sources defined, substitution gives the current at every angle:

$$\boxed{I(\phi) = \left(\mathbf{Y}^T(\phi) - \mathbf{Y}^T(\phi)\mathbb{Z}\mathbb{F}^{-1}\mathbb{Y}\right)\mathbf{V}} \tag{1.26}$$

The Fourier coefficients are readily found from (1.26). Noting that

$$Y_q(\phi) = \sum_{n=-\infty}^{\infty} \left(Y_n' e^{-jn\phi_q}\right) e^{jn\phi} \tag{1.27}$$

we have

$$I(\phi) = \sum_{n=-\infty}^{\infty} \left[Y_n'\left(e^{-jn\phi_1}, e^{-jn\phi_2}, \ldots\ e^{-jn\phi_M}\right)\left(\mathbb{I} - \mathbb{Z}\mathbb{F}^{-1}\mathbb{Y}\right)\mathbf{V}\right] e^{jn\phi} \tag{1.28}$$

Letting

$$\mathbf{E} \equiv e^{-jn\phi_q} \quad \text{and} \quad \mathbb{G} \equiv \mathbb{I} - \mathbb{Z}\mathbb{F}^{-1}\mathbb{Y} \tag{1.29}$$

allows the simpler notation

$$I(\phi) = \sum_{n=-\infty}^{\infty} \left[Y_n'\ \mathbf{E}^{\mathbf{T}}\mathbb{G}\mathbf{V}\right] e^{jn\phi} \equiv \sum_{n=-\infty}^{\infty} \left[\mathbf{Y}_{\mathbf{n}}^{\mathbf{T}}\mathbf{V}\right] e^{jn\phi} \tag{1.30}$$

and identifies the Fourier coefficients:

$$\boxed{I_n = \mathbf{Y}_{\mathbf{n}}^{\mathbf{T}}\mathbf{V}} \tag{1.31a}$$

These coefficients are not symmetric around $n = 0$ because **E** is not symmetric around $n = 0$. This imposes an important burden when determining the currents of a loaded loop. Therefore,

$$I(\phi) = \sum_{n=-\infty}^{\infty} I_n e^{jn\phi} = \sum_{n=0}^{\infty} \varepsilon_n \left(I_n e^{jn\phi} + I_{(-n)} e^{-jn\phi}\right) \tag{1.31b}$$

The input impedance and admittance for a loaded loop make no sense if more than one voltage source is present, or if the remaining source is not at $\phi = 0$. In this case, the voltage source vector **V** reduces to a single voltage and the matrix $\mathbb{G}$ reduces to a vector.

$$\mathbf{V} \to V_1 = V_s \quad \text{and} \quad \mathbb{G} \to \mathbf{G} = \begin{bmatrix} 1 \\ 0 \\ \vdots \\ 0 \end{bmatrix} - \sum_{k=1}^{\infty} \mathbb{Z}_{qk} \sum_{p=1}^{\infty} \left(\mathbb{F}^{-1}\right)_{kp} Y_{p1} \tag{1.32}$$

The input admittance and impedance are then given by

$$\boxed{Y_{in} \equiv \frac{I_\phi(0)}{V_s} = \sum_{n=-\infty}^{\infty} Y'_n\,\mathbf{E}^{\mathrm{T}}\mathbf{G} \equiv \frac{1}{Z_{in}}} \tag{1.33}$$

1.1.4 Radiation from a driven thin-wire loop antenna

The far fields were recently derived by Nagar *et al.* [26] in the case of asymmetric currents when driven by a single source. These are given by

$$\begin{aligned} E_\theta &\approx j\frac{\xi_0 e^{-jk_0 r}\cot\theta}{2r}\sum_{n=1}^{\infty} nj^n\left(I_n e^{jn\phi} - I_{-n}e^{-jn\phi}\right)\mathbf{J}_n(k_b\sin\theta) \\ E_\phi &\approx -\frac{\xi_0 e^{-jk_0 r}}{2r}\sum_{n=0}^{\infty}\varepsilon_n j^n\left(I_n e^{jn\phi} + I_{-n}e^{-jn\phi}\right)\mathbf{J}'_n(k_b\sin\theta) \end{aligned} \tag{1.34}$$

The power radiated can now be calculated using the usual prescription:

$$P_{rad} = \int_0^{2\pi}\int_0^{\pi}\frac{\left(E_\theta^* E_\theta + E_\phi^* E_\phi\right)}{2\xi} r^2 \sin\theta d\theta d\phi \tag{1.35}$$

Integrating leads to integrals of the form

$$\mathcal{Q}^{(p)}_{m,n}(k_b) = \int_0^{\pi/2}\mathbf{J}_m(k_b\sin\theta)\mathbf{J}_n(k_b\sin\theta)\sin^p\theta d\theta \tag{1.36}$$

an auxiliary function suggested by Savov in 2002 [13]. Assuming a single source, V_s at $\phi = 0$, the result [24] is

$$\begin{aligned} P_{rad} &= \frac{\xi_0\pi k_b^2}{2}|V_s|^2 \\ &\times\sum_{n=0}^{\infty}\varepsilon_n\left(|Y_n|_{in}^2 + |Y_{(-n)}|_{in}^2\right)\left[\frac{1}{2}\mathcal{Q}^{(1)}_{n-1,n-1}(k_b) + \frac{1}{2}\mathcal{Q}^{(1)}_{n+1,n+1}(k_b) - \frac{n^2}{k_b^2}\mathcal{Q}^{(1)}_{n,n}(k_b)\right] \end{aligned} \tag{1.37}$$

This makes use of the helpful recurrence relation when $n = m$:

$$\mathcal{Q}^{(-1)}_{n,n}(k_b) = \frac{k_b^2}{4n^2}\left[\mathcal{Q}^{(1)}_{n-1,n-1}(k_b) + 2\mathcal{Q}^{(1)}_{n-1,n+1}(k_b) + \mathcal{Q}^{(1)}_{n+1,n+1}(k_b)\right] \tag{1.38}$$

We also note that [24]

$$\mathcal{Q}^{(1)}_{n,n}(k_b) = \frac{1}{k_b}\sum_{m=0}^{\infty}\mathbf{J}_{2n+2m+1}(2kb) \tag{1.39}$$

which may be used to evaluate the three terms inside the square brackets of (1.37).

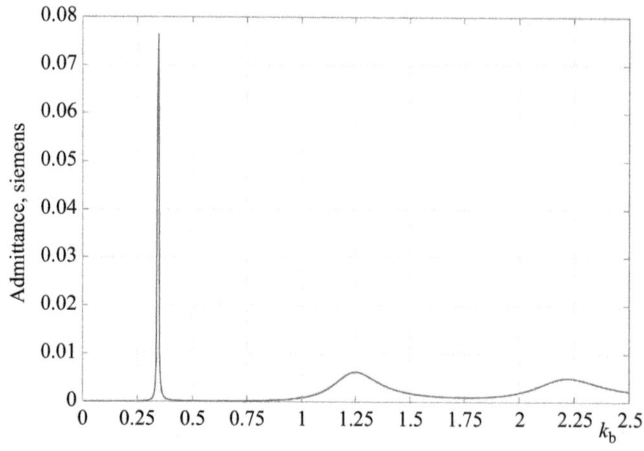

Figure 1.3 The sub-wavelength resonance of a thin-wire, PEC loop, loaded with a single capacitor, $l_\varepsilon = 1.0$ at $\phi = 0$. The resonance in this case occurs at about $k_b = 0.3$.

1.1.5 The sub-wavelength resonance of loops and rings

The gap in a split-ring meta-atom, often used in meta-materials, acts like a capacitance, which, with the natural inductance of the loop, creates a high-Q resonance in the region below the first anti-resonance at 0.46λ (see Figure 1.3). This is true for good metals out to about 150 GHz. The resonance then experiences wavelength scaling as noted in Section 1.1.1.

The phenomenon can be studied by placing a single capacitive load in the periphery of the loop and using (1.33). Taking $M = 1$, $\phi_1 = 0$ with the source $V_1 = V_s$, the load matrix has one element $Z_1 = -jX_C$, where X_C is the capacitive reactance, the input admittance of the loop becomes

$$Z_{in} = Z_{cl} - jX_C \tag{1.40}$$

This is probably evident because the capacitor is in series with the input port source and therefore its reactance should add to that of the closed loop.

If a particular resonance is desired, a capacitor can be picked to produce it. Expanding (1.40)

$$Z_{in} = R_{cl} + j\left(X_{L,cl} - X_{C,cl} - X_C\right) \tag{1.41}$$

At resonance, the imaginary term must go to 0. Therefore,

$$X_C = X_{L,cl} - X_{C,cl} \tag{1.42}$$

Select your resonance and calculate its corresponding k_{br} for the loop with the desired radius, b. Then calculate l_ε using

$$X_C = \frac{k_{br}}{\xi_0 l_\varepsilon} \tag{1.43}$$

The value of the required capacitor is

$$C = \varepsilon_0 b l_\varepsilon \tag{1.44}$$

This is well and good for loop sizes that are large enough to handle a lumped capacitor, as in the MHz region. Otherwise, one needs to use a gap filled with a dielectric material that allows some tuning, perhaps through doping [34].

Exactly how much reactance a free space gap has depends upon the size and shape of the wire and how much the end effects are taken into account. It can be shown that for thicker wires the flat-plate capacitor model is a reasonable estimate [29].

1.2 Analytical theory of mutual coupling in nanoloops

The previous sections detailed the theory of loop antennas in isolation. However, the coupling between multiple loop antennas is also of interest in a variety of areas including mine communications [35] and geological characterization [36]. In addition, coupling plays a critical role in the analysis and design of arrays of loops, most notably the directive Yagi–Uda configuration [37]. In the optical and THz regimes, nanoloop arrays have been considered for their potential use for light-trapping in solar cells [38] and for highly directive emission [39]. A full-wave simulation is typically used when designing these structures, but this approach can be prohibitive in terms of computational resources and time required. In this section, the induced EMF method [5] will be employed to evaluate the mutual coupling between nanoloop array elements. This approach has previously been used to study the mutual coupling in arrays of dipoles [40] and slots [41]. However, due to the complex integrals which arise with loop antennas [42], no analytical solutions have been developed for the mutual coupling of arrays of loops of arbitrary sizes and at arbitrary locations with respect to each other. In 1955, Wait [43] developed expressions for coplanar electrically small loops above a homogeneous ground through a quasi-static approximation. This was extended in 1972 by Fuller and Wait [44] to include the effects of an inhomogeneous ground, but the loops were still required to be electrically small and coplanar. Efficient integral-equation solutions have been developed for two coaxial loops of the same radius [45] and of different radii [42]. A completely general integral-equation based formulation where the loops can have arbitrary radii, orientation and location was developed in 2005 [46]. While these expressions have proven to be valuable in the analysis and design of loop arrays, they lack the computational efficiency and physical insight provided by fully analytical expressions. This section will present two formulations for the coupling between nanoloop arrays: fully analytical expressions when the loops are relatively far apart and pseudo-analytical expressions when the loops are extremely closely spaced. Both formulations offer advantages when compared to full-wave solvers in terms of computational efficiency, while also providing physical intuition and insight.

1.2.1 Theory

In order to accurately calculate the mutual coupling for closely-spaced loops, the near-zone electric field must be used. Werner derived exact closed-form expressions for the near-zone fields in 1996 [22]. The geometry under consideration is shown in Figure 1.4, where the near-zone electric field at an arbitrary field point (r, θ, ϕ) must be calculated based on a source point $(b, \pi/2, \phi')$. The distance between source point and field point can be represented by:

$$R' = \sqrt{R^2 - 2b\sin(\theta)\cos(\phi - \phi')} \tag{1.45a}$$

$$R = \sqrt{r^2 + b^2} \tag{1.45b}$$

The near-zone electric field components in spherical coordinates are given by:

$$E_r(r,\theta,\phi) = \frac{\xi_0 k_0^2 b\sin(\theta)}{4}\left\{\sum_{m=1}^{\infty}\sum_{\substack{n=1\\ m-n=2k\\ k=0,1,\ldots}}^{m}\left(\frac{k_0^2 br\sin(\theta)}{2}\right)^{m-1}\sin(n\phi)\times\left[(k_0b)^2 C_{mn}^3\frac{h_{m+1}^{(2)}(k_0R)}{(k_0R)^{m+1}} - C_{mn}^4\frac{h_m^{(2)}(k_0R)}{(k_0R)^m}\right]\right\} \tag{1.46a}$$

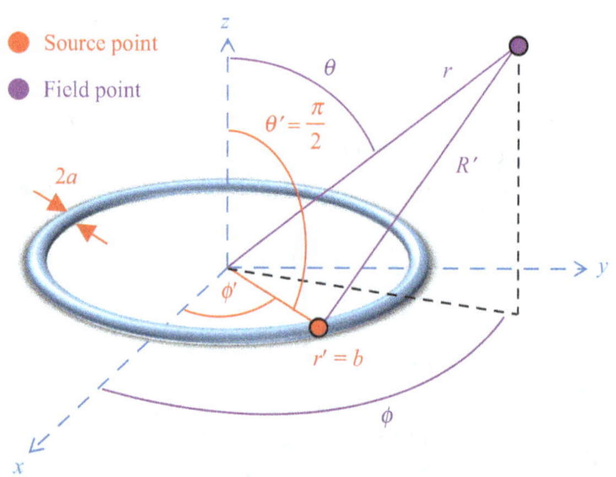

Figure 1.4 Nanoloop geometry.

$$E_\theta(r,\theta,\phi) = \frac{\xi_0 k_0^2 b\cos(\theta)}{4}\left\{\sum_{m=1}^{\infty}\sum_{\substack{n=1\\ m-n=2k\\ k=0,1,\ldots}}^{m}\left(\frac{k_0^2 br\sin(\theta)}{2}\right)^{m-1}\sin(n\phi)\right.$$
$$\left.\times\left[C_{mn}^5\frac{h_m^{(2)}(k_0R)}{(k_0R)^m} - C_{mn}^3\frac{h_{m-1}^{(2)}(k_0R)}{(k_0R)^{m-1}}\right]\right\} \tag{1.46b}$$

$$E_\phi(r,\theta,\phi) = \frac{\xi_0 k_0^2 b}{4}\left\{\sum_{m=1}^{\infty}\sum_{\substack{n=0\\ m-n=2k\\ k=0,1,\ldots}}^{m}\left(\frac{k_0^2 br\sin(\theta)}{2}\right)^{m-1}\cos(n\phi)\right.$$
$$\left.\times\left[C_{mn}^6\frac{h_m^{(2)}(k_0R)}{(k_0R)^m} - C_{mn}^2\frac{h_{m-1}^{(2)}(k_0R)}{(k_0R)^{m-1}}\right]\right\} \tag{1.46c}$$

$$C_{mn}^2 = I_n\frac{m}{[(m-n)/2]![(m+n)/2]!} \tag{1.47a}$$

$$C_{mn}^3 = I_n\frac{n}{[(m-n)/2]![(m+n)/2]!} \tag{1.47b}$$

$$C_{mn}^4 = (m+1)C_{mn}^3 \tag{1.47c}$$

$$C_{mn}^5 = mC_{mn}^3 \tag{1.47d}$$

$$C_{mn}^6 = nC_{mn}^3 \tag{1.47e}$$

where $h_m^{(2)}$ are spherical Hankel functions of the second kind.

The geometry which will be considered for this section is shown in Figure 1.5. An unprimed coordinate system (x, y, z) is defined with origin at the center of the loop i which is assumed to be active. A loop j which is assumed to be passive is placed at $(x = x_0, y = y_0, z = z_0)$. A primed coordinate system (x', y', z') has its origin at the center of loop j. Loop i has loop radius b_i and wire radius a_i whereas loop j has loop radius b_j and wire radius a_j. The vector $\left(\widehat{r}_0, \widehat{\theta}_0, \widehat{\phi}_0\right)$ points from the center of loop i to the center of loop j while the vector $\left(\widehat{r}, \widehat{\theta}, \widehat{\phi}\right)$ points from the

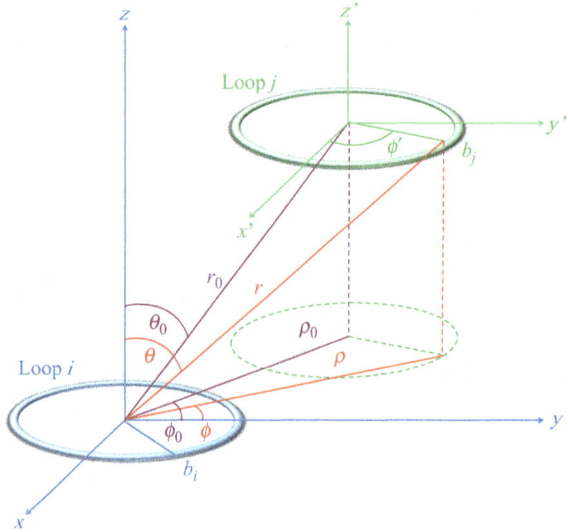

Figure 1.5 Nanoloop coupling geometry.

center of loop i to an arbitrary point on loop j. Loop j is projected onto the xy-plane and shown as a dashed line in Figure 1.5. The associated projected vectors $\left(\widehat{\rho}_0, \pi/2, \widehat{\phi}_0\right)$ and $\left(\widehat{\rho}, \pi/2, \widehat{\phi}\right)$ are defined in a similar manner. With these geometrical parameters defined, the induced EMF method is used to calculate the current at arbitrary angle $\mathbf{\Phi}'$ on the passive loop j due to radiation from the active loop i, as follows:

$$I_j(\mathbf{\Phi}') = b_j \int_0^{2\pi} \left[\vec{E}_i \cdot \widehat{t}\right] \left[\sum_{p=0}^{\infty} Y'_{p,j} \cos(p(\mathbf{\Phi}' - \phi'))\right] d\phi' \tag{1.48}$$

In this expression, the first term in the integrand is the induced voltage on loop j due to loop i where $\vec{E}_i$ is the electric field due to loop i and $\widehat{t}$ is the tangential vector in the coordinate system of loop j. The second term is related to the modal admittances $Y'_{p,j}$ of mode p for loop j which is given in (1.10). The tangential vector is explicitly given by:

$$\widehat{t} = -\sin\phi' \widehat{x} + \cos\phi' \widehat{y} \tag{1.49}$$

The corresponding dot products required in (1.48) are given by:

$$\widehat{r} \cdot \widehat{t} = \frac{1}{r}(-x \sin\phi' + y \cos\phi') \tag{1.50a}$$

$$\hat{\theta}\cdot\hat{t} = \frac{z}{r\rho}(-x\ \sin\phi' + y\ \cos\phi') \tag{1.50b}$$

$$\hat{\varphi}\cdot\hat{t} = \frac{1}{\rho}(y\ \sin\phi' + x\ \cos\phi') \tag{1.50c}$$

These results can be combined to derive an expression for the induced current which involves the single integral from 0 to 2π of (1.48). When implementing this equation, the fully analytical near-zone electric fields are used but numerical integration is required; for this reason, this formulation will be referred to as a pseudo-analytical representation. No approximations are made in this case, and the induced current is valid for any inter-loop separation. This formulation is much more computationally efficient compared to a full-wave solver, but it does not necessarily provide the physical insight which can be gleaned from a fully analytical representation. Furthermore, a sophisticated numerical integration routine such as global adaptive quadrature is required for accurate calculations. When using the exact near-zone fields, an analytical representation has not been found. However, using some approximations which assume the loops are moderately spaced, a fully analytical form can be derived. Once the induced current has been found, the self and mutual admittances can be determined. In particular, the self-admittance of loop i is given by:

$$Y_{ii} = \sum_{p=0}^{\infty} Y'_{p,j} \tag{1.51}$$

The mutual admittance Y_{ji} is derived by applying a voltage source V_i to loop i placed at $\Phi' = 0$ and shorting loop j:

$$Y_{ji} = \frac{I_j(\Phi' = 0)}{V_i} \tag{1.52}$$

Given these expressions, the total current at the input terminals of loop j can be computed for an array of N loops:

$$I_j = \sum_{j=1}^{N} Y_{ji} V_i \tag{1.53}$$

Since the final expression is a summation of sinusoids and cosinusoids, the far-zone expressions derived earlier can be used along with array factor theory to compute the resulting far-field from an array of loops. For an array of N loops where the ith loop is located at $(x_{0,i}, y_{0,i}, z_{0,i})$ the far-zone electric field is given by:

$$E(\theta,\phi) = \sum_{i=1}^{N} E_i(\theta,\phi) e^{jk_0\left[x_{0,i}\ \sin\ \theta\ \cos\ \phi + y_{0,i}\ \sin\ \theta\ \sin\ \phi + z_{0,i}\ \cos\ \theta\right]} \tag{1.54}$$

1.2.1.1 Pseudo-analytical representation

Combining (1.46) and (1.48–1.50) results in the following expression:

$$I_j(\Phi') = \frac{\xi_0 k_0^2 b_i b_j}{4} \sum_{p=0}^{\infty} Y'_{p,j} \sum_{m=1}^{\infty} \sum_{\substack{n=1 \\ m-n=2k \\ k=0,1,\ldots}}^{m} \left(\frac{k_0^2 b_i}{2}\right)^{m-1} F_{mnp}(\Phi') \tag{1.55}$$

The summation over p involves the modal admittance for loop j while the summations over m and n involve the induced voltage at the terminals of loop j due to radiation from loop i. The C_{mn} terms in the near-zone expressions involve the modal current coefficients. The notation $C_{mn,i}$ will be used to explicitly specify the modal coefficients corresponding to loop i. The expression $F_{mnp}(\Phi')$ involves numerical integration and is given explicitly by:

$$\begin{aligned} F_{mnp}(\Phi') = & \int_0^{2\pi} \cos(p(\Phi' - \phi'))(r \sin\theta)^{m-1} \\ & \times \left[\frac{1}{r} \sin(\theta)\sin(n\phi)(-x \sin\phi' + y \cos\phi') \left(k_0^2 b_i^2 C_{mn,i}^3 \frac{h_{m+1}^{(2)}(k_0 R)}{(k_0 R)^{m+1}} - C_{mn,i}^4 \frac{h_m^{(2)}(k_0 R)}{(k_0 R)^m} \right) \right. \\ & + \frac{z}{r\rho} \cos(\theta)\sin(n\phi)(-x \sin\phi' + y \cos\phi') \left(C_{mn,i}^5 \frac{h_m^{(2)}(k_0 R)}{(k_0 R)^m} - C_{mn,i}^3 \frac{h_{m-1}^{(2)}(k_0 R)}{(k_0 R)^{m-1}} \right) \\ & \left. + \frac{1}{\rho} \cos(n\phi)(y \sin\phi' + x \cos\phi') \left(C_{mn,i}^6 \frac{h_m^{(2)}(k_0 R)}{(k_0 R)^m} - C_{mn,i}^2 \frac{h_{m-1}^{(2)}(k_0 R)}{(k_0 R)^{m-1}} \right) \right] d\phi' \end{aligned} \tag{1.56}$$

where

$$x = x_0 + b_j \cos\phi' \tag{1.57a}$$

$$y = y_0 + b_j \sin\phi' \tag{1.57b}$$

$$r = \sqrt{x^2 + y^2 + z_0^2} \tag{1.57c}$$

$$\rho = \sqrt{x^2 + y^2} \tag{1.57d}$$

1.2.1.2 Fully analytical representation

An analytical representation of the exact integral in (1.55) could not be found. However, for intermediate inter-loop spacings a number of approximations can be made and an analytical representation can be found. The exact limits of when these approximations can be applied will be discussed later. It has been found from numerical experimentation that the far-zone representations of E_θ and E_ϕ can be used to accurately calculate the mutual coupling but that the far-zone assumption that $E_r = 0$ cannot be used. Instead, a series of approximations will be made which will result in an "intermediate-zone" representation of the radial electric field. The major difficulty with the integral is that many of the parameters depend on the variable of integration ϕ'. In order to make the integral tractable, the following assumptions for the spherical components of the electric field are made:

$$\rho \approx \begin{cases} \rho_0, & \text{for amplitude terms} \\ \rho_0\left(1 + \dfrac{b_j}{\rho_0}\cos(\phi_0 - \phi')\right), & \text{for phase terms} \end{cases} \tag{1.58a}$$

$$r \approx \begin{cases} \sqrt{\rho_0^2 + z_0^2}, & \text{for amplitude terms} \\ \sqrt{\rho_0^2 + z_0^2}\left(1 + \dfrac{b_j\rho_0}{r_0^2}\cos(\phi_0 - \phi')\right), & \text{for phase terms} \end{cases} \tag{1.58b}$$

In addition, the following approximations are made:

$$R \approx r \tag{1.59a}$$

$$\theta \approx \theta_0 \tag{1.59b}$$

When the loops are non-coaxial, the approximation $\phi \approx \phi_0$ will also be used. The coaxial case will be discussed later. For the radial component, the large argument asymptotic limit of the spherical Hankel function of the second kind will be employed:

$$\lim_{x\to\infty} h_m^{(2)}(x) = j^{m+1}\frac{e^{-jx}}{x} \tag{1.60}$$

The argument in this case is k_0R which will be very large in the RF or optical regimes. Using these assumptions, the near-zone expression of (1.46) reduces to the following:

$$E_r(r,\theta,\phi) \approx \frac{\xi_0 b_i \sin\theta_0}{4r^2}\sum_{m=1}^{\infty}\sum_{\substack{n=1 \\ m-n=2k \\ k=0,1,\ldots}}^{m} \sin(n\phi_0)\left[\frac{k_0 b_i \sin\theta_0}{2}\right]^{m-1} \times j^{m+1}\left[\frac{jk_0 b_i^2}{r}C_{mn,i}^3 - C_{mn,i}^4\right]e^{-jk_0 r} \tag{1.61}$$

Defining $D_i = \frac{k_0 b_i \rho_0}{r_0}$ and $D_j = \frac{k_0 b_j \rho_0}{r_0}$ the fully analytical expression for the induced current is given by:

$$
\begin{aligned}
I_j(\Phi') = &-\frac{\xi_0\pi}{r_0}e^{-jk_0r_0}\sum_{p=0}^{\infty}Y'_{p,j}(-1)^p(j)^{p-1} \\
&\times\Bigg\{\sum_{n=0}^{\infty}\Bigg[j^nI_{n,i}\frac{z_0^2}{\rho_0^2k_0}np\sin(n\phi_0)J_p(D_j)J_n(D_i)\sin(p(\Phi'-\phi_0)) \\
&+k_0b_ib_j\left(j\frac{b_j}{\rho_0}J_p(D_j)-J'_p(D_j)\right)J'_n(D_i)\cos(n\phi_0)\cos(p(\Phi'-\phi_0))\Bigg] \\
&-\frac{\rho_0b_i}{2r_0^2k_0}\sum_{m=1}^{\infty}\sum_{\substack{n=1\\ m-n=2k\\ k=0,1,\dots}}^{m} j^{m+1}\sin(n\phi_0)\left(\frac{k_0b_i\sin\theta_0}{2}\right)^{m-1} \\
&\times\left(\frac{jk_0b_i^2C^3_{mn,i}}{r_0}-C^4_{mn,i}\right)pJ_p(D_j)\sin(p(\Phi'-\phi_0))\Bigg\}
\end{aligned}
\tag{1.62}
$$

For the coaxial configuration, $\phi = \phi'$ and the induced current can be expressed as:

$$
\begin{aligned}
I_j(\Phi') = &-\frac{\xi_0k_0b_ib_j\pi}{2r}e^{-jk_0r} \\
&\times\left[\sum_{p=0}^{\infty}j^pI_{p,i}Y'_{p,j}J'_p\left(\frac{k_0b_ib_j}{r}\right)\cos(p\Phi')+I_{0,i}Y'_{0,j}J'_0\left(\frac{k_0b_ib_j}{r}\right)\right]
\end{aligned}
\tag{1.63}
$$

where $r = \left(r_0^2 + b_j^2\right)^{1/2}$, which is independent of ϕ' in the coaxial case.

1.2.2 Results

The fully analytical and pseudo-analytical expressions derived in the previous section were implemented in MATLAB® [47] and comparisons were made with the full-wave integral equation solver FEKO [64]. The tests were made on a Dual Intel Xeon processor with 10 cores. FEKO used all 10 cores in parallel, while the MATLAB® code used only a single core. Table 1.4 shows a summary of the computational resources required by each method for two gold nanoloops with circumference 3,000 nm evaluated at 51 discrete frequency points from $k_b = 0.01$ to $k_b = 2.5$. The time and memory required is independent of the distance between the loops since the integral equation method only meshes the surface of the loops and not the air region in between. For the MATLAB® code, the specific case of loops in a co-planar configuration was implemented separately since the expression

Table 1.4 Comparison of computational resources required.

Method	Time	Memory
Full-wave	5.9 h	8.2 GB
Pseudo-analytical	33 s	34 MB
Analytical (general)	11 s	20 MB
Analytical (stacked)	0.1 s	2 MB

for the induced current can be represented in a particularly simple form. As can be seen, the full-wave solver required 5.9 h and 8.2 GB of memory, while the pseudo-analytical solver only requires 33 s and 34 MB, resulting in a more than 600× speed improvement and 200× memory improvement. The fully analytical solution in general required 11 s and 20 MB. The simple form derived for the stacked configuration required only 0.1 s and 2 MB, a more than 200,000× speed improvement and 4,000× memory improvement.

First the analytical expressions derived in the previous section will be validated for the coupling between two identical thin ($\Omega = 12$) PEC loops with $\Phi' = 0$ and $2\pi b = 3,000$ nm for a variety of geometrical configurations. Figure 1.6 shows a comparison between the full-wave, analytical and pseudo-analytical current calculations for three different cases. As can be seen, the full-wave and pseudo-analytical results are nearly identical for all cases, while the analytical results start to deviate when the loops are closely spaced. This is due to the approximations made with this calculation.

As shown explicitly in (1.48), there are two major components of the induced current: the induced voltage due to radiation from the active loop and a component proportional to the input admittance of the passive loop. Before examining the total induced current, these two components can be examined in more detail. While the exact expression is an integral of the two components over ϕ', examining the magnitudes of the individual components can give insight into the maximum values the induced currents can achieve.

Figure 1.7 shows the magnitude of the current of an active loop, which for $\Phi' = 0$, is identical to the second term. The value is plotted versus k_b on the x-axis and ϕ' on the y-axis. As can be seen, the PEC loop has its first two resonances slightly above $k_b = 1$ and $k_b = 2$, while the Au loop has its first two resonances around $k_b = 0.4$ and $k_b = 0.78$. This shows that in a gold loop, the effective wavelength is shortened, which agrees with the theory of Novotny [32]. Note that the magnitude of the PEC loop's second resonance is much lower than its first resonance, while the gold loop's resonances fall off gradually. This plot shows that

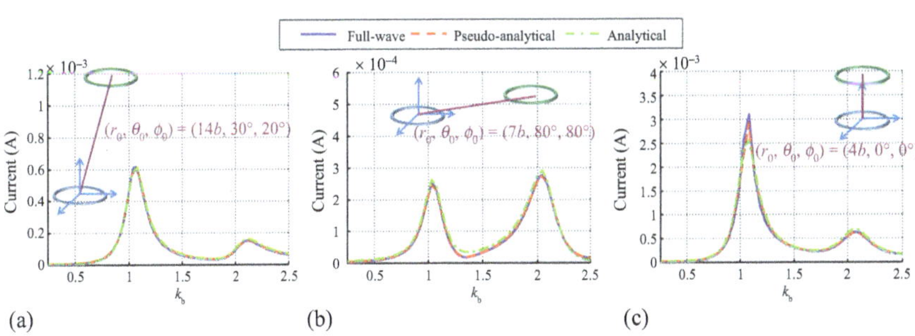

Figure 1.6 Comparison between passive current derived from full-wave, pesudeo-analytical and fully-analytical calculations for two PEC loops for three different geometrical configurations.

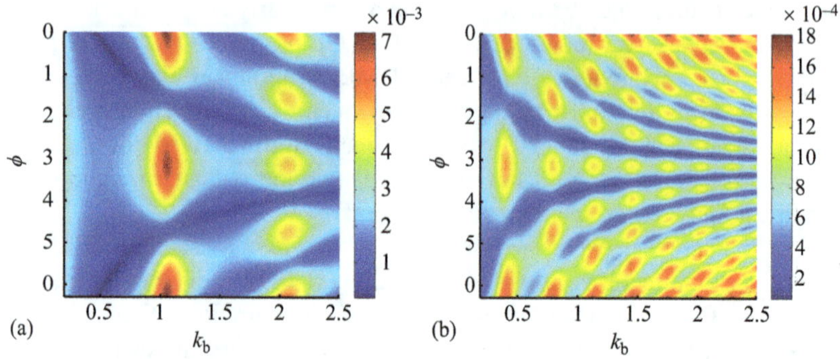

Figure 1.7 Magnitude of active current for (a) PEC and (b) Au loop.

the largest coupling will potentially occur at these resonances; however, the actual magnitude depends on the near-zone electric field and the location of the passive loop. Figure 1.8 displays the near-fields in dB of the PEC loop at $k_b = 1$ in the plane of the loop, where (a) shows E_x and (b) shows E_y. As can be seen, the E_x component of the field is very small along $\phi_0 = 0°$ and falls off rapidly along $\phi_0 = 90°$. In contrast, E_y is large along $\phi_0 = 0°$ and falls off slowly along $\phi_0 = 90°$. At the second resonance, the E_x component of the field is very small along $\phi_0 = 0°$ and is relatively constant along $\phi_0 = 90°$. In contrast, E_y is large along $\phi_0 = 0°$ and falls off extremely rapidly along $\phi_0 = 90°$. Similar plots are shown in Figure 1.9 for the gold loop. For both resonances, the E_x component falls off rapidly along $\phi_0 = 0°$ and falls off slower along $\phi_0 = 90°$, while the E_y component falls off slowly in both directions.

Finally, the current induced on a passive PEC loop is shown in Figure 1.10 along two directions, while the current induced on a passive Au loop is shown in Figure 1.11 along the same directions. For the PEC loop, the magnitude of the active current at the first resonance is much higher than the second, a trend which can also be observed in the passive current. The magnitude of E_x in the $\phi_0 = 0°$ is much smaller than the $\phi_0 = 90°$ direction, leading to a smaller induced current. The first resonance shows a strong E_y component in the $\phi_0 = 90°$ direction, leading to a relatively large induced current even for small distances. However, the E_y component at the second resonance falls off quickly in this direction, leading to much smaller magnitudes of induced current for this resonance. For the gold loop, the large active current of the first resonance leads to a large induced current at this resonance. The other resonances have magnitudes which fall off gradually. For the first two resonances, both E_x and E_y fall off relatively quickly in the $\phi_0 = 0°$ direction, leading to a small induced magnitude even when the loops are closely spaced. Both components fall off slowly in the $\phi_0 = 90°$ direction, leading to a larger induced current.

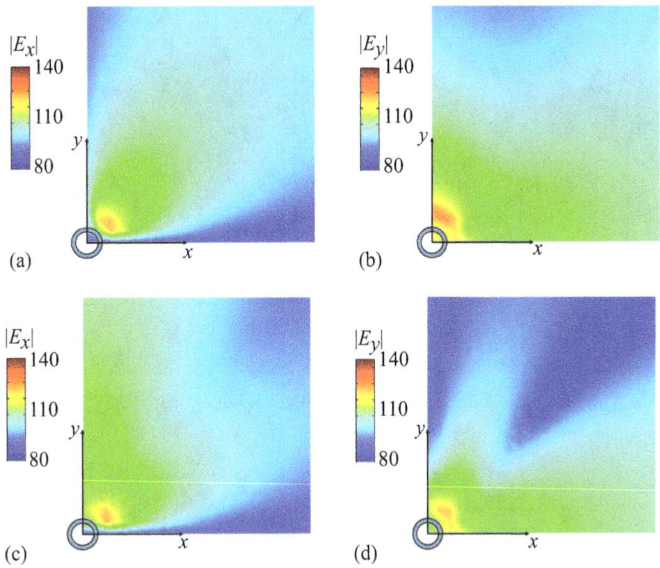

Figure 1.8 *Magnitude of near-zone electric field of PEC loop. (a)–(b) is at the first resonance while (c)–(d) is at the second resonance.*

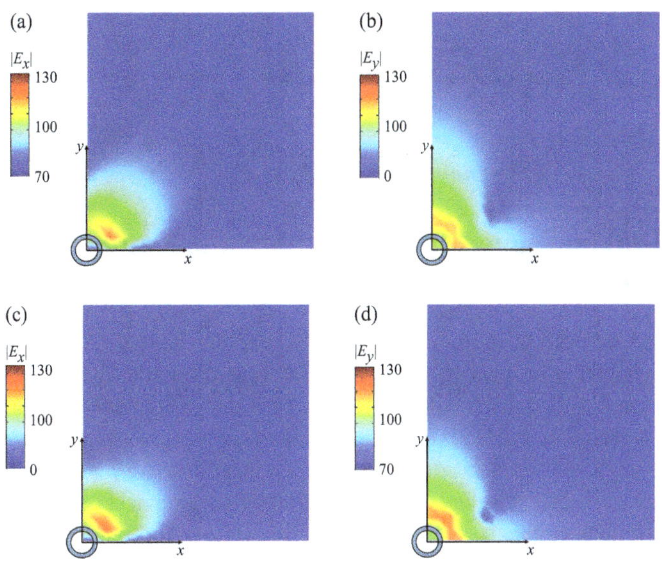

Figure 1.9 *Magnitude of near-zone electric field of Au loop. (a)–(b) is at the first resonance while (c)–(d) is at the second resonance.*

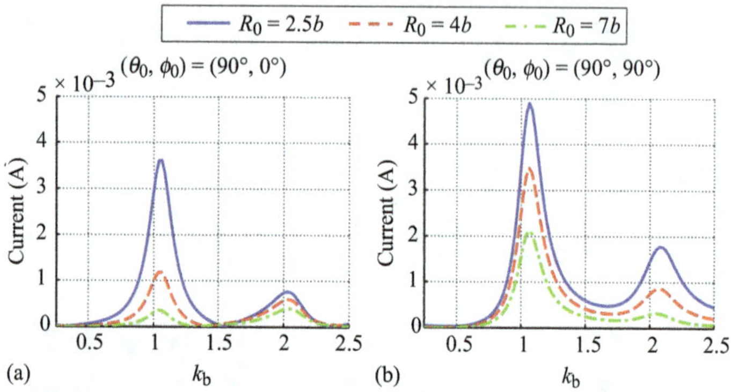

Figure 1.10 Magnitude of passive current for PEC loop.

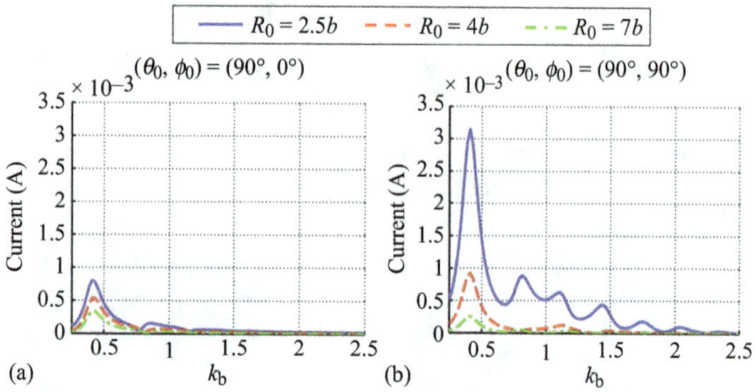

Figure 1.11 Magnitude of passive current for Au loop.

1.3 Broadband superdirective radiation modes in nanoloops

As it was pointed out at the beginning of the chapter, the theoretical modeling of nanoloops enables an intensive exploration of their properties with low computational resources. In this way, the authors of [48] discovered that single thin-wire nanoloops present superdirective radiation (also referred to as superdirectivity) for appropriate material composition and size, thereby offering a potential solution to the problem of the short-range communication of nanodevices. In the context of electrically small antennas, such as those with sizes in the range of hundreds of nanometers and operating at frequencies up to the optical regime, superdirective radiation appears for a significant enhancement of the directionality of the radiated field, in comparison to the directionality of a Hertzian or ideal point dipole. In a similar way, in the context of arrays of antennas, any directivity higher than that obtained with an array of the same length where the elements are excited by currents with uniform amplitude and phase is considered superdirectivity [49]. In practice, superdirective antennas or arrays tend to be narrowband and challenging to manufacture. These are sufficient reasons for considering superdirectivity a challenge in current antenna theory. The physical explanation for features of superdirectivity in nanoloops can be given in two ways: either by an analysis of the electric currents, or by a spherical multipole decomposition of the radiated fields.

Suppose we consider a thin-wire gold nanoloop of circumference $C = 2\pi b = 3{,}000$ nm and $\Omega = 12$ located in the XY plane. Figure 1.12 shows the directivity versus frequency along three directions calculated by means of equations provided in Section 1.1. Surprisingly, the $(\theta, \phi) = (90^\circ, 180^\circ)$ direction yields a directivity of 4.5 (6.53 dBi) at 118 THz in the far-infrared regime (i.e., at a normalized wavelength $k_b = 1.18$). Another relevant fact is the large fractional bandwidth, around 10% for values larger than 4. It is 25% for values larger than 3. Those fractional bandwidths are greater than the ones reported for two-element arrays (only 1–2% [50]). Similarly for other single-element nanoantennas such as dielectric nanodisks and nanospheres (9% and 14%, respectively [51,52], for the case of directivities over 3).

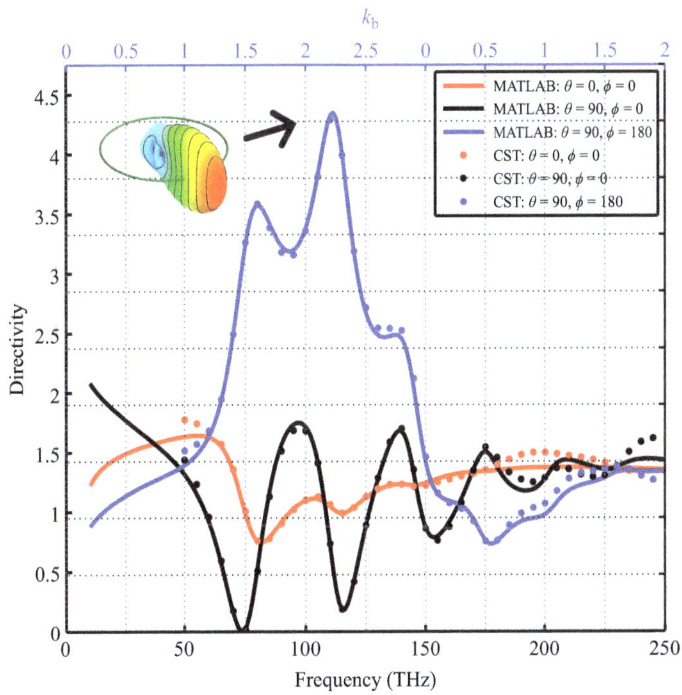

Figure 1.12 *Directivity of a thin-wire gold nanoloop of circumference* $C = 2\pi b = 3{,}000$ nm *located in the XY plane along three directions of interest. Inset shows the 3D radiation pattern in the (90,180) direction at 118 THz* ($\lambda_0 = 2{,}540$ nm) *or at* $k_b = 2\pi b/\lambda_0 = 1.18$.

From the point of view of electric sources, Figure 1.13 depicts the magnitude and phase of the current distribution in the nanoloop at the frequency of maximum directivity. At this frequency, the thin-wire gold nanoloop operates as a resonant loop of length 3 times the effective wavelength of propagation [32]. Two remarkable points are inferred from Figure 1.13: (a) there is a near 180° phase difference between the feed point and the opposite point on the nanoloop, corresponding, respectively, to $\phi = 0°$ and $\phi = 180°$; and (b) the magnitude of the current at the feeding point is around a 30% higher than the opposite point. Those properties are similar to the relationship between the feeding point currents at a superdirective two-element array of parallel linear dipoles. According to [53], there is a simple relationship describing the maximum directivity of two parallel linear dipoles of length L separated by a distance d, such as those shown in the right-hand side of Figure 1.14. This relationship is presented in Figure 1.15. Furthermore, the left-

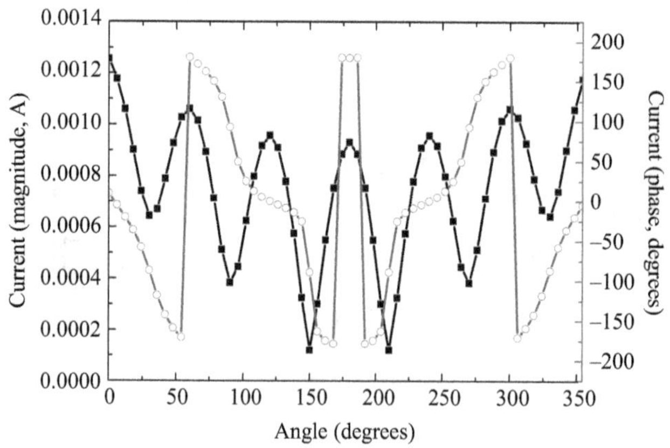

Figure 1.13 Current distribution (magnitude and phase) for the nanoloop of Figure 1.12 operating at 118 THz ($\lambda_0 = 2.54$ μm). Reprinted, with permission from [48]. © 2017 American Chemical Society.

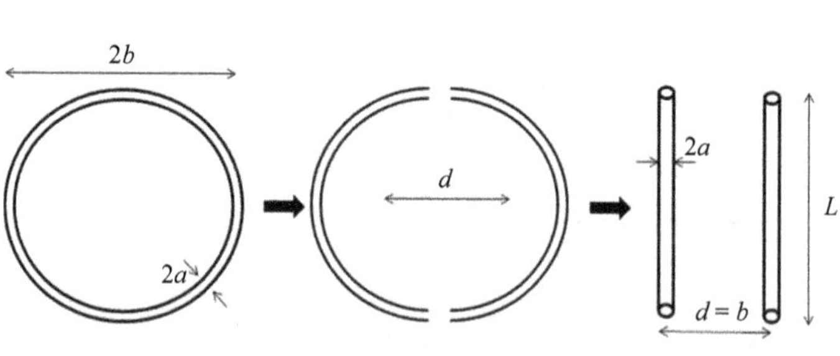

Figure 1.14 Analogy between a thin-wire nanoloop and a two-element nanodipole array. Note that this interpretation only holds when the circumference of the loop is an integer multiple of λ_{eff}. Reprinted, with permission, from [48]. © 2017 American Chemical Society.

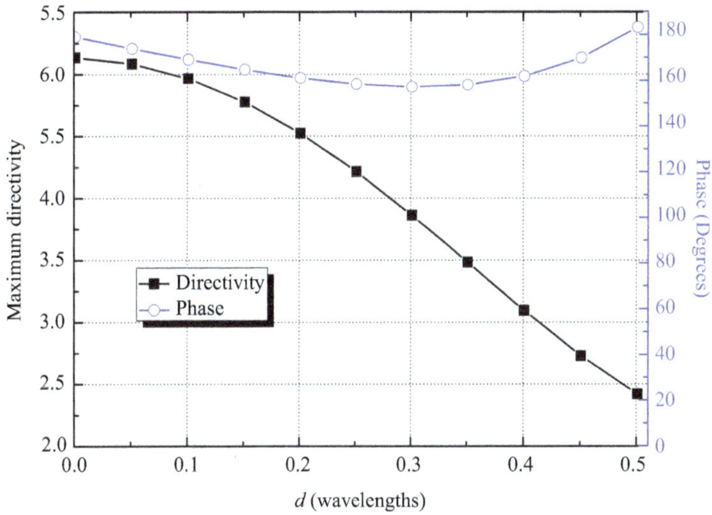

Figure 1.15 Maximum directivity and required phase delay in an array of two thin-wire $3\lambda/2$ dipoles vs. inter-element spacing. Reprinted, with permission, from [48]. © 2017 American Chemical Society.

hand side of Figure 1.14 depicts the conceptual model of the nanoloop as a 2-element array of dipoles. The separation distance d between the elements is considered roughly to be the same as the radius of the nanoloop b, which corresponds to 0.19λ at the frequency of interest. Since $C = 3\lambda_{eff}$ for the loop, our simplified model will consist of two dipoles of length $L = 3\lambda/2$. Using those dimensions, Figure 1.15 predicts a maximum directivity of approximately 4.5 in the $(\theta, \phi) = (90°, 180°)$ direction, in concordance with those obtained in the nanoloop. The phenomenon of superdirectivity is thus confirmed by inspection of electric currents.

However, to explain broadband superdirectivity we need to analyze the far-field multipole coefficients [54]. Assuming knowledge either of electric or magnetic fields surrounding the antenna, the radiated fields can be expanded in spherical coordinates in terms of multipole coefficients calculated [55] as:

$$a_M(l,m) = \frac{1}{E_0 h_l^{(1)}(kr) j^l \sqrt{\pi(2l+1)}} \int \vec{X}^*_{lm} \cdot \vec{E} d\Omega \tag{1.64}$$

$$a_E(l,m) = \frac{kr}{E_0 h_l^{(1)}(kr) j^{l+1} \sqrt{l(l+1)} \sqrt{\pi(2l+1)}} \int Y^*_{lm} E_r d\Omega \tag{1.65}$$

in the case of known electric fields, or:

$$a_E(l,m) = \frac{Z_0}{E_0 h_l^{(1)}(kr) j^{l-1} \sqrt{\pi(2l+1)}} \int \vec{X}^*_{lm} \cdot \vec{H} d\Omega \tag{1.66}$$

$$a_M(l,m) = \frac{kr}{E_0 h_l^{(1)}(kr) j^l \sqrt{l(l+1)}\sqrt{\pi(2l+1)}} \int Y_{lm}^* H_r d\Omega \tag{1.67}$$

when magnetic fields are available. In the above expressions, Y_{lm} corresponds to the scalar spherical harmonics; $\vec{X}_{lm}$ are the vector spherical harmonics; $h_l^{(1)}$ are spherical Hankel functions of the first kind; k is the wavenumber; Z_0 is the impedance of free space; a_E and a_M are the electric and magnetic multipole coefficients, respectively. In the results presented here, (1.34) are used to compute the electric fields. The magnetic fields are derived by applying the far-field identities $H_\phi = E_\theta/Z_0$ and $H_\theta = -E_\phi/Z_0$, while (1.65) and (1.67) are used to compute the electric and magnetic multipole coefficients.

Directive radiation is achieved when either magnetic or electric multipole coefficients are out of phase and create constructive interference in some particular direction [56]. For the 3,000 nm thin-wire gold nanoloop, the in-phase superposition of the $(l,m) = (3,\pm 3)$ electric modes coupled with the out-of-phase superposition of the $(l,m) = (2,\pm 2)$ electric modes are responsible for the superdirective radiation (Figure 1.16). Also, the electric multipole coefficients are larger than the magnetic ones, a fundamental difference from the superdirectivity reported for dielectric nanospheres and nanodisks [57]. A further analysis of these single-element nanoantennas shows that dielectric structures present high-order magnetic multipoles which contribute to a high directivity over a narrow bandwidth. On the other hand, thin-wire metallic nanoloops are dominated by low-order electric multipoles which results in a very large frequency bandwidth at the expense of a low radiation efficiency, a drawback which can be circumvented by increasing the wire thickness. Figure 1.17 illustrates the influence of the thickness for the 3,000 nm gold nanoloops of $\Omega = 12$ and $\Omega = 8$.

In addition to studying the directivity of these antennas, it is necessary to also investigate their radiation efficiency and gain. It is these properties that determine the communication range for such structures as well as the practicality of the device in general. Though the thin $\Omega = 12$ loop is superdirective, it suffers from a low radiation efficiency as evidenced by the gain plot in Figure 1.17(a). However, if the thickness of the loop is increased, the loop will radiate with high gain over a wide bandwidth. This is shown for the case of the thicker $\Omega = 8$ loop in Figure 1.17(b). As the loop becomes thicker, the directive band is shifted to a higher frequency regime, which no longer classifies as superdirective. Though not superdirective, this stable, directive, broadband gain radiated by thick nanoloops is novel as other nanoantennas such as dielectric nanospheres and nanodisks suffer from narrowband resonances. This suggests nanoloops could play an important role in next generation communication devices.

Finally, it is worth remarking that superdirective modes appear for thin-wire nanoloops of different constituent materials and sizes. In Figure 1.18 the red regions mark the superdirective radiation regime for Au and Al thin-wire nanoloops of varying size. At low frequencies the electrical size of the nanoloops are too small to create the superdirective condition, while for very high frequencies the electrical

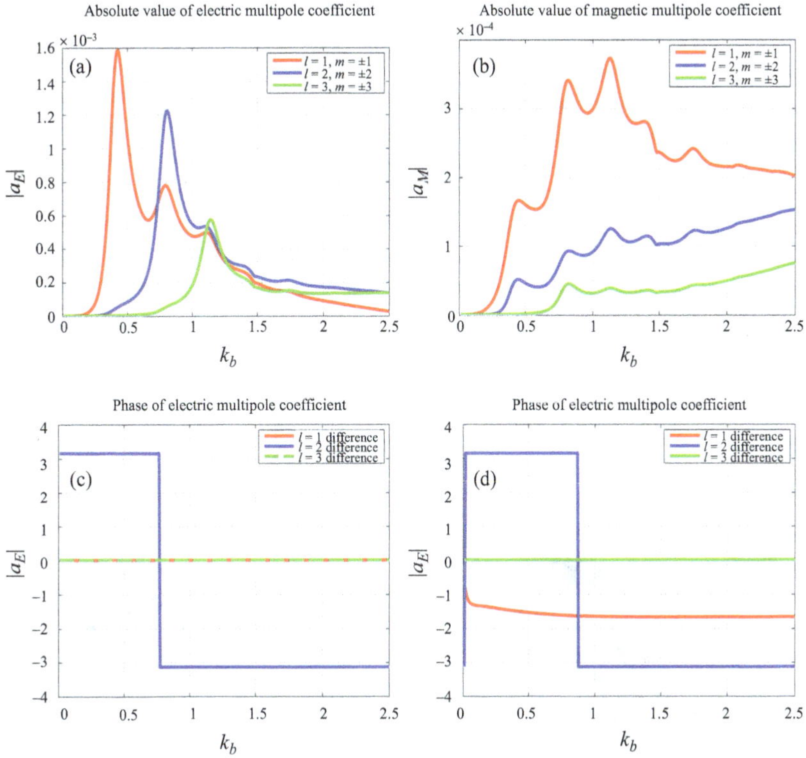

Figure 1.16 Amplitude of the (a) electric and (b) magnetic multipole coefficients and phase differences of the (c) electric and (d) magnetic multipole coefficients corresponding to the radiated fields for the 3,000 nm thin-wire gold nanoloop. Reprinted, with permission, from [48]. © 2017 American Chemical Society.

size is too large and an excessive attenuation avoids the superdirectivity. At the intermediate part of the frequency spectrum, the lower refractive index of gold enables superdirectivity at lower frequencies because of the lower phase velocity (Figure 1.19). The role of the extinction coefficient, depicted for the same figure, is more related to the bandwidth of the superdirective regime, which is higher for aluminum because the extinction coefficient has a shorter dynamic range compared to that of gold.

1.4 Trade-offs in electrical size, directivity, and gain for nanoloops

As shown in Section 1.3, a single, unloaded loop antenna of a specific size and material composition will radiate superdirectively in the endfire direction. A

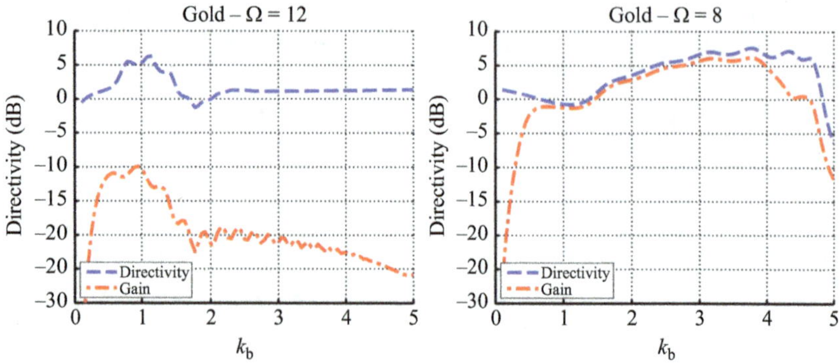

Figure 1.17 Directivity and gain for 3,000 nm circumference gold nanoloops of $\Omega = 12$ (left) and $\Omega = 8$ (right). Reprinted, with permission, from [48]. © 2017 American Chemical Society.

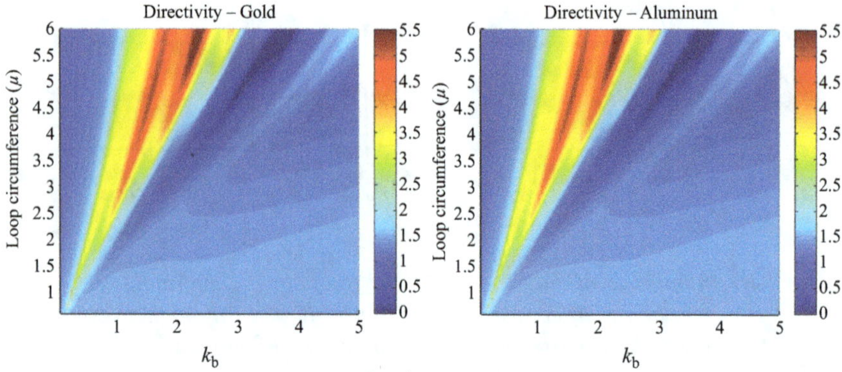

Figure 1.18 Contour plots of directivity versus loop circumference and normalized frequency for (left) gold, (right) aluminum. Reprinted, with permission, from [48]. © 2017 American Chemical Society.

natural question to consider is what trade-offs exist between directivity and electrical size for loops composed of different materials. Additionally, one might ask whether the loop's current can be modified in any way to further improve this trade-off as well as achieve this superdirective radiation pattern in other directions.

An efficient way to find these trade-offs is to use multi-objective optimization (MOO). This optimization method, compared with single-objective optimization, will return a Pareto front of solutions rather than just a single solution. This Pareto front consists of the set of solutions that are non-dominated. The concept of dominance is best illustrated graphically, as shown in Figure 1.20.

The result is a set of solutions that record the trade-offs between different objectives. The optimizer BORG will be used throughout this section [58]. BORG is a multi-objective genetic algorithm with a variety of mutation operators that the

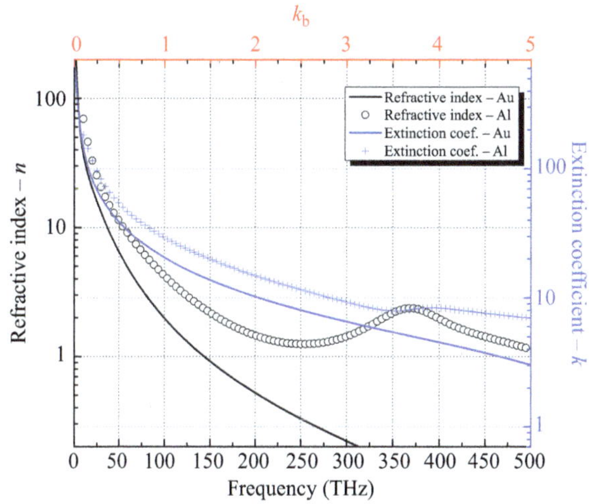

Figure 1.19 Refractive index versus frequency for gold and aluminum. Reprinted, with permission, from [48]. © 2017 American Chemical Society.

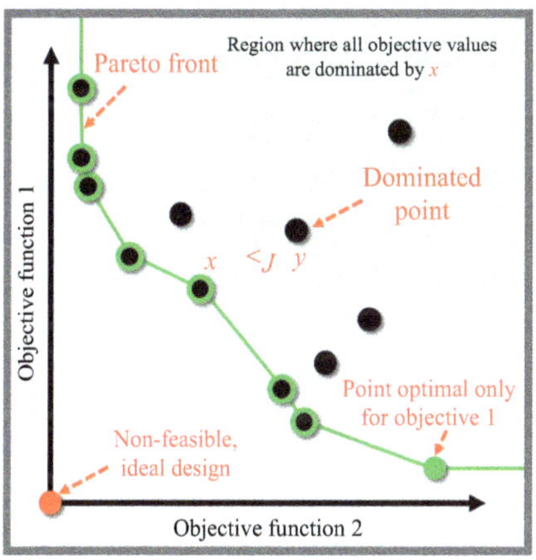

Figure 1.20 Example Pareto front.

algorithm will adaptively tune throughout the course of the optimization based upon each operator's success. Additionally, BORG introduces the concept of epsilon-dominance when converging to the Pareto front. This ensures the solutions on the Pareto front are diversified and not bunched up in one location of the objective space.

1.4.1 Optimizing a single nanoloop

The first question can be answered by comparing the Pareto fronts, representing the trade-offs between electrical size and endfire directivity, for loops comprised of different materials. In this case a comparison is made using Au, Ag, Al, and PEC materials. In all of the following studies, the antenna directivities will also be compared to the classical Harrington and Geyi limits for electrically small antennas [59,60]. These place theoretical bounds on the radiation emanating from a given volume. The limiting case of the Geyi limit reduces to the radiation pattern of a Huygen's source. The parameters are constrained such that $k_b \in [0.1,\ 2.5]$, $b \in [47.7\ \text{nm},\ 636\ \text{nm}]$, and $\Omega \in [8,\ 12]$. The results of the comparison are shown in Figure 1.21. It is clear that the superdirective radiation properties also manifest for loops comprised of different metals due to the similar dispersive behavior. Thus, given a desired operating size, the proper metal can be chosen in order to yield the largest directivity. Additionally, the PEC loop of RF frequencies, shows a different trend as a symmetric lower directivity endfire pattern becomes redirected broadside as the first resonance ($k_b = 1$) is approached.

In order to answer the second question posed above, a variety of MOO studies were conducted on a gold, loaded nanoloop antenna. This is the subject of a recent paper [26]. One well-known method used for tuning the current distribution on an antenna is to load the antenna with inductive and capacitive elements. Following the theory presented in Section 1.1, the same analysis can now be performed including the effects of lumped loads. In the optical regime, lumped loads such as inductors and capacitors are not readily available. However, in [61], Engheta has shown that the use of positive and negative permittivity materials can act as capacitive and inductive optical circuit elements, respectively. Though the load has loss in the form of a parallel resistance, the loss will be neglected for the purposes of this study. The load impedance is defined as a slab with its center located in the periphery of the loop at $\phi = \phi_L$. The impedance of the load is defined as

$$Z_L = -\frac{jd}{\omega\varepsilon_0\varepsilon_r\pi a^2} \tag{1.68}$$

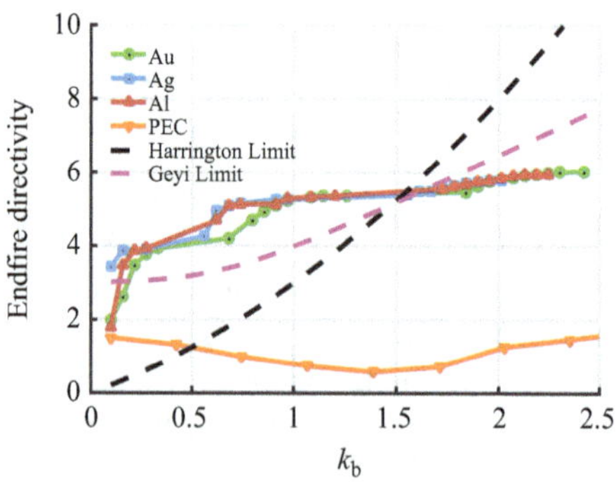

Figure 1.21 Pareto fronts displaying the trade-offs between electrical size and endfire directivity for loops comprised of Au, Ag, and Al in comparison to the Harrington and Geyi directivity limits. The directivity of a PEC loop is also plotted for the purposes of comparison.

where d is the arc length of a slab of angular width α. For this optimization the angular width and location of the loads are optimized in addition to the loop radius and thickness factor Ω. The parameters are constrained within $k_b \in [0.1,\ 1]$ for the directivity optimization and $k_b \in [0.1,\ 3]$ for the gain optimization. As a result, $b \in [47.7\ \text{nm},\ 636\ \text{nm}]$, and $\Omega \in [8,\ 12]$, $\phi_L \in [0°,\ 360°], \alpha \in [0°,\ 200°]$. The large dynamic range of α is used only for the purpose of modeling the impedance. In practice, an angularly small slab of larger permittivity can be used in place of an angularly large slab of low permittivity. The Pareto fronts for the case of no loads, one load, and two loads are shown below for the directivity optimization in four different directions. From Figure 1.22, it is clear that loading can both improve the trade-off between electrical size and directivity as well as enable superdirectivity in other directions within the plane of the loop. The solutions corresponding to $k_b = 1$ with the desired maximum directivity in the $(90°,\ 0°)$ direction are shown in Table 1.5 with their associated currents analyzed in Figure 1.23. In order to radiate

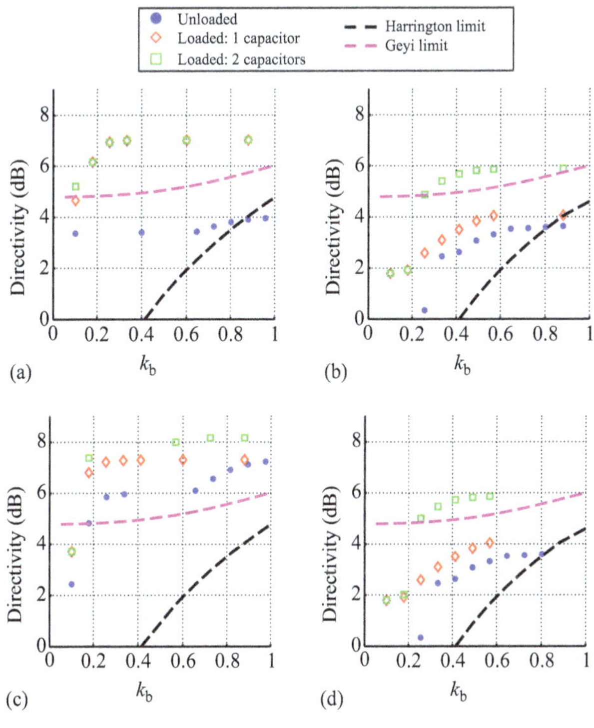

Figure 1.22 Pareto fronts of an unloaded gold nanoloop and a gold nanoloop loaded with one or two capacitors. The objectives are minimizing k_b and maximizing directivity along (a) $(\theta, \phi) = (90°,\ 0°)$, (b) $(\theta, \phi) = (90°,\ 90°)$, (c) $(\theta, \phi) = (90°,\ 180°)$, (d) $(\theta, \phi) = (90°,\ 270°)$. © 2019 IEEE. Reprinted, with permission, from [26].

Table 1.5 Results of optimization for $(\theta, \phi) = (90°, 0°)$ *© 2019 IEEE. Reprinted, with permission, from [26].*

Number of loads	b (nm)	Ω	$\phi_L(1)$ (deg)	$\alpha(1)$ (deg)	$\phi_L(2)$ (deg)	$\alpha(2)$ (deg)
0	506	11.5				
1	406	11.8	180	0.131		
2	72.5	8.8	60	10.65	300	10.65

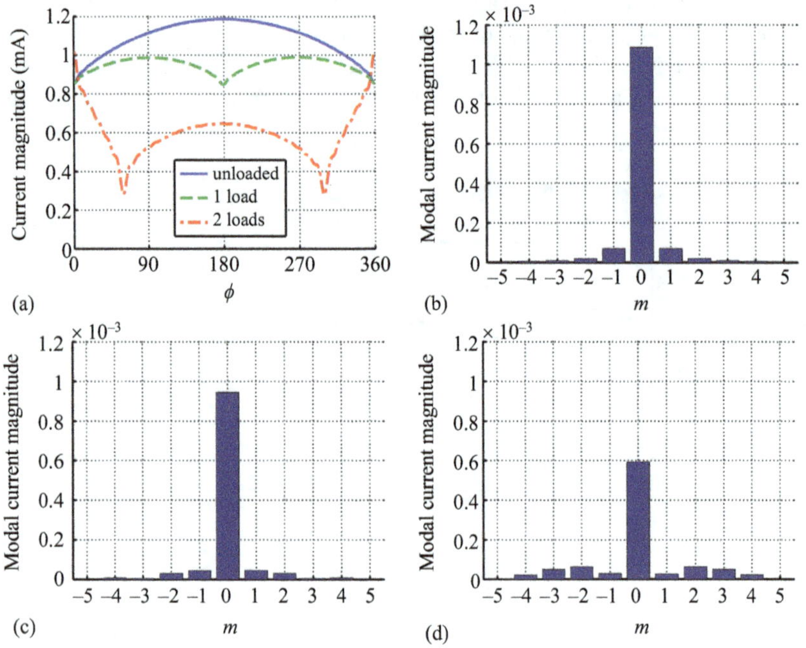

Figure 1.23 Optimized results for maximum directivity along $(\theta, \phi) = (90°, 0°)$ *at* $k_b = 0.1$. *(a) Current distributions for the optimized unloaded and loaded gold nanoloops. (b)–(d) show the modal current magnitudes for the unloaded, loaded with one capacitor and loaded with two capacitor cases, respectively. © 2019 IEEE. Reprinted, with permission, from [26].*

in the desired direction, the current distributions must be azimuthally symmetric and the modal currents must be symmetric with respect to its mode order. Interestingly, the modal currents of the two-load solution do not monotonically decrease away from the zeroth order mode. Rather, the magnitudes of the second and third modes are increased such that they are larger than the first mode. This distribution of the modal currents is only possible for the case of two loads due to the symmetry required to radiate along the (90°, 0°) direction. The result is that the doubly loaded solution outperforms both the unloaded and singly loaded solutions. However, in order for the loop to radiate strongly in either the (90°, 0°) or (90°, 270°) directions, the current distribution must now be asymmetric since the desired radiation direction is no longer symmetric with respect to the feed located at $\phi = 0°$. Thus, it is not surprising that the loaded loop has a better performance than the unloaded loop as seen in Figure 1.22(b) and (d).

Due to the symmetry of the standing wave within an unloaded loop, the unloaded loop is constrained to radiate only a bidirectional pattern in this case, rather than a unidirectional pattern as in the case of the loaded loop. This can be seen in the full 3D directivity plots shown in Figure 1.24. Similarly, from the modal current distributions in Figure 1.25, it is evident that the $m = -2$ mode is enhanced when the loop is loaded. The beam can be steered oppositely towards the (90, 270) by mirroring the loads. This results in an enhancement of the $m = +2$ current mode.

Though a majority of the loops discussed in this section are superdirective, with respect to the Geyi and Harrigton limits, they suffer from a poor efficiency since many of the loops must be thin in order to support the necessary current modes. As a result, it is instructive to also examine the trade-offs between the electrical size and gain, including the effects of the radiation efficiency. The Pareto fronts of these studies are shown in Figure 1.26. In this case, the loops are unable to radiate with supergain, i.e., gain greater than that predicted by the Geyi limit. However, the capacitive loading still improves the loop's performance.

As shown in the above examples, the analytical equations describing the current and radiation of the circular loop enables both a greater understanding of the underlying physics involved as well as highly efficient design cycles. These examples demonstrate only a small portion of the capabilities of this theory. Additionally, both unloaded and loaded nanoloops can be designed and optimized to satisfy various objectives, such as polarization conversion or higher modal scattering.

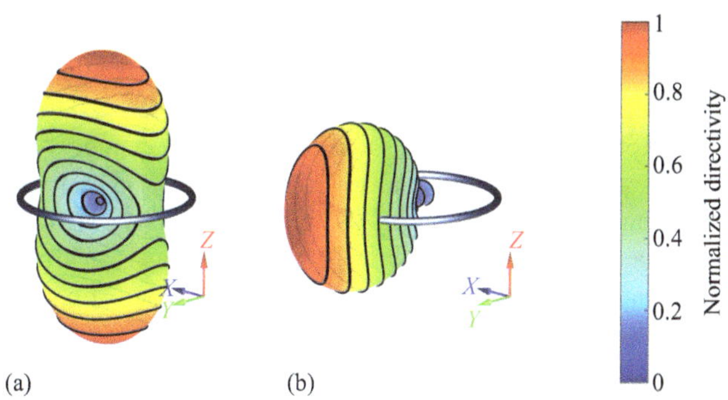

Figure 1.24 Normalized directivity patterns for the (a) unloaded loop and (b) loaded loop optimized for maximum directivity along $(\theta, \phi) = (90°, 90°)$.

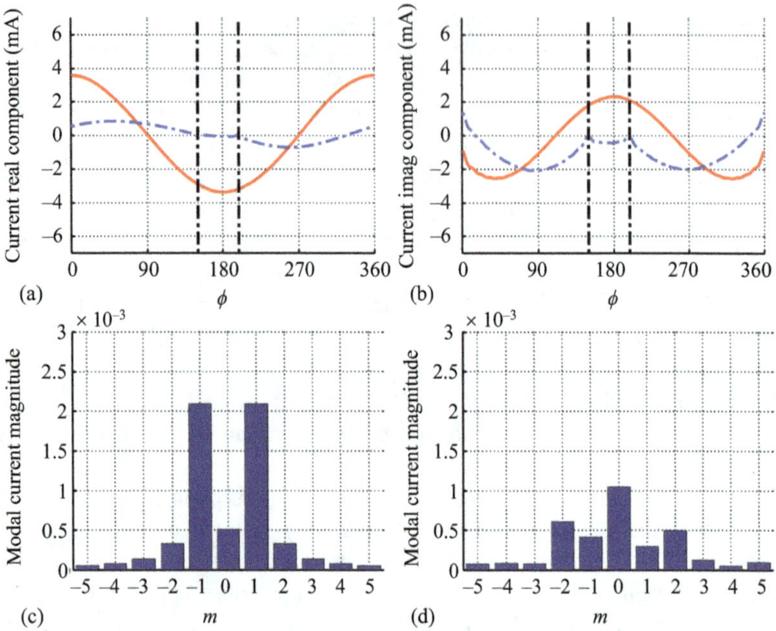

Figure 1.25 Comparison of a loaded and unloaded gold nanoloop optimized for maximum directivity along $(\theta, \phi) = (90^\circ, 90^\circ)$ *at* $k_b = 0.88$. *The solid red and dashed blue lines, respectively, correspond to the unloaded and loaded loops. The black dashed lines show the load locations. (a) Real and (b) imaginary components of the current. (c) and (d) show the modal current magnitudes for the unloaded and loaded cases, respectively. © 2019 IEEE. Reprinted, with permission, from [26].*

1.4.2 Optimizing arrays of nanoloops

In practice, loop antennas are often employed within an array of similar elements. One parasitic array configuration that is commonly used is the Yagi–Uda antenna. In this configuration, a single element is fed by the source, while a reflector and several directors function to create a highly directive radiation pattern. A particular configuration of interest is the loop Yagi–Uda in which the elements of the antenna are loops of comparable size on the order of λ. Though the loop antenna theory, in its current form, cannot be directly used to optimize such an antenna, it can help establish bounds for the various geometric parameters as well as highlight regions of interest within the frequency band under study. The trade-off between the total electrical size of the antenna, the number of directors, and the antenna directivity can similarly be studied for the case of the Yagi–Uda antenna. These trade-offs for PEC Yagi–Uda antennas have been well documented in the literature [62]. However, such trade-offs for the case of dispersive nanoloop Yagi–Uda antennas are not well known. Again, BORG is used to determine this trade-off. This study is the subject of [63]. The general testbed antenna design is shown in Figure 1.27 in

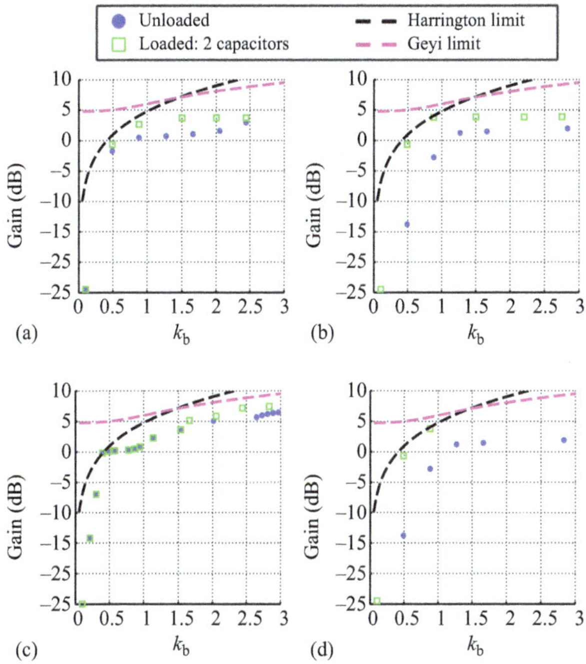

Figure 1.26 Pareto fronts of an unloaded gold nanoloop and a gold nanoloop loaded with one or two capacitors. The objectives are minimizing k_b and maximizing gain along (a) $(\theta, \phi) = (90°, 0°)$, (b) $(\theta, \phi) = (90°, 90°)$, (c) $(\theta, \phi) = (90°, 180°)$, (d) $(\theta, \phi) = (90°, 270°)$. © 2019 IEEE. Reprinted, with permission, from [26].

which the element radii, wire radii, as well as the spacing between elements are varied as parameters in the optimization. The antenna is arranged in a broadside, co-axial configuration, with the loop normal vector lying parallel to the antenna axis. The analysis is completed for the number of directors ranging from 0 to 5 using the commercial method of moments solver FEKO. The first study investigates this trade-off at a frequency of 340 THz since a single nanoloop with radius $b = 0.1635$ μm exhibits a large broadside directivity at this frequency. The results are shown in Figure 1.29. Similar to the PEC Yagi–Uda, the broadside directivity increases as the number of directors is increased. However, this increase in the directivity comes at the expense of requiring an electrically larger antenna. This trade-off is best visualized by examining two solutions, as shown in the same figure. These solutions are the Yagi–Uda composed of 0 directors with a directivity of 7.2 dB and the antenna composed of 4 directors with a maximum directivity of 14.1 dB. Additionally, a similar study involving the gain is also instructive. The results of this study are shown in Figure 1.29 revealing a comparable trade-off trend. However, the gain begins to saturate as the number of directors is increased, unlike the directivity trade-off.

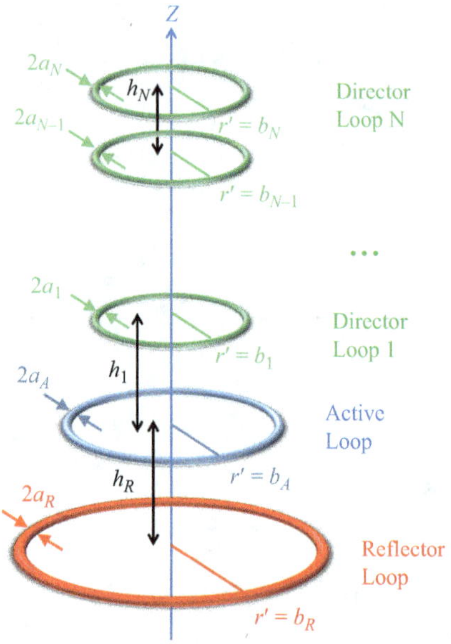

Figure 1.27 Broadside Yagi–Uda nanoloop array geometry. © 2017 IEEE. Reprinted, with permission, from [63].

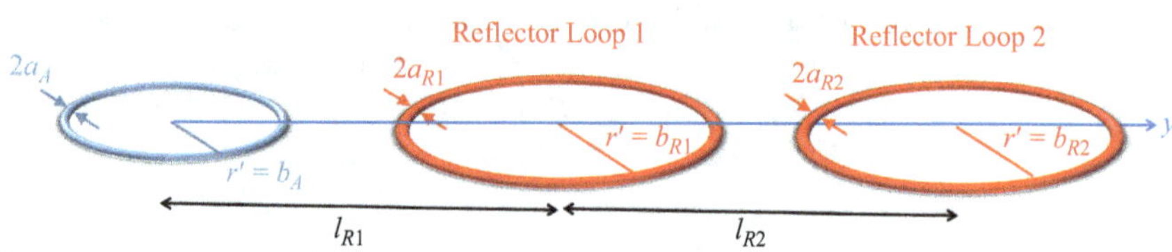

Figure 1.28 Endfire Yagi–Uda nanoloop array geometry. © 2017 IEEE. Reprinted, with permission, from [63].

Lastly, loops in a co-planar Yagi–Uda configuration are studied as shown in Figure 1.28. In this case, the loops lie in a straight line parallel to the antenna axis and radiate in the endfire direction. An operating frequency of 250 THz is chosen based on the highly endfire directive properties of a single $b = 0.4775$ μm loop at this frequency. The results are shown in Figure 1.30. Interestingly, the radiation mechanism of the endfire Yagi–Uda is different than the broadside configuration as no directors and multiple reflectors are necessary in order to achieve the endfire radiation. A similar trend concerning the trade-off between gain and size is found as the gain saturates if the number of the reflectors exceeds 2.

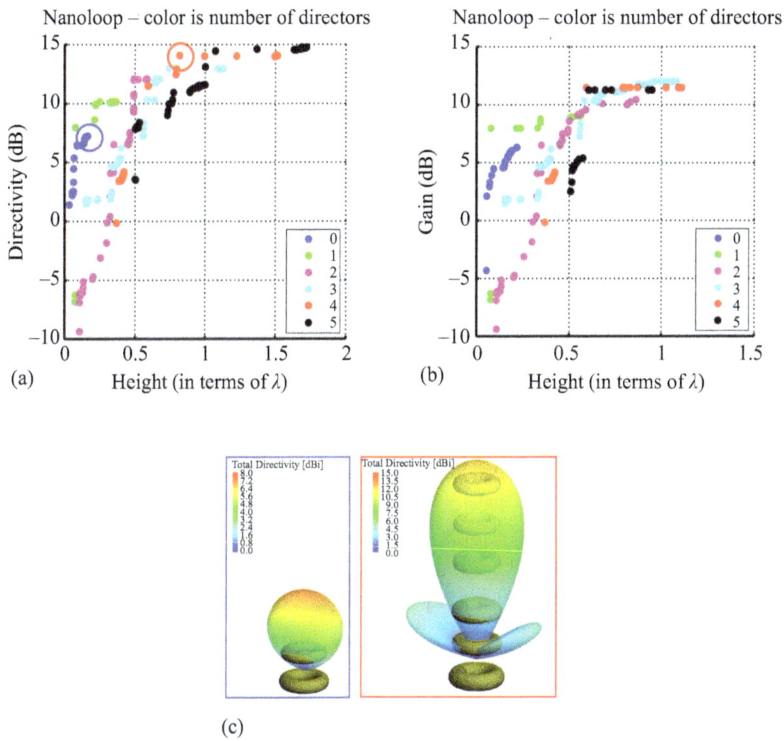

Figure 1.29 Pareto fronts with the goals of maximizing (a) directivity, (b) gain, and minimizing height for the broadside nanoloop Yagi–Uda array for 0, 1, 2, 3, 4, *and* 5 *director elements. (c) Two example solutions composed of* 1 *and* 5 *directors. © 2017 IEEE. Reprinted, with permission, from [63].*

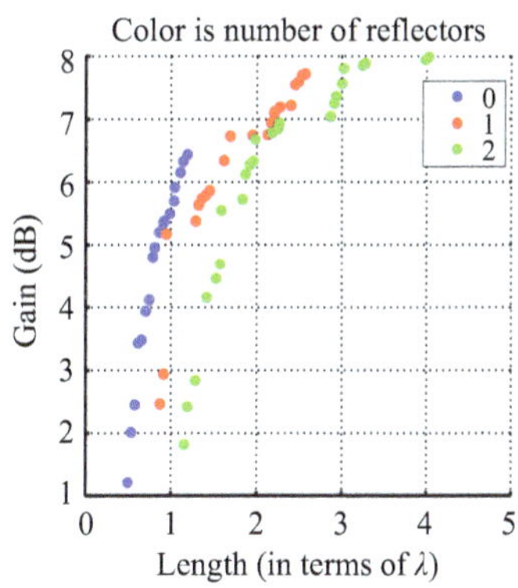

Figure 1.30 Pareto fronts of the endfire nanoloop Yagi–Uda array for 0–2 reflector elements. © 2017 IEEE. Reprinted, with permission, from [63].

1.5 Elliptical nanoloops

A more general case of the circular loop is the elliptical loop. The elliptical loop now has two radii, b_x and b_y, in addition to the wire radius a as depicted in Figure 1.31. This provides an extra parameter for tuning the currents and thereby the radiated fields. In 1958 and 1959, Loh and Wong derived closed-form analytical expressions for the far-zone radiating fields [65] and radiation resistance [66] of an elliptical loop antenna assuming a constant current, an assumption which is valid only when the loop is electrically very small. This theory was further extended to include more complicated current distributions in 1991 [67] by employing asymptotic antenna theory. A variety of numerical methods have been employed to study the characteristics of elliptical loop antennas, most notable the method of moments [68] and the theory of characteristic modes [69]. Following the recent work by Werner in [70], this section will present an analytical approach for calculating the far-zone radiated fields produced by an elliptical loop of arbitrary electrical size valid in the RF, assuming the current takes the form of a Fourier cosine series of the general azimuthal coordinate u. This loop is parametrized in (1.69) and (1.70) as

$$\left.\begin{aligned} x &= b_x \cos\theta \\ y &= b_y \sin\theta \end{aligned}\right\} -\pi \le u \le \pi \tag{1.69}$$

$$r'(u) = \sqrt{b_x^2\cos^2(u) + b_y^2\sin^2(u)} \tag{1.70}$$

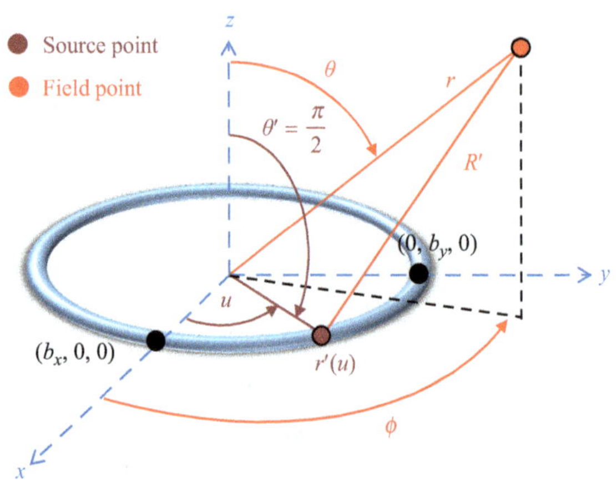

Figure 1.31 Elliptical loop antenna geometry.

Further, the vector potential $\mathbf{A} = A_x\widehat{a}_x + A_y\widehat{a}_y$, where

$$A_x(r,\theta,\phi) = -\frac{\mu b_x}{4\pi}\frac{e^{-jk_0 r}}{r} \times \int_{-\pi}^{\pi} I(u)e^{jk_0\left[b_x \cos u \sin\theta \cos\phi + b_y \sin u \sin\theta \sin\phi\right]} \sin u \, du \tag{1.71a}$$

$$A_y(r,\theta,\phi) = \frac{\mu b_y}{4\pi}\frac{e^{-jk_0 r}}{r} \times \int_{-\pi}^{\pi} I(u)e^{jk_0\left[b_x \cos u \sin\theta \cos\phi + b_y \sin u \sin\theta \sin\phi\right]} \cos u \, du \tag{1.71b}$$

Similar to the case of the circular loop, the current is now decomposed into a Fourier series as a function of the general azimuthal coordinate u.

$$I(u) = \sum_{n=0}^{\infty} I_n \cos(nu) \tag{1.72}$$

The components of the vector potential $\mathbf{A}$ now take the following form:

$$A_x(r,\theta,\phi) = \frac{\mu b_x}{4\pi}\frac{e^{-jk_0 r}}{r}\sum_{n=0}^{\infty} I_n g_n^x(\theta,\phi) \tag{1.73a}$$

$$A_y(r,\theta,\phi) = \frac{\mu b_y}{4\pi}\frac{e^{-jk_0 r}}{r}\sum_{n=0}^{\infty} I_n g_n^y(\theta,\phi) \tag{1.73b}$$

The g_n functions contain the θ and ϕ dependence of the vector potential and are expressed below.

$$g_n^x(\theta,\phi) = \int_{-\pi}^{\pi} e^{j\xi \cos(u-\gamma)} \cos(nu)\sin(u)\, du \tag{1.74a}$$

$$g_n^y(\theta,\phi) = \int_{-\pi}^{\pi} e^{j\xi \cos(u-\gamma)} \cos(nu)\cos(u)\, du \tag{1.74b}$$

where the quantities γ and ξ are defined as

$$\gamma = \arctan\left(\frac{b_y}{b_x}\tan\phi\right) \tag{1.75a}$$

$$\xi = k_0 b_x \sin\theta\sqrt{\cos^2\phi + \left(\frac{b_y}{b_x}\right)^2 \sin^2\phi} = \frac{k_0 b_x \sin\theta}{\sqrt{\cos^2\gamma + \left(\frac{b_x}{b_y}\right)^2 \sin^2\gamma}} \tag{1.75b}$$

Further, the Jacobi–Anger representation can be used to expand the integrands in (1.74a) and (1.74b) as shown below.

$$e^{j\xi\cos(u-\gamma)} = \sum_{m=0}^{\infty} \varepsilon_m (j)^m \mathbf{J}_m(\xi)\cos[m(u-\gamma)] \tag{1.76a}$$

$$\varepsilon_m = \begin{cases} 1, & m = 0 \\ 2, & m > 0 \end{cases} \tag{1.76b}$$

Upon computing the integrals in (1.74a) and (1.74b) and transforming to spherical coordinates, the electric far-field is expressed as

$$E_\theta(r,\theta,\phi) = -\frac{k_0 b_y \xi_0}{2}\frac{e^{-jk_0 r}}{r}\frac{\cos\theta}{F(\gamma)}S_2(\xi,\gamma) \tag{1.77a}$$

$$E_\phi(r,\theta,\phi) = -\frac{k_0 b_x \xi_0}{2}\frac{e^{-jk_0 r}}{r}\left\{F(\gamma)S_1(\xi,\gamma) + \left[\left(\frac{b_y}{b_x}\right)^2 - 1\right] \times \frac{\cos\gamma\,\sin\gamma}{F(\gamma)}S_2(\xi,\gamma)\right\} \tag{1.77b}$$

where

$$S_1(\xi,\gamma) = \sum_{n=0}^{\infty} I_n (j)^n \mathbf{J}'_n(\xi)\cos(n\gamma) \tag{1.78a}$$

$$S_2(\xi,\gamma) = \sum_{n=0}^{\infty} I_n (j)^n n\frac{\mathbf{J}_n(\xi)}{\xi}\sin(n\gamma) \tag{1.78b}$$

$$F(\gamma) = \sqrt{\sin^2\gamma + \left(\frac{b_y}{b_x}\right)^2\cos^2\gamma} \tag{1.78c}$$

1.5.1 Special cases

The first special case to consider is the $n = 0$ modal solution, the case of constant current. For this case $I(u) = I_0$, $S_1 = -I_0\mathbf{J}_0(\xi)$, $S_2 = 0$. As a result, the electric field only possesses a ϕ component, defined as

$$E_\phi(r,\theta,\phi) = \frac{k_0 b_x I_0}{2}\frac{e^{-jk_0 r}}{r}F(\gamma)\mathbf{J}_1(\xi) \tag{1.79}$$

This agrees with the solution reported by Loh and Wong in [65]. The second case considers a cosinusoidal current distribution with $I(u) = I_p\cos(pu)$. In this case the far-field expressions reduce to

$$E_\theta(r,\theta,\phi) = -\frac{(j)^p k_0 b_y \xi_0 I_p}{2}\frac{e^{-jk_0 r}}{r}\frac{\cos\theta}{F(\gamma)}p\frac{\mathbf{J}_p(\xi)}{\xi}\sin(p\gamma) \tag{1.80a}$$

$$E_\phi(r,\theta,\phi) = -\frac{(j)^p k_0 b_x \xi_0 I_p}{2}\frac{e^{-k_0 r}}{r}\left\{F(\gamma)\mathbf{J}'_p(\xi)\cos(p\gamma) + \left[\left(\frac{b_y}{b_x}\right)^2 - 1\right]\right.$$
$$\left.\times\frac{\cos\gamma\sin\gamma}{F(\gamma)}p\frac{\mathbf{J}_p(\xi)}{\xi}\sin(p\gamma)\right\} \tag{1.80b}$$

These electric field expressions are implemented in MATLAB® and their validity is studied in comparison to the commercial method of moments solver FEKO [64]. This study is completed for a variety of loops with different ratios, b_x:b_y, the circular loop being the loop in which b_x:b_y = 1:1. It is expected that a PEC loop with circumferences of λ and 2λ will have purely cosinusoidal current distributions of $\cos u$ and $\cos(2u)$, respectively. A comparison between the MATLAB® implementation and FEKO for different cuts of the electric field radiated by loops of circumference λ is shown in Figure 1.32. Here, it is demonstrated that there is full agreement between the MATLAB® implementation and FEKO for the cases

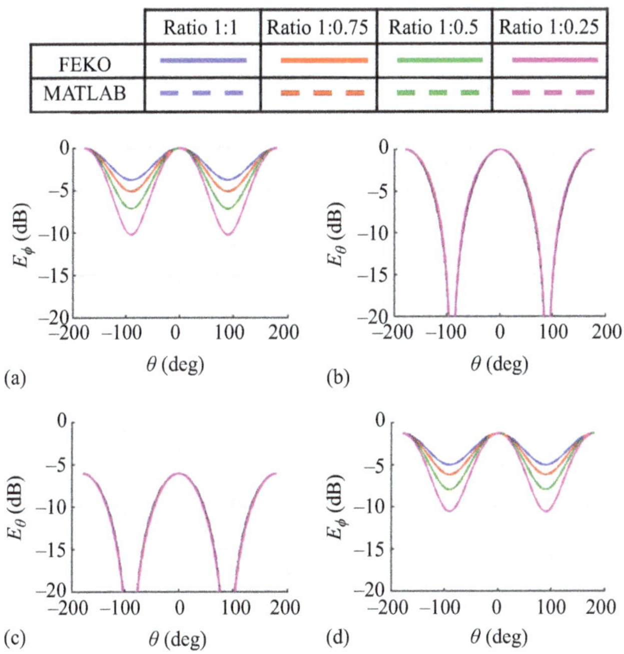

Figure 1.32 Cuts of the normalized far-zone electric field pattern in decibels as computed by FEKO and MATLAB® for a current of $I(u) = \cos u$ on the four elliptical loop antenna geometries under consideration, where each antenna is assumed to have a perimeter of λ. (a) E_ϕ along $\phi = 0°$, (b) E_θ along $\phi = 90°$, (c) E_θ along $\phi = 30°$, (d) E_ϕ along $\phi = 30°$. © 2018 IEEE. Reprinted, with permission, from [70].

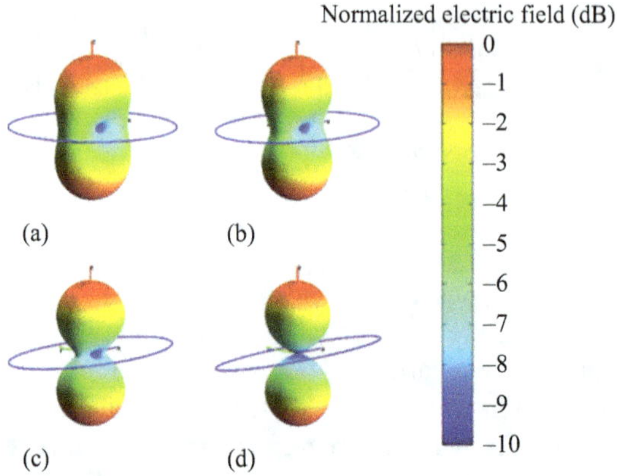

Figure 1.33 Normalized far-zone electric field pattern in decibels as computed by MATLAB® for a current of $I(u) = \cos u$ on an elliptical loop antenna with ratio $b_x : b_y$ of (a) 1:1, (b) 1:0.75, (c) 1:0.5, (d) 1:0.25. Note that each antenna, in this case has a perimeter of λ. © 2018 IEEE. Reprinted, with permission, from [70].

shown. Additionally, 3D normalized electric far-field plots for the different loops are shown in Figure 1.33. According to both of these figures, it is evident that the pattern becomes more directive broadside as the ratio between b_x and b_y is increased. This is an interesting phenomenon inherent to the elliptical loop.

The same studies are completed for the loops of circumference 2λ. The results are shown in Figures 1.34 and 1.35. Again, there is full agreement between the MATLAB® implementation and FEKO. Similarly, upon examining the 2D cuts and the 3D electric far-field pattern, a similar enhancement trend is again noticed as the ratio between b_x and b_y is increased. In the case of the circular loop, *i.e.* $b_x : b_y = 1$, the pattern possesses four-fold symmetry with the maximum field occurring in each of the $\phi = 0°, 90°, 180°, 270°$ directions. However, for the case of the largest ratio between b_x and b_y, the beam shows a maximum field occuring in the $\phi = 0°, 180°$ directions. As a result, it is evident that the act of squeezing a circular loop into an elliptical loop of the same perimeter enhances the far-fields radiated by the loop.

Though only results for the elliptical PEC loop were discussed here, full-wave analysis predicts that the same enhanced radiated field directionality trend is also true for elliptical loops comprised of dispersive materials. Thus, elliptical nanoloops will likewise produce an enhanced form of the superdirective properties of circular loops, as shown in Section 1.3, highlighting an exciting future area of research.

1.5.2 The electrically small elliptical loop antenna

In this section, we further consider the important special case of an electrically small elliptical loop antenna (i.e. when $n = 0$). For this degenerate case, $E_\theta = 0$ and the expression for the total radiated power given in (1.35) reduces to

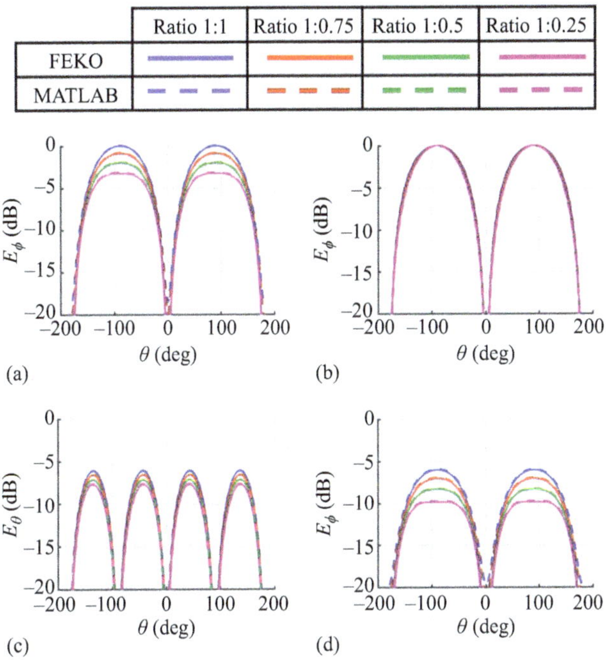

Figure 1.34 Cuts of the normalized far-zone electric field pattern in decibels as computed by FEKO and MATLAB® for a current of $I(u) = \cos(2u)$ *on the four elliptical loop antenna geometries under consideration, where each antenna is assumed to have a perimeter of 2λ. (a)* E_ϕ *along* $\phi = 0°$*, (b)* E_ϕ *along* $\phi = 90°$*, (c)* E_θ *along* $\phi = 30°$*, (d)* E_ϕ *along* $\phi = 30°$*. © 2018 IEEE. Reprinted, with permission, from [70].*

$$P_{rad} = \int_0^{\pi}\int_0^{\pi} \frac{|E_\phi|^2}{\xi_0} r^2 \sin\theta d\theta d\phi \tag{1.81}$$

Since $n = 0$, the electric field component E_ϕ represented by (1.77b) reduces to the form

$$E_\phi = \xi_0 I_0 \frac{k_0 b_x}{2} \frac{e^{-jk_0 r}}{r} F(\gamma)\mathbf{J}_1(\xi) \tag{1.82}$$

Substituting (1.82) into (1.81) yields

$$P_{rad} = \xi_0 |I_0|^2 \left(\frac{k_0 b_x}{2}\right)^2 \int_0^{\pi}\int_0^{\pi} F^2(\gamma)\mathbf{J}_1^2(\xi)\sin\theta d\theta d\phi \tag{1.83}$$

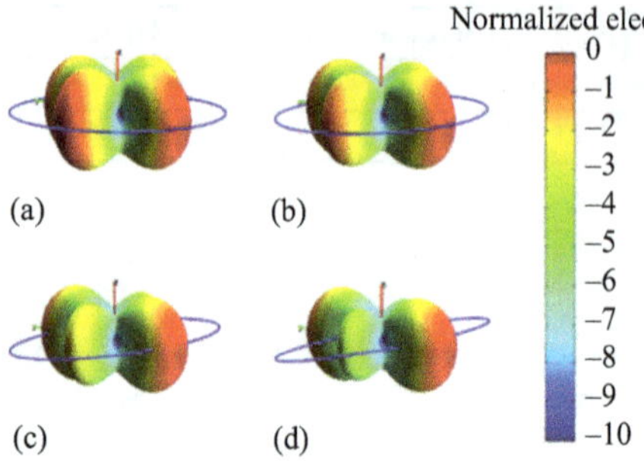

Figure 1.35 Normalized far-zone electric field pattern in decibels as computed by MATLAB® for a current of $I(u) = \cos(2u)$ on an elliptical loop antenna with ratio $b_x:b_y$ of (a) 1:1, (b) 1:0.75, (c) 1:0.5, (d) 1:0.25. Note that each antenna, in this case has a perimeter of 2λ. © 2018 IEEE. Reprinted, with permission, from [70].

The double integral in (1.83) may be evaluated, which leads to the following exact expression for the total radiated power

$$P_{rad} = \xi_0 |I_0|^2 (k_0 b_x)^2 (k_0 b_y)^2 \sum_{q=0}^{\infty} \frac{(-1)^q (k_0 b_x)^{2q}}{(q+1)(2q+3)(2q+4)[q!]^2} \Xi_q(\nu) \tag{1.84}$$

where

$$\Xi_q(\nu) = \int_0^{\pi} \left[\cos^2\phi + \nu^2 \sin^2\phi\right]^q d\phi = \frac{\pi}{2^{2q}} \sum_{k=0}^{q} C(k,q) \nu^{2q-2k} \tag{1.85}$$

and

$$C(k,q) = \frac{(2k)!(2q-2k)!}{[k!]^2[(q-k)!]^2} \tag{1.86a}$$

$$\nu = (b_y/b_x) \tag{1.86b}$$

Next, we consider the radiation resistance for an electrically small elliptical loop antenna, which may be represented as

$$R_{rad} = \frac{2P_{rad}}{|I_0|^2} = 2\xi_0 (k_0 b_x)^2 (k_0 b_y)^2 \sum_{q=0}^{\infty} \frac{(-1)^q (k_0 b_x)^{2q}}{(q+1)(2q+3)(2q+4)[q!]^2} \Xi_q(\nu) \tag{1.87}$$

If $k_0 b_x \ll 1$ then it follows that

$$R_{rad} \approx \frac{1}{6} \xi_0 (k_0 b_x)^2 (k_0 b_y)^2 \Xi_0(\nu) \tag{1.88}$$

But

$$\Xi_0(\nu) = \int_0^{\pi} d\phi = \pi \tag{1.89}$$

Hence, (1.88) reduces to the result

$$R_{rad} = \frac{\xi_0 \pi}{6} (k_0 b_x)^2 (k_0 b_y)^2 \tag{1.90}$$

If $b_x = b_y = b$, then (1.90) becomes

$$R_{rad} = \frac{\xi_0 \pi}{6} (k_0 b)^4 \tag{1.91}$$

which is the well-known result for an electrically small circular loop.

1.6 Summary

The over-arching theory of nanoloop antennas has been laid out in Section 1.1. Here it was shown that the RF loop antenna theory could be extended up through the optical regime by representing the metallic loop as a frequency-dependent surface impedance. This impedance modified the definition of the loop's modal admittance and associated current distribution. As a result, both the near-field and far-field radiation properties of loop antennas were likewise extended to dispersive frequency regimes. Additionally, the theory was also developed to include lumped impedances placed along the periphery of the loop, providing a means to further control the radiation from such structures. In Section 1.2, it was shown how nanoloop theory could be integrated with the induced EMF method to investigate the coupling between nanoloops set in various configurations. Despite the geometric simplicity of nanoloops, Section 1.3 showed that a single nanoloop of the correct size and material composition is remarkably able to radiate superdirectively. Further, nanoloops that are sufficiently thick are capable of directive, high gain, broadband radiation, surpassing that of other nanoantennas such as dielectric nanodisks and nanospheres. Section 1.4 showcased the advantage of the closed-form nature of the nanoloop theory, as it enables both rapid parametric studies as well as numerical optimizations for specific performance objectives. It was shown that it is possible to optimize a loaded nanoloop to achieve superdirectivity in various directions as well as to achieve increased gain from the antenna. Finally, Section 1.5 highlighted an exciting new area of research, in which the concepts of a circular loop were generalized to those of an elliptical loop in which there are now additional degrees of freedom to be exploited for future studies.

References

[1] Schulte J. Nanotechnology: global strategies, industry trends and applications. Chichester, UK: John Wiley & Sons; 2005.

[2] Biagioni P, Huang JS, Hecht B. Nanoantennas for visible and infrared radiation. Reports on Progress in Physics. 2012;75(2):024402.

[3] Johnson PB, Christy RW. Optical constants of the noble metals. Physical Review B. 1972;6(12):4370.

[4] Novotny L, Hecht B. Principles of nano-optics. Cambridge, UK: Cambridge University Press; 2012.

[5] Balanis CA. Antenna theory: analysis and design. Hoboken, NJ: John Wiley & Sons; 2016.

[6] Yosif B, Samra A, Modeling of optical nanoantennas. Physics Research International. 2012;2012(321075):1–10.

[7] Ansys. Ansys Electronics Desktop; 2016. https://www.ansys.com/products/electronics/ansys-hfss.

[8] Alù A, Engheta N. Input impedance, nanocircuit loading, and radiation tuning of optical nanoantennas. Physical Review Letters. 2008;101(4):043901/1–4.

[9] Panaretos A, Yuwen Y, Werner D, *et al.* Tuning the optical response of a dimer nanoantenna using plasmonic nanoring loads. Scientific Reports. 2015;5(1):9813–9826.

[10] Ahmadian D, Ghobadi C, Nourinia J. Ultra-compact two-dimensional plasmonic nano-ring antenna array for sensing applications. Optical and Quantum Electronics. 2014;46(9):1097–1106.

[11] Etezadi D, Cetin AE, Altug H. Plasmonic nanoantennas on nanopedestals for ultra-sensitive vibrational IR-spectroscopy. In: Conference on Lasers and Electro-Optics (CLEO). IEEE; 2015. p. 1–2.

[12] McKinley AF, White TP, Catchpole KR. Designing nano-loop antenna arrays for light-trapping in solar cells. In: 39th Photovoltaic Specialists Conference (PVSC). IEEE; 2013. p. 1894–1896.

[13] Savov SV. An efficient solution of a class of integrals arising in antenna theory. IEEE Antennas and Propagation Magazine. 2002;44(5):98–101.

[14] Mahony J. A comment on Q-type integrals and their use in expressions for radiated power. IEEE Antennas and Propagation Magazine. 2003;45(3):127–128.

[15] Hallen E. Theoretical investigations into the transmitting and receiving qualities of antennae. Nova Actae Regiae SocSci Upsaliensis. 1938;Ser. IV, 11(4):1–44.

[16] Storer JE. Impedance of the thin wire loop. Transactions of the American Institute of Electrical Engineers. 1956;75:606–619.

[17] Wu TT. Theory of the thin circular loop antenna. Journal of Mathematical Physics. 1962;3(6):1301–1304.

[18] Iizuka K. The circular loop antenna multiloaded with positive and negative resistors. IEEE Transactions on Antennas and Propagation. 1965;13(1):7–20.

[19] Pocklington HC. Electrical oscillations in wires. Proceedings of the Cambridge Philosophical Society. 1897;9:324.

[20] Harrington RF, Ryerson JL. Electromagnetic scattering by loaded wire loops. Radio Science. 1966;1:347–352.

[21] Rao BA. Far field patterns of large circular loop antennas: Theoretical and experimental result. IEEE Transactions on Antennas and Propagation. 1968;16(2):269–270.

[22] Werner DH. An exact integration procedure for vector potentials of thin circular loop antennas. IEEE Transactions on Antennas and Propagation. 1996;44(2):157–165.

[23] McKinley AF, White TP, Catchpole KR. Theory of the circular closed loop antenna in the terahertz, infrared, and optical regions. Journal of Applied Physics. 2013;114(4):044317.

[24] Lu BQ, Nagar J, Yue T, *et al.* Closed-form expressions for the radiation properties of nanoloops in the terahertz, infrared and optical regimes. IEEE Transactions on Antennas and Propagation. 2017;65(1):121–133.

[25] Nagar J, McKinley AF, Werner DH. Far-field radiation properties of impedance-loaded loop antennas from RF to optical frequencies. In: 12th European Conference on Antennas and Propagation (EuCAP). IEEE; 2018. p. 1–5.

[26] Nagar J, Chaky RJ, Pantoja MF, *et al.* Optimization of far-field radiation from impedance-loaded nanoloops accelerated by an exact analytical formulation. IEEE Transactions on Antennas and Propagation. 2019;67 (3):1448–1458.

[27] Pendry JB, Holden AJ, Robbins DJ, *et al.* Magnetism from conductors and enhanced nonlinear phenomena. IEEE Transactions on Microwave Theory and Techniques. 1999;47(11):2075–2084.

[28] McKinley AF. Theory of impedance loaded loop antennas and nanorings from RF to optical wavelengths. IEEE Transactions on Antennas and Propagation. 2017;65(5):2276–2281.

[29] McKinley A. The analytical foundations of loop antennas and nano-scaled rings. Signals and Communication Technology. Springer, Singapore; 2019.

[30] Stratton JA. Electromagnetic theory. New York and London: McGraw-Hill Book Company; 1941.

[31] Hanson GW. On the applicability of the surface impedance integral equation for optical and near infrared copper dipole antennas. IEEE Transactions on Antennas and Propagation. 2006;54(12):3677–3685.

[32] Novotny L. Effective wavelength scaling for optical antennas. Physical Review Letters. 2007;98(26):266802.

[33] McKinley AF, White TP, Maksymov IS, *et al.* The analytical basis for the resonances and anti-resonances of loop antennas and meta-material ring resonators. Journal of Applied Physics. 2012;112(9):094911.

[34] Maksymov IS, Miroshnichenko AE, Kivshar YS. Actively tunable bistable optical Yagi–Uda nanoantenna. Optics Express. 2012;20(8):8929–8938.

[35] Large D, Ball L, Farstad A. Radio transmission to and from underground coal mines–theory and experiment. IEEE Transactions on Communications. 1973;21:194–202.

[36] Fuller JA, Wait JR. Mutual electromagnetic coupling of coaxial loops in a borehole. Radio Science. 1973;8(5):453–457.
[37] Yadava R. Antenna and wave propagation. PHI Learning Pvt. Ltd.; 2011.
[38] Simovski C, Morits D, Voroshilov P, *et al.* Enhanced efficiency of light-trapping nanoantenna arrays for thin-film solar cells. Optics Express. 2013;21(104):A714–A725.
[39] Ahmadi A, Mosallaei H. Plasmonic nanoloop array antenna. Optics Letters. 2010;35(21):3706–3708.
[40] Hansen R. Formulation of echelon dipole mutual impedance for computer. IEEE Transactions on Antennas and Propagation. 1972;20(6):780–781.
[41] Kakatkar SS, Ray KP. Evaluation of mutual coupling between slots from dipole expressions. Progress In Electromagnetics Research. 2009;9:123–138.
[42] Ito S, Inagaki N, Sekiguchi T. An investigation of the array of circular-loop antennas. IEEE Transactions on Antennas and Propagation. 1971;19(4):469–476.
[43] Wait JR. Mutual electromagnetic coupling of loops over a homogeneous ground. Geophysics. 1955;20(3):630–637.
[44] Fuller JA, Wait JR. High-frequency electromagnetic coupling between small coplanar loops over an inhomogeneous ground. Geophysics. 1972;37 (6):997–1004.
[45] Iizuka K, King R, Harrison C. Self and mutual admittances of two identical circular loop antennas in a conducting medium and in air. IEEE Transactions on Antennas and Propagation. 1966;14(4):440–450.
[46] Krishnan S, Li LW, Leong MS. Entire-domain MoM analysis of an array of arbitrarily oriented circular loop antennas: a general formulation. IEEE Transactions on Antennas and Propagation. 2005;53(9):2961–2968.
[47] MathWorks. MATLAB; 2014. https://www.mathworks.com/products/matlab.html.
[48] Pantoja M, Nagar J, Lu B, *et al.* Existence of superdirective radiation modes in thin-wire nanoloops. ACS Photonics. 2017;4(3):509–516.
[49] IEEE. IEEE Standard Definitions of Terms for Antennas, IEEE Std 145-1983. IEEE: New York, USA; 1983.
[50] Haskou A, Sharaiha A, Collardey S. Design of small parasitic loaded superdirective end-fire antenna arrays. IEEE Transactions on Antennas and Propagation. 2015;63:5456–5464.
[51] Taminiau TH, Stefani FD, Segerink FB, *et al.* Optical antennas direct single-molecule emission. Nature Photonics. 2008;2:234–237.
[52] Krasnok AE, Filonov DS, Simovski CR, *et al.* Experimental demonstration of superdirective dielectric antenna. Applied Physics Letters. 2014;104 (13):133502.
[53] Altshuler EA, O'Donnell TH, Yaghjian AD, *et al.* A monopole super-directive array. IEEE Transactions on Antennas and Propagation. 2005;53: 2053–2060.
[54] Jackson JD. Classical electrodynamics. Wiley: New York, USA; 1999.

[55] Grahn P. Multipole excitations in optical meta-atoms. Aalto University, Aalto, Finland; 2012.
[56] Campbell SD, Ziolkowski RW. Near-field directive beams from passive and active asymmetric optical nanoantennas. IEEE Journal of Selected Topics in Quantum Electronics. 2015;21:312–323.
[57] Krasnok AE, Simovski CR, Belov PA, *et al.* Superdirective dielectric nanoantennas. Nanoscale. 2014;6:7354–7361.
[58] Hadka D, Reed P. Borg: an auto-adaptive many-objective evolutionary computing framework. Evolutionary Computation. 2013;21(2):231–259.
[59] Geyi W. Physical limitations of antenna. IEEE Transactions on Antennas and Propagation. 2003;51(8):2116–2123.
[60] Chu L. Physical limitations of omni-directional antennas. Journal of Applied Physics. 1948;19(12):1163–1175.
[61] Engheta N, Salandrino A, Alù A. Circuit elements at optical frequencies: nanoinductors, nanocapacitors, and nanoresistors. Physical Review Letters. 2005;95(9):095504/1–4.
[62] Shen L, Raffoul G. Optimum design of Yagi array of loops. IEEE Transactions on Antennas and Propagation. 1974;22(6):829–830.
[63] Nagar J, Campbell S, Ren Q, *et al.* Multi-objective optimization aided metamaterials-by-design with application to highly directive nano-devices. IEEE Journal on Multiscale and Multiphysics Computational Techniques. 2017;2(1):147–158.
[64] FEKO Suite 7.0, EM Software and Systems (www.feko.info), 2015.
[65] Loh S, Wong J. Radiation field of an elliptic loop antenna with a constant current. Canadian Journal of Physics. 1958;36(6):672–676.
[66] Wong JY, Loh S. Calculated radiation resistance of an elliptical loop antenna with constant current. Journal of the British Institution of Radio Engineers. 1959;19(2):89–91.
[67] Zuffada C, Yang FC. Currents and fields of thin circular and elliptical loops. IEEE Transactions on Electromagnetic Compatibility. 1991;33(2):97–104.
[68] Rothwell E, Gharsallah N. Determination of the natural frequencies of a thin wire elliptical loop. IEEE Transactions on Antennas and Propagation. 1987;35(11):1319–1324.
[69] Cabedo-Fabres M, Valero-Nogueira A, Ferrando-Bataller M. Systematic study of elliptical loop antennas using characteristic modes. In: Antennas and Propagation Society International Symposium. IEEE; 2002. p. 156–159.
[70] Werner DH. Exact expressions for the far-zone electromagnetic fields radiated by thin elliptical loop antennas of arbitrary size. IEEE Transactions on Antennas and Propagation. 2018;66(12):6844–6850.

Chapter 2

Passive and active nano cylinders for enhanced and directive radiation and scattering phenomena

Samel Arslanagić[1] and Richard W. Ziolkowski[2]

2.1 Introduction and chapter overview

Future technologies are in a desperate need of increasingly higher frequencies as the desire for smaller devices, higher speeds and larger bandwidths continues its ever dramatic increase. To properly address these challenging demands, our ability to effectively tailor and control wave-matter interactions on the nano scale is of paramount importance. Indeed, significant accomplishments in this broad nano-photonics research field have been reported in recent years owing to the advent and development of metamaterials, metasurfaces, metamaterial-inspired structures, and plasmonics [1–4]. The large number of reported configurations may have a profound impact on several nano-photonics fields of utmost societal importance, e.g., nano-lasers and sensors, energy harvesting, wireless communication at the nano-scale, and medicine [5–11].

Nano-antennas (NAs) are the key enablers to realize effective interactions on the nano-scale. They empower efficient conversion of incident light into a strongly localized field (receiving mode of operation), as well as the efficient conversion of the emissions from weak sources into intense and directed propagating waves (transmitting mode of operation) [3,4,12]. Realizations of NAs may be grouped into two main categories. The first category is comprised of traditional NAs which simply are scaled down versions of their well-known microwave counterparts such as, e.g., dipoles, bow-ties, Yagi–Uda antennas and a variety of antenna arrays [3,13–16]. The majority of the first traditional NAs were made from metals (plasmonic materials), and their over-all size is either comparable to or larger than the operating wavelength. The second category consists of a wide range of NAs based on nano-particles (NPs) supporting the excitation of different multipoles, even for NPs of electrically small size. The vast majority of the initial realizations of the

[1]Department of Electrical Engineering, Technical University of Denmark, Kongens Lyngby, Denmark
[2]Faculty of Engineering and Information Technology, Global Big Data Technologies Centre, University of Technology Sydney, Ultimo, NSW, Australia

NP-based NAs (which also will, interchangeably, be referred to as simply NPs) likewise utilized plasmonic materials [3,4,12,17–22]. The physics underlying their operation is ultimately tied to one of the well-known effects associated with optics at surfaces, namely the occurrence of surface plasmons [23]. While plasmonic-based NAs offer a number of advantages, including profound field enhancement and localization with deeply sub-wavelength configurations, their performance is limited by the plasmonic losses at optical frequencies. As a means of alleviating the loss issue in these plasmonic-based NAs, active materials were proposed and used recently in a variety of active NPs [24–28]. Due to the intense interaction of the localized field in active NAs and the gain material, the resulting radiation and scattering characteristics are significantly enhanced relative to their passive counterparts. As such, these active NAs are often said to be super-resonant and to possess super-resonances, see also [29] and the references therein. As an alternative low-loss route towards efficient designs at optical frequencies, Mie-resonances were recently exploited in a variety of all-dielectric structures. These include not only all-dielectric NAs (e.g., silicon-based), but also complex structures such as all-dielectric metamaterials and metasurfaces [30–35]. The key behind the success of such all-dielectric NAs lies in the use of high-permittivity materials which enable strong and well-isolated excitations of not only the electric, but even more importantly, the magnetic dipole and as well as higher order electric and magnetic responses.

Many of the initial realizations of NP-based NAs involved simple configurations at sub-wavelength sizes. Hence, they scattered and radiated fields as though they were small dipole elements. However, many emerging applications are in strong need of highly directive sub-wavelength NAs, i.e., nano-devices that enable strong power flows in preferred directions. In configurations with symmetric material arrangements (canonical shapes), the directivity can be enhanced through the simultaneous excitation of electric and magnetic dipoles (Huygens sources) in layered NPs [36–38], or through the excitation of higher order modes [39–44]. On the other hand, the introduction of asymmetric material arrangements (non-canonical shapes) in layered NPs has likewise been shown to enhance the directivities of such subwavelength structures [45–49]. The asymmetries stimulate the excitation of the required higher order modes; they can be achieved by introducing appropriate perforations [45,46,48,49] or eccentricities [47] into the NP layers. Interesting optical properties of plasmonic-based NPs possessing broken symmetries were also considered in [50,51]. Moreover, super-directive NP-based NAs with profound needle-like radiation patterns were recently demonstrated in multi-layered spherical [52] and cylindrical [53] geometries. In these cases, the analytically developed scattering coefficients were designed to follow ideal Dirac-delta and other tailored realistic distributions to give rise to increased directivities by overlapping several multipolar electric and magnetic responses. The available number of these higher order modes is associated with the number of layers in the NA structure and their optimized material and size parameters.

In this chapter our main objective is to review our previous and current research efforts in this promising area of directive NP-based NAs. Focus will be placed on both symmetric and asymmetric two-dimensional (2D) and three-dimensional (3D) passive and active (with constant gain model) cylindrical NPs with appropriate layer designs, perforations or eccentricities. We will illustrate how perforations essentially only enhance the near-field radiation properties, while asymmetries introduced through the layer eccentricities also enhance the far-field radiation behaviors. We will also describe how symmetric multilayered cylindrical NPs with tailored scattering coefficients lead to super-directive and needle-like radiation patterns in forward as well as backward directions. Our results will be illustrated by various near-and far-field distribution results obtained either analytically and/or numerically with full-wave commercial packages. The canonical problem solutions prove not only to be essential reference results to build intuition and knowledge, but also to guide and verify more complex numerical simulations.

This chapter is organized as follows. Section 2.2 introduces all the configurations to be discussed, their material and gain models, the pertaining analysis methods and all quantities used to quantify the resulting performance characteristics. The results obtained for the symmetric and asymmetric 2D configurations are summarized in Section 2.3, while those related to their 3D counterparts are found in Section 2.4. Section 2.5 discusses the symmetric and multilayered cylindrical configurations with tailored scattering coefficients that lead to super-directive and needle-like behavior. In each section, except Section 2.5, the results are reported predominantly for active configurations. Section 2.6 summarizes the chapter. The chapter concludes with two appendixes. Appendix A includes the main analytical steps that lead to the solution of the symmetric 2D configurations. Appendix B includes the corresponding details for the eccentrically-layered configurations. Throughout the chapter, the time-factor $\exp(j\omega t)$, with ω being the angular frequency and t being the time, is assumed and suppressed.

2.2 Configurations, materials, gain model, and analysis methods

Both analytical and numerical solutions have been obtained for a rather large set of 2D and 3D, passive and active, radiating and scattering problems. Selected examples will illustrate these solution processes. The presented results will highlight the realized radiation and scattering physics and engineering fundamentals. When available, the associated, more exotic outcomes will be emphasized.

2.2.1 Configurations

The various nano-cylinder configurations featured in this chapter are shown in Figure 2.1. Most of them are 2D, i.e., they are of infinite extent along one axis. However, some have finite counterparts, i.e., are finite along that axis. These 3D configurations will also be described.

We will begin with the *symmetric* 2D-*coated nanoparticle* (CNP) configuration of which a cross-section is shown in Figure 2.1(a). The CNP consists of a nanocore (region 1) covered concentrically by a nano-shell (region 2). The specific choices of the materials employed for these two regions will be discussed later on.

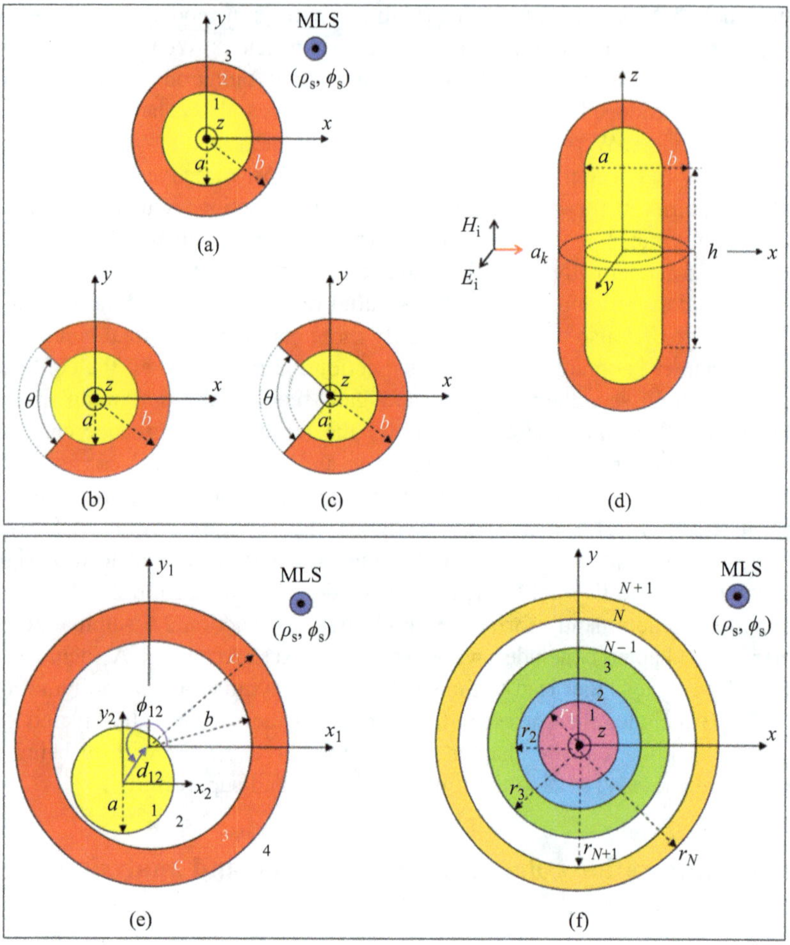

Figure 2.1 The featured nano-cylinder radiating and scattering configurations: (a) 2D core-shell; (b) 2D holey core-shell; (c) 2D cake core-shell; (d) 3D core-shell; (e) 2D core-shell with interior offset cylinder; and (f) 2D multilayered core-shell. These cylindrical structures are driven by either a line source located in their interior or exterior regions or by a plane wave.

The radius of the nano-core is denoted by a, while the outer radius of the nano-shell is denoted by b. The CNP is immersed in free space (region 3). A rectangular coordinate system (x, y, z) is introduced so that the center-axis of the CNP coincides with the z-axis. Thus, the CNP is infinite along the z-axis. The associated circular cylindrical coordinate system (ρ, ϕ, z) is also introduced; the coordinates of the observation point are denoted by (ρ, ϕ).

The symmetric 2D CNPs will initially be contrasted to the *asymmetric* 2D CNPs shown in Figure 2.1(b) and (c). These are the so-called *holey* and *cake* CNPs, respectively. They follow at once from their symmetric counterpart through angular sector perforations of the nano-shell and then also of the nano-core. The amount of the sector removed from the CNP is specified by the cut-out angle θ, cf., Figure 2.1(b) and (c). This choice and the configuration determine the overall degree of asymmetry. Since the asymmetric 2D CNPs are derived from the corresponding symmetric ones, they too are infinite along the z-direction. Moreover, their material composition and their cross-sectional size (as specified by the radii a and b) remain unchanged relative to the symmetric case.

To approach a more realistic scenario, 3D counterparts of both the symmetric and asymmetric 2D CNPs will also be addressed. These 3D CNPs are obtained by truncating the infinitely long CNPs so that their height becomes finite. A rectangular (x, y, z) coordinate system has also been introduced to help connect the geometry to the features of the plane wave excitation. Explicitly, the symmetric 3D CNP consists of a cylindrical core-shell of height h and its cross-sectional size in the xy-plane remains specified by the radii a and b, which remain unchanged from the corresponding 2D CNP case. As shown in Figure 2.1(d), the cylindrical part is completed, on the top and bottom, with a core-shell semi-spherical part of the same material composition and core-shell dimensions as the cylindrical part to facilitate the numerical computations by eliminating any edges and the corresponding field singularities. We note that for $h = \infty$, this 3D CNP goes over to the previously described 2D CNP, while for $h = 0$, it recovers a spherical 3D CNP.

Another way of introducing asymmetry is depicted in Figure 2.1(e). It shows a cross-section of a 2D three-zone (three-region) nano-particle (3Z-NP). It consists of an offset nano-core (region 1) of radius a located in a cylinder of radius b (region 2). A nano-shell (region 3) of the outer radius c encloses both regions 1 and 2 and is concentric with region 2. The 3Z-NP is immersed in free space (region 4, exterior). Two rectangular, (x_i, y_i, z_i), and the associated cylindrical, (ρ_i, ϕ_i, z_i), coordinate systems, $i = 1$ and 2, are introduced so that the center-axis of the nano-core coincides with the z_2-axis and the center-axis of regions 2 and 3 coincide with the z_1-axis. The z_2-axis is parallel to the z_1-axis. The 3Z-NP is thus a 2D configuration, i.e., it is infinite in the z_1-direction. The eccentricity of the 3Z-NP, and thus its degree of asymmetry, is characterized by the amount of the offset of the nano-core with respect to the nano-shell. The center of the nano-core is described by the distance d_{12} and the angle ϕ_{12}, cf., Figure 2.1(e). The observation points in regions 1–3 will be denoted by their (ρ_2, ϕ_2) coordinates, while those in region 4 will be denoted by their (ρ_1, ϕ_1) coordinates. The performance of this asymmetric 3Z-NP will be compared to the corresponding symmetric (concentric) 3Z-NP (i.e., when the z-axes of the two coordinate systems coincide) as well as to the simpler 2D CNPs.

The cross-section of the remaining nano-cylinder configuration is shown in Figure 2.1(f). It is a symmetric, circular multi-layer 2D nano-particle (ML-NP) that consists of a cylindrical nano-core of radius r_1 (region 1) covered concentrically with $N - 1$ layers immersed in free space, region $N + 1$. Each layer is labeled as region j and has an outer radius $r_j, j = 2, \ldots, N$. Note that N is an arbitrary integer. The z-axis of the rectangular, (x, y, z), and cylindrical, (ρ, ϕ, z), coordinate systems coincide. The length of the ML-NP in the z-direction is infinite, i.e., it is a

2D configuration. The coordinates of the observation point in all regions are denoted by (ρ, ϕ).

2.2.2 Materials

All nano cylinders are surrounded by free space with permittivity ε_0 and permeability μ_0, and thus, the wave number $k_0 = \omega\sqrt{\varepsilon_0\mu_0} = 2\pi/\lambda$ and the intrinsic impedance $\eta_0 = \sqrt{\mu_0/\varepsilon_0}$. The quantity $\lambda = c/f$ denotes the free-space wavelength, where $c = 1/\sqrt{\varepsilon_0\mu_0}$ is the speed of light and f is the frequency. For all but the ML-CNP case, the nano-core of the 2D and 3D CNPs in Figure 2.1 is made of silica (silicon dioxide, SiO_2), a dielectric material. It has the permittivity $\varepsilon_{nc} = 2.05\,\varepsilon_0$ and permeability $\mu_{nc} = \mu_0$. Their nano-shells are a plasmonic (metallic) medium and are taken to be silver (Ag)* with the complex permittivity $\varepsilon_{ns} = \varepsilon'_{ns} - j\varepsilon''_{ns}$ and permeability $\mu_{ns} = \mu_0$. Region 2 of the 3Z-NP is taken to be free space. The actual nano-cylinder dimensions (i.e., the radii and the height) will be specified in the subsections in which their behaviors are discussed. However, in all cases, the silver nano-shell is always 6-nm thick.

This small, nanometer thickness requires, however, rather accurate modeling of the underlying optical properties which includes the size dependence of the Ag permittivity. Using the empirically determined permittivities of silver [56,57] along with a permittivity model decomposed into a size-dependent Drude response and an inter-band transition response [24], accurate size-dependent permittivity models of the plasmonic nano-shells have been established for a range of nano-shell thicknesses. With this in mind, Figure 2.2 shows the real part of the relative permittivity,

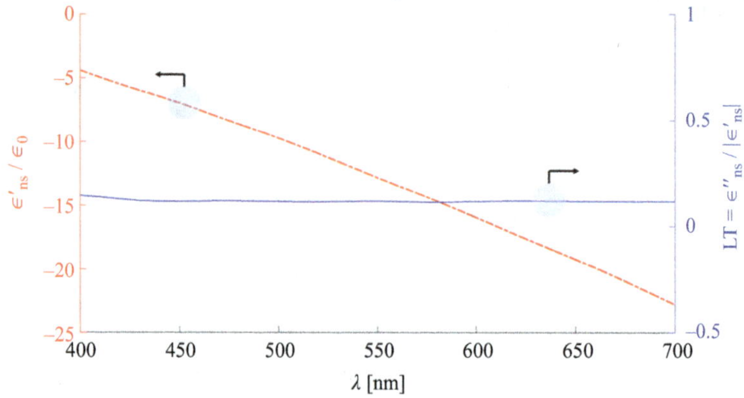

Figure 2.2 The permittivity, ε_{ns}, of the silver 6-nm-thick nano-shell. The left axis shows the real part of the relative permittivity, $\varepsilon'_{ns}/\varepsilon_0$, while the right axis shows the loss-tangent, $\mathrm{LT} = \varepsilon''_{ns}/|\varepsilon'_{ns}|$.

*Only silver is emphasized in this work as the plasmonic material since the use of other choices, e.g., gold or copper, in the nano-shells yields similar behaviors [29, 47–49, 54, 55].

$\varepsilon_{ns}^{'}/\varepsilon_0$ of Ag, as well as its loss tangent, LT $= \varepsilon_{ns}^{''}/|\varepsilon_{ns}^{'}|$ for a 6-nm-thick nano-shell as functions of the excitation wavelength. These results are provided over the visible range of wavelengths, 400–700 nm, which will suffice in illustrating all of the interesting results for the light-driven nano-cylinder configurations examined herein. It is clear that the real part of Ag's permittivity attains negative values, as well as large loss tangent values, in this range of wavelengths.

For the symmetric ML-NP shown in Figure 2.1(f), its region i, $i = 1, \ldots, N$, is characterized by a permittivity $\varepsilon_i = \varepsilon_i^{'} - j\varepsilon_i^{''}$ and a permeability $\mu_i = \mu_0$, and thus a wave number, $k_i = \omega\sqrt{\varepsilon_i\mu_i}$, and an intrinsic impedance, $\eta_i = \sqrt{\mu_i/\varepsilon_i}$. These regions are taken to be either a double positive (DPS, $\varepsilon > 0, \mu > 0$) or an epsilon negative (ENG, $\varepsilon < 0, \mu > 0$) material. The specific radii and material parameters will be associated with a specific choice of scattering coefficients or determined by a numerical optimization process to approach the derived theoretical bounds. Only passive ML-NP configurations are considered.

2.2.3 Gain model

As shown in Figure 2.2, Ag is very lossy in the wavelength region of interest. This feature of the nano-shells has notable effects on the potential resonant properties of the nano-cylinder configurations. In order to achieve enhanced performance characteristics, these losses must be overcome. This can be accomplished by impregnating the CNPs with a gain material. We will use the so-called constant frequency gain model which we include into the silica nano-core to illustrate the resulting active characteristics. The nano-core permittivity will thus read:

$$\varepsilon_{nc} = \varepsilon_0\left(n^2 - \kappa^2 - 2jn\kappa\right), \tag{2.1}$$

where $n = \sqrt{2.05}$ is the refractive index of silica (maintained at its value for the lossless case). The parameter κ represents the loss/gain coefficient, and thus determines the nature of the nano cylinders according to the following classification:

$\kappa = 0$ and $\kappa > 0$: *Passive* nano cylinders,
$\kappa < 0$: *Active* nano cylinders.

In relation to the Ag-based nano cylinders, our focus here is almost exclusively on the active cases due to the obvious inefficiency of the passive ones. Consequently, the latter will be addressed only briefly to illustrate the realization of enhanced and directive radiation and scattering properties of the former. We will also emphasize that realistic gain materials are dispersive and have their own frequency dependence. Our choice of using a constant frequency gain model simplifies tuning the model in order to achieve its maximum performance. It corresponds to the concept that in practice one would chose the gain material to have its maximum value occur at a specific frequency (wavelength) of interest or redesign the structure to have its resonance occur at the frequency at which the maximum gain does.

2.2.4 Analysis methods

2.2.4.1 2D nano cylinders

The 2D nano cylinders illustrated in Figure 2.1(a–c), (e), and (f) are all excited by an infinitely long magnetic line source (MLS) driven with a constant magnetic current I_m [V]. This excitation type corresponds to the transverse electric (TE) polarization. The MLS is parallel to the z-axis of the employed coordinate systems in all cases. Its position is specified by the coordinates (ρ_s, ϕ_s). The field due to the MLS constitutes the known incident field in these problems; its electric and magnetic fields are denoted by $\boldsymbol{E}_{MLS}$ and $\boldsymbol{H}_{MLS}$, respectively. Upon the excitation of the nano cylinders, scattered electric and magnetic fields are created in each of the regions. The MLS is, with no loss of generality, located in the region exterior to the nano cylinders in the numerical examples. In the analytical formulation, the total field in the region in which the source exists is the sum of the scattered fields and the source/incident field. In all other regions, the total and scattered fields are the same. In all regions, the total electric, $\boldsymbol{E}$, and magnetic, $\boldsymbol{H}$, fields take on the form:

$$\boldsymbol{E} = \boldsymbol{a}_\rho E_\rho + \boldsymbol{a}_\phi E_\phi = \boldsymbol{a}_x E_x + \boldsymbol{a}_y E_y, \tag{2.2a}$$

$$\boldsymbol{H} = \boldsymbol{a}_z H_z, \tag{2.2b}$$

respectively. In these equations, E_ρ and E_ϕ denote the ρ- and ϕ-components of the electric fields, while E_x and E_y denote its x- and y-components. The quantity H_z denotes the z-component of the magnetic field.

The unknown fields are determined analytically in the symmetric 2D CNP cases. The incident and the unknown scattered/total fields are expanded in multipolar cylindrical waves. The unknown scattered field expansion coefficients are readily determined by enforcing the boundary conditions at the interfaces between the various regions of the nano-cylinder configurations. This well-known approach is briefly reviewed in Appendix A. It is easily extended in order to cope with the symmetric ML-NP configuration shown in Figure 2.1(f); its main results are likewise summarized in Appendix A. Moreover, both the concentric and eccentric 3Z-NP configurations in Figure 2.1(e) are solved in a similar way. However, in the eccentric case, a specific addition theorem for the cylindrical wave functions is required in order to make the use of the boundary conditions practical by translating the cylindrical waves from one coordinate system to another. This solution procedure is briefly outlined in Appendix B for the eccentric 3Z-NP problem.

For the asymmetric holey and cake 2D CNPs (Figure 2.1(b) and (c)), a numerical solution has been established using the full-wave Comsol Multiphysics modeling software, which is based on the finite-element method. To this end, a 2D design domain is utilized. A graphical visualization of the model is given in Figure 2.3(a).

The CNP in question is immersed in a free-space circular host domain, which is completed with a perfectly matched layer (PML) shell. The inner surface of the PML constitutes the so-called integration surface for a number of derived quantities to be introduced below. Its outer surface is replaced with a scattering boundary to

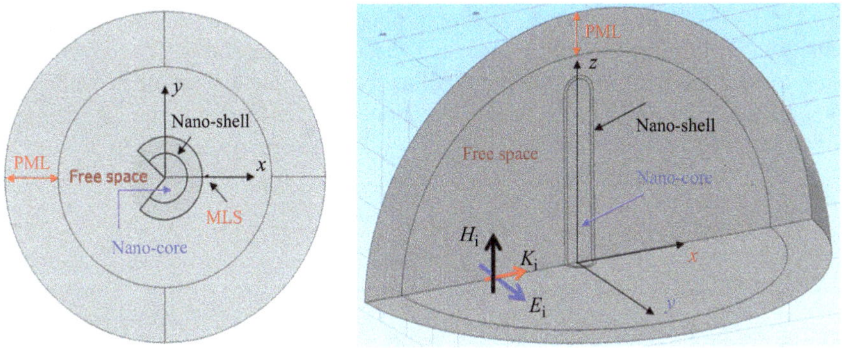

(a) (b)

Figure 2.3 (a) Numerical (Comsol Multiphysics [58]) model of the MLS excited asymmetric (cake) 2D CNPs. The model is representative also for the holey 2D CNP. (b) Numerical model of the plane wave excited symmetric 3D CNP. Perfect electric (xz-plane) and perfect magnetic (xy-plane) conductor symmetry planes are used to reduce the problem size. The model is representative also for the corresponding holey and cake 3D CNPs with the exception that the symmetry only exists with respect to the xy-plane in those cases.

further suppress any fields reflected from it. The distance to the PML (90 nm) and its thickness (45 nm), along with the mesh-fineness, particularly on the integration surface and in the CNP region, are all selected in order to ensure the convergence and accuracy of the final numerical solution.

Once the unknown fields in the 2D nano-cylinder configurations have been determined, either analytically or numerically, a number of quantities can be derived in order to quantify their performance. It is implicit that closed-form expressions exist in those cases where the fields are determined analytically (see Appendixes A and B for further details).

The power flow density is determined with the well-known Poynting vector, $\boldsymbol{S}(\rho, \phi)$. Its magnitude S and direction $\boldsymbol{a}_S$ are expressed as:

$$\boldsymbol{S}(\rho, \phi) = \boldsymbol{a}_S S = \frac{1}{2}\mathrm{Re}\{\boldsymbol{E} \times \boldsymbol{H}^*\}. \tag{2.3}$$

While the near-field behavior of the 2D CNPs is nicely illustrated by (2.2) and (2.3), their far-field properties are expressed through the normalized radiation resistance, NRR, and the directivity, $D(\phi)$. The NRR is the total power, P_t, radiated by the MLS in presence of the CNP, normalized by the power, P_{MLS}, radiated by the MLS alone in free space. Expressed in dB, it takes on the following form:

$$\mathrm{NRR} = 10\,\log_{10}\frac{P_t}{P_{MLS}}. \tag{2.4}$$

The total radiated power P_t is simply obtained as the flux of the power flow density (2.3) through a surface enclosing the CNP. This surface is conveniently taken to be a circle of infinitely large radius for the analytical solution, while it is taken as the finite integration surface mentioned above in the numerical solution. The power P_{MLS} is simply $P_{MLS} = I_m^2 k_0/8\eta_0$. The directivity is defined as:

$$D(\phi) = \frac{U(\phi)}{P_t/2\pi}, \tag{2.5}$$

where the radiation intensity is given by:

$$U(\phi) = \lim_{\rho \to \infty} \boldsymbol{S}(\rho, \phi) \cdot \rho \boldsymbol{a}_\rho. \tag{2.6}$$

2.2.4.2 3D nano cylinders

All of the 3D CNPs are studied under their excitation with the uniform plane wave indicated in Figure 2.1(d). The TE polarized electric and magnetic fields of the incident wave are given by:

$$\boldsymbol{E}_i = \boldsymbol{a}_y E_0 e^{-jk_0 x}, \tag{2.7a}$$

$$\boldsymbol{H}_i = \boldsymbol{a}_z (E_0/\eta_0) e^{-jk_0 x}, \tag{2.7b}$$

respectively, where E_0 [V/m] is the magnitude of the incident electric field. The presence of the nano cylinders gives rise to scattered fields in all regions. The total fields, obtained as the appropriate sum of the scattered and incident fields, are designated as in the 2D case: $\boldsymbol{E}$ for the electric field and $\boldsymbol{H}$ for the magnetic field. Note, however, that they may possess all three field components, e.g., the x-, y-, and z-components.

All of the 3D CNPs reported herein were also analyzed using the full-wave Comsol Multiphysics modeling software. A graphical visualization of the corresponding 3D design domain is illustrated in Figure 2.3(b) for the symmetric 3D CNP. The figure actually shows only a quarter of the entire configuration. The symmetry of the problem was exploited in order to reduce the computational domain and time. To this end, symmetry planes were utilized, i.e., perfect electric (xz-plane) and perfect magnetic (xy-plane) conductor planes were introduced in accordance with image theory. A similar numerical model is established for the holey and cake 3D CNPs. However, symmetry exists only with respect to the xy-plane in those cases. Thus, only the perfect magnetic conductor boundary condition is applied. The CNPs are again located in a free space domain that is terminated with a PML shell. Its inner and outer surfaces are again, respectively, the integration surface and a radiating boundary surface. In all cases, the distance to the PML was 120 nm in the xy-plane and varying in the xz-plane. Its thickness (70 nm) and the mesh-fineness, particularly on the integration surface and within the CNP region, were selected in order to ensure the convergence and accuracy of the final solution.

With the unknown fields determined in all regions, several derived quantities can be introduced to assess the 3D CNPs. To complement the near-field analysis done in terms of the total fields in all regions, the power flow density (2.3) is used; it is understood that this power flow density for 3D CNPs is a function of all three spatial variables. Under plane wave illumination, the far-field properties are taken into account through the total scattering cross section (SCS), σ_t, which is defined as:

$$\sigma_t = \frac{P_s}{S_i}, \tag{2.8}$$

where P_s is the total scattered power obtained simply as the flux of the scattered power flow density through a surface (integration surface) enclosing the 3D CNP and $S_i = |E_0|^2/2\eta_0$ is the power flow density of the incident field. The total SCS will be stated in dB as:

$$\sigma_t = 10\log_{10}(\sigma_t/A)[\text{dBcs}] \quad (\text{3D CNP}), \tag{2.9}$$

where dBcs denotes the decibels relative to the cross-sectional area of the 3D CNP, $A = \pi b^2$.

The general definition of the total SCS in (2.8) also applies to the 2D configurations. However, the total SCS in 2D will be expressed as:

$$\sigma_t = 10\log_{10}(\sigma_t/C)[\text{dBc}] \quad (\text{2D CNP}), \tag{2.10}$$

where dBc denotes dB relative to the circumference $C = 2\pi b$ of the 2D CNP.

Finally, we re-emphasize that the analytical results in the symmetric 2D and 3D CNP (for $h = 0$) cases have been obtained to enhance our understanding of the associated light-matter interactions, as well as to provide appropriate reference cases in order to validate the numerical models.

2.3 Symmetric and asymmetric CNPs

2.3.1 Dipole-based symmetric 2D CNPs

Substantial efforts have been devoted to analyzing and understanding symmetric 2D CNP problems. These investigations have included not only the Ag–SiO_2 versions but also SiO_2 nano-cores coated with gold (Au) and copper (Cu) nano-shells. We have theoretically shown that when active, these CNPs are super-resonant, i.e., they have dipole resonances that produce significantly enhanced radiation and scattering characteristics [29,57,58]. The enhanced exterior field responses result from their strongly resonant interior fields overlapping the gain region. As noted, the gain inclusion facilitates overcoming the losses in the plasmonic nano-shells. The dimensions of the symmetric 2D CNPs in our examples are chosen to be:

$a = 24$ nm (nano-core radius),
$b = 30$ nm (overall CNP radius),

which yields the specified 6-nm-thick plasmonic nano-shell. Figure 2.4(a) compares the NRR for the symmetric, passive and active, Ag-based 2D CNPs. In both cases, the source of excitation is a MLS with $I_m = 1$ V; it is located outside the CNPs along the x-axis at 35 nm. Very low NRR values are obtained in the passive, lossy case ($\kappa = 0$), the maximum NRR = 2.60 dB being attained at 593.5 nm. When the gain parameter $\kappa = -0.175$, the active CNP has a significantly enhanced NRR value, a remarkable NRR = 50.1 dB being attained at the wavelength 577.7 nm.

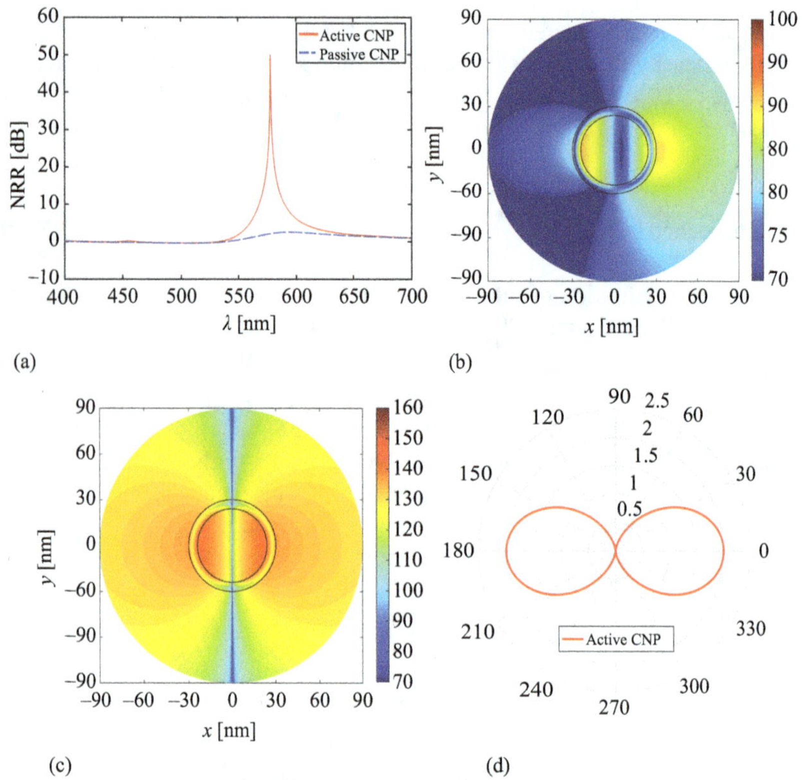

Figure 2.4 Near- and far-field results for symmetric Ag-based 2D CNPs. The normalized radiation resistance, NRR, as a function of the wavelength λ is shown in (a) for both passive and active cases. The magnetic field, $20 \log |\boldsymbol{H}/1[\mathrm{A/m}]|$, corresponding to the peak NRRs in (a) are shown in (b) for passive, and (c) for active cases. The resulting directivity, D, of the super-resonant active case is included in (d). In all cases, the MLS is located outside the CNP along the x-axis at 35 nm. The results in (b) and (c) are shown in a circular region with a 90 nm radius. Curves representing the surfaces of the CNPs are included in those figures.

This phenomenon of significant enhancement of the total radiated power in the active case is termed a super-resonance [29]. The very large NRR in the active case implies that large power radiated by the MLS is efficiently channeled into the far-field region. The active CNPs can thus be said to play the role of a natural matching network that transduces the MLS power coupled to the gain region into free-space propagating fields. The values of the NRR, gain parameter κ, and the resonant wavelength λ_0, are summarized below for quick reference:

$$\text{NRR} = 50.1\ \text{dB}, \kappa = -0.175, \lambda_0 = 577.7\ \text{nm} \quad \textit{Symmetric active Ag-based 2D CNP}$$

The far-field NRR results are nicely correlated with the near-field distributions. To this end, Figure 2.4(b) and (c), respectively, show the magnetic field $20\log|\boldsymbol{H}/1[\text{A/m}]|$ for the passive and active cases. The near-field distribution associated with the passive CNP is clearly rather weak, thereby confirming the lack of any radiated power enhancements and the extremely low NRR values. On the other hand, the near-field distribution for the active CNP displays a very strong resonant dipole mode response with a maximum response along the x-axis. This dipole mode is at the very heart of the extremely high NRR values.

Although the symmetric active Ag-based 2D CNPs can be designed to be super-resonant, the significantly enhanced radiation characteristics are not very directive since they arise from their fundamental dipole mode. While this can be inferred already from the near-field distribution shown in Figure 2.4(c), a clear confirmation is provided by the directivity pattern given in Figure 2.4(d), as determined by (2.5). A clear symmetric dipole pattern is in evidence with a maximum directivity of 2 attained in the forward ($\phi = 0^\circ$) and backward ($\phi = 180^\circ$) directions. A useful figure of merit for the assessment of the directional nature of both the near- and far-field properties of the examined nano cylinders is so-called the front-to-back ratio (FTBR). In the near-field we examine:

$$\text{FTBR}_{\text{NF}} = 10\log_{10}\left[|\boldsymbol{S}|_{+x}/|\boldsymbol{S}|_{-x}\right], \tag{2.11a}$$

which is the ratio of the power flow density $\boldsymbol{S}$, defined by (2.3), evaluated in the forward direction (observation points along the $+x$-axis) to the backward direction (observation points along the $-x$-axis). In the far-field we examine:

$$\text{FTBR}_{\text{FF}} = 10\log_{10}[D(\phi = 0^\circ)/D(\phi = 180^\circ)], \tag{2.11b}$$

where D is the directivity. Due to the symmetry in the near- and far-field responses, $\text{FTBR}_{\text{NF}} = \text{FTBR}_{\text{FF}} = 0$ dB, clearly confirming the weakly directional nature of this symmetric 2D CNP.

A number of nano-technological applications rely strongly on our ability to capture and enable efficient flows of electromagnetic energy in prescribed directions. As viewed from this perspective, the symmetric active 2D CNPs are not effective. Thus, it becomes obvious that other routes must be pursued towards the design of highly directive NA configurations.

2.3.2 Why go for something else? Asymmetric, holey and cake, active 2D CNPs

The asymmetric, holey, and cake active 2D CNPs provide a means to achieve higher directivities. With reference to Figure 2.1(b) and (c), the two radii of the asymmetric 2D CNPs remain the same as in the symmetric case: $a = 24$ nm and $b = 30$ nm. The holey and cake active Ag-based 2D CNPs can likewise be designed to be super-resonant. Simply, the gain constant needs to be adjusted to compensate for the weaker field localization within the nano-structure, i.e., the losses are higher because the fields leak out of the interior through the aperture in the shell. On the other hand, the edges of the aperture and the associated singular behaviors of the fields near them provide access to higher order modes. The result is rather directive near-field beams and significant near-field FTBRs. In our analysis, specific attention has been devoted to the influence of the degree of asymmetry, as measured by the cut-out angle, θ, on the resonant wavelength λ_0 and the gain parameter κ. The present section is a summary of [48]. As stated before, only the results for the Ag-based active CNPs are included in here. When relevant, they will be contrasted to those of the corresponding symmetric active CNPs.

Figure 2.5 shows the NRR as a function of the wavelength λ for the holey (Figure 2.5(a)) and cake (Figure 2.5(b)) active Ag-based 2D CNPs. The MLS is again located outside of the CNPs along the x-axis at 35 nm and driven by a constant magnetic current with $I_m = 1$ V. Because of the very time-consuming numerical optimizations, only the results at, and slightly around the resonant wavelength, are shown. The cut-out angles are chosen as: $\theta = 1^o, 5^o, 10^o, 22.5^o, 45^o, 90^o$, i.e., the shape of the CNPs ranges from being almost symmetric to being very asymmetric.

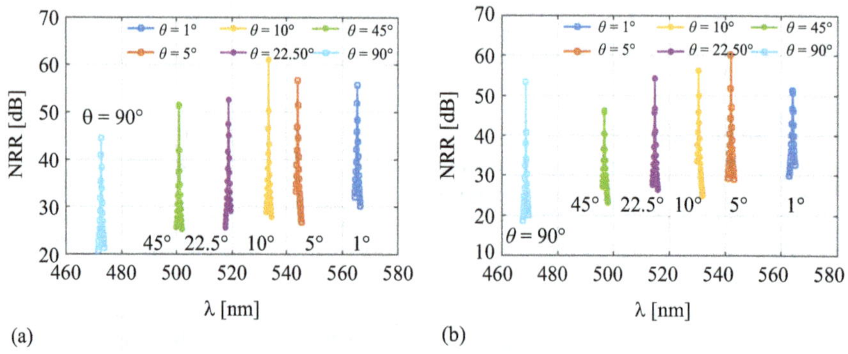

Figure 2.5 The normalized radiation resistance, NRR, as a function of the wavelength λ for asymmetric (a) holey and (b) cake active Ag-based 2D CNPs. The cut-out angles are chosen as: $\theta = 1^o, 5^o, 10^o, 22.5^o, 45^o, 90^o$. In all cases, the MLS is located outside the CNP along the x-axis at 35 nm.

It is evident that significant enhancements of the NRR, comparable to those in the symmetric case, are obtained for both the holey and cake active CNPs for all values of the cut-out angle θ. Thus, the inclusion of gain in the nano-cores does overcome the plasmonic losses and the losses induced by the open geometry. For a given θ, similar NRR values are obtained for both the holey and cake CNPs. Nevertheless, some variations in them are evident as θ varies. It is observed that the super-resonances get shifted to lower wavelengths (blue-shift) as the CNP asymmetry increases. This is expected since an increased asymmetry leads to a more profound excitation of the higher order modes arising from the edge singularities. The higher order modes generally appear at lower wavelengths (higher frequencies), which explains the observed behavior. This is additionally illustrated by the results in Figure 2.6 which compares the resonant wavelengths and the gain parameter κ for the holey and cake CNPs as functions of θ. As illustrated, less gain is needed overall as the holey and cake CNP asymmetry is increased (except for $\theta = 5^{\circ}$). As larger portions of the silver nano-shell are removed, the larger plasmonic losses decrease, even though the open geometry losses increase. Thus, less gain is required with an increasing asymmetry. Note that for the cake CNPs, more gain region is likewise removed as θ increases. Nevertheless, its cross-sectional area is smaller than that of the corresponding missing portion of the Ag nano-shell. Moreover, the edge and corner induced higher order modes lead to larger fields localized in the interior which leads to a more efficient field-gain region interaction. Figure 2.6 also indicates that slightly less gain is needed for the holey CNPs than for the cake CNPs simply because the gain region of the former is physically larger than that of the latter. We can also note that for relatively small cut-out angles, both the holey and cake CNPs require more gain than the corresponding symmetric CNPs. The loss reduction due to the removal of silver nano-shell portions in those cases does not really compensate the open geometry losses, thereby requiring slightly more gain than for the symmetric configurations.

Table 2.1 summarizes the relevant values (the NRR, the resonant wavelength λ_0, and the gain parameter κ) for the holey and cake 2D CNPs for the range of

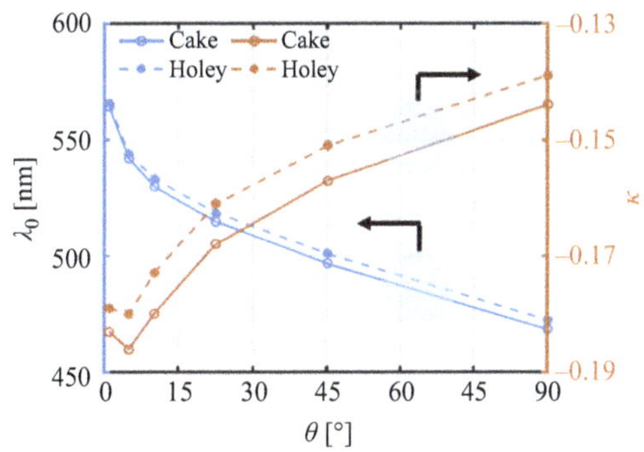

Figure 2.6 The resonant wavelength, λ_0, and the gain parameter, κ, as a function of the cut-out angle θ for holey and cake Ag-based 2D CNPs.

Table 2.1 A summary of various parameters for the super-resonant holey and cake Ag-based 2D CNPs.

θ [°]	Holey CNP			Cake CNP		
	NRR [dB]	κ	λ_0 [nm]	NRR [dB]	κ	λ_0 [nm]
0	50.1	−0.175	577.7	50.1	−0.175	577.7
1	55.6	−0.179	565.4	50.7	−0.183	564.1
5	56.6	−0.180	544.1	60.2	−0.186	542.0
10	60.9	−0.173	533.4	56.2	−0.180	530.5
22.5	52.5	−0.161	518.7	54.2	−0.168	514.9
45	51.5	−0.151	501.0	46.3	−0.157	496.7
90	44.5	−0.139	472.5	53.3	−0.144	468.5

considered cut-out angles θ. The results for the symmetric CNP ($\theta = 0°$) are included for reference purposes.

We next shed more light onto the super-resonant behaviors of the asymmetric Ag-based 2D CNPs by investigating their near-field behaviors. This is done by visualizing the total magnetic field, $20 \log |\boldsymbol{H}/1[\mathrm{A/m}]|$, and the associated power flow density, $10 \log |\boldsymbol{S}/1[\mathrm{W/m^2}]|$. Specifically, for $\theta = 1°, 10°, 45°, 90°$, Figure 2.7(a), (c), (e), and (g) shows the magnetic field distributions for the super-resonant holey CNP while the corresponding power flow densities are given in Figure 2.7(b), (d), (f), and (h). Please note that the colors of the power density results represent the magnitude of the complex Poynting vector, while the arrows simply show the direction of the power flow.

First, we note that rather intense local fields and power flows are obtained in all cases, particularly inside the gain impregnated nano-core region and just outside in the exterior of the CNP structures. In the small asymmetry case ($\theta = 1°$), a dominant dipole mode is evident, similar to the symmetric CNP case, cf., Figure 2.4(c). The asymmetry in the response is, however, present. This behavior is revealed in the power flow density result in Figure 2.7(b). As the asymmetry increases, the responses are no longer due simply to a dominant dipole mode, but rather due to a mixture of it with the higher order modes. As a consequence, the magnetic field gets more directional along the forward direction (positive x-axis), particularly when the observation points are close to the CNP. On the other hand, its response is suppressed in other directions, this being particularly so in the backward direction (negative x-axis). This behavior is not associated with any reactive and non-radiating near-fields as confirmed by the power flow density results. All of them exhibit rather intense and directed power flows in the forward direction as opposed to what happens in the backward direction. In fact, one observes the formation of localized cold spots with extremely low field and power flow density values in the backward direction.

The appearance of more directive near-field beams is further illustrated in Figure 2.8. It shows planar sections of the magnetic field distributions in Figure 2.7 when the radial distance ρ to the observation point is set to 40 nm in Figure 2.8(a)

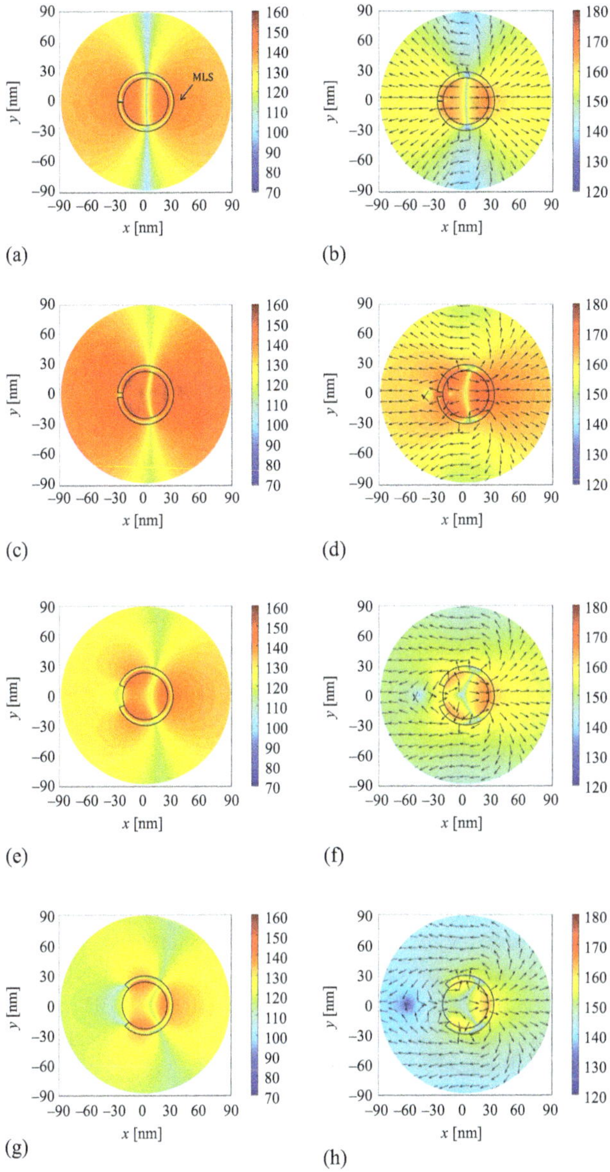

Figure 2.7 The magnetic field, $20 \log |\boldsymbol{H}/1[\mathrm{A/m}]|$*, of the super-resonant Ag-based holey 2D CNPs for the cut-out angles* θ *equal to (a)* 1°*, (c)* 10°*, (e)* 45°*, and (g)* 90°*. The associated power flow densities (color = magnitude, arrows = direction) are shown in Figure 2.7(b), (d), (f), and (h), respectively. The MLS is located outside the CNP along the x-axis at 35 nm. The results are shown in a circular region with a 90 nm radius. Curves representing the surfaces of the holey CNPs are included in the figure.*

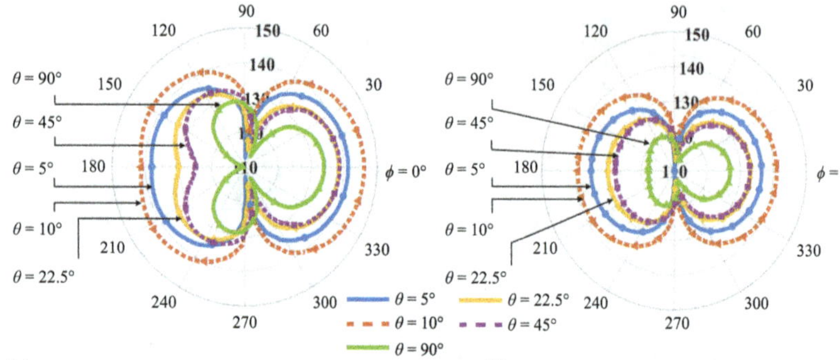

(a) (b)

Figure 2.8 The planar sections of the magnetic field, $20 \log |\boldsymbol{H}/1[\mathrm{A/m}]|$, *as a function of the observation angle* ϕ *for (a)* $\rho = 40$ *nm and (b)* $\rho = 80$ *nm for the super-resonant holey Ag-based 2D CNPs for the indicated cut-out angles* θ.

and at 80 nm in Figure 2.8(b). When the observation point is close to the CNP (Figure 2.8(a)), i.e., when we observe the immediate near-field of the nano-cylinder configuration, the response exhibits an asymmetric, directional pattern. However, as the observation point is moved further away, only the case with the largest cut-out angle $\theta = 90^{\circ}$ leads to any noticeable directional response. The other cases show a more symmetric, dipole mode type of pattern. One finds that the fields of the higher order modes excited by the asymmetries in the CNPs simply die out well before the onset of the far-field region, i.e., they are evanescent fields.

As a final confirmation of the ability of the examined active holey CNPs to act as directional near-field radiators, Figure 2.9 shows their near-field front-to-back ratio values determined by (2.11a). Very large $\mathrm{FTBR_{NF}}$ values are obtained in all of the cases. As the asymmetry increases, the $\mathrm{FTBR_{NF}}$ increases. This holds true even as the observation point is moved further away from the CNP in its near-field region. To exemplify, the power flow density at 60 nm is 30 dB more intense along the positive x-axis than it is along the negative x-axis for the largest asymmetry case, $\theta = 90^{\circ}$. While Figures 2.7–2.9 illustrated the behaviors of the super-resonant holey 2D CNPs, the related cake configurations exhibit very similar results and, therefore, will not be included here.

We have noted in our discussion of the results pertaining to Figure 2.8(b) that a symmetric response is expected in the far-field region due to extinction of the higher order modes before they reach this region. To firmly establish this, the directivity for the super-resonant holey (Figure 2.10(a)) and cake (Figure 2.10(b)) cases is shown in Figure 2.10. Except for some minor deviations along the y-axis ($\phi = 90^{\circ}$ and 270°) for the $\theta = 90^{\circ}$ case, the obtained patterns are overall symmetric and resemble closely that of an electric dipole. Their maxima are, however, somewhat lower than 2. This decrease arises from the open nature of the configurations. Some energy is tied to the evanescent fields of the higher order modes.

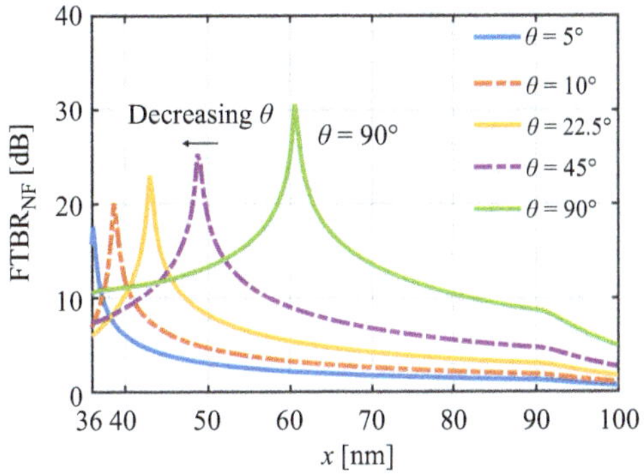

Figure 2.9 The near-field front-to-back-ratio, $FTBR_{NF}$, as a function of the observation point along the x-axis for the super-resonant holey Ag-based 2D CNPs for the indicated cut-out angles θ.

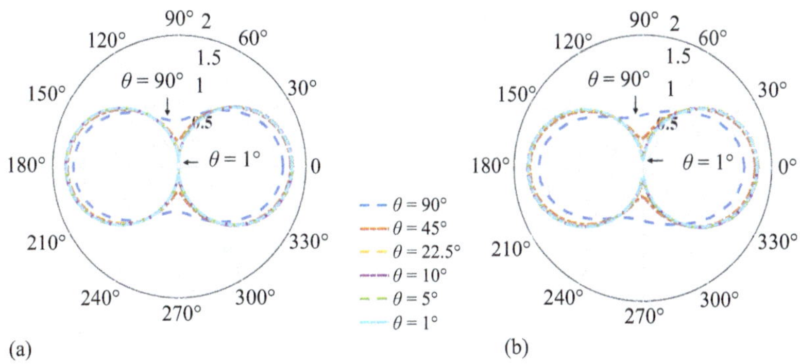

Figure 2.10 The directivity, D, as a function of the observation angle ϕ for asymmetric (a) holey and (b) cake active Ag-based 2D CNPs. The cut-out angels are $\theta = 1^o, 5^o, 10^o, 22.5^o, 45^o, 90^o$. In all cases, the MLS is located outside the CNP along the x-axis at 35 nm.

Consequently, the observed enhancements of the total radiated power (far-field) remain only weakly directive and dipolar. The associated far-field front-to-back-ratio given by (2.11b) is $\text{FTBR}_{\text{FF}} \approx 0$ dB.

Before proceeding to the 3D cases, we remind the reader that the above results were all reported for a particular location of the MLS. Figure 2.11 shows the NRR values as a function of the MLS location, x_s, along the positive and negative x-axes for the holey active CNP case. The vertical dashed lines indicate the location of the CNP boundaries. The response is, of course, not symmetric with respect to x_s due to the asymmetry of the CNP. Nevertheless, it is clear that significant enhancements are found for many other source locations and that the largest NRR values occur when the MLS is close to the core-shell interface, i.e., near to $x_s = 24$ nm. It is also

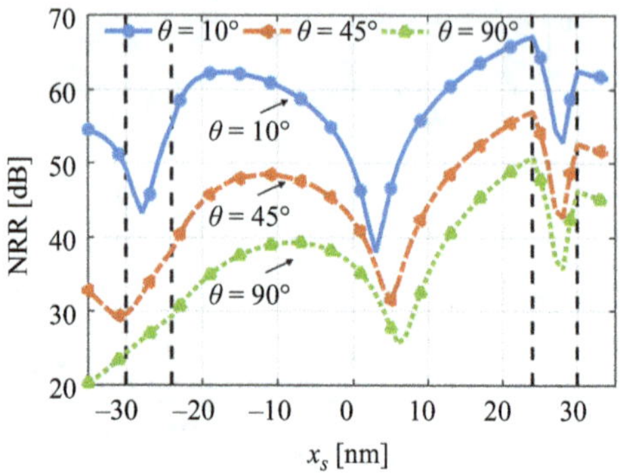

Figure 2.11 The normalized radiation resistance, NRR, as a function of the MLS location, $\rho_s = x_s$, along the x-axis for the super-resonant holey Ag-based 2D CNP for the indicated cut-out angles θ.

very interesting to observe that the NRR in Figure 2.11 never drops to 0 dB. This is in sharp contrast to the corresponding symmetric active CNPs for which no enhancements of the radiated power are obtained, i.e., NRR = 0 dB, even when the source is at $x_s = 0$ nm. This source location only excites the monopole mode and none of the higher order modes [29,54,55]. Consequently, one finds that the monopole mode is strongly suppressed in the asymmetric CNPs, even when the source is at the origin.

The results of the present section have shown that appropriate perforations of the symmetric 2D CNPs may give rise to rather directive radiation beams from these electrically small structures. The examined holey and cake Ag-based 2D CNPs can be super-resonant while at the same time enabling access to the higher order modes necessary for the radiation enhancement in preferred directions. This, however, only happens in the near-field region since the higher order modes excited in these asymmetric configurations die out before they reach the far-field region.

2.3.3 Eccentric three-region CNPs

As shown in Section 2.3.2, the asymmetric holey and cake active 2D CNPs do possess higher order modes which, however, die out before they reach the far-field region. Consequently, such configurations result only in directive and enhanced near-field patterns and front-to-back ratios. A variety of applications require, however, NAs to possess directive and enhanced responses not only in the near-field, but also in their far-field region. In order to accomplish this, NAs capable of supporting higher order modes also in the far-field region are required. One way of achieving this is by introducing asymmetry through layer eccentricity, as opposed to cutting, as shown in Figure 2.1(e) for our 3Z-NP. The eccentricity is introduced by displacing the nano-core relative to the surrounding regions. We will identify particular 3Z-NPs with comparable and simultaneous (at the same wavelength)

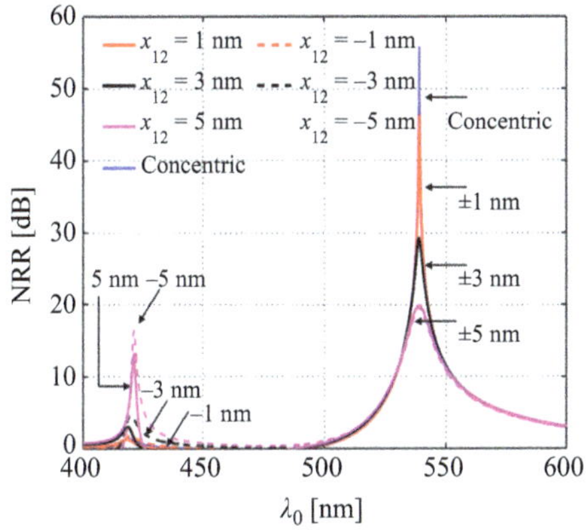

Figure 2.12 The normalized radiation resistance, NRR, as a function of the wavelength λ_0 for the 3Z-NP. The results are obtained for $\kappa = -0.4409$, *which is the gain parameter value that optimizes the concentric 3Z-NP. The influence of layer eccentricity is also shown through the indicated nano-core displacements.*

excitation of dipole and quadrupole modes with large NRR values and associated directive and enhanced near- and far-field patterns. Even more notably, we will show that the main beam directions of these patterns can be effectively tailored through variations of both the direction and the amount of nano-core displacement. Only active designs are emphasized. The included results are analytical, and the necessary details of the analytical solution are included in Appendix B. See also [47] for further details.

First, the concentric Ag-based 3Z-NP is considered, and then the influence of the eccentricity (asymmetry) is considered as measured by the offset of the center of the nano-core described by the distance d_{12} and the angle ϕ_{12}. Specifically, nano-core displacements are considered along the positive and negative x_1-axis ($d_{12} \equiv x_{12} = \pm x$ nm; $\phi_{12} = 0^o$) and along the positive y_1-axis ($d_{12} \equiv y_{12} = +y$ nm; $\phi_{12} = 90^o$).

As shown in Figure 2.12, a super-resonant state exists also in the case of a concentric 3Z-NP. It is attained for the gain parameter $= -0.4409$; the resulting NRR $= 55.8$ dB and the super-resonant wavelength $\lambda_0 = 538.7$ nm. As for the symmetric active 2D CNP considered in Section 2.3.1, the super-resonant state is due to a strong dipole mode excitation in the 3Z-NP. This is confirmed by the results in Figure 2.13(a) where the total power spectrum of the n-th mode is shown for the concentric 3Z-NP. The total power spectrum is essentially the total power in the $n = 0, 1, 2, 3,$ and 4 modes, normalized by the power in the $n = 1$ (dipole) mode. It is obvious that the vast majority of the power is contained in the dipole mode (the bar is invisible!) with only a negligible portion being contained in the other modes. This is also confirmed by the magnetic field result ($20 \log |\boldsymbol{H}/1[\text{A/m}]|$) in Figure 2.14(a), which shows a clear and strong dipole field response.

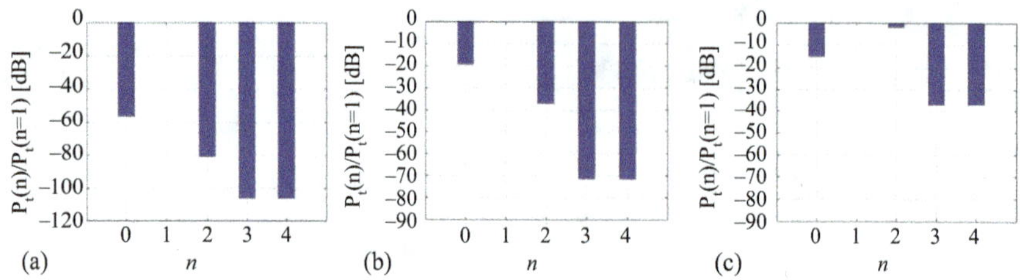

Figure 2.13 *The total power spectrum, $P_t(n)/P_t(n=1)$, of the nth mode, $n = 0, 1, 2, 3$, and 4, for (a) the concentric 3Z-NP ($\kappa = -0.4409, \lambda = 538.7$ nm), (b) the eccentric 3Z-NP ($\kappa = -0.4409, \lambda = 538.5$ nm) and (c) the eccentric 3Z-NP ($\kappa = -0.4409, \lambda = 421.0$ nm). For the eccentric 3Z-NPs, the nano-core displacement is $x_{12} = -5$ nm.*

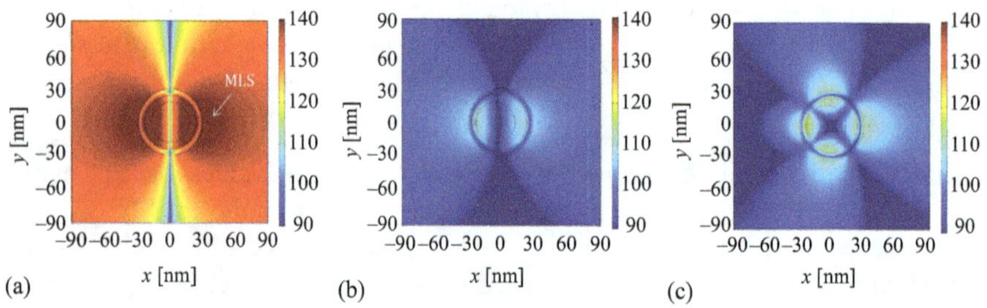

Figure 2.14 *The magnetic field, $20 \log |\boldsymbol{H}/1[\text{A/m}]|$, of (a) the concentric 3Z-NP ($\kappa = -0.4409, \lambda = 538.7$ nm), (b) the eccentric 3Z-NP ($\kappa = -0.4409, \lambda = 538.5$ nm), and (c) the eccentric 3Z-NP ($\kappa = -0.4409, \lambda = 421.0$ nm). This gain parameter value optimizes the concentric 3Z-NP. For the eccentric 3Z-NPs, the nano-core displacement is $x_{12} = -5$ nm. The MLS is located outside the 3Z-NP along the x-axis at 35 nm. The results are shown in a square region with a side length of 90 nm. Curves representing the surfaces of the 3Z-NP are included in the figure.*

We next assess the influence of the eccentricity on the resonant properties of the investigated 3Z-NP. To this end, Figure 2.12 shows the NRR for the nano-core displacements along the x_1-axis by an amount $x_{12} = \pm(1, 3, 5)$ nm; the gain parameter is still that of the super-resonant concentric case, i.e., $\kappa = -0.4409$. In comparison to the concentric 3Z-NP, the NRR is found to decrease with increasing nano-core displacements. On the other hand, the resonant wavelength remains essentially the same in all eccentric cases as in the concentric one. To exemplify, NRR $= 19.4$ dB at $\lambda = 538.5$ nm for the nano-core displacement of $x_{12} = -5$ nm. The corresponding total power spectrum is shown in Figure 2.13(b). While it is clear that the dipole mode is still the dominant mode, other modes are found to contribute more than in the concentric case, cf., Figure 2.13(a). Thus, the eccentricity causes a somewhat stronger excitation of these other modes while also diminishing the dominance of the dipole mode. This behavior is also observed in the magnetic field distribution for the $x_{12} = -5$ nm nano-core displacement shown in Figure 2.14(b). The response is still mostly dipolar, but its magnitude is reduced and its symmetry is broken, clearly signifying the increased presence of the other modes.

Moreover, the eccentricity leads to a few other interesting phenomena. As can be observed in Figure 2.12, the increasing nano-core displacement induces additional peaks in the NRR at much shorter wavelengths. This is more pronounced for the largest nano-core displacements and particularly those along the negative x_1-axis. For instance, one obtains NRR = 16.4 dB at $\lambda = 421.0$ nm for the displacement $x_{12} = -5$ nm. The total power spectrum of this case is shown in Figure 2.13(c). While the dipole mode is still the largest contributor, the $n = 0$ (monopole) mode is enhanced. Even more notable, a very strong $n = 2$ (quadrupole) mode is induced at this shorter wavelength whose strength is comparable to the dipole mode. This combined excitation of the dipole and quadrupole modes in the eccentric 3Z-NP leads to asymmetric and rather directive magnetic near-field distributions with the main beam directed along the positive x_1-axis. This is thus opposite to the direction in which the nano-core is displaced. This feature is clear from Figure 2.14(c) which shows the corresponding magnetic field results.

The above results are next supported with an account of the directivity properties of the investigated 3Z-NPs. Figure 2.15(a) shows the directivity of the concentric and eccentric 3Z-NPs for wavelengths corresponding to the peaks in the NRR around the longer wavelength (539 nm) in Figure 2.12. The nano-core displacements in the eccentric cases are given by $x_{12} = -(1, 3, 5)$ nm. The displacements along the positive x_1- and y_1-axes yield similar outcomes. The obtained directivities are in line with the near-field distribution of the magnetic fields shown in Figure 2.14(a) and (b). A symmetric pattern is found in the concentric case with the maximum directivity $D_{\max} = 2$ attained along the $\pm x_1$-direction. In contrast, the pattern is slightly asymmetric in the eccentric case, but is still dominantly dipolar. As to the directivities attained for wavelengths corresponding to the peaks in the NRR around the shorter wavelength (421.0 nm) in Figure 2.12 (where the dipole and quadrupole modes are almost equally excited), the results are shown in Figure 2.15(b) and (c) for the nano-core displacements along the positive and negative x_1-axis, respectively. Clearly, an increasingly distinct main beam occurs in the direction opposite to the displacement direction as the nano-core displacement increases with a maximum directivity $D_{\max} = 3$ for the largest nano-core displacements of $x_{12} = \pm 5$ nm. Thus, the directivity in this case is enhanced relative to that of the dipole mode alone in Figure 2.15(a). For completeness, Figure 2.15(d) shows the corresponding directivity (around the shorter wavelength 421.0 nm) for the nano-core displacements along the positive y_1-axis. Overall, a similar trend is in evidence here as well. The maximum directivity, however, exceeds 3 for the largest displacement. Moreover, the main beam direction is pointed somewhat off from the direction opposite to the displacement direction. This effect is caused by the asymmetric MLS location with respect to the 3Z-NP.

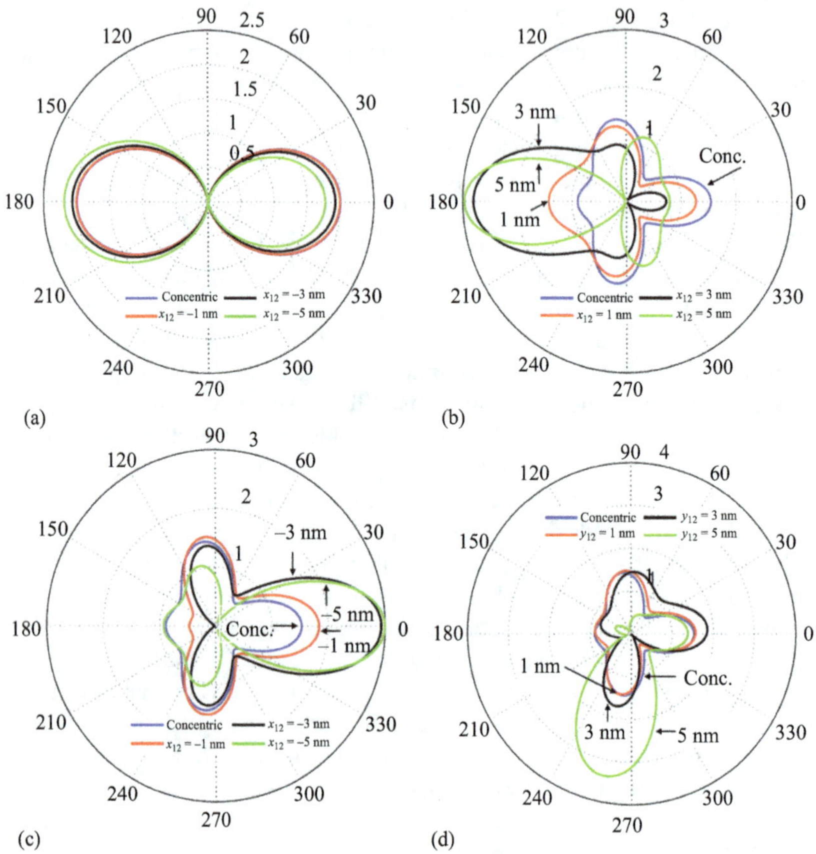

Figure 2.15 The directivity, D, as a function of the observation angle ϕ for the concentric and eccentric 3Z-NPs. In all cases, the gain parameter $\kappa = -0.4409$, which is the value that optimizes the concentric 3Z-NP. The results in (a) are attained around the longer wavelength, 539 nm (where the dipole mode is dominant), while the results in (b)–(d) are attained around the shorter wavelength, 421 nm (where the dipole and quadrupole modes are comparably excited). In (b)–(d), the directivity is shown for the indicated amounts of nano-core displacements along the $\pm x_1$- and y_1-axes.

Our investigations have shown that the eccentricity lowers the NRR at the dipole mode resonance (around 539 nm) when it is introduced into a super-resonant concentric 3Z-NP (with the gain parameter $\kappa = -0.4409$). At the same time it induces an additional resonance (around 421 nm) where the dipole and quadrupole modes are excited with a comparable strength. At the latter, interesting directive near- and far-field patterns were obtained by tailoring the nano-core displacements,, cf., Figures 2.14 and 2.15. For the $x_{12} = -5$ nm nano-core displacement,

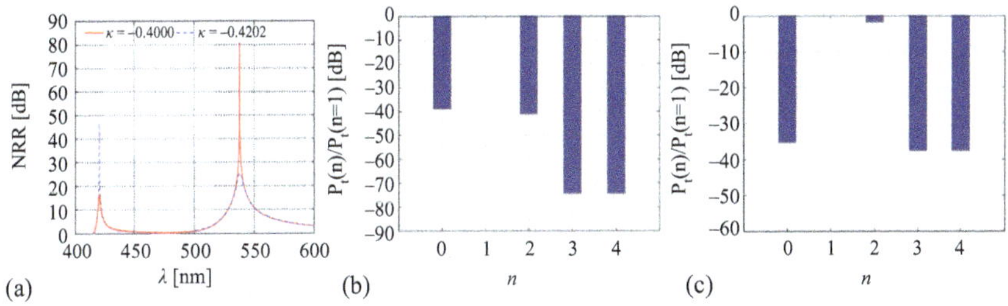

Figure 2.16 *The normalized radiation resistance, NRR, as a function of the wavelength λ for the eccentric 3Z-NP with $x_{12} = -5$ nm. One result is optimized around 539 nm, and the other is optimized around 421 nm. The corresponding total power spectrum, $P_t(n)/P_t(n=1)$, of the n'th mode, $n = 0, 1, 2, 3,$ and 4, is shown in (b) for $\kappa = -0.4000$, $\lambda = 538.7$ nm, and in (c) for $\kappa = -0.4202$, $\lambda = 420.7$ nm.*

NRR = 19.4 dB at $\lambda = 538.5$ nm and NRR = 16.4 dB at $\lambda = 421.0$ nm. Since these results were obtained with the gain parameter which optimized the NRR of the concentric 3Z-NP, they are not optimal for the eccentric cases. This is evidenced by the diminished NRRs and field magnitudes, cf., Figures 2.12, 2.14(b) and (c). It is therefore interesting to note that each of these eccentric 3Z-NPs can be gain optimized to further enhance their radiation characteristics around both the longer (539 nm) and shorter (421.0 nm) wavelengths.

These results indicate that the shorter wavelength designs pave a way for eccentric 3Z-NPs with enhanced and directive near- and far-field radiation characteristics. This is illustrated in Figure 2.16(a), which shows the NRR as a function of the wavelength λ for the largest nano-core displacement, $x_{12} = -5$ nm, for two cases: one providing the optimized response around the longer wavelength, and the other providing the optimized response around the shorter wavelength. More specifically, NRR $\approx$ 80 dB at $\lambda = 538.7$ nm (longer wavelength) for $\kappa = -0.4000$ and NRR $\approx$ 46 dB at $\lambda = 420.7$ nm (shorter wavelength) for $\kappa = -0.4202$. The associated total power spectra are shown in Figure 2.16(b) and (c). While the dipole mode is dominant at 538.7 nm (Figure 2.16(b)), it effectively combines with the quadrupole mode at 420.7 nm (Figure 2.16(c)), where the two modes are comparably excited.

The magnetic fields corresponding to the peaks at the longer and shorter wavelengths in Figure 2.16(a) are shown in Figure 2.17(a) and (b), respectively. The overall shapes of these responses resemble those already seen in Figure 2.14(b) and (c). However, since the present eccentric 3Z-NPs have been gain optimized, the field magnitudes are significantly enhanced. This reflects the significantly enhanced NRR values in Figure 2.16(a). We again emphasize that the highly directive near-field response attained at 420.7 nm (Figure 2.17(b)) for the nano-core displacement of $x_{12} = -5$ nm yields a main beam in the direction opposite to the nano-core displacement direction. As noted previously, the direction of the main beam can be effectively tailored by changing the direction of the nano-core displacement. This is illustrated clearly with the magnetic field results shown in Figure 2.17(c) (obtained for $x_{12} = 5$ nm) and Figure 2.17(d) (obtained for $y_{12} = 5$ nm). In all cases, the main beam direction clearly occurs in the direction opposite (or close to it) to the direction in which the nano-core is displaced.

It is also useful to note that the directivities of the eccentric 3Z-NPs optimized to provide the maximum NRR in Figure 2.16(a) around 421.0 nm resemble those of the non-optimized cases shown in Figure 2.15(b–d), for the three considered directions of the nano-core displacements. They are, therefore, not included here.

The results of the present section have demonstrated that eccentric 3Z-NP can be designed with enhanced and directive near- and far-field radiation properties. Since they are due to the simultaneous excitation of dipole and quadrupole modes in such structures, they can be effectively tailored (enhanced, reshaped, and steered) through appropriate changes of both the amount and the direction of the nano-core displacements. As such, the proposed 3Z-NPs, in which the asymmetry

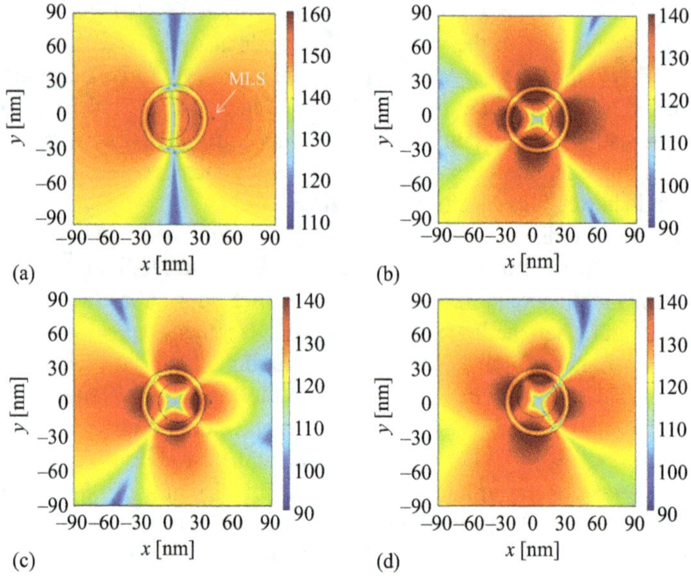

Figure 2.17 *The magnetic field,* 20 log $|\boldsymbol{H}/1[\mathrm{A/m}]|$, *of the eccentric 3Z-NP optimized (a) at 538.7 nm for* $\kappa = -0.4000$, *and (b)-(d) at 420.7 nm for* $\kappa = -0.4202$. *The fields in (a) and (b) correspond to the NRR peaks observed in Figure 2.16(a). The nano-core displacement in (a) and (b) is* $x_{12} = -5$ *nm, while it is* $x_{12} = +5$ *nm in (c). It is* $y_{12} = +5$ *nm in (d). The MLS in all cases is located outside the 3Z-NP along the x-axis at 35 nm. The results are shown in a square region with a side length of 90 nm. Curves representing the surfaces of the 3Z-NP are included in the figure.*

is introduced through the layer eccentricity, may be interesting candidates as NAs with enhanced and directive radiation characteristics.

2.4 Symmetric and asymmetric active 3D CNPs

Our review in Section 2.3 was devoted to the two and three region 2D geometries. They were all excited by an infinitely long MLS and, hence, were TE polarization configurations. Our aim in what follows is to describe systematically the outcomes for the corresponding finite, 3D symmetric and asymmetric (holey and cake) active CNPs. We recall that the 3D CNPs are obtained through an appropriate truncation of the corresponding 2D CNPs. Thus, their material composition, as well as their cross-sectional size in the *xy*-plane, remains unchanged relative to their 2D counterparts. As a consequence, the shells of the 3D CNPs are all Ag-based as well. The results of the present section have been previously reported in [49].

2.4.1 Symmetric active 3D CNPs

The diameter of the 3D CNP in the cross-sectional plane (*xy*-plane) is $D_{\mathrm{CNP}} = 2b = 60$ nm. Recall that the finite cylindrical structure is terminated at both ends with a core-shell semi-spherical part of the same material composition and core-shell dimensions. The total length H is thus the sum of the finite vertical length h of its cylinder portion and twice the radius of its endcap hemispheres, i.e., $H = h + D_{\mathrm{CNP}}$. The length h of the symmetric 3D CNPs discussed below were chosen to be: $h = \{1D_{\mathrm{CNP}}, 5D_{\mathrm{CNP}}, 10D_{\mathrm{CNP}}\}$. Therefore, their overall lengths are $H = 2D_{\mathrm{CNP}}$, $6D_{\mathrm{CNP}}$, and $11D_{\mathrm{CNP}}$. All of the 3D CNPs are illuminated by a TE polarized plane wave $(\boldsymbol{E}_{\mathrm{i}}, \boldsymbol{H}_{\mathrm{i}})$ as shown in Figure 2.1(d) with the magnitude of the incident electric field $E_0 = 1$ [V/m].

Extensive numerical optimizations of the symmetric active Ag-based 3D CNPs have shown that they too can be super-resonant and lead, in those cases, to significantly enhanced scattered fields as evidenced by the very large values of the total scattering cross section σ_{t} as determined by (2.8). Table 2.2 summarizes these values along with the gain parameter, κ, and the resonant wavelength, λ_0, for three lengths of the super-resonant 3D CNPs. For reference purposes, the results for the corresponding 2D CNPs ($h = \infty$) and the 3D spherical CNP (i.e., with $h = 0$) are also included. Our 3D CNP numerical model recovers, as expected, the super-resonant behaviors of the symmetric active 2D CNPs as well as of the spherical active CNPs results previously reported [24]. These comparisons thereby confirm the validity of the established numerical model.

Table 2.2 shows that very large, similar values of σ_{t} are obtained in all of these 3D active CNP cases. Thus, as in the 2D cases, the plasmonic losses in these CNPs have been overcome by the inclusion of gain inside their nano-cores. To emphasize the remarkable orders of magnitude of these enhancements, we note that a symmetric passive 3D CNP with $h = 5D_{\mathrm{CNP}}$ exhibits a total cross section of only $\sigma_{\mathrm{t}} = 12.8$ dBcs. This is more than 50 dBcs smaller than the active cases. Table 2.2 also shows that the amount of gain required for the super-resonance increases as the

Table 2.2 A summary of various parameters for the symmetric super-resonant Ag-based 3D CNPs.

h **[nm]**	σ_t **[dBcs]**	κ	λ_0 **[nm]**
∞	47.7 [dBc]	−0.175	577.7
$10D_{CNP}$	66.4	−0.186	577.7
$5D_{CNP}$	67.9	−0.187	576.6
$1D_{CNP}$	62.6	−0.196	549.3
0	68.0	−0.245	502.1

length of the particle decreases. This is readily explained in terms of the smaller physical volume of the gain region as the length decreases. Moreover, the super-resonances are shifted towards smaller wavelengths (blue-shift) as the particle length decreases. This behavior is also expected simply from scaling the CNP size relative to the wavelength. With the same reasoning, longer active CNPs (more needle-like shapes) attain their super-resonant behavior at longer wavelengths and require less gain in contrast to shorter CNPs. This is very useful knowledge towards the successful impregnation of these particles with realistic gain materials, e.g., quantum-dot materials. With the proper fine-tuning of the height of the 3D CNPs, their resonances can be made to spectrally overlap the emission lines of the quantum-dots, thereby attaining realistic super-resonant systems in line with the presented results.

As with the 2D CNP cases, the super-resonances of the 3D CNPs are easily associated with the magnetic field distribution, $20 \log |\boldsymbol{H}/1[\mathrm{A/m}]|$. The results are shown in Figure 2.18.

The upper row (Figure 2.18(a)–(c)) shows the field distributions in the xy-plane for the $h = 1D_{CNP}$, $5D_{CNP}$, and $10D_{CNP}$ cases, respectively. The lower row (Figure 2.18(d)–(f)) shows the corresponding results in the xz-plane. The insets in Figure 2.18(a) and (d) depict the propagation direction of the incident field, $\boldsymbol{a}_k = \boldsymbol{a}_x$. The responses in the xy-plane all display a strong dipole mode. They are found to be very similar to those reported in Figure 2.4(c) for the corresponding 2D CNP cases. Notice, however, that there is a difference in the scales/magnitudes between the 2D and 3D cases. The results in the xz-plane likewise show a strong field response, particularly close to the surfaces of the nano-shell part of the structure. While the response of the $h = 1D_{CNP}$ case is large along the entire structure, there is a notable decrease in the field away from the center of the particle along the longer 3D CNPs. This is particularly noticeable in the $h = 10D_{CNP}$ case. The total length of the 3D CNP is 660 nm in the latter. This length is a bit more than the wavelength at which this configuration resonates (577.7 nm, cf., Table 2.2). The ends of these longer 3D CNPs seem, therefore, not to be fully exploited in these wave-matter interactions. This observation indicates there are better and worse placements for possible quantum-dot media that would be introduced into these 3D CNPs in a realistic scenario. Also note that the further the observation point is moved away from the CNPs, the more the field in the xz-plane is found to be omnidirectional for all cases.

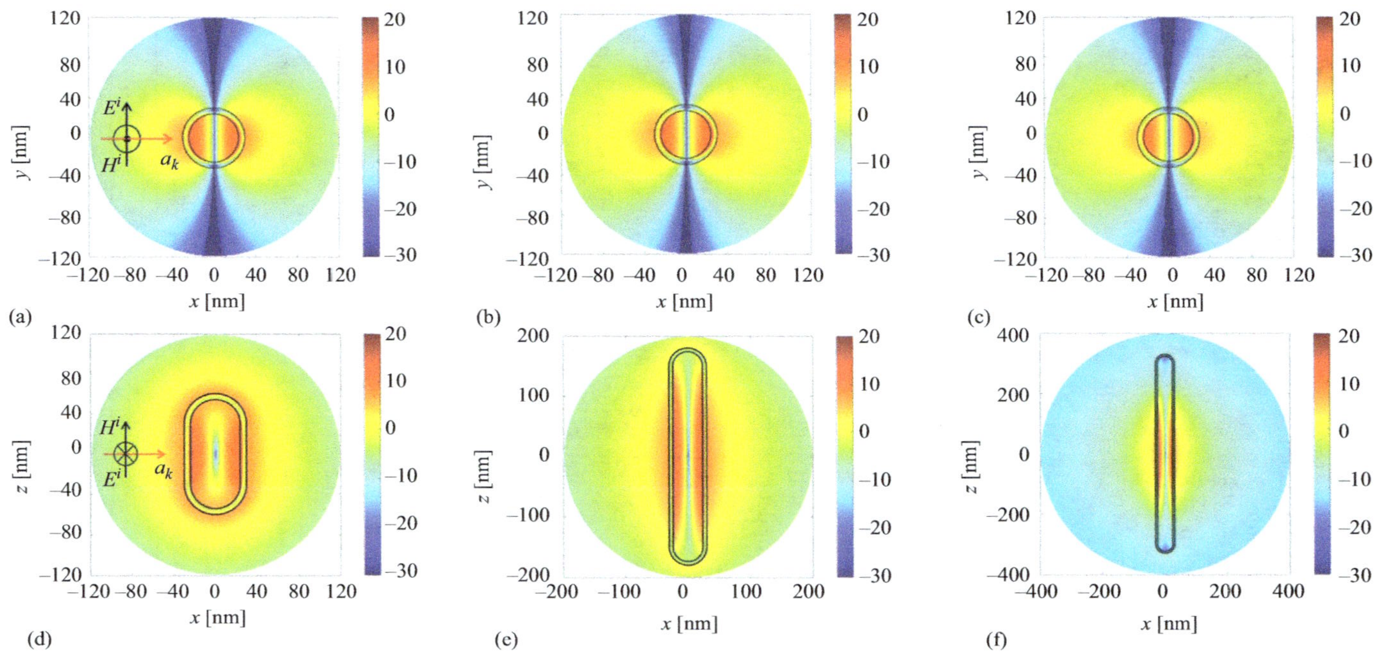

Figure 2.18 *The magnetic field,* $20 \log |\boldsymbol{H}/1[\mathrm{A/m}]|$, *of the super-resonant Ag-based 3D CNPs for* $h = 1D_{\mathrm{CNP}}$, *(a) and (d);* $h = 5D_{\mathrm{CNP}}$, *(b) and (e); and* $h = 10D_{\mathrm{CNP}}$, *(c) and (f). The upper row depicts the distributions in the xy-plane in a circular region with a 90 nm radius. The lower row shows them in the xz-plane (in a circular region with a varying radius). Curves representing the surfaces of the 3D CNPs are included in each figure.*

2.4.2 Asymmetric, holey and cake, active 3D CNPs

Because the dipole mode dominates their behavior, the symmetric active 3D CNPs are not very directive in either of their two cross-sectional planes. Consequently, in analogy with the 2D cases, we have examined the corresponding asymmetric holey and cake active 3D CNPs. Our attention is focused here on the Ag-based 3D CNP with $h = 5D_{\text{CNP}}$. Because very small asymmetries, i.e., small cut-out angles θ, have little impact on their directional properties, we consider only the larger cut-out angle cases:

$$\theta = 22.5^\circ, 45^\circ, \text{and } 90^\circ.$$

Table 2.3 summarizes the total scattering cross section σ_t , the gain parameter κ, and the resonant wavelength λ_0, for both holey and cake active Ag-based 3D CNPs. The results for the symmetric 3D CNP case ($\theta = 0^\circ$) are also included for reference. The table shows that these asymmetric active 3D CNPs can also be optimized to be super-resonant and provide significantly enhanced scattering responses. Comparing the resonant wavelengths in Table 2.3 with those in Table 2.1, it is interesting to find that the resonant wavelengths for the asymmetric 3D and 2D CNPs are almost equal to each other. Moreover, the same overall trend with increasing asymmetry is observed in the 3D case as previously reported for the 2D case, i.e., the amount of gain decreases and the resonant wavelength gets blue-shifted.

The magnetic field distributions ($20 \log |\boldsymbol{H}/1[\text{A/m}]|$) for the holey super-resonant Ag-based 3D CNP when $h = 5D_{\text{CNP}}$ are shown in Figure 2.19. The upper row (Figure 2.19(a)–(c)) shows them in the xy-plane for the $\theta = 22.5^\circ$, $\theta = 45^\circ$, and $\theta = 90^\circ$, respectively, cases. The lower row (Figure 2.19(d)–(f)) shows them in the xz-plane for the same cut-out angles. The responses in all cases are rather strong. Moreover, it is clear that an increasing asymmetry leads to a more asymmetric and directive response, as was found with the asymmetric 2D cases. This is particularly clear when the observation points are in the vicinity of the CNPs in the half-space opposite to that in which the cut-outs and plane wave source occur. The insets in Figure 2.19(a) and (d) depict the propagation direction of the incident field, $\boldsymbol{a}_k = \boldsymbol{a}_x$.

Table 2.3 A summary of various parameters for the super-resonant holey and cake Ag-based 3D CNPs.

θ [°]	3D Holey CNP			3D Cake CNP		
	σ_t [dBcs]	κ	λ_0 [nm]	σ_t [dBcs]	κ	λ_0 [nm]
0	67.9	−0.187	576.6	67.9	−0.187	576.6
22.5	67.0	−0.170	518.7	59.7	−0.178	514.8
45	59.5	−0.157	500.8	61.0	−0.164	496.3
90	53.7	−0.142	472.1	48.2	−0.147	468.0

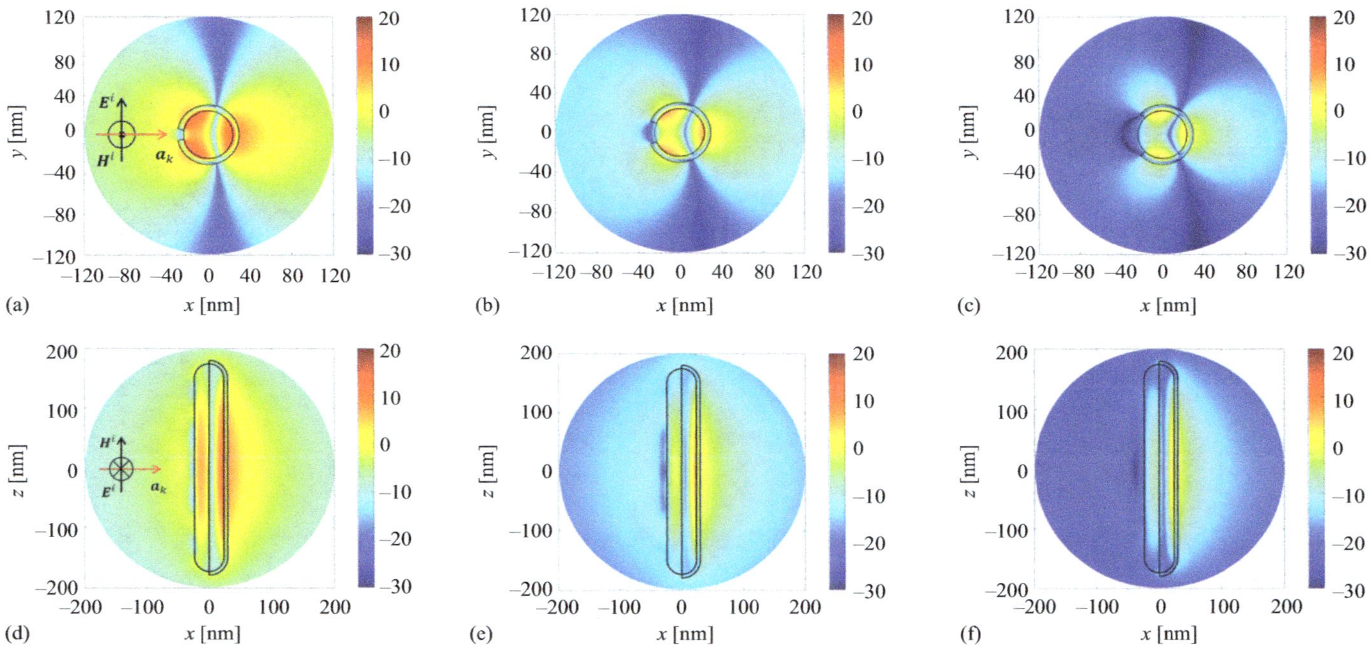

Figure 2.19 *The magnetic field ($20 \log |\boldsymbol{H}/1[\mathrm{A/m}]|$) of the super-resonant Ag-based, 3D holey CNP for $\theta = 22.5^{\circ}$, (a) and (d); $\theta = 45^{\circ}$, (b) and (e); and $\theta = 90^{\circ}$, (c) and (f). In all cases, $h = 5D_{\mathrm{CNP}}$. The upper row depicts the distributions in the xy-plane in a circular region with a 90 nm radius. The lower row shows them in the xz-plane (in a circular region with a varying radius). Curves representing the surfaces of the 3D CNPs are included in each figure.*

To further account for the directional nature of the near-field radiation exhibited by the holey super-resonant Ag-based 3D CNPs, the planar sections of the magnitude of the magnetic field in the *xy*-plane ($z = 0$) normalized to its maximum value as a function of the azimuthal angle ϕ are shown in Figure 2.20 when the radial distance ρ of the observation point is set to 40 nm (Figure 2.20(a)), 80 nm (Figure 2.20(b)), and 120 nm (Figure 2.20(c)). The corresponding planar sections of the magnitude of the power flow density $\boldsymbol{S}$ normalized to its maximum value are shown in Figure 2.20(d–f). The response is rather asymmetric and directional for observation points near the structure. The near-field $FTBR_{NF}$ values are very similar to those obtained for their 2D counterparts.

Only the largest asymmetry case ($\theta = 90°$) displays some noticeable directional response for more distant observation points in this plane. This feature is particularly noticeable in the power flow density plot in Figure 2.20(f). Thus, the higher order modes, which are abundantly present in the near-field regions in these 3D CNP cases, also die out and never reach too far away from the CNP. Due to their electrically small cross-section in the *xy*-plane, these asymmetric 3D CNPs

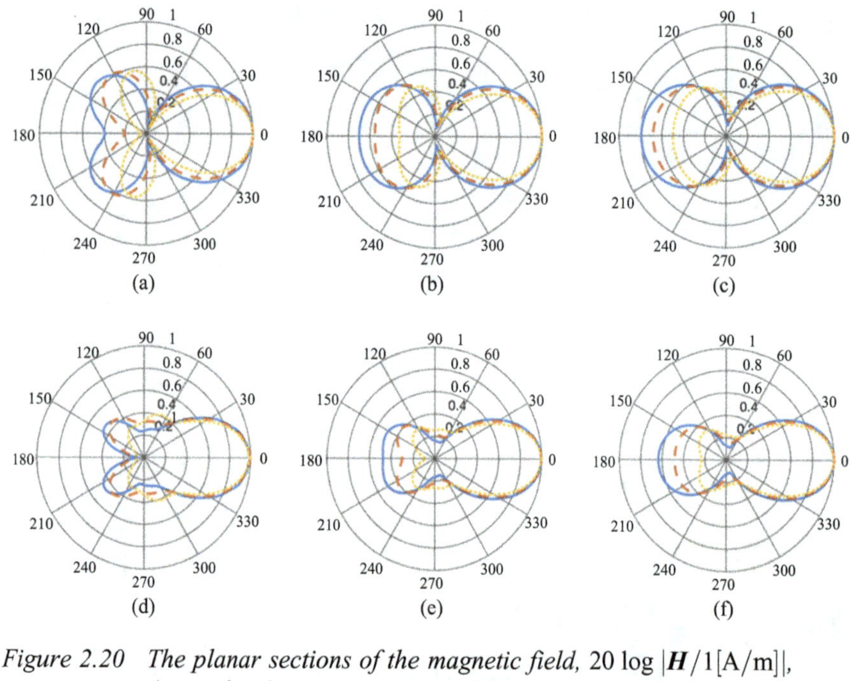

Figure 2.20 The planar sections of the magnetic field, $20 \log |\boldsymbol{H}/1[\mathrm{A/m}]|$, *(normalized to its maximum value) shown in the xy-plane ($z = 0$) for the super-resonant holey Ag-based 3D CNPs as a function of the observation angle ϕ for (a) $\rho = 40$ nm, (b) $\rho = 80$ nm, and (c) $\rho = 120$ nm. The respective planar sections of the normalized magnitude of the power flow density $\boldsymbol{S}$ normalized to its maximum value are shown in (d)–(f). The cut-out angles $\theta = 22.5°$ (solid-blue), 45° (dash-red), and 90° (dot-orange).*

also operate in their fundamental dipole mode and exhibit a dipolar response in that plane similar to those reported for the asymmetric 2D CNPs. Thus, their $FTBR_{FF} \approx 0$ dB. Finally, we emphasize that similar results are obtained for the corresponding super-resonant cake 3D CNPs and, as a result, are not presented here.

The results of this section have elucidated several important things. Truncated cylindrical NAs can be designed to be super-resonant just as their 2D counterparts considered back in Section 2.3.2 are. The super-resonances in these 3D cases require, however, increased gain parameter values, and they are found to occur at blue-shifted wavelengths. Introducing asymmetry through perforations had basically the same effect as in the 2D-cases; the higher order modes were in evidence only in the near-field regions where they gave rise to rather directive near-field beams. These were found to approach the standard dipole mode form for farther distances, typical of electrically small configurations.

2.5 Symmetric multi-layer NPs

The results in Section 2.3.3 have shown that directive radiation can be achieved not only in the near-field but also in the far-field region provided that the structure supports higher order modes in the latter. This was the case with the eccentric 3Z-NPs in which the dipole and quadrupole modes were excited at the same wavelength. In the present section we extend these ideas to achieve superdirective radiation and scattering behaviors in electrically small symmetric structures. To this end, the metamaterial-inspired paradigm [59] that employs near field resonant parasitic (NFRP) elements to enhance and control the emissions from a driven source has been combined with the concept of having those elements generate higher order modes (HOM) [52]. In this regard the symmetric ML-NP configuration from Figure 2.1(f) is considered. As opposed to the array of elemental Huygens multipole radiations examined in [52], our ML-NP is the actual NFRP element that is driven by the nearby MLS in order to achieve the desired super-directive and needle-like characteristics. The directivity enhancement is directly tied to the number of HOMs excited in this structure. If N HOMs are desired, they can be obtained if there are a corresponding number of layers in the ML-NP and if the source can efficiently excite them. The superdirective performance, in which the cylindrical waves from the MLS are transduced into a highly collimated beam, is demonstrated with the simultaneous excitation of the dipole and HOMs in ML-NP at the same wavelength [53].

The problem at hand has been solved analytically. Its solution procedure is a straightforward extension of the 2D CNP results summarized briefly in Appendix A. We will consider two N-layer scenarios for achieving a super-directive behavior, which are also briefly summarized in Appendix A. In one scenario, the expansion coefficients A_n^{N+1} of the scattered field in the exterior $(N+1)$ region are forced to follow the Dirac-delta distribution (A.15), while in the other scenario they are forced to follow a more realistic, binomial distribution (A.16). After discussing some general results obtained in these two cases, including the theoretical bound on the maximum directivity of 2D systems, several 5-layer ML-NPs possessing specific material and geometrical parameters are discussed and their super-directive behaviors are described.

By forcing the expansion coefficients A_n^{N+1} to recover a Dirac-delta distribution, and by truncating the cylindrical multipole expansion to $N+1$ terms, the directivity takes on the following form [53]:

$$D_N(\phi) = \frac{\left|1 + 2\sum_{n=1}^{N} \cos(n\psi)\cos(\phi - \phi_s)\right|^2}{1 + 2\sum_{n=1}^{N} |\cos(n\psi)|^2}, \tag{2.12}$$

where $\psi = \phi_{max} - \phi_s$, with ϕ_{max} denoting the desired direction for maximum directivity. With no loss of generality we set $\phi_s = \pi$ (the MLS is on the $-x$-axis) and $\phi_{max} = 0$ (maximum directivity along the $+x$-axis). One then finds that (2.12) becomes:

$$D_N(\phi) = \frac{\left|1 + 2\sum_{n=1}^{N} \cos(n\phi)\right|^2}{1 + 2N}. \tag{2.13}$$

As shown in [53], one finds immediately that as $N \to \infty$, $D_N(\phi) \propto \delta(\phi)$, and needle radiation in the $\phi = 0$ direction is obtained. It was also proved that the maximum 2D directivity for a finite number of layers is N :

$$D_{N,max}(\phi) = \frac{\left|1 + 2\sum_{n=1}^{N} 1\right|^2}{1 + 2N} = 1 + 2N. \tag{2.14}$$

Figure 2.21 shows the directivity (2.13) (in dB) as a function of the observation angle ϕ obtained for $N = 5, 10, 100$, and 1,000 layer cases. The maximum directivity in the respective cases is confirmed to be 11, 21, 201, and 2,001, this being in line with (2.14). The resulting far-field FTBR for each case is, respectively, 20.83, 26.44, 46.06, and 66.02 dB. Moreover, the $N = 1{,}000$ layer case clearly demonstrates a needle-like behavior.

From the standard antenna theory [60], we know that the amplitude and/or phase distributions between the elements of an antenna array can be designed in order to control its radiation characteristics. Some common amplitude distributions for broadside radiating and uniformly spaced arrays are the uniform (largest directivity), Dolph–Tschebyscheff (specified flat side-lobe level) and binomial (few or no sidelobes). In order to demonstrate that the expansion coefficients of the scattered field for our ML-NP can in fact be weighted to produce other beams tailored by choice, a binomial distribution was also considered [53] (see also (A.16)). The directivities resulting from both the Dirac-delta and binomial-distribution based scattered field coefficients are shown in Figure 2.22(a) for $N = 10$ layers.

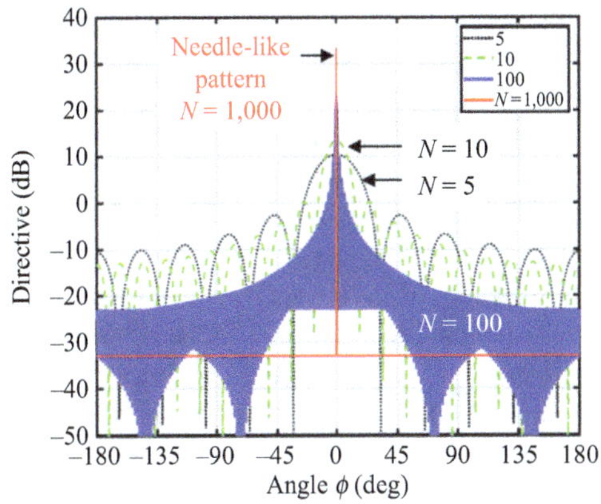

Figure 2.21 The directivity as a function of the observation angle ϕ associated with the $N = 5, 10, 100,$ and 1000 ML-NPs. In all cases, the directivity is generated by the Dirac-delta-based expansion coefficients.

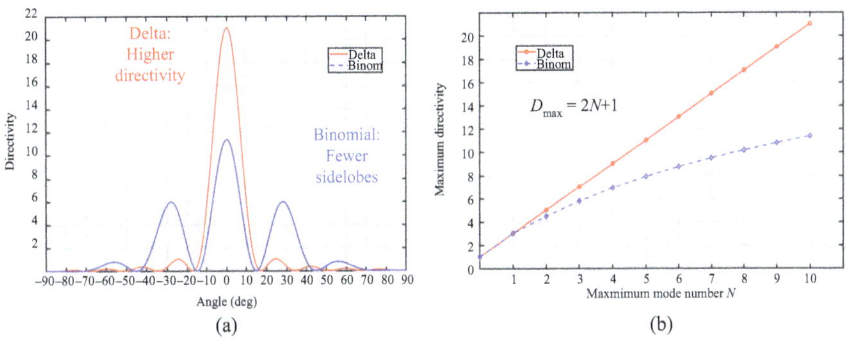

Figure 2.22 The directivity obtained with both the Dirac-delta and binomial-distribution-based scattered field coefficients for N = 10 (a), and the maximum directivity as a function of the number N (b). Both figures are in linear scale.

As expected, the Dirac-delta based coefficients produce a more directive and more tightly confined main beam with low sidelobes. On the other hand, the binomial-based case leads to a broader main beam and fewer sidelobes. The corresponding maximum directivities with those scattered field coefficients are found to increase as the number of layers N (hence, modes) increases as Figure 2.22(b) indicates. The $D_{N,\max}$values achieve their upper bounds $1 + 2N$ in each of the Dirac-delta cases. On the other hand, the binomial coefficient-based beam clearly begins to saturate as the number of layers (hence, modes) increases. As opposed to the equal power weighting associated with each Dirac-delta mode of the field expansions, the binomial weights decrease as the mode number increases. These coefficient choices clearly demonstrate the ability to customize the output beam of the system simply by tailoring the scattered field coefficients through the geometry and material choices in analogy with amplitude tapering in a classic antenna array.

In order to illustrate the potential of our ML-NP to achieve these high directivities, we have considered a number of specific configurations. As noted above, the number of modes that can be generated and controlled in such structures is intimately related to the number of layers and materials from which these layers are made. A ML-NP with N layers can, in principle, be optimized to generate at least $N+1$ modes of significance with the desired amplitudes and phases. As we have shown [53], the remaining HOMs of the cylindrical wave expansion contribute little to the overall scattering since they are not designed to resonate with the source in any way. In all cases, the mode sums are thus truncated to $n = 0, 1, \ldots, N$ terms.

As a first example, we have investigated a highly subwavelength, MLS-excited, $N = 5$ layer, lossless ML-NP whose exterior radius was set to $r_5 = \lambda/10$, implying a transverse effective width of $W_{\text{eff}} = 2\,r_5 = \lambda/5$. The MLS was located at $(\rho_s, \phi_s) = (1.05\,r_5, 180°)$ which is sufficiently away from the origin to produce several HOMs. As derived in [53], the 2D directivity bound of the analogous uniformly excited 2D antenna system is $D_{\text{2D}} = 2\pi W_{\text{eff}}/\lambda$. Consequently, any 5-layer ML-NP with a directivity greater than $D_{\text{2D}} = 1.257$ is superdirective [60], whereas it is needle-like if it approaches the maximum value of $D_{N,\text{max}} = 11 = 8.75\,D_{\text{2D}}$. Since all parameters are defined in terms of λ, our results scale to any frequency. In particular, for an optical source with $\lambda = 500$ nm, the ML-NP nano-antenna has a 50 nm radius.

The electrical and geometrical parameters of the examined ML-NPs were tailored by numerical optimization to produce scattering coefficients yielding the desired superdirective pattern. The details of the optimization process may be found in [53]. The specific choice of the materials took advantage of the known miniaturization capabilities associated with the juxtaposition of positive and negative material parameter regions [1], of which the latter are freely available in the optical regime.

In the first example, the resulting relative permittivities of the 5-layer ML-NP were found to be: $\varepsilon_{1r} = 4.98852$, $\varepsilon_{2r} = 0.15285$, $\varepsilon_{3r} = -2.02885$, $\varepsilon_{4r} = 7.46530$, and $\varepsilon_{5r} = -8.400$, when the radii of the respective layers were specified as: $r_1 = 0.01\lambda$, $r_2 = 0.03\lambda$, $r_3 = 0.06\lambda$, $r_4 = 0.08\lambda$, and $r_5 = 0.10\lambda$. The first $n = 0, 1, \ldots, 5$ scattered field coefficients were obtained numerically and used to calculate the resulting directivity. It is shown in Figure 2.23(a), which also includes the directivity based on the Dirac-delta scattered field coefficients for this 5-layer ML-NP. The maximum $D_{N=5,\text{max}} = 6.368 = 5.07\,D_{\text{2D}}$ is found along the $\phi = 0°$ direction with the associated FTBR = 10.03 = 10.01 dB. These results clearly demonstrate the ability of this ML-NP to act as a superdirective lens which effectively transduces the cylindrical waves from the MLS into a superdirective beam.

To shed further light on the behavior of this ML-NP, the total magnetic field ($10\log|\boldsymbol{H}_{\text{tot}}(x,y)/1[\text{A/m}]|[\text{dB}]$) is shown in Figure 2.23(b). It is clear that the $N = 5$ mode is rather dominant in the vicinity of the ML-NP. Moreover, these

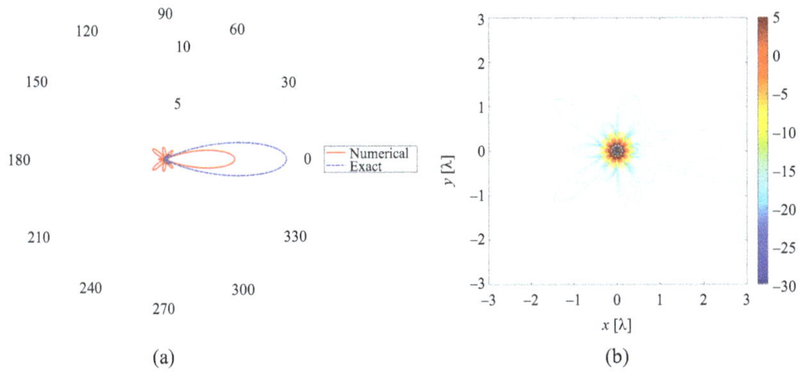

Figure 2.23 The numerical and the Dirac-delta based directivities (linear scale) obtained with only $n = 0, 1, \ldots, 5$ for a MLS excited 5-layer ML-NP (a), and the associated total magnetic field ($10\log|\boldsymbol{H}_{\text{tot}}(x,y)/1[\text{A/m}]|$) (b). The MLS is located at $(\rho_s, \phi_s) = (1.05\, r_5, 180°)$.

mid-range results show a main beam in the $\phi = 0°$ direction which becomes increasingly dominant as the sidelobes begin to diminish away from the structure. It is the $N = 5$ mode that is at the heart of the large $D_{\max}$ value, whereas the lower order modes with the correct weighting are instrumental for the observed reduction of the sidelobe levels. We note that the observed superdirective behavior is obtained with only one set of modes (TE^z) as the complementary TM^z set is not present. Moreover, we reiterate that only the modes $n = 0, 1, \ldots, 5$ were used and superimposed to obtain the above results. Additional HOMs would be required and available for contributions to the directivity only if more layers were included in the ML-NP.

The ML-NP considered above is electrically small. The superdirective behavior can also be obtained by allowing its size to be larger. Moreover, the electrically larger ML-NPs can be tuned to yield maximum scattering not only in the forward direction, but also in the backward direction [53]. With the exterior radius set to $r_5 = 1.0\lambda$, the optimization process determined the relative permittivities to be: $\varepsilon_{1r} = 3.0, \varepsilon_{2r} = 3.0, \varepsilon_{3r} = -0.3350, \varepsilon_{4r} = 0.4915$, and $\varepsilon_{5r} = -15.40$, with the remaining radii of the layers equal to: $r_1 = 0.2\lambda$, $r_2 = 0.4\lambda$, $r_3 = 0.6\lambda$, and $r_4 = 0.8\lambda$. The numerical and the Dirac-delta based directivity results are shown in Figure 2.24(a). The maximum directivity $D_{N=5,\max} = 7.303$ clearly occurs along the $\phi = 0°$ direction where the associated FTBR = 9.11 = 9.60 dB. Since $D_{2D} = 2\pi r_5/\lambda = 6.28$ and $D_{N=5,\max} = 1.16\, D_{2D}$ this ML-NP NA is superdirective as well. The corresponding magnetic field distributions, which reaffirms the superdirective behavior with enhanced forward scattering and tiny sidelobes, are shown in Figure 2.24(b).

The ML-NP results demonstrate that a superdirective NA could be realistically achieved by superimposing several HOMs at the same wavelength. The ML-NP

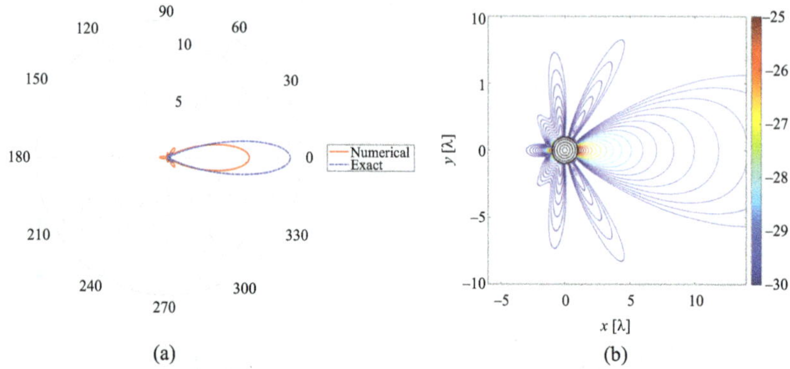

Figure 2.24 The numerical and the Dirac-delta based directivities (linear scale) obtained with only $n = 0, 1, \ldots, 5$ for a MLS excited positive-permittivity 5-layer ML-NP (a), and the associated total magnetic field ($10\log|\boldsymbol{H}_{\text{tot}}(x, y)/1[\text{A/m}]|$) (b). The MLS is located at $(\rho_s, \phi_s) = (1.05\, r_5, 180°)$.

scattered field coefficients were targeted to obey the theoretical Dirac-delta distribution to achieve needle-like radiation and the more realistic binomial distribution to tailor the number of sidelobes. The number of modes supported by these structures was intimately related to the number of their layers. The ideal needle-like radiation representing the theoretical 2D directivity bound was demonstrated. Moreover, optimized designs yielding superdirective behavior that approached the 2D directivity bound were demonstrated for forward scattering examples.

As a final remark we recall that our ML-NPs were made of both positive and negative permittivities, some of which assumed rather large values. The negative permittivities are naturally available at optical frequencies from both metals and resonant polaritonic materials. However, as explained in Section 2.2.1, the plasmonic materials are generally very lossy. While the losses do not impact the ratio of the radiated powers, they do, however, degrade the structure's ability to promote the HOMs which are imperative for the superdirective behavior. Thus, any losses will have a negative impact on the overall directivity. Nevertheless, as demonstrated with the other cylindrical core-shell structures, the introduction of gain into a ML-NP could mitigate these deleterious effects. With this in mind, a practical demonstration of superdirective ML-NP will target those optical frequencies. The resulting optical superdirective ML-NPs would have a variety of photonics applications, e.g., sensors, microscopy, and integrated optical circuits. On the other hand, natural as well as artificial microwave and mm-wave low-loss materials are now available with very small and large positive and negative permittivities. The ability to achieve super directivity from electrically small structures at these lower frequencies would have many practical applications in current and future wireless systems. Therefore, the present results are also of relevance to those lower frequency bands.

2.6 Conclusions and summary

In this chapter we have reviewed our previous and ongoing research efforts in the promising area of directive NP-based NAs. Focus was devoted to both symmetric and asymmetric 2D and 3D active, and to some extent passive cylindrical NPs with appropriate layer designs, perforations or eccentricities. The asymmetric arrangements were obtained through the introduction of appropriate perforations or layer eccentricities in the examined NPs. More specifically, we have investigated symmetric and asymmetric (holey and cake) 2D and 3D active CNPs, concentric and eccentric 2D active 3Z-NPs, as well as symmetric 2D passive ML-NPs. All but the latter case utilized a combination of silicon dioxide nano-core and a silver nano-shell in their configurations. The material parameters of the ML-NP were obtained through a numerical optimization procedure. All 2D structures were excited with a near-by MLS, while the 3D configurations were excited with a uniform plane wave. The active designs were represented through the constant frequency gain model. Our results were illustrated by various near-and far-field distribution results obtained either analytically and/or numerically with full-wave commercial packages.

The main messages conveyed by this chapter are the following. Both perforated (holey and cake designs) and eccentric (3Z-NPs) active NPs may be super-resonant with very large radiated power levels clearly exceeding those of their passive counterparts. However, while the asymmetries introduced through perforations increase only the directionality of the near-field radiation properties, the eccentricity-based asymmetries lead to enhanced directional responses also in the far-field. In the former case, the excited higher order modes die out before they ever reach the far-field. In the latter case, eccentric configurations were identified with simultaneous and comparable excitations of their dipole and quadrupole modes that also appeared in their far-field spectra. These combinations lead to enhanced directional responses in both their near- and far-field regions. Moreover, they can be tuned (enhanced, reshaped, and steered) via appropriate eccentricities and relative source locations. In addition to these active NP results, we have also shown theoretically how a symmetric, passive ML-NP with tailored scattering coefficients will lead to super-directive radiation patterns in forward (as well as backward) directions. Furthermore, we have demonstrated that ideal needle-like radiation representing the theoretical 2D directivity bound for the highest order mode index is attained and that needle radiation is reached in the limit of infinitely many layers.

The NA configurations described in this chapter may serve not only as alternative candidates to the large plethora of existing NAs, but also as a new generation of probes with significantly enhanced and directional radiation and scattering properties. As such, they may have a wide range of interesting nano-photonics applications of utmost societal importance.

Appendix A Analytical details related to the 2D CNP and ML-NP solutions

The main steps of the analytical solution pertaining to the symmetric 2D CNP configuration illustrated in Figure 2.1(a) are first summarized. They are then generalized to the ML-NP configuration shown in Figure 2.1(f) and its results.

The symmetric 2D CNP is excited with an infinitely long MLS driven by a constant magnetic current I_m [V]. The MLS is located in the exterior of the CNP and its position is specified by the coordinates (ρ_s, ϕ_s). The field due to the MLS constitutes the known incident field; its electric and magnetic fields are denoted by $\boldsymbol{E}_{\mathrm{MLS}}$ and $\boldsymbol{H}_{\mathrm{MLS}}$, respectively. In terms of cylindrical waves originating at $\rho = 0$, the incident magnetic field reads [61]:

$$\boldsymbol{H}_{\mathrm{MLS}} = -\boldsymbol{a}_z I_{\mathrm{m}} \frac{\omega\varepsilon_0}{4} \begin{cases} \sum_{n=0}^{\infty} \tau_n J_n(k_0\rho) H_n^{(2)}(k_0\rho_s) \cos[n(\phi - \phi_s)] & \rho \le \rho_s, \\ \sum_{n=0}^{\infty} \tau_n J_n(k_0\rho_s) H_n^{(2)}(k_0\rho) \cos[n(\phi - \phi_s)] & \rho \ge \rho_s, \end{cases} \tag{A.1}$$

where $J_n(\cdot)$ is the Bessel function of order n, $H_n^{(2)}(\cdot)$ is the Hankel function of second kind and order n, and where $\tau_0 = 1$ and $\tau_n = 2$ for $n \neq 0$. The unknown fields in regions $i = 1, 2,$ and 3 are also expanded into cylindrical waves as follows:

$$\boldsymbol{H}_{\mathrm{Reg}\ i} = -\boldsymbol{a}_z I_{\mathrm{m}} \frac{\omega\varepsilon_i}{4} \sum_{n=0}^{\infty} \tau_n h_n(\rho) \cos[n(\phi - \phi_s)], \tag{A.2}$$

where

$$h_n(\rho) = \begin{cases} A_n^1 J_n(k_1\rho), & \text{Region 1: } \rho \le a \\ A_n^2 J_n(k_2\rho) + A_n^3 Y_n(k_2\rho), & \text{Region 2: } a \le \rho \le b \\ A_n^4 H_n^{(2)}(k_0\rho), & \text{Region 3: } \rho \ge b \end{cases} \tag{A.3}$$

with $Y_n(\cdot)$ denoting the Neumann function of order n. The total fields in the three regions thus read:

$$\begin{Bmatrix} \boldsymbol{H}_1 \\ \boldsymbol{H}_2 \\ \boldsymbol{H}_3 \end{Bmatrix} = \begin{cases} \boldsymbol{H}_{\mathrm{Reg}\ 1} & \text{(Region 1)}, \\ \boldsymbol{H}_{\mathrm{Reg}\ 2} & \text{(Region 2)}, \\ \boldsymbol{H}_{\mathrm{Reg}\ 3} + \boldsymbol{H}_{\mathrm{MLS}} & \text{(Region 3)}, \end{cases} \tag{A.4}$$

The electric fields associated with (A.4) follow at once through the application of the Maxwell-Ampere's law $\boldsymbol{E}_i = \nabla \times \boldsymbol{H}_i / j\omega\mu_i = \boldsymbol{a}_\rho E_\rho + \boldsymbol{a}_\phi E_\phi$. They are not included in here for the sake of brevity. The quantities $A_n^1 \ldots A_n^4$ in (A.3) are the unknown expansion coefficients. They are determined through the satisfaction of the interface boundary conditions. These require the continuity of the tangential components of the total electric and magnetic fields at the two cylindrical interfaces of the 2D CNP at $\rho = a$ and $\rho = b$. More explicitly, this process establishes the following equations:

$$\varepsilon_1 A_n^1 J_n(k_1 a) - \varepsilon_2 A_n^2 J_n(k_2 a) - \varepsilon_2 A_n^3 Y_n(k_2 a) = 0, \tag{A.5a}$$

$$k_1 A_n^1 J_n'(k_1 a) - k_2 A_n^2 J_n'(k_2 a) - k_2 A_n^3 Y_n'(k_2 a) = 0, \tag{A.5b}$$

at $\rho = a$ and:

$$\varepsilon_2 A_n^2 J_n(k_2 b) + \varepsilon_2 A_n^3 Y_n(k_2 b) - \varepsilon_0 A_n^4 H_n^{(2)}(k_0 b) = \varepsilon_0 J_n(k_0 a) H_n^{(2)}(k_0 \rho_s), \tag{A.5c}$$

$$k_2 A_n^2 J_n'(k_2 b) - k_2 A_n^3 Y_n'(k_2 b) - k_0 A_n^4 H_n^{(2)'}(k_0 b) = k_0 J_n'(k_0 a) H_n^{(2)}(k_0 \rho_s). \tag{A.5d}$$

at $\rho = b$. The equations in (A.5) represent a 4×4 linear system of equations:

$$\overleftarrow{\boldsymbol{M}}_n \boldsymbol{C}_n = \boldsymbol{\Lambda}_n, \tag{A.6}$$

where the matrix $\overleftarrow{\boldsymbol{M}}_n$ contains the information about the cylindrical waves at the two interfaces, the vector $\boldsymbol{\Lambda}_n$ contains the information related to the MLS excitation, and the vector $\boldsymbol{C}_n$ contains the unknown expansion coefficients $A_n^1 \ldots A_n^4$. The linear system of equations in (A.6) can be inverted:

$$\boldsymbol{C}_n = \left[\overleftarrow{\boldsymbol{M}}_n\right]^{-1} \boldsymbol{\Lambda}_n, \tag{A.7}$$

thereby determining the expansion coefficients $A_n^1 \ldots A_n^4$. With these coefficients in hand, the fields in all regions of the 2D CNP problem are known. As a consequence, several quantities can be calculated analytically for its further characterization.

The total radiated power, P_t, is defined as:

$$P_t = \lim_{\rho \to \infty} \int_{\phi=0}^{2\pi} \boldsymbol{S}(\rho, \phi) \cdot \boldsymbol{a}_\rho \rho d\phi, \tag{A.8}$$

where $\boldsymbol{S}(\rho, \phi)$ is the power flow density defined in (2.3). For the considered MLS excitation (TE polarization), the expression in (A.8) gives:

$$P_t = \lim_{\rho \to \infty} \frac{\eta_0}{2} \int_{\phi=0}^{2\pi} |\boldsymbol{H}_3|^2 \rho d\phi. \tag{A.9}$$

The far-field expression of $\boldsymbol{H}_3$ can be obtained easily. Then (A.9) leads to the final expression:

$$P_t = \frac{I_m^2}{8} \frac{k_0}{\eta_0} \sum_{n=0}^{\infty} \tau_n |J_0(k_0 \rho_s) + A_n^4|^2. \tag{A.10}$$

The total power radiated by the MLS alone in free space is readily obtained from (A.10) by setting $A_n^4 = 0$:

$$P_{\mathrm{MLS}} = \frac{I_{\mathrm{m}}^2}{8}\frac{k_0}{\eta_0}. \tag{A.11}$$

When the power expressions (A.10) and (A.11) are introduced into the definition of the normalized radiation resistance (NRR) (2.4), one obtains:

$$\mathrm{NRR} = 10\,\log_{10}\sum_{n=0}^{\infty}\tau_n|J_0(k_0\rho_{\mathrm{s}}) + A_n^4|^2. \tag{A.12}$$

To determine the directivity (2.5), we first note that the radiation intensity (2.6) can be written as:

$$U(\phi) = \lim_{\rho\to\infty}\frac{\eta_0}{2}|\boldsymbol{H}_3|^2\rho. \tag{A.13}$$

Its closed-from expression follows from the far-field representation of $\boldsymbol{H}_3$. Substituting (A.10) and (A.13) into (2.5), one obtains the final expression for the directivity:

$$D(\phi) = \frac{2\pi U(\phi)}{P_{\mathrm{t}}} = \frac{\left|\sum_{n=0}^{\infty}\tau_n j^n\left(J_n(k_0\rho_{\mathrm{s}}) + A_n^4\right)\cos[n(\phi-\phi_{\mathrm{s}})]\right|^2}{\sum_{n=0}^{\infty}\tau_n|J_0(k_0\rho_{\mathrm{s}}) + A_n^4|^2}. \tag{A.14}$$

Since the total radiated power (A.10) and the directivity (A.14) are far-field results, their closed-form expressions depend on the expansion coefficients A_n^4 of the scattered field in the region exterior to the CNP.

It is straightforward to generalize these results to the case of the ML-NP depicted in Figure 2.1(e). In this $N+1$ region problem, with N denoting the number of layers of the ML-NP, let us denote the expansion coefficients of the scattered field in the region exterior to the ML-NP by A_n^{N+1}. The resulting total radiated power P_{t} and the directivity $D(\phi)$ are thus obtained from (A.10) and (A.14) simply by replacing A_n^4 with A_n^{N+1}. It is understood that the coefficients A_n^{N+1} are obtained from a $2N \times 2N$ linear system of equations similar in form to (A.6).

As noted in Sections 2.1 and 2.5, we will consider ML-NPs exhibiting superdirective behavior obtained by forcing the expansion coefficients A_n^{N+1} to follow the Dirac-delta and binomial distributions. The Dirac-delta-based expansion coefficients $A_{n,\delta}^{N+1}$ are defined as:

$$A_{0,\delta}^{N+1} = \frac{1}{2\pi} - J_0(k_0\rho_{\mathrm{s}}), \tag{A.15a}$$

$$A_{n,\delta}^{N+1} = \frac{1}{\tau_n \pi j^n} \cos(n\psi) - J_n(k_0\rho_s), n > 0 \tag{A.15b}$$

where $\psi = \phi_{\max} - \phi_s$ with $\phi_{\max}$ is the desired maximum directivity direction. The resulting analytical expression for the directivity is easily obtained and is given by (2.12) in Section 2.5. The binomial-distribution-based expansion coefficients $A_{n,\text{binom}}^{N+1}$ are obtained as:

$$A_{n,\text{binom}}^{N+1} = j^{-n} \frac{p_n^N}{\max p_n^N} \cos(n\psi) - J_n(k_0\rho_s), \tag{A.16}$$

where $p_n^N = (N-1)!/[n!(N-1-n)!]$, $n = 0, 1, \ldots, N-1$, are the weights of the binomial distribution with N modes. The resulting directivity, obtained numerically, is discussed in Section 2.5.

Appendix B Analytical details of the eccentric 3Z-NP solution

The main steps of the analytical solution pertaining to the eccentric 3Z-NP configuration shown in Figure 2.1(e) follow. Our analytical approach applies, however, to a configuration which is slightly more general than the one depicted in Figure 2.1(e). Not only can the nano-core be displaced relative to the nano-shell, but the middle-layer can be as well. Thus, we introduce three rectangular, (x_i, y_i, z_i), and the associated cylindrical, (ρ_i, ϕ_i, z_i), coordinate systems, $i =$ 1, 2, and 3 such that their z_i-axes coincide, respectively, with the center of the nano-shell ($i = 1$), middle layer ($i = 2$), and nano-core ($i = 3$). The center of the nano-core (and thus its displacement) is described by the distance d_{13} and the angle ϕ_{13} with respect to the x_1-axis. The center of the middle layer is described by the distance d_{12} and the angle ϕ_{12} with respect to the x_1-axis. The center of the outer layer is taken to be true coordinate origin.

The eccentric 3Z-NP is excited with an infinitely long MLS driven by a constant magnetic current I_m [V]. The MLS is located in the region exterior to the 3Z-NP. In terms of cylindrical waves originating at $\rho_1 = 0$, the incident magnetic field $\boldsymbol{H}_{\text{MLS}}$ is given by:

$$\boldsymbol{H}_{\text{MLS}}(\rho_1, \phi_1) = -\boldsymbol{a}_z I_m \frac{\omega\varepsilon_0}{4} \begin{cases} \sum_{n=-\infty}^{\infty} J_n(k_0\rho_1) H_n^{(2)}(k_0\rho_s)\, e^{jn(\phi-\phi_s)} & \rho_1 \le \rho_s, \\ \sum_{n=-\infty}^{\infty} J_n(k_0\rho_s) H_n^{(2)}(k_0\rho_1)\, e^{jn(\phi-\phi_s)} & \rho_1 \ge \rho_s. \end{cases} \tag{B.1}$$

The unknown fields in regions $i =$ 1, 2, 3, and 4 are also expanded into cylindrical waves as follows:

$$\boldsymbol{H}_{\text{Reg } i} = -\boldsymbol{a}_z I_m \frac{\omega\varepsilon_i}{4} \sum_{n=-\infty}^{\infty} h_{i,n}(\rho, \phi), \tag{B.2}$$

where

$$h_{i,n}(\rho,\phi) = \begin{cases} A_n^1 J_n(k_1\rho_3)e^{jn\phi_3}, & \text{Region 1: } \rho_3 \le a \\ \left[A_n^2 J_n(k_2\rho_3) + B_n^2 Y_n(k_2\rho_3)\right]e^{jn\phi_3}, & \text{Region 2: } a \le \rho_3 \le b \\ \left[A_n^3 J_n(k_3\rho_3) + B_n^3 Y_n(k_3\rho_3)\right]e^{jn\phi_3}, & \text{Region 3: } b \le \rho_3 \le c \\ B_n^4 H_n^{(2)}(k_0\rho_1)e^{-jn\phi_s}e^{jn\phi_1}, & \text{Region 4: } \rho_1 \ge c \end{cases} \tag{B.3}$$

The total magnetic fields in the various regions thus read:

$$\begin{Bmatrix} \boldsymbol{H}_1(\rho_3,\phi_3) \\ \boldsymbol{H}_2(\rho_3,\phi_3) \\ \boldsymbol{H}_3(\rho_3,\phi_3) \\ \boldsymbol{H}_4(\rho_1,\phi_1) \end{Bmatrix} = \begin{cases} \boldsymbol{H}_{\text{Reg 1}} & \text{Region 1} \\ \boldsymbol{H}_{\text{Reg 2}} & \text{Region 2} \\ \boldsymbol{H}_{\text{Reg 3}} & \text{Region 3} \\ \boldsymbol{H}_{\text{Reg 4}} + \boldsymbol{H}_{\text{MLS}} & \text{Region 4} \end{cases} \tag{B.4}$$

The corresponding electric fields are obtained from Maxwell's equation: $\boldsymbol{E}_i = \nabla \times \boldsymbol{H}_i / j\omega\mu_i$, $i = 1, 2, 3$, and 4.

The quantities $A_n^1 \ldots A_n^3$ and $B_n^2 \ldots B_n^4$ in (B.3) are the unknown expansion coefficients that must be determined. As in Appendix A, they are determined through the application of the boundary conditions which require the continuity of the tangential components of the total electric and magnetic fields at the three cylindrical interfaces of the 3Z-NP at $\rho_1 = c$, $\rho_2 = b$, and $\rho_3 = a$. However, for the analytical expressions to be practical at the interfaces at $\rho_1 = c$ and $\rho_2 = b$, the fields in regions 2 and 3, initially expressed in terms of the $i = 3$ coordinate system, must be translated to the $i = 1$ and $i = 2$ coordinate systems. This can be accomplished through the application of the addition theorem for cylindrical wave functions of order n [62]. In this way, the following equations are established:

$$\varepsilon_1 A_n^1 J_n(k_1 a) - \varepsilon_2 A_n^2 J_n(k_2 a) - \varepsilon_2 B_n^2 H_n^{(2)}(k_2 a) = 0, \tag{B.5a}$$

$$k_1 A_n^1 J_n'(k_1 a) - k_2 A_n^2 J_n'(k_2 a) - k_2 B_n^2 H_n^{(2)'}(k_2 a) = 0, \tag{B.5b}$$

at $\rho = a$ (nano-core interface) and:

$$\begin{aligned} &\varepsilon_2 \sum_{n=-\infty}^{\infty} J_{q-n}(k_2 d_{23})\left[A_n^2 J_q(k_2 b) + B_n^2 H_q^{(2)}(k_2 b)\right]e^{-j(q-n)\phi_{23}} \\ &= \varepsilon_3 \sum_{n=-\infty}^{\infty} J_{q-n}(k_3 d_{23})\left[A_n^3 J_q(k_3 b) + B_n^3 H_q^{(2)}(k_3 b)\right]e^{-j(q-n)\phi_{23}}, \end{aligned} \tag{B.6a}$$

$$\begin{aligned} &k_2 \sum_{n=-\infty}^{\infty} J_{q-n}(k_2 d_{23})\left[A_n^2 J_q'(k_2 b) + B_n^2 H_q^{(2)'}(k_2 b)\right]e^{-j(q-n)\phi_{23}} \\ &= k_3 \sum_{n=-\infty}^{\infty} J_{q-n}(k_3 d_{23})\left[A_n^3 J_q'(k_3 b) + B_n^3 H_q^{(2)'}(k_3 b)\right]e^{-j(q-n)\phi_{23}}, \end{aligned} \tag{B.6b}$$

at $\rho = b$ (inner nano-shell interface) and:

$$\begin{aligned} &\varepsilon_3 \sum_{n=-\infty}^{\infty} J_{q-n}(k_3 d_{13}) \left[A_n^3 J_q(k_3 c) + B_n^3 H_q^{(2)}(k_3 c) \right] e^{-j(q-n)\phi_{13}} \\ &\quad = \varepsilon_0 \left[H_q^{(2)}(k_0 \rho_s) J_q(k_0 c) + B_q^4 H_q^{(2)}(k_0 c) \right] e^{-jq\phi_s}, \end{aligned} \tag{B.7a}$$

$$\begin{aligned} &k_3 \sum_{n=-\infty}^{\infty} J_{q-n}(k_3 d_{13}) \left[A_n^3 J_q'(k_3 c) + B_n^3 H_q^{(2)'}(k_3 c) \right] e^{-j(q-n)\phi_{13}} \\ &\quad = k_0 \left[H_q^{(2)}(k_0 \rho_s) J_n'(k_0 c) + B_q^4 H_q^{(2)'}(k_0 c) \right] e^{-jq\phi_s}, \end{aligned} \tag{B.7b}$$

at $\rho = c$ (outer nano-shell interface).

Equations (B.5a) and (B.5b) are used to eliminate the coefficients A_n^1. By lengthy, but straightforward manipulations, the following expression is obtained:

$$\tau_{1n} A_n^2 + \tau_{2n} B_n^2 = 0, \tag{B.8}$$

where

$$\tau_{1n} = \frac{\varepsilon_2}{\varepsilon_1} \frac{J_n(k_2 a)}{J_n(k_1 a)} - \frac{k_2}{k_1} \frac{J_n'(k_2 a)}{J_n'(k_1 a)}, \tag{B.9a}$$

$$\tau_{2n} = \frac{\varepsilon_2}{\varepsilon_1} \frac{H_n^{(2)}(k_2 a)}{J_n(k_1 a)} - \frac{k_2}{k_1} \frac{H_n^{(2)'}(k_2 a)}{J_n'(k_1 a)}. \tag{B.9b}$$

We next consider the inner interface of the nano-shell ($\rho = b$). To this end, we note that (B.6a) and (B.6b) can be written as:

$$\sum_{n=-\infty}^{\infty} \left[-\chi_1^2(q,n) A_n^2 - \chi_2^2(q,n) B_n^2 + \chi_1^3(q,n) A_n^3 + \chi_2^2(q,n) B_n^3 \right] = 0, \tag{B.10a}$$

$$\sum_{n=-\infty}^{\infty} \left[-\chi_3^2(q,n) A_n^2 - \chi_4^2(q,n) B_n^2 + \chi_3^3(q,n) A_n^3 + \chi_4^2(q,n) B_n^3 \right] = 0, \tag{B.10b}$$

where

$$\chi_1^{\nu}(q,n) = \varepsilon_{\upsilon} J_{q-n}(k_{\upsilon} d_{23}) J_q(k_{\upsilon} b) e^{-j(q-n)\phi_{23}}, \tag{B.11a}$$

$$\chi_2^{\nu}(q,n) = \varepsilon_{\upsilon} J_{q-n}(k_{\upsilon} d_{23}) H_q^{(2)}(k_{\upsilon} b) e^{-j(q-n)\phi_{23}}, \tag{B.11b}$$

$$\chi_3^{\nu}(q,n) = k_{\upsilon} J_{q-n}(k_{\upsilon} d_{23}) J_q'(k_{\upsilon} b) e^{-j(q-n)\phi_{23}}, \tag{B.11c}$$

$$\chi_4^{\nu}(q,n) = k_{\upsilon} J_{q-n}(k_{\upsilon} d_{23}) H_q^{(2)'}(k_{\upsilon} b) e^{-j(q-n)\phi_{23}}, \tag{B.11d}$$

with $\nu = 2, 3$.

We finally consider the outer nano-shell interface ($\rho = c$). By tedious manipulations of (B.7a) and (B.7b), the expansion coefficient B_n^1 can be eliminated. The following relation is obtained:

$$\sum_{n=-\infty}^{\infty} \left[\psi_1(q,n)A_n^3 + \psi_2(q,n)B_n^3\right] = \zeta_q, \tag{B.12}$$

where

$$\psi_1(q,n) = J_{q-n}(k_2 d_{13})\left[\frac{\varepsilon_3}{\varepsilon_0}\frac{J_q(k_3 c)}{H_q^{(2)}(k_0 c)} - \frac{k_3}{k_0}\frac{J_q'(k_3 c)}{H_q^{(2)'}(k_0 c)}\right]e^{jq\phi_s}e^{-j(q-n)\phi_{13}}, \tag{B.13a}$$

$$\psi_2(q,n) = J_{q-n}(k_3 d_{13})\left[\frac{\varepsilon_3}{\varepsilon_0}\frac{H_q^{(2)}(k_3 c)}{H_q^{(2)}(k_0 c)} - \frac{k_3}{k_0}\frac{H_q^{(2)'}(k_3 c)}{H_q^{(2)'}(k_0 c)}\right]e^{jq\phi_s}e^{-j(q-n)\phi_{13}}, \tag{B.13b}$$

$$\zeta_q = H_q^{(2)}(k_0 c)\left[\frac{J_q(k_0 c)}{H_q^{(2)}(k_0 c)} - \frac{J_q'(k_0 c)}{H_q^{(2)'}(k_0 c)}\right]. \tag{B.13c}$$

Equations (B.8), (B.10), and (B.12) represent a system of linear equations to be solved for the unknown expansion coefficients. They can be written compactly as:

$$\begin{bmatrix} \overline{\psi}_1(q,n) & \overline{\psi}_2(q,n) & \overline{\mathbf{0}} & \overline{\mathbf{0}} \\ -\overline{\boldsymbol{\chi}}_1^2(q,n) & -\overline{\boldsymbol{\chi}}_2^2(q,n) & \overline{\boldsymbol{\chi}}_1^3(q,n) & \overline{\boldsymbol{\chi}}_2^3(q,n) \\ -\overline{\boldsymbol{\chi}}_3^2(q,n) & -\overline{\boldsymbol{\chi}}_4^2(q,n) & \overline{\boldsymbol{\chi}}_3^3(q,n) & \overline{\boldsymbol{\chi}}_4^3(q,n) \\ \overline{\mathbf{0}} & \overline{\mathbf{0}} & \overline{\boldsymbol{\tau}}_{1n} & \overline{\boldsymbol{\tau}}_{2n} \end{bmatrix} \begin{bmatrix} \boldsymbol{A}_n^2 \\ \boldsymbol{B}_n^2 \\ \boldsymbol{A}_n^3 \\ \boldsymbol{B}_n^3 \end{bmatrix} = \begin{bmatrix} \boldsymbol{\zeta}_q \\ \mathbf{0} \\ \mathbf{0} \\ \mathbf{0} \end{bmatrix}, \tag{B.14}$$

where $\overline{\psi}_p$, $p = 1, 2$, is a matrix formulation of (B.13a) and (B.13b); $\overline{\boldsymbol{\chi}}_i^{\upsilon}$, $\upsilon = 2, 3$, $i = 1, 2, 3, 4$, is a matrix formulation of (B.11); $\overline{\boldsymbol{\tau}}_{pn}$, $p = 1, 2$, is a diagonalized matrix containing (B.9); $\boldsymbol{A}_n^{\upsilon}$ and $\boldsymbol{B}_n^{\upsilon}$, $\upsilon = 2, 3$, are vectors containing the expansion coefficients; and $\boldsymbol{\zeta}_q$ is a vector containing (B.13c). The overall form of (B.14) resembles the one in (A.6) derived for the symmetric 2D CNP. It can thus be inverted, as in (A.7), in order to determine the unknown coefficients in regions 2 and 3 of the 3Z-NP. With this result in hand, the expansion coefficients in regions 1 and 4 are obtained from (B.5a) and (B.7a), respectively, as:

$$A_n^1 = \frac{\varepsilon_2}{\varepsilon_1}\frac{J_n(k_2 a)A_n^2 + H_n^{(2)}(k_2 a)B_n^2}{J_n(k_1 a)}, \tag{B.15a}$$

$$\begin{aligned} B_n^4 = {} & \frac{1}{H_q^{(2)}(k_0 c)} \\ & \times \left[\frac{\varepsilon_3}{\varepsilon_0}\sum_{n=-\infty}^{\infty}\left\{J_{q-n}(k_3 d_{13})\left[A_n^3 J_q(k_3 c) + B_n^3 H_q^{(2)}(k_3 c)\right]e^{jq\phi_s}e^{-(q-n)\phi_{13}}\right\}\right. \\ & \left. - H_q^{(2)}(k_0\rho_s)J_q(k_0 c)\right]. \end{aligned} \tag{B.15b}$$

When all expansion coefficients have been determined, the derived quantities, such as the normalized radiation resistance and the directivity, can be easily obtained through the same general equations as those used in Appendix A.

References

[1] N. Engheta and R. W. Ziolkowski, Eds., *Metamaterials: Physics and Engineering Explorations*, Piscataway, NJ, IEEE Press, 2006.
[2] W. Cai and V. M. Shalaev, *Optical Metamaterials*, New York, Springer, 2010.
[3] L. Novotny and B. Hecht, *Principles of Optics*, 2nd ed., Cambridge University Press, UK, 2012.
[4] M. Agio and A. Alú, *Optical Antennas*, Cambridge University Press, USA, 2013.
[5] M. T. Hill, "Status and prospects of metallic and plasmonic nano-lasers," *J. Opt. Soc. Am. B*, vol. 27, pp. 36–44, 2010.
[6] I. Choi and Y. Choi, "Plasmonic nanosensors: Review and prospect," *IEEE J. Sel. Top. Quant. Electron.*, vol. 18, pp. 1110–1121, 2012.
[7] C. Wu, A. B. Khanikaev, R. Adato, N. Arju, A. A. Yanik, H. Altug, and G. Shvets, "Fano-resonant asymmetric metamaterials for ultrasensitive spectroscopy and identification of molecular monolayers," *Nat. Mater.*, vol. 11, 69, 2012.
[8] H. A. Atwater and A. Polman, "Plasmonics for improved photovoltaic devices," *Nat. Mater.*, vol. 9, pp. 205–213, 2013.
[9] A. Alú and N. Engheta, "Wireless at the nanoscale: Optical interconnects using matched nanoantennas," *Phys. Rev. Lett.*, vol. 104: 213902, 2010.
[10] D. Sikdar, I. D. Rukhlenko, W. Cheng and M. Premaratne, "Optimized gold nanoshell ensembles for biomedical applications," *Nanoscale Res. Let.*, vol. 8, pp. 142–146, 2013.
[11] N. J. Halas, "Plasmonics – an emerging field fostered by nano letters," *Nano Lett.*, vol. 10, pp. 3816–3822, 2010.
[12] L. Novotny and N. van Hulst, "Antennas for light," *Nat. Photon.*, vol. 5, pp. 83–90, 2011.
[13] T. H. Taminiau, F. D. Stefani, and N. F. van Hulst, "Enhanced directional excitation and emission of single emitters by a nano-optical Yagi–Uda antenna," *Opt. Express*, vol. 16, pp. 10858–10866, 2008.
[14] T. Kosako, Y. Kadoya, and H. F. Hofmann, "Directional control of light by a nano-optical Yagi–Uda antenna," *Nat. Photonics*, vol. 4, 312, 2010.
[15] Z. Liu, A. Boltasseva, R. H. Pedersen, R. Bakker, A. V. Kildishev, V. P. Drachev, and V. M. Shalaev, "Plasmonic nanoantenna arrays for the visible," *Metamaterials*, vol. 2, pp. 45–51, 2008.
[16] A. Ahmadi and H. Mosallaei, "Plasmonic nanolopp array antenna," *Opt. Lett.*, vol. 35, pp. 3706–3709, 2010.

[17] S. J. Oldenburg, G. D. Hale, J. B. Jackson, and N. J. Halas, "Light scattering from dipole and quadrupole shells," *Appl. Phys. Let.*, vol. 75, pp. 1063–1065, 1999.
[18] S. Kühn, U. Håkanson, L. Rogobete, and V. Sandoghdar, "Enhancement of single-molecule fluorescence using a gold nanoparticle as an optical nanoantennas," *Phys. Rev. Lett.*, vol. 97, 017402, 2006.
[19] B. S. Luk'yanchuk and V. Trenovsky, "Light scattering by a thin wire with a surface-plasmon resonance: Bifurcations of the Poynting vector field," *Phys. Rev. B*, vol. 73, 235432, 2006.
[20] M. Agio, "Optical antennas as nanoscale resonators," *Nanoscale*, vol. 4, 692, 2012.
[21] N. A. Mirin, and N. J. Halas, "Light-bending nanoparticles," *Nano Lett.*, vol. 9, pp. 1255–1259, 2009.
[22] A. Alú and N. Engheta, "Theory, modeling and features of optical nanoantennas," *IEEE Trans. Antennas Propag.*, vol. 61, 1508, 2013.
[23] S. Maier, *Plasmonics*: *Fundamentals and Applications*, Springer, New York, 2007.
[24] J. A. Gordon and R. W. Ziolkowski, "The design and simulated performance of a coated nano-particle laser," *Opt. Express*, vol. 15, pp. 2622–2653, 2007.
[25] M. Noginov, G. Zhu, A. M. Belgrave, R. Bakker, V. M. Shalaev, E. E. Narimanov, S. Stout, E. Herz, T. Suteewong, and U. Wiesner, "Demonstration of a spaser-based nanolasers," *Nature*, vol. 460, 1110, 2009.
[26] S. Y. Liu, J. Li, F. Zhou, L. Gan, and Z.-Y. Li, "Efficient surface plasmon amplification from gain-assisted gold nanorods," *Opt. Lett.*, vol. 36, pp. 1296–1298.
[27] I. E. Protsenko, A. V. Uskov, O. A. Zaimidoroga, V. N. Samoilov, and E. P. O'Reilly, "Dipole nanolaser," *Phys. Rev. A*, vol. 71, 063812, 2015.
[28] S. D. Campbell and R. W. Ziolkowski, "The performance of active coated nanoparticles based on quantum dot media," *Adv. Optoelectron.*, vol. 2012, 368786, 2012.
[29] S. Arslanagić and R. W. Ziolkowski, "Cylindrical and spherical active coated nanoparticles as nanoantennas," *IEEE Ant. Propagat. Magazine*, vol. 59, pp. 14–29, 2017.
[30] R. G. Medina, B. G. Cámara, I. S. Lacalle, F. Gonzáles, F. Moreno, M. N. Vesperinas, and J. J. Sáenz, "Electric and magnetic dipolar response of germanium nanospheres: Interference effects, scattering anisotropy, and optical forces," *J. Nanophoton.*, vol. 5, 053512, 2011.
[31] A. I. Kuznetsov, A. E. Miroshnichenko, Y. H. Fu, J. B. Zhang, and B. Luk'yanchuk, "Magnetic light," *Sci. Rep.*, vol. 2, 492, 2012.
[32] A. B. Evlyukhin, S. M. Novikov, U. Zywietz, R. L. Eriksen, C. Reinhardt, S. I. Bozhevolnyi, and B. N. Chichkov, "Demonstration of magnetic dipole resonances of dielectric nanospheres in the visible region," *Nano Lett.*, vol. 12, pp. 3749–3755, 2012.
[33] A. E. Krasnok, A. E. Miroshnichenko, P. A. Belov, and Y. S. Kivshar, "All-dielectric optical nanoantennas," *Opt. Exp.*, vol. 20, pp. 20599–20604, 2012.

[34] S. Jahani and Z. Jacob, "All-dielectric metamaterials," *Nat. Technolog.*, vol. 11, pp. 23–36, 2016.

[35] A. I. Kuznetsov, A. E. Miroshnichenko, M. L. Brongersma, Y. S. Kivshar, and B. Luk'yanchuk, "Optically resonant dielectric strucures," *Science*, vol. 354, aag2472, 2016.

[36] Y. H. Fu, A. E. Krasnok, A. E. Miroshnichenko, Y. F. Yu, and B. Luk'yanchuk, "Directional visible light scattering by silicon nanoparticles," *Nat. Commun.*, vol. 4, 1527, 2013.

[37] S. Campbell and R. W. Ziolkowski, "Simultaneous excitation of electric and magnetic dipole modes in a resonant core-shell particle at infrared frequencies to achieve minimal backscattering," *IEEE J. Sel. Top. Quantum Electron.*, vol. 19, 4700209, 2013.

[38] I. Liberal, I. Ederra, R. Gonzalo, and R. W. Ziolkowski, "Induction theorem analysis of resonant nanoparticles: Design of a Huygens source nanoparticle laser," *Phys. Rev. Appl.*, vol. 1, 044002, 2014.

[39] W. Liu, J. Zhang, B. lei, H. Ma, W. Xie, and H. Hu, "Ultradirectional forward scattering by individual core-shell nanoparticles," *Opt. Exp.*, vol. 22, 16178, 2014.

[40] D. Sikdar, W. Cheng, and M. Premaratne, "Optically resonant magneto-electric cubic nanoantennas for ultra-directional light scattering," *J. Appl. Phys.*, vol. 117, 083101, 2015.

[41] A. Alù and E. Engheta, "Enhanced directivity from subwavelength infrared/optical nano-antennas loaded with plasmonic materials or metamaterials," *IEEE Trans. Antennas Propagat.*, vol. 55, pp. 3027–3039, 2007.

[42] B. G. Cámara, J. M. Saiz, F. Gonzáles, and F. Moreno, "Nanoparticles with unconventional scattering properties: size effects," *Opt. Commun.*, vol. 283, pp. 490–496, 2010.

[43] Z. Ruan and S. Fan, "Superscattering of light from subwavelength nanostructures," *Phys. Rev. Let.*, vol. 105, 013901, 2010.

[44] Z. Ruan and S. Fan, "Design of subwavelength superscattering nanospheres," *Appl. Phys. Lett.*, vol. 98, 04101, 2011.

[45] A. E. Krasnok, C. R. Simovski, P. A. Belov, and Y. S. Kivshar, "Superdirective dielectric nanoantennas," *Nanoscale*, vol. 6, pp. 7354–7361, 2014.

[46] S. D. Campbell and R. W. Ziolkowski, "Near-field directive beams from passive and active asymmetric optical nanoantennas," *IEEE J. Sel. Top. Quantum Electron.*, vol. 21, 4800112, 2014.

[47] R. Ø. Thorsen and S. Arslanagić, "Eccentrically-layered active coated nanoparticles for directive near- and far-field radiation," *Photonics*, vol. 2, pp. 773–794, 2015.

[48] S. Arslanagić and R. Ø. Thorsen, "Asymmetric active nano-particles for directive near-field radiation," IEEE Int. Symposium on Antennas and Propagation and USNC-URSI National Radio Science Meeting, Puerto Rico, USA, June/July 2016.

[49] S. Arslanagić and R. E. Jacobsen, "Active coated nano rod antennas for enhanced and directive scattering phenomena," *EPJ Appl. Metamat.*, vol. 6, 19, 2019.
[50] M. W. Knight and N. J. Halas, "Nanoshells to nanoeggs to nanocups: optical properties of reduced symmetry core-shell nanoparticles beyond the quasi-static limit," *New J. Phys.*, vol. 10, 105006, 2008.
[51] J. B. Lassiter, M. W. Knight, N. A. Mirin, and N. J. Halas, "Reshaping the plasmonic properties of an individual nanoparticle," *Nano Lett.*, vol. 9, pp. 4326–4332, 2009.
[52] R. Ziolkowski, "Using Huygens multipole arrays to realize unidirectional needle-like radiation," *Phys. Rev. X*, vol. 7, 031017, 2017.
[53] S. Arslanagić and R. W. Ziolkowski, "Highly subwavelength, superdirective cylindrical nanoantennas," *Phys. Rev. Lett.*, vol. 120, 237401, 2018.
[54] S. Arslanagić and R. W. Ziolkowski, "Impact of the excitation source and plasmonic material on cylindrical active coated nano-particles," *Sensors*, vol. 11, pp. 9109–9120, 2011.
[55] S. Arslanagić and R. W. Ziolkowski, "Directive properties of active coated nano-particles," *Adv. Electromagn.*, vol. 1, pp. 57–64, 2012.
[56] P. B. Johnson and R. W. Christy, "Optical constants of noble metals," *Phys. Rev. B*, vol. 6, 4370, 1972.
[57] N. W. Ashcroft and N. D. Mermin, *Solid State Physics*, Holt, Rinehart and Winston, NY, 1976.
[58] COMSOL Multiphysics®, https://www.comsol.com/
[59] R. Ziolkowski, P. Jin, and C.-C. Lin, "Metamaterial-inspired engineering of antennas," *Proc. IEEE*, vol. 99, 1720, 2011.
[60] C. A. Balanis, *Antenna Theory: Analysis and Design*, 4th ed., Wiley, New York, 2016.
[61] C. A. Balanis, *Advanced Engineering Electromangetics*, 2nd ed., Wiley, New York, 2012.
[62] J. A. Stratton, *Electromagnetic Theory*, McGraw-Hill, New York and London, 1941.

Chapter 3
Coherent control of light scattering

Alex Krasnok[1] *and Andrea Alú*[1]

Scattering of electromagnetic waves, i.e., changing the direction of wave propagation and power flow caused by elastic (preservation of frequency) interactions with an object or scatterer, lies at the heart of most experimental techniques over the entire optical spectrum. The first experimental investigations of electromagnetic wave scattering were performed in the years 1886–1889 by Heinrich Hertz in his famous experiments that proved the theory of electromagnetic waves established by James Clerk Maxwell 20 years earlier [1]. The theory and experiments have convincingly proven that light is an electromagnetic wave and since then our research in the field of light phenomena has not ceased. Another example of light scattering that we encounter in nature is the reddening of the Sun at sunset and the blue color of the daytime sky, explained by Lord Rayleigh in 1871, well before the rigorous theory of such scattering was published in 1908 by Gustav Mie [2].

In everyday life, we meet simple scattering processes, such as specular reflection, absorption, diffuse scattering, focusing, and lensing. However, progress in nanofabrication and material science made it feasible to design scatterers demonstrating unusual scattering effects, including non-radiating states [3], cloaking [4,5], bound states in the continuum [6] and exceptional points (EPs) [7]. Understanding these scattering effects plays a fundamental role in modern optical physics and is vital for many applications relying on enhanced light-matter interactions, e.g., energy harvesting [8], sensing [9] and optical data processing [10], to mention just a few.

These examples show that the design of single scatterers can lead to the appearance of unique scattering properties and exotic light-matter interaction effects. About fifteen years ago, it has been proved that combining subwavelength scatterers into an architected material, i.e., a *metamaterial*, may allow the realization of electromagnetic properties much different than those of the constituent materials [11–20]. Such artificial materials provide exotic electromagnetic responses leveraging on the unusual scattering properties of collections of plasmonic or high-index nanoparticles to enhance light-matter interactions at subwavelength scales and gain control over linear and nonlinear fields [21–28]. 2D

[1]Photonics Initiative, Advanced Science Research Center, City University of New York, New York, NY, USA

planarized metamaterials, *metasurfaces*, have allowed obtaining specific optical properties on demand by controlling the local field distribution on a thin surface for various applications, including most notably wavefront engineering [12,26,29–37].

It has been shown recently that another fundamental property of light – *coherence* – can provide one more degree of freedom in scattering engineering and control. Nowadays this research area is a subject of significant interest. The examples of coherent scattering effects include *coherent perfect absorption* (CPA), *virtual perfect absorption* (VPA), *CPA lasing* and *coherently enhanced wireless power transfer* (CWPT). In this chapter, we discuss these coherent scattering phenomena in photonics. We first establish a unified description of such exotic scattering effects in electromagnetics and show that the origin of all these effects can be traced back to singularities of the scattering matrix. Then we discuss these effects paying particular attention to their physical essence and coherent nature.

3.1 Poles and zeros of the $\widehat{S}$ matrix

To begin with, here we derive a formulation for the general problem of light scattering. It consists in finding a solution of Maxwell's equations describing spatial and temporal evolution of total (impinging and scattered) electric $\mathbf{E}(\mathbf{r}, t)$ and magnetic $\mathbf{H}(\mathbf{r}, t)$ fields in one, two, or three dimensions

$$\nabla \times \mathbf{E}(\mathbf{r}, t) = -\dot{\mathbf{B}}(\mathbf{r}, t), \nabla \cdot \mathbf{D}(\mathbf{r}, t) = \rho(\mathbf{r}, t), \tag{3.1a}$$

$$\nabla \times \mathbf{H}(\mathbf{r}, t) = \dot{\mathbf{D}}(\mathbf{r}, t) + \mathbf{j}(\mathbf{r}, t), \nabla \cdot \mathbf{B}(\mathbf{r}, t) = 0, \tag{3.1b}$$

with corresponding boundary conditions. Here $\mathbf{D} = \varepsilon_0 \mathbf{E} + \mathbf{P} = \widehat{\varepsilon}\mathbf{E}$, $\mathbf{B} = \mu_0 (\mathbf{H} + \mathbf{M}) = \widehat{\mu}\mathbf{H}$, ρ and $\mathbf{j}$ are the free charge density and current, c is the speed of light in vacuum, ε_0 is the dielectric constant, and $\widehat{\varepsilon}$ and $\widehat{\mu}$ are permittivity and permeability (tensors and/or integral operators, in general). The general boundary conditions can be written as $\mathbf{n} \times (\mathbf{E}_2 - \mathbf{E}_1) = 0$, $\mathbf{n} \cdot (\mathbf{D}_2 - \mathbf{D}_1) = \rho$, $\mathbf{n} \times (\mathbf{H}_2 - \mathbf{H}_1) = \mathbf{J}$ and $\mathbf{n} \cdot (\mathbf{B}_2 - \mathbf{B}_1) = 0$, with $\mathbf{n}$ being the normal vector from medium "1" to medium "2".

Hereinafter, we consider only uncharged, linear, nonmagnetic, and isotropic materials when $\widehat{\varepsilon}$ is reduced to one complex valued permittivity $\varepsilon(\omega)$. We also assume that the host medium permittivity ε_h is a real valued constant. Moreover, we imply the time convention $\exp(-i\omega t)$. Under these assumptions, we derive the Helmholtz equation from (3.1)

$$\nabla \times \nabla \times \mathbf{E}(\omega, \mathbf{r}) - \varepsilon(\omega)\frac{\omega^2}{c^2}\mathbf{E}(\omega, \mathbf{r}) = 0. \tag{3.2}$$

The total electric field can be divided into the incident $\mathbf{E}_i$ and scattered $\mathbf{E}_s$ components, $\mathbf{E} = \mathbf{E}_i + \mathbf{E}_s$. Hence, the solution of (3.2) can be formally found in the form $\mathbf{E}_s = \widehat{S}\mathbf{E}_i$, with $\widehat{S}$ being the *scattering matrix* (in operator form) [38].

Let us discuss the structure and the main properties of the $\widehat{S}$ matrix. For this, we first chose a basis of certain incoming and outgoing waves, or *channels*, as

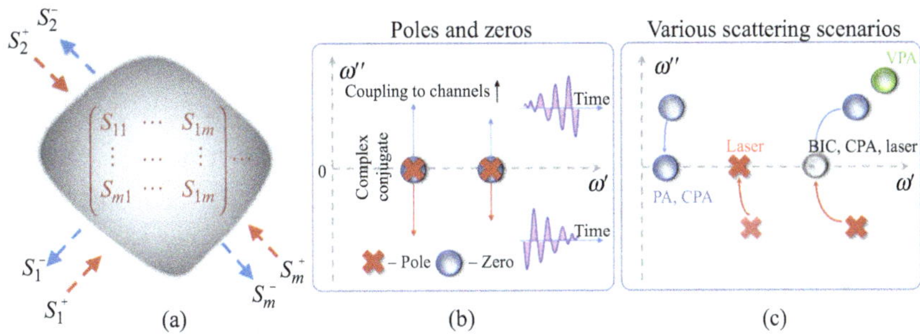

Figure 3.1 Scattering matrix and its zeros and poles: (a) Description of a linear electromagnetic structure by a scattering ($\widehat{S}$) matrix, which links the amplitudes of the ingoing (s_m^+) and outgoing (s_m^-) waves; m stands for the number of channels. (b), (c) Various cases of light scattering scenario and their relation to poles and zeros location. (b) Hermitian system. The zeros and poles are related through complex conjugation. In the limit case of almost no coupling to channels, pole and zero lie very close to the real axis. With increase in coupling to channels (radiative losses), the pole (zero) goes to upper (lower) complex plane. (c) Non-Hermitian scenarios, including lasing, perfect absorption (PA), CPA. BIC stands for bound state in the continuum, the situation when the zero and pole of a Hermitian system coalesce at the real axis. VPA stands for virtual perfect absorption. CPA-laser regime is possible in a system with balanced loss and gain.

shown in Figure 3.1(a), and represent the incident and scattered waves in the form $\mathbf{E}_i = \sum_m s_m^+ \mathbf{E}_m^+$, and $\mathbf{E}_s = \sum_m s_m^- \mathbf{E}_m^-$, with corresponding amplitudes $\mathbf{s}^+ = \{s_1^+, s_2^+, ...\}$ and $\mathbf{s}^- = \{s_1^-, s_2^-, ...\}$, respectively. Hence, the formal solution of (3.2) in the basis of chosen channels reads

$$\mathbf{s}^- = \widehat{S}\mathbf{s}^+, \tag{3.3}$$

where $\hat{S}$ is the scattering matrix of the system in a chosen basis. Thus, amplitudes of the outgoing waves $\mathbf{s}^-$ are linear combinations of the incoming wave amplitudes $\mathbf{s}^+$, which is established by the $\widehat{S}$ matrix of the system. Usually, the amplitudes are normalized such that $|s_m^+|^2$ and $|s_m^-|^2$ correspond to the energy of the incoming and outgoing waves in a certain channel m. For example, in a single port system, the $\widehat{S}$ matrix coincides with the reflection coefficient (r). In a two-port system, the scattering matrix is $\widehat{S} = \begin{pmatrix} r_{11} & t_{12} \\ t_{21} & r_{22} \end{pmatrix}$, where r_{ii} and t_{ij} stand for corresponding reflection and transmission coefficients. The channels exist outside the system and represent freely propagating solutions of the wave equation in the absence of the scattering object. The specific choice of channels often depends on the exact geometry of the system. For instance, in a planar slab the basis of plane waves is the obvious choice, whereas for a spherical particle it is convenient to use a basis of vector spherical harmonics.

Now, we discuss the general properties of scattering matrix. If a system is *reciprocal* [39–44], it implies that the scattering matrix is symmetric, $\widehat{S} = \widehat{S}^T$. Next, usually the $\widehat{S}$ matrix can be reduced to its diagonal form $\widehat{S} = \mathrm{diag}(d_1, d_2, \cdots, d_m)$, where each $\widehat{S}$ matrix *eigenvalue* $d_p(\omega)$ is associated with its incoming amplitudes (eigenvectors) $\mathbf{s}_p^+$, $\widehat{S}\mathbf{s}_p^+ = d_p\mathbf{s}_p^+$. In a *Hermitian system*, i.e., a system with no gain and material loss (the permittivity is real everywhere), the scattering eigenvalues d_p *at real frequency* ω' are unimodular, i.e., $|d_p(\omega')| = 1$. This constraint is imposed by the unitarity of the scattering matrix, i.e., $\widehat{S}\widehat{S}^+ = \widehat{I}$, which is associated with a Hermitian Hamiltonian $\widehat{H}_0 = \widehat{H}_0^+$ of the scatterer [45,46]. The eigenvalues of $\widehat{S}$ matrix of a Hermitian system can be presented in the exponential form $d_p(\omega') = \exp[2\iota\delta_p(\omega')]$, which shows that a corresponding eigenvector s_p^+ experiences a phase shift $\delta_p(\omega')$ upon scattering on the system.

When the frequency ω is analytically continued to the complex plane, $\omega = \omega' + i\omega''$, the scattering eigenvalues $|d_p(\omega)|$ may take any values between ∞ and 0. The values ∞ and 0 correspond to the poles and zeros of the $\widehat{S}$ matrix, respectively. The poles relate to the *eigenmodes* (self-sustained solutions) [47], while the latter is associated with the *perfect absorbing states* (no scattering). A fundamental property of the $\widehat{S}$ matrix eigenvalues d_p is that they can be expanded into a special product via the *Weierstrass factorization theorem* [48–50]:

$$d_p(\omega) = A_p \exp(iB_p\omega) \prod_m \frac{\omega - \omega_{\mathrm{zr,m}}}{\omega - \omega_{\mathrm{pl,m}}}, \tag{3.4}$$

where A_p and B_p are constants, and the product is taken over all resonances which match the symmetry of d_p. As one can directly see, these expressions include only information about positions of poles $\omega_{\mathrm{pl,m}}$ and zeros $\omega_{\mathrm{zr,m}}$ in the complex plane. Therefore, the Weierstrass expansion shows that the scattering process is *fully described by the positions of poles and zeros*. This fact makes it possible to describe and classify a variety of scattering phenomena in terms of the position of zeros and poles.

In lossless systems, poles $\omega_{\mathrm{pl},n}$ and zeros $\omega_{\mathrm{zr},n}$ occur in *complex conjugate pairs*, i.e., $\omega_{\mathrm{zr},n} = \omega_{\mathrm{pl},n}^*$ and restricted to specific parts of the complex frequency plane. Resonant states of a passive system must decay in time due to leakage of the energy from the system into free space. With the time dependence defined through the factor $\exp(-i\omega t)$, such attenuation corresponds to the complex frequencies $\omega_{pl,n}$ in the lower half-plane. Correspondingly, the zeros $\omega_{zr,n} = \omega_{pl,n}^*$ sit symmetrically in the upper half-plane (Figure 3.1(b)). In the limit of zero coupling, poles coincide with the eigenfrequencies of the close-system Hamiltonian, $\widehat{H}_0$. When the coupling between the scatterer and the channels increases, the pairs of pole and zero start to move apart from the real axis of the complex frequency plane (Figure 3.1(b)). Note that the coalescence of a pole and zero can happen not only in a closed system but also in so-called bound state in the continuum (BIC) (Figure 3.1(c)), which is an eigenmode of the scatterer with a real frequency in the continuum of radiation modes [6,51–53].

If a system is non-Hermitian ($\widehat{H}_0 \neq \widehat{H}_0^+$) the poles and zeros are located asymmetrically in the complex frequency plane and their position can be controlled by material gain and loss parameters, which *break the time-reversal symmetry* $\mathcal{T}$. In pure lossy scatterers ($|d_m(\omega)|^2 < 1$, for each channel), the complex zeros move down from the upper-frequency plane as dissipation increases. At some critical value of dissipation, the scattering zero crosses the real axis at some frequency leading to perfect absorption (PA) of an appropriate incoming waveform (Figure 3.1(c)). This waveform is determined by the eigenstate of $\widehat{S}(\omega)$ corresponding to its zero eigenvalue. In the case of several ports, it gives rise to the coherent perfect absorption (CPA), the time-reversed counterpart of a laser for a structure with gain [54,55]. If the system has only one port, this condition corresponds to a common perfect absorption regime.

For a given system with some amount of losses, the increase in gain allows compensation of the energy loss. In the complex plane, it corresponds to pushing a pole closer to the real axis (Figure 3.1(c)). As the material part of loss is compensated, the further growth of gain allows *amplification regime* and a signal passing through the system gets amplified. The *lasing threshold* can be reached when further gain enhancement compensates the entire losses (material and radiation) in the system. This is accompanied by the phase transition to a coherent light emission [56]. Thus, in the complex plane, lasing corresponds to the presence of a pole at the real axis [47] (Figure 3.1(c)). It is worth mentioning that although the *laser generation is essentially a nonlinear process*, the threshold pumping for lasing can be found within a linear approximation [47,57] because the amplitude of the laser self-oscillation is equal to zero below the lasing threshold and the problem can be treated linearly.

The balance of gain and loss leads to the $\mathcal{PT}$-symmetric system, whose evolution is governed by a Hamiltonian symmetrical upon reversal of space $\mathcal{P}$ and time $\mathcal{T}$. The $\mathcal{PT}$-symmetric optical system has balanced absorption and amplification as long as the relative permittivity of the medium satisfies $\varepsilon(\mathbf{r}) = \varepsilon^*(-\mathbf{r})$. Such $\mathcal{PT}$-symmetric Hamiltonians support the phase transition of eigenvalue spectrum from entirely real to complex ones, which occur as the so-called EPs [58–63]. EPs emerge when two (or more) eigenvalues and their corresponding eigenstates coalesce simultaneously so that the Hamiltonian becomes defective. In an the interesting subclass associated with wave scattering due to $\mathcal{PT}$-symmetric targets, like the Hermitian case, poles and zeros also occur in complex conjugate pairs, as seen from the interchange between the secular equation for the poles and zeros after performing the combined $\mathcal{PT}$ symmetry operation. Nevertheless, as the strength of the non-Hermiticity varies, pairs of pole and zero will move symmetrically in the complex frequency plane, Figure 3.1(c). Eventually, a pole and a zero of the $\widehat{S}$ matrix may coincide on the real axis and a *CPA-laser* [64–66] point emerges, when the system behaves simultaneously as a laser oscillator and a CPA.

Note that, besides the poles and zeros of the $\widehat{S}$ matrix, one can use poles and zeros of other physical quantities for analysis and designing scattering at will. For example, in the 1D case, the reflection coefficient also has these singularities in the complex frequency plane. Interestingly, poles of the scattering amplitude coincide with those of $\widehat{S}$ matrix eigenvalues. However, in general, zeros do not coincide with zeros of $\widehat{S}$ matrix since they depend on excitation geometry (angle of incidence, number of excitation beams). This gives rise to the *coherent control of scattering* when one uses two or more coherent light beams in a two (or more)-channel configuration.

There is another example of scattering poles and zeros we meet in 2D and 3D cases. For the analysis of such structures, we often use the basis of spherical or cylindrical channels (harmonics) with the corresponding scattering coefficients. Let us consider here the case of the 3D object of spherical symmetry of a radius R. In this case, Mie scattering theory yields the following equation for the scattering cross section [67,68]

$$SCS = \frac{\lambda^2}{2\pi}\sum_{l=1}(2l+1)\left(|c_l^{\mathrm{TM}}|^2 + |c_l^{\mathrm{TE}}|^2\right), \tag{3.5}$$

with c_l^{TM} and c_l^{TE} being the scattering coefficients (Mie coefficients), which can be calculated analytically [67]. In (3.5), l defines the order of a scattering channel and λ is the wavelength of impinging light. The scattering coefficients can be expressed as $c_l^{TE,\,TM} = -U_l^{TE,\,TM}/(U_l^{TE,\,TM} - jV_l^{TE,\,TM})$ [69], for TE and TM polarized waves, respectively. The quantities $U_l^{TE,\,TM}$ and $V_l^{TE,\,TM}$ are defined by a combination of spherical Bessel and Neumann functions [70]. The scattering coefficients indicate the contribution of the different scattering channels, and the superscript TM or TE indicates whether the vector spherical harmonic has magnetic or electric field orthogonal to the radial direction.

If a scatterer is subwavelength ($kR \leq 1$), we may cut (3.5) at the first dipole term ($l = 1$). In this case, the only one dipole scattering amplitude $c_1^i = -U_1^i/(U_1^i - jV_1^i)$ governs the scattering properties. This amplitude has a *scattering zero* when $U_1 = 0$ gives rise to the *anapole state* [3], which is relative to the cloaking effect [5] for a subwavelength scatterer. The anapole state is of great interest in modern photonics and has been employed for boosting nonlinear effects in dielectric nanoparticles, such as third harmonic generation and four-wave mixing [71–74]. In its turn, the *scattering poles* (eigenmodes) are realized when the denominator turns to zero, $U_1^i - jV_1^i = 0$. In such a structure, adding material gain can also push the pole to the real frequency axis leading to realization of lasing regime (nanolasers, spacers) [75]. The pole and zero of the 3D structure can also coalesce at the real axis by tailoring the geometry, inducing the BIC scattering regime [52].

3.2 Coherent perfect absorption

Scattering of any linear structure is determined by two basic kinds of peculiarities in the complex frequency plane, zeros and poles of the $\widehat{S}$ matrix eigenvalues located in the upper and lower half-planes, respectively; see for example Figure 3.2(a).

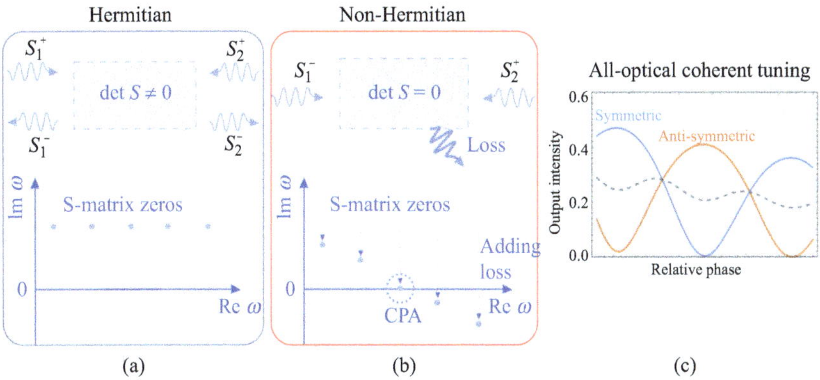

Figure 3.2 Coherent perfect absorbers: (a) Hermitian two-port system (e.g., Fabry–Perot resonator). The $\widehat{S}$ matrix zeros of such a system lie in the upper complex plane. Adding material loss to the system pushes the zeros down towards the real axis (b). As one of them gets to the real axis, the system becomes CPA at the corresponding frequency. (c) Tuning of phase or amplitude of one of the port signals allows tuning the output intensity of the system.

These singularities correspond to perfectly absorbing and lasing solutions and can never be achieved by *monochromatic excitations* in a Hermitian system. However, the presence of material losses inevitably breaks the time-reversal symmetry $\mathcal{T}$ and a specific amount of material loss can get one of the zeros to the real frequency axis (Figure 3.2(b)). In this case, the system would work as a perfect absorber (PA) for excitation by monochromatic waves in the corresponding *eigenvector configuration*. In some sense, this regime can be considered as time-reversed lasing; when adding gain to the system pushes poles towards the real axis until one of them get to the real axis with the formation of self-sustained oscillations of the electromagnetic field without incident field at the same frequency [47,76].

The vast majority of perfectly absorbing systems utilizes the one port excitation geometry when the incident energy is delivered to the system through a single channel. This situation corresponds to the most practical case of an absorber illuminated from one side by, e.g., the Sun, radar or a lab light source. For example, classical one-port PAs, *Salisbury* and *Dallenbach screens* [77,78] utilize the destructive self-interference of a plane wave excitation on one side of an absorber. More recent examples include PAs based on diffraction gratings and metasurfaces [79–88]. In order to enable destructive interference, an absorbing material should be placed above a mirror or covered with an anti-reflective coating [89]. Thus, achieving one-sided perfect absorption without backing mirrors is a challenging task, especially in the visible range.

In a system with two or more excitation channels, the perfect absorption concept is generalized to CPA. In this scenario, if the system is excited through two or more channels with appropriate amplitudes and phases (eigenvectors of $\widehat{S}$ matrix), it perfectly absorbs all energy [54,55,90–92]. This phenomenon can be illustrated by the consideration of a two-port planar structure, Figure 3.2. The inputs and outputs of such system are related via the $\widehat{S}$ matrix (for normal incidence), $\begin{pmatrix} s_1^- \\ s_2^- \end{pmatrix} = \widehat{S}\begin{pmatrix} s_1^+ \\ s_2^+ \end{pmatrix}$, where $\widehat{S} = \begin{pmatrix} r_{11} & t_{12} \\ t_{21} & r_{22} \end{pmatrix}$. Here s_i^+ and s_i^-, respectively, denote the input and output wave amplitudes in the i-th channel; r_{ii} are the reflection coefficients in each port and t_{ij} are the transmission coefficients. This $\widehat{S}$ matrix has the following eigenvalues and eigenvectors

$$d_{1,2} = \frac{1}{2}\left(r_{11} + r_{22} \pm \sqrt{r_{11}^2 - 2r_{11}r_{22} + r_{22}^2 + 4t_{12}t_{21}}\right), \tag{3.6a}$$

$$\mathbf{s}_{1,2} = \left\{1, -\frac{-r_{11} + r_{22} \pm \sqrt{r_{11}^2 - 2r_{11}r_{22} + r_{22}^2 + 4t_{12}t_{21}}}{2t_{21}}\right\}, \tag{3.6b}$$

If $\widehat{S}$-matrix is constrained by optical reciprocity [41], then $t_{12} = t_{21} = t$. If in addition a two-port structure is symmetric under mirror reflection, $r_{11} = r_{22} = r$, the corresponding eigenvalues and eigenvectors become $d_{1,2} = r \pm t$ and $\mathbf{s}_{1,2} = \{1, \pm 1\}$ representing *symmetric and antisymmetric inputs* of equal intensity.

If the system is lossless, all zeros of $d_{1,2} = r \pm t$ are located in the upper complex plane in Figure 3.2(a). However, adding a certain amount of material losses can make $r \pm t = 0$ at the real frequency axis in Figure 3.2(b). Then illumination of it with the corresponding CPA eigenmode $\mathbf{s}_{1,2} = \{1, \pm 1\}$ will lead to complete absorption. The absorption in a CPA turns out to be highly sensitive to variations in the conditions of illumination, in particular, in the relative phase between the components of the exciting field. To characterize a CPA it is convenient to define the *joint absorption* [93]

$$\mathcal{A} \equiv 1 - \frac{|s_1^-|^2 + |s_2^-|^2}{|s_1^+|^2 + |s_2^+|^2}. \tag{3.7}$$

Here, $|s_1^+|^2 + |s_2^+|^2$ is proportional to the total input intensity, and $|s_1^-|^2 + |s_2^-|^2$ to the total output intensity; thus a CPA has $\mathcal{A}$=1. Figure 3.2(c) features a typical plot of $1 - \mathcal{A}$ for the symmetric (blue curve) and anti-symmetric (orange curve) modes of a uniform dielectric slab near a CPA frequency. CPA occurs whenever one of the curves falls to zero. An intriguing feature of this plot is that at the CPA frequency, while one of the incident modes yields total absorption, the other mode strongly scatters due to constructive interference outside the system, thus suppressing the absorption. By adjusting the relative phase of the inputs, one can switch to the regime of suppressed absorption caused by the constructive interference of exiting waves [54,91,93].

Zeros of different eigenvalues of $\widehat{S}$ matrix typically occur at distinct frequencies, meaning that if a certain incident wave experiences CPA, the orthogonal wave does not. However, there are examples of doubly degenerate CPA [94,95], when both eigenvalues of $\widehat{S}$ can be zero at the same frequency. In this case the absorption becomes insensitive to the relative phase of the two input waves so that even a single input wave incident on either port would be perfectly absorbed.

CPAs have been studied and realized in a wide variety of geometries, including planar structures (slabs, diffraction gratings, metasurfaces, and thin films) [54,90,91,96] and guided-mode structures [97–99] compatible with integrated nanophotonics applications. Different techniques have been devised for increasing or decreasing the absorption bandwidth [100–102] and for achieving perfect absorption under strongly-asymmetric beam intensities [103]. Also CPA has been extended to systems exhibiting strong light-matter interaction [104]. Recently, CPA has been generalized to disordered systems [92].

The phenomenon of CPA may be useful in a variety of applications, ranging from all-optical data processing to enhanced photocurrent generation [55]. CPAs can be highly sensitive to the parameters of the input waves, enabling attractive opportunities for the flexible control of light scattering and absorption [91,105,106]. For instance, in a two-port CPA, one input can be regarded as a signal beam, and the other as a control beam. A coherent control beam serves as a resource enabling efficient control of optical absorption, even in the linear regime. This opens the door to novel low-energy logical circuits and small signal amplifiers [105,106] without the use of non-linear materials that rely on high intensity of incident radiation to achieve non-linear response. Moreover, CPA can be employed for processing binary images and pattern recognition [107]. Coherent absorption effects may prove very useful for improving photocurrent generation in photo-electrochemical systems [108]. Finally, the ability to control and remotely trigger CPA of single- or few-photon states [109,110] could be promising for quantum information processing.

3.3 Virtual perfect absorption

Absorbing electromagnetic waves is crucial to many problems of applied science, including sensing and molecular detection, RF cloaking, photovoltaics and photo-detection. Craftily designed PAs allow light absorption in an ultimately effective fashion boosting the performance of devices relying on light absorption, e.g., ultrathin solar cells. CPA is a two or more port absorber with, as in the case of common PAs, zeros of $\widehat{S}$ matrix eigenvalues at the real frequency axis. Although, CPAs allow perfect absorption and all-optical coherent light manipulation, the energy is eventually dissipated into heat, which is undesirable for some applications in active light manipulation such as low-energy memories and optical modulators. For these applications, it is important to design a system capable of storing the absorbed energy for an arbitrary amount of time and releasing it at will, rather than its dissipation. Conventional high-Q lossless cavities such as microdisks [111–115], microspheres [116], Bragg reflector microcavities [117] and photonic crystals [118–120] with extremely high-Q factors ($\sim 10^3 - 10^6$) are capable of storing energy for a relatively long time; however, the process of charging these cavities is highly inefficient. In these systems, at every cycle of the external excitation, a large amount of energy is reflected and only a tiny part of the illumination penetrates into the cavity, resulting in a very inefficient storing process. Although, to perfectly couple to an optical resonator the so-called *critical coupling* (often called as *Q*-factor matching) approach can be used, this requires presence of losses (material or radiative), which is very undesirable in applications. In other words, to critically couple to a mode with a large *Q*-factor one need to add losses into it and hence spoil the cavity.

Recently, the concept of CPA has been generalized to the *complex frequencies* and *Hermitian systems*. Instead of adding loss in order to push the complex zeros of the system towards the real frequency axis, it is feasible to engage the complex zeros of a lossless system by tailoring a signal with the corresponding complex frequency [121]. It was shown that if a lossless dielectric slab is coherently excited by signals with these complex frequencies (Figure 3.3(a)), the energy of the incoming waves will get stored in the slab *without any outgoing energy from the system* (Figure 3.3(b)). Figure 3.3(c) shows the stored energy during this temporal process. Figures 3.3(d) and (e) show the electric field distribution in the system during the excitation (d) and right after excitation breaking (e). Hence, during this excitation by a complex signal the structure looks like perfectly absorbing. As soon as the illuminating signals stop, all the energy stored in the system will get released through its electromagnetically open channels.

Thus, instead of adding loss to the system to push the complex zero towards the real axis, it is possible to tailor the incident field in time, such that its temporal profile matches the exponentially diverging mode associated with the complex zero over a finite interval of time, giving rise to VPA [121]. The effect can be reproduced in any lossless (or with small loss) electromagnetic structure provided that the spectral position of its scattering zero is known and the corresponding spatial and temporal profile of the incident field can be created. It is also robust to the

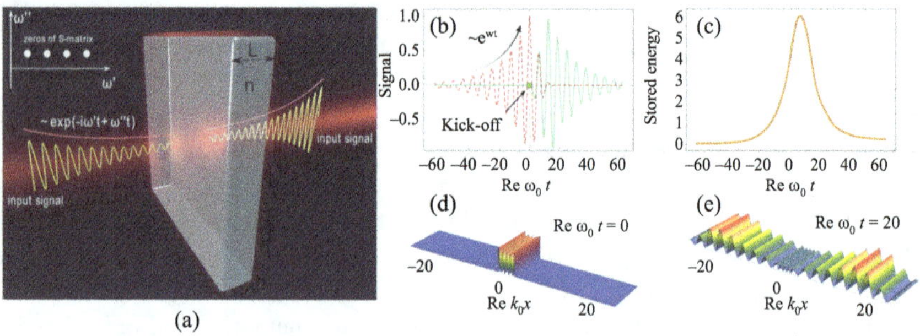

Figure 3.3 Virtual perfect absorption: (a) Dielectric slab resonator is coherently excited with signals corresponding to one of the zeros in the upper complex plane; (b) During the excitation by a signal corresponding to one of the zeros (red curve), the total scattering to all ports is absent (green curve). As the excitation stops abruptly, the stored energy gets released; (c) Stored energy during this temporal process; (d) Electric field distribution in the system during the excitation; and (e) right after excitation breaking.

inevitable material dissipation and material dispersion because the position of poles and zeros entails information about this dispersion.

3.4 Coherently enhanced wireless power transfer

An antenna is a crucial element for many important wireless technologies, including communications and power transfer [122]. First antennas emerged at the same time with the discovery of electromagnetic waves by H. Hertz in 1888 and since then antennas have been developing with human civilization often being a catalyst for its development. A plethora of antennas have been invented for the radio, microwave, THz and even optical frequency range, where they become unique elements for quantum optics and interconnections on a chip [123–125].

While wireless communications are rather established, the wireless power transfer (WPT), proposed at the beginning of the twentieth century by Tesla [126], is experiencing a rebirth. It was caused by an experimental demonstration that the WPT efficiency, i.e. amount of energy received by a reviving antenna related to the total amount of the emitted energy by transmitter (usually a magnetic coil), can be drastically enhanced in the so-called near-field WPT regime, when the power is transferred via resonant coupling [127]. Transfer over the distance of 2 m with high efficiency in the kHz range via strongly coupled magnetic resonances between two metallic coils has been achieved [127], which gave rise to research on ways of using this effect for novel technologies. Examples include electric vehicles, implanted medical devices and consumer electronics. Since then, significant research efforts have been devoted to exploring the ways to achieve as high WPT efficiency as possible [128], whereas the vast majority of them has been concentrated on optimizing of the resonators' geometry, a surrounding and their relative arrangement.

In the previous subsections, it has been shown that electromagnetic absorption and scattering can be effectively controlled through tailoring of coherent temporal shaping of the incident electromagnetic field. For example, a CPA is a linear electromagnetic system in which perfect absorption is achieved with two or more incident waves, creating constructive interference inside an absorbing structure. Here, we show that the same principle that underlies the operation of CPAs can be employed to improve the efficiency of antenna energy transfer and as a result improve the efficiency of WPT systems [129]. More specific, by coherently exciting the receiving antenna [see Figure 3.4(a)] with an auxiliary signal, tuned in sync with the impinging signal from the transmitting antenna, it is possible to enhance the robustness of the system largely. This additional signal improves energy transfer through constructive interference with the impinging wave, compensating any imbalance in the antenna coupling without having to modify the load.

The analysis of this system can be performed by the temporal coupled mode theory (TCMT) [130–132]. We assume that the dipole receiving antenna has a single mode with the real Eigen frequency ω_0 and the mode amplitude a, normalized such that $|a|^2$ is the energy of the mode. The dipole antenna couples to the waveguide and free space radiation with the coupling constants $|\kappa\rangle = \{\kappa_w, \kappa_f\}$.

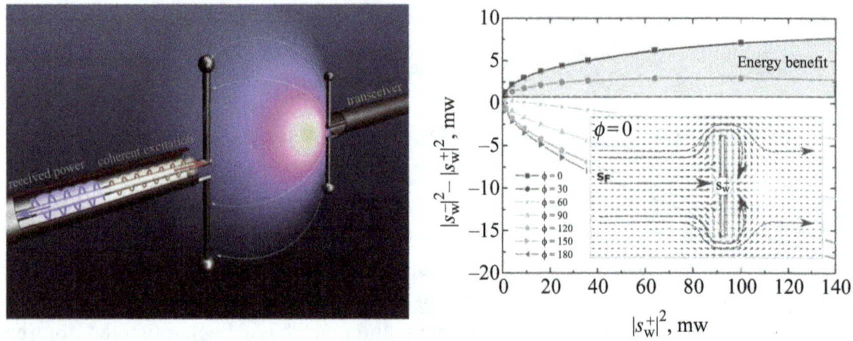

Figure 3.4 Coherently enhanced wireless power transfer: (a) Schematic of a CWPT system consisting of a transceiver antenna and a receiver antenna driven by an additional coherent excitation. (b) Net extracted energy as a function of the auxiliary signal intensity for different relative phase values. Filled area indicates the region of energy benefit, where the coherently assisted energy balance exceeds that for $s_w^+ = 0$. *Inset: Poynting vector distribution around the antenna with the auxiliary signal* s_w^+ *and relative phase of 0 deg.*

The excited antenna mode can decay to both channels with the total dumping rate, $1/\tau = 1/\tau_w + 1/\tau_f$. As a result, we can write the first equation of the TCMT describing the evolution of the dipole antenna mode amplitude in the form

$$\frac{da}{dt} = \left(i\omega_0 - \frac{1}{\tau}\right)a + \langle\kappa^*|s_+\rangle, \quad (3.8)$$

where $|s_+\rangle$ is the vector of input amplitudes. Now we assume that the vector of input amplitudes consists of not only the field of the transceiver s_f but also the additional field of coherent excitation s_w^+, Figure 3.4(a). As follows from (3.1), without excitation ($|s_+\rangle = 0$) the initially excited dipole mode decays in time as $a \sim e^{i\omega_0 t}e^{-t/\tau}$, i.e., exponentially with the complex frequency, $\omega = \omega' + i\omega''$, where $\omega' = \omega_0$ and $\omega'' = -1/\tau$.

In its turn, the output amplitude vector $|s_-\rangle$ is related to the input vector and the mode amplitude via

$$\Big|s_-\rangle = \widehat{C}|s_+\rangle + a|\kappa\rangle \quad (3.9)$$

where $\widehat{C}$ is the matrix of *direct scattering*. We assume the waveguide is highly mismatched with the free space and hence $\widehat{C} = -1$.

The results of the TCMT analysis yield the following expression for the scattered energy back to the receiver (extracted field) [129]

$$s_w^- = -s_w^+ + \kappa_w \frac{\kappa_w s_w^+ + \kappa_f s_f}{i(\omega - \omega_0) + 1/\tau}. \quad (3.10)$$

If $s_f = 0$ (no radiation from the transceiver), this expression gives the reflection parameter (s_{11}), which can be measured or calculated. The typical results that this equation yields for the net extracted energy are shown in Figure 3.4(b). The area where the net extracted energy exceeds that for $s_w^+ = 0$ (filled area) corresponds to the energy benefit. In these regimes, the antenna receives more energy from the transceiver than it does without coherent excitation even after subtraction of this additional energy. Interesting that in such a regime of positive net extracted energy the Poynting vector distribution around the antenna demonstrates many flow lines ending by the dipole antenna, as shown in the inset of Figure 3.4. Otherwise, when the relative phase between s_w^+ and s_f is 180 deg., there are a few of Poynting vector lines flowing into the antenna that operates in the radiation regime (radiates more than receives). In a practical WPT device, the amplitude and phase of the additional coherent signal can be controlled in real time to adjust the antenna as a function of changes in the environment, temperature changes in the load and distance relative to the transmitter.

3.5 Conclusions

Scattering of electromagnetic waves lies at the heart of all experimental techniques in radiophysics, visible and X-ray optics. Hence, a deep insight into the basis of scattering theory and understanding the peculiar features of electromagnetic scattering are necessary for the correct interpretation of experimental data and an understanding of the underlying physics. Recently, a broad spectrum of exceptional scattering effects, including bound states in the continuum, EPs in PT-symmetrical non-Hermitian systems, and many others attainable in wisely suitably engineered structures have been predicted and demonstrated. Among these scattering effects, those that rely on coherence properties of light are of a particular interest today. In this chapter, we have discussed coherent scattering effects in photonics. CPA generalizes the concept of perfect absorption to systems with several excitation channels. This effect allows all-optical light manipulation by tuning of parameters of coherent excitation. In its turn, the VPA effect generalizes the concept of CPA to complex frequencies and Hermitian systems. In contrast to real energy dissipation in PA and CPA, in VPA the energy is trapped in a resonator till excitation continuous exponentially growth and gets released when it breaks. Additionally, VPA provides a way to perfectly couple to any even lossless high-Q resonators. Finally, employing principles of coherent excitation can boost the performance of traditional WPT systems via CWPT effect.

Acknowledgement

This work was supported by the Office of Naval Research, the Air Force Office of Scientific Research, the Department of Defense, the Simons Foundation and the National Science Foundation.

References

[1] J. C. Maxwell, *A Treatise on Electricity and Magnetism*. Cambridge, UK: Cambridge University Press, 2010.

[2] G. Mie, "Beiträge zur optik trüber medien speziell kolloidaler metallösungen," *Ann. Phys.*, vol. 330, pp. 377–445, 1908.

[3] K. V. Baryshnikova, D. A. Smirnova, B. S. Luk'yanchuk, and Y. S. Kivshar, "Optical anapoles: concepts and applications," *Adv. Opt. Mater.*, vol. 1801350, 1801350, 2019.

[4] A. Alù, and N. Engheta, "Achieving transparency with plasmonic and metamaterial coatings," *Phys. Rev. E*, vol. 72, no. 1, 016623, 2005.

[5] R. Fleury, F. Monticone, and A. Alù, "Invisibility and cloaking: origins, present, and future perspectives," *Phys. Rev. Appl.*, vol. 4, no. 3, 037001, 2015.

[6] C. W. Hsu, B. Zhen, A. D. Stone, J. D. Joannopoulos, and M. Soljačić, "Bound states in the continuum," *Nat. Rev. Mater.*, vol. 1, no. 9, 16048, 2016.

[7] M.-A. Miri, and A. Alù, "Exceptional points in optics and photonics," *Science (80).*, vol. 363, no. 6422, eaar7709, 2019.

[8] J. A. Schuller, E. S. Barnard, W. Cai, Y. C. Jun, J. S. White, and M. L. Brongersma, "Plasmonics for extreme light concentration and manipulation," *Nat. Mater.*, vol. 9, no. 4, pp. 368–368, 2010.

[9] M. E. Stewart, C. R. Anderton, L. B. Thompson, *et al.*, "Nanostructured plasmonic sensors," *Chem. Rev.*, vol. 108, no. 2, pp. 494–521, 2008.

[10] H. Kwon, D. Sounas, A. Cordaro, A. Polman, and A. Alù, "Nonlocal metasurfaces for optical signal processing," *Phys. Rev. Lett.*, vol. 121, no. 17, 173004, 2018.

[11] N. Engheta and R. W. Ziolkowski, *Metamaterials: Physics and Engineering Explorations*. Hoboken, NJ, USA: John Wiley & Sons, Inc., 2006.

[12] F. Monticone, and A. Alu, "Metamaterial, plasmonic and nanophotonic devices," *Reports Prog. Phys.*, vol. 80, no. 3, 036401, 2017.

[13] R. C. McPhedran, I. V. Shadrivov, B. T. Kuhlmey, and Y. S. Kivshar, "Metamaterials and metaoptics," *NPG Asia Mater.*, vol. 3, no. 11, pp. 100–108, 2011.

[14] A. Baev, P. N. Prasad, H. Ågren, M. Samoć, and M. Wegener, "Metaphotonics: an emerging field with opportunities and challenges," *Phys. Rep.*, vol. 594, pp. 1–60, 2015.

[15] F. Monticone, and A. Alù, "Metamaterials and plasmonics: from nanoparticles to nanoantenna arrays, metasurfaces, and metamaterials," *Chinese Phys. B*, vol. 23, no. 4, 047809, 2014.

[16] S. B. Glybovski, S. A. Tretyakov, P. A. Belov, Y. S. Kivshar, and C. R. Simovski, "Metasurfaces: from microwaves to visible," *Phys. Rep.*, vol. 634, pp. 1–72, 2016.

[17] N. I. Zheludev, and Y. S. Kivshar, "From metamaterials to metadevices," *Nat. Mater.*, vol. 11, no. 11, pp. 917–924, 2012.

[18] Y. Xu, Y. Fu, and H. Chen, "Planar gradient metamaterials," *Nat. Rev. Mater.*, vol. 1, no. 12, 16067, 2016.

[19] C. Della Giovampaola, and N. Engheta, "Digital metamaterials," *Nat. Mater.*, vol. 13, no. 12, pp. 1115–1121, 2014.

[20] M. Lapine, I. V. Shadrivov, and Y. S. Kivshar, "Colloquium: nonlinear metamaterials," *Rev. Mod. Phys.*, vol. 86, no. 3, pp. 1093–1123, 2014.

[21] L. Gui, S. Bagheri, N. Strohfeldt, *et al.*, "Nonlinear refractory plasmonics with titanium nitride nanoantennas," *Nano Lett.*, vol. 16, no. 9, pp. 5708–5713, 2016.

[22] J. Y. Suh, and T. W. Odom, "Nonlinear properties of nanoscale antennas," *Nano Today*, vol. 8, no. 5, pp. 469–479, 2013.

[23] S. Bin Hasan, F. Lederer, and C. Rockstuhl, "Nonlinear plasmonic antennas," *Mater. Today*, vol. 17, no. 10, pp. 478–485, 2014.

[24] V. Giannini, A. I. Fernández-Domínguez, S. C. Heck, and S. A. Maier, "Plasmonic nanoantennas: fundamentals and their use in controlling the radiative properties of nanoemitters," *Chem. Rev.*, vol. 111, no. 6, pp. 3888–3912, 2011.

[25] R. Adato, S. Aksu, and H. Altug, "Engineering mid-infrared nanoantennas for surface enhanced infrared absorption spectroscopy," *Mater. Today*, vol. 18, no. 8, pp. 436–446, 2015.

[26] S. He, Y. Cui, Y. Ye, P. Zhang, and Y. Jin, "Optical nano-antennas and metamaterials," *Mater. Today*, vol. 12, no. 12, pp. 16–24, 2009.

[27] U. Guler, V. M. Shalaev, and A. Boltasseva, "Nanoparticle plasmonics: going practical with transition metal nitrides," *Mater. Today*, vol. 18, no. 4, pp. 227–237, 2015.

[28] H. Wei, and H. Xu, "Plasmonics in composite nanostructures," *Mater. Today*, vol. 17, no. 8, pp. 372–380, 2014.

[29] A. Krasnok, M. Caldarola, N. Bonod, and A. Alú, "Resonant dielectric nanoparticles: spectroscopy and biosensing with optically resonant dielectric nanostructures (advanced optical materials 5/2018)," *Adv. Opt. Mater.*, vol. 6, no. 5, 1870018, 2018.

[30] A. M. Shaltout, A. V. Kildishev, and V. M. Shalaev, "Evolution of photonic metasurfaces: from static to dynamic," *J. Opt. Soc. Am. B*, vol. 33, no. 3, pp. 501–510, 2016.

[31] A. Alù, "Metamaterials: prime time," *Nat. Mater.*, vol. 15, no. 12, pp. 1229–1231, 2016.

[32] N. Meinzer, W. L. Barnes, and I. R. Hooper, "Plasmonic meta-atoms and metasurfaces," *Nat. Photonics*, vol. 8, no. 12, pp. 889–898, 2014.

[33] M. Decker, I. Staude, M. Falkner, *et al.*, "High-efficiency dielectric Huygens' surfaces," *Adv. Opt. Mater.*, vol. 3, no. 6, pp. 813–820, 2015.

[34] K. Fan, and W. J. Padilla, "Dynamic electromagnetic metamaterials," *Mater. Today*, vol. 18, no. 1, pp. 39–50, 2015.

[35] C. Caloz, "Perspectives on EM metamaterials," *Mater. Today*, vol. 12, no. 3, pp. 12–20, 2009.

[36] B. N. Zheludev, "Obtaining optical properties on demand," *Science*, vol. 348, no. 6238, pp. 973–974, 2015.
[37] L. Sun, C. Y. Wang, A. Krasnok, *et al.*, "Separation of valley excitons in a MoS2 monolayer using a subwavelength asymmetric groove array," *Nat. Photonics*, vol. 13, no. 3, pp. 180–184, 2019.
[38] M. Gell-Mann, Murray and Goldberger, "The formal theory of scattering," *Phys. Rev.*, vol. 91, 398, 1953.
[39] A. B. Khanikaev, and A. Alù, "Optical isolators: nonlinear dynamic reciprocity," *Nat. Photonics*, vol. 9, no. 6, pp. 359–361, 2015.
[40] Y. Hadad, J. C. Soric, and A. Alu, "Breaking temporal symmetries for emission and absorption," *Proc. Natl. Acad. Sci.*, vol. 113, no. 13, pp. 3471–3475, 2016.
[41] R. J. Potton, "Reciprocity in optics," *Reports Prog. Phys.*, vol. 67, no. 5, pp. 717–754, 2004.
[42] N. A. Estep, D. L. Sounas, J. Soric, and A. Alù, "Magnetic-free non-reciprocity and isolation based on parametrically modulated coupled-resonator loops," *Nat. Phys.*, vol. 10, no. 12, pp. 923–927, 2014.
[43] D. L. Sounas, C. Caloz, and A. Alù, "Giant non-reciprocity at the sub-wavelength scale using angular momentum-biased metamaterials," *Nat. Commun.*, vol. 4, 2407, 2013.
[44] D. L. Sounas, and A. Alù, "Non-reciprocal photonics based on time modulation," *Nat. Photonics*, vol. 11, no. 12, pp. 774–783, 2017.
[45] H. M. Nussenzveig, *Causality and Dispersion Relations.* Academic Press, San Diego, 1972.
[46] R. G. Newton, *Scattering Theory of Waves and Particles.* Springer, New York, 1982.
[47] L. Ge, Y. D. Chong, and A. D. Stone, "Steady-state ab initio laser theory: generalizations and analytic results," *Phys. Rev. A - At. Mol. Opt. Phys.*, vol. 82, no. 6, 063824, 2010.
[48] V. Grigoriev, A. Tahri, S. Varault, *et al.*, "Optimization of resonant effects in nanostructures via Weierstrass factorization," *Phys. Rev. A*, vol. 88, no. 1, 011803, 2013.
[49] G. B. Arfken, and H. J. Weber, *Mathematical Methods for Physicists.* Academic Press, San Diego, 2001.
[50] B. Stout, and R. McPhedran, "Egocentric physics: just about Mie," *EPL (Europhysics Lett.)*, vol. 119, no. 4, 44002, 2017.
[51] S. Lannebère, and M. G. Silveirinha, "Optical meta-atom for localization of light with quantized energy," *Nat. Commun.*, vol. 6, 8766, 2015.
[52] F. Monticone, and A. Alù, "Embedded photonic eigenvalues in 3D nanostructures," *Phys. Rev. Lett.*, vol. 112, no. 21, 213903, 2014.
[53] A. Krasnok, and A. Alu, "Embedded scattering eigenstates using resonant metasurfaces," *J. Opt. (United Kingdom)*, vol. 20, no. 6, 064002, 2018.
[54] Y. D. Chong, L. Ge, H. Cao, and A. D. Stone, "Coherent perfect absorbers: time-reversed lasers," *Phys. Rev. Lett.*, vol. 105, no. 5, 053901, 2010.

[55] D. G. Baranov, A. Krasnok, T. Shegai, A. Alù, and Y. Chong, "Coherent perfect absorbers: linear control of light with light," *Nat. Rev. Mater.*, vol. 2, no. 12, 17064, 2017.

[56] P. Meystre, G. Khitrova, D. Boggavarapu, H. M. Gibbs, R. Jin, and M. Sargent, "Photon-number correlations near the threshold of microcavity lasers in the weak-coupling regime," *Phys. Rev. A*, vol. 49, no. 5, pp. 4038–4042, 2002.

[57] H. Haken, *Light: Laser Light Dynamics*. North-Holland Publishing Company, Amsterdam, 1986.

[58] K. G. Makris, R. El-Ganainy, D. N. Christodoulides, and Z. H. Musslimani, "Beam dynamics in PT symmetric optical lattices," *Phys. Rev. Lett.*, vol. 100, no. 10, 103904, 2008.

[59] C. E. Rüter, K. G. Makris, R. El-Ganainy, D. N. Christodoulides, M. Segev, and D. Kip, "Observation of parity-time symmetry in optics," *Nat. Phys.*, vol. 6, no. 3, pp. 192–195, 2010.

[60] Y. Liu, T. Hao, W. Li, J. Capmany, N. Zhu, and M. Li, "Observation of parity-time symmetry in microwave photonics," *Light Sci. Appl.*, vol. 7, no. 1, 38, 2018.

[61] S. Longhi, "Parity-time symmetry meets photonics: a new twist in non-Hermitian optics," *Epl*, vol. 120, no. 6, 64001, 2017.

[62] R. El-Ganainy, K. G. Makris, M. Khajavikhan, Z. H. Musslimani, S. Rotter, and D. N. Christodoulides, "Non-Hermitian physics and PT symmetry," *Nat. Phys.*, vol. 14, no. 1, pp. 11–19, 2018.

[63] S. V. Suchkov, A. A. Sukhorukov, J. Huang, S. V. Dmitriev, C. Lee, and Y. S. Kivshar, "Nonlinear switching and solitons in PT-symmetric photonic systems," *Laser Photonics Rev.*, vol. 10, no. 2, pp. 177–213, 2016.

[64] S. Longhi, and L. Feng, "PT-symmetric microring laser-absorber," *Opt. Lett.*, vol. 39, no. 17, pp. 5026–5029, 2014.

[65] J. E. Dyson, T. W. Hartquist, M. T. Malone, and S. D. Taylor, "Boundary layers and highly supersonic molecular hydrogen flows," *Rev. Mex. Astron. y Astrofis. Ser. Conf.*, vol. 1, no. 9, pp. 119–127, 1995.

[66] S. Longhi, "PT-symmetric laser absorber," *Phys. Rev. A - At. Mol. Opt. Phys.*, vol. 82, no. 3, 031801(R), 2010.

[67] C. F. Bohren, and D. R. Huffman, *Absorption and Scattering of Light by Small Particles*. Weinheim, Germany, Germany: Wiley-VCH Verlag GmbH, 1998.

[68] A. Krasnok, M. Caldarola, N. Bonod, and A. Alú, "Spectroscopy and biosensing with optically resonant dielectric nanostructures," *Adv. Opt. Mater.*, vol. 6, no. 5, 1701094, 2018.

[69] A. Aì, and N. Engheta, "Polarizabilities and effective parameters for collections of spherical nanoparticles formed by pairs of concentric double-negative, single-negative, andor double-positive metamaterial layers," *J. Appl. Phys.*, vol. 97, no. 9, 094310, 2005.

[70] M. Abramowitz, I. A. Stegun, and R. H. Romer, "Handbook of mathematical functions with formulas, graphs, and mathematical tables," *Am. J. Phys.*, vol. 56, no. 10, pp. 958 958, 1988.

[71] G. Grinblat, Y. Li, M. P. Nielsen, R. F. Oulton, and S. A. Maier, "Enhanced third harmonic generation in single germanium nanodisks excited at the anapole mode," *Nano Lett.*, vol. 16, no. 7, pp. 4635–4640, 2016.

[72] T. Shibanuma, G. Grinblat, P. Albella, and S. A. Maier, "Efficient third harmonic generation from metal-dielectric hybrid nanoantennas," *Nano Lett.*, vol. 17, no. 4, pp. 2647–2651, 2017.

[73] S. A. Maier, M. P. Nielsen, G. Grinblat, Y. Li, and R. F. Oulton, "Efficient third harmonic generation and nonlinear subwavelength imaging at a higher-order anapole mode in a single germanium nanodisk," *ACS Nano*, vol. 11, no. 1, pp. 953–960, 2016.

[74] G. Grinblat, Y. Li, M. P. Nielsen, R. F. Oulton, and S. A. Maier, "Degenerate four-wave mixing in a multiresonant germanium nanodisk," *ACS Photonics*, vol. 4, no. 9, pp. 2144–2149, 2017.

[75] R. M. Ma, and R. F. Oulton, "Applications of nanolasers," *Nat. Nanotechnol.*, vol. 14, no. 1, pp. 12–22, 2019.

[76] A. Mostafazadeh, "Spectral singularities of complex scattering potentials and infinite reflection and transmission coefficients at real energies," *Phys. Rev. Lett.*, vol. 102, no. 22, 220402, 2009.

[77] R. L. Fante, and M. T. McCormack, "Reflection properties of the Salisbury screen," *IEEE Trans. Antennas Propag.*, vol. 36, no. 10, pp. 1443–1454, 1988.

[78] W. Salisbury, "Absorbent body for electromagnetic waves," US Patent 2599944, 1952.

[79] E. Popov, D. Maystre, R. C. McPhedran, M. Nevière, M. C. Hutley, and G. H. Derrick, "Total absorption of unpolarized light by crossed gratings," *Opt. Express*, vol. 16, no. 9, 6146, 2008.

[80] N. I. Landy, S. Sajuyigbe, J. J. Mock, D. R. Smith, and W. J. Padilla, "Perfect metamaterial absorber," *Phys. Rev. Lett.*, vol. 100, no. 20, p. 207402, 2008.

[81] X. Liu, T. Starr, A. F. Starr, and W. J. Padilla, "Infrared spatial and frequency selective metamaterial with near-unity absorbance," *Phys. Rev. Lett.*, vol. 104, no. 20, 207403, 2010.

[82] V. S. Asadchy, I. A. Faniayeu, Y. Ra'di, S. A. Khakhomov, I. V. Semchenko, and S. A. Tretyakov, "Broadband reflectionless metasheets: frequency-selective transmission and perfect absorption," *Phys. Rev. X*, vol. 5, no. 3, 031005, 2015.

[83] N. Yu, and F. Capasso, "Flat optics with designer metasurfaces," *Nat. Mater.*, vol. 13, no. 2, pp. 139–150, 2014.

[84] Y. C. Chang, A. V. Kildishev, E. E. Narimanov, and T. B. Norris, "Metasurface perfect absorber based on guided resonance of a photonic hypercrystal," *Phys. Rev. B*, vol. 94, no. 15, 155430, 2016.

[85] Y. Yao, R. Shankar, M. A. Kats, *et al.*, "Electrically tunable metasurface perfect absorbers for ultrathin mid-infrared optical modulators," *Nano Lett.*, vol. 14, no. 11, pp. 6526–6532, 2014.

[86] M. A. Kats, R. Blanchard, P. Genevet, and F. Capasso, "Nanometre optical coatings based on strong interference effects in highly absorbing media," *Nat. Mater.*, vol. 12, no. 1, pp. 20–24, 2013.

[87] Y. Ra'di, C. R. Simovski, and S. A. Tretyakov, "Thin perfect absorbers for electromagnetic waves: theory, design, and realizations," *Phys. Rev. Appl.*, vol. 3, no. 3, 037001, 2015.
[88] M. A. Kats, and F. Capasso, "Optical absorbers based on strong interference in ultra-thin films," *Laser Photonics Rev.*, vol. 10, no. 5, pp. 735–749, 2016.
[89] H. K. Raut, V. A. Ganesh, A. S. Nair, and S. Ramakrishna, "Anti-reflective coatings: a critical, in-depth review," *Energy Environ. Sci.*, vol. 4, no. 10, pp. 3779–3804, 2011.
[90] W. Wan, Y. Chong, L. Ge, H. Noh, A. D. Stone, and H. Cao, "Time-reversed lasing and interferometric control of absorption," *Science*, vol. 331, no. 6019, pp. 889–892, 2011.
[91] J. Zhang, K. F. MacDonald, and N. I. Zheludev, "Controlling light-with-light without nonlinearity," *Light Sci. Appl.*, vol. 1, e18, 2012.
[92] K. Pichler, M. Kühmayer, J. Böhm, *et al.*, "Random anti-lasing through coherent perfect absorption in a disordered medium," *Nature*, vol. 567, no. 7748, pp. 351–355, 2019.
[93] L. Baldacci, S. Zanotto, G. Biasiol, L. Sorba, and A. Tredicucci, "Interferometric control of absorption in thin plasmonic metamaterials: general two port theory and broadband operation," *Opt. Express*, vol. 23, no. 7, 9202, 2015.
[94] J. R. Piper, and S. Fan, "Total absorption in a graphene monolayer in the optical regime by critical coupling with a photonic crystal guided resonance," *ACS Photonics*, vol. 1, no. 4, pp. 347–353, 2014.
[95] J. R. Piper, V. Liu, and S. Fan, "Total absorption by degenerate critical coupling," *Appl. Phys. Lett.*, vol. 104, no. 25, 251110, 2014.
[96] J. W. Yoon, G. M. Koh, S. H. Song, and R. Magnusson, "Measurement and modeling of a complete optical absorption and scattering by coherent surface plasmon-polariton excitation using a silver thin-film grating," *Phys. Rev. Lett.*, vol. 109, no. 25, 257402, 2012.
[97] R. Bruck, and O. L. Muskens, "Plasmonic nanoantennas as integrated coherent perfect absorbers on SOI waveguides for modulators and all-optical switches," *Opt. Express*, vol. 21, no. 23, 27652, 2013.
[98] J. M. Rothenberg, C. P. Chen, J. J. Ackert, *et al.*, "Experimental demonstration of coherent perfect absorption in a silicon photonic racetrack resonator," *Opt. Lett.*, vol. 41, no. 11, pp. 2537–2540, 2016.
[99] A. I. Ignatov, I. A. Nechepurenko, and D. G. Baranov, "Anisotropy-assisted non-scattering coherent absorption of surface plasmon-polaritons," *Ann. Phys.*, vol. 528, no. 7–8, pp. 537–542, 2016.
[100] S. Li, J. Luo, S. Anwar, *et al.*, "Broadband perfect absorption of ultrathin conductive films with coherent illumination: superabsorption of microwave radiation," *Phys. Rev. B – Condens. Matter Mater. Phys.*, vol. 91, no. 22, 220301, 2015.
[101] S. Yu, X. Piao, J. Hong, and N. Park, "Progress toward high-Q perfect absorption: a fano antilaser," *Phys. Rev. A – At. Mol. Opt. Phys.*, vol. 92, no. 1, 011802(R), 2015.

[102] L. N. Pye, M. L. Villinger, S. Shabahang, W. D. Larson, L. Martin, and A. F. Abouraddy, "Octave-spanning coherent perfect absorption in a thin silicon film," *Opt. Lett.*, vol. 42, no. 1, pp. 151–154, 2017.

[103] H. Zhao, W. S. Fegadolli, J. Yu, *et al.*, "Metawaveguide for asymmetric interferometric light-light switching," *Phys. Rev. Lett.*, vol. 117, no. 19, 193901, 2016.

[104] S. Zanotto, F. P. Mezzapesa, F. Bianco, *et al.*, "Perfect energy-feeding into strongly coupled systems and interferometric control of polariton absorption," *Nat. Phys.*, vol. 10, no. 11, pp. 830–834, 2014.

[105] X. Fang, K. F. MacDonald, and N. I. Zheludev, "Controlling light with light using coherent metadevices: all-optical transistor, summator and invertor," *Light Sci. Appl.*, vol. 4, no. 5, e292, 2015.

[106] M. Papaioannou, E. Plum, J. Valente, E. T. Rogers, and N. I. Zheludev, "Two-dimensional control of light with light on metasurfaces," *Light Sci. Appl.*, vol. 5, no. 4, e16070, 2016.

[107] M. Papaioannou, E. Plum, and N. I. Zheludev, "All-optical pattern recognition and image processing on a metamaterial beam splitter," *ACS Photonics*, vol. 4, no. 2, pp. 217–222, 2017.

[108] S. F. Liew, S. M. Popoff, S. W. Sheehan, *et al.*, "Coherent control of photocurrent in a strongly scattering photoelectrochemical system," *ACS Photonics*, vol. 3, no. 3, pp. 449–455, 2016.

[109] T. Roger, S. Vezzoli, E. Bolduc, *et al.*, "Coherent perfect absorption in deeply subwavelength films in the single-photon regime," *Nat. Commun.*, vol. 6, no. 1, 7031, 2015.

[110] T. Roger, S. Restuccia, A. Lyons, *et al.*, "Coherent absorption of N00N States," *Phys. Rev. Lett.*, vol. 117, no. 2, 023601, 2016.

[111] D. K. Armani, T. J. Kippenberg, S. M. Spillane, and K. J. Vahala, "Ultra-high-Q toroid microcavity on a chip," *Nature*, vol. 421, no. 6926, pp. 925–928, 2003.

[112] N. Zhang, Y. Wang, W. Sun, *et al.*, "High-Q and highly reproducible microdisks and microlasers," *Nanoscale*, vol. 10, no. 4, pp. 2045–2051, 2018.

[113] N. Kryzhanovskaya, Y. Polubavkina, E. Moiseev, *et al.*, "Enhanced light outcoupling in microdisk lasers via Si spherical nanoantennas," *J. Appl. Phys.*, vol. 124, no. 16, 163102, 2018.

[114] E. I. Moiseev, N. Kryzhanovskaya, Y. S. Polubavkina, *et al.*, "Light outcoupling from quantum dot-based microdisk laser via plasmonic nanoantenna," *ACS Photonics*, vol. 4, no. 2, pp. 275–281, 2017.

[115] S. L. McCall, A. F. J. Levi, R. E. Slusher, S. J. Pearton, and R. A. Logan, "Whispering-gallery mode microdisk lasers," *Appl. Phys. Lett.*, vol. 60, no. 3, pp. 289–291, 1992.

[116] V. V. Vassiliev, V. L. Velichansky, V. S. Ilchenko, M. L. Gorodetsky, L. Hollberg, and A. V. Yarovitsky, "Narrow-line-width diode laser with a high-Q microsphere resonator," *Opt. Commun.*, vol. 158, no. 1–6, pp. 305–312, 1998.

[117] M. Koschorreck, R. Gehlhaar, V. G. Lyssenko, M. Swoboda, M. Hoffmann, and K. Leo, "Dynamics of a high-Q vertical-cavity organic laser," *Appl. Phys. Lett.*, vol. 87, no. 18, 181108, 2005.

[118] M. Lončar, T. Yoshie, A. Scherer, P. Gogna, and Y. Qiu, "Low-threshold photonic crystal laser," *Appl. Phys. Lett.*, vol. 81, no. 15, pp. 2680–2682, 2002.

[119] B.-S. Song, S. Noda, T. Asano, and Y. Akahane, "Ultra-high-Q photonic double-heterostructure nanocavity," *Nat. Mater.*, vol. 4, no. 3, pp. 207–210, 2005.

[120] O. Painter, R. K. Lee, A. Scherer, *et al.*, "Two-dimensional photonic band-gap defect mode laser," *Science (80-)*, vol. 284, no. 5421, pp. 1819–1821, 1999.

[121] D. G. Baranov, A. Krasnok, and A. Alù, "Coherent virtual absorption based on complex zero excitation for ideal light capturing," *Optica*, vol. 4, no. 12, 1457, 2017.

[122] C. A. Balanis, *Antenna Theory: Analysis and Design.* New York; Brisbane: J. Wiley, 1997.

[123] A. Alù, and N. Engheta, "Wireless at the nanoscale: optical interconnects using matched nanoantennas," *Phys. Rev. Lett.*, vol. 104, no. 21, 213902, 2010.

[124] A. Alu, and N. Engheta, "Theory, modeling and features of optical nanoantennas," *IEEE Trans. Antennas Propag.*, vol. 61, no. 4, pp. 1508–1517, 2013.

[125] L. Novotny, and N. Van Hulst, "Antennas for light," *Nat. Photonics*, vol. 5, no. 2, pp. 83–90, 2011.

[126] W. C. Brown, "The history of power transmission by radio waves," *IEEE Trans. Microw. Theory Tech.*, vol. 32, no. 9, pp. 1230–1242, 1984.

[127] A. Kurs, A. Karalis, R. Moffatt, J. D. Joannopoulos, P. Fisher, and M. Soljacic, "Wireless power transfer via strongly coupled magnetic resonances," *Science (80).*, vol. 317, no. 5834, pp. 83–86, 2007.

[128] M. Song, P. Belov, and P. Kapitanova, "Wireless power transfer inspired by the modern trends in electromagnetics," *Appl. Phys. Rev.*, vol. 4, no. 2, 021102, 2017.

[129] A. Krasnok, D. G. Baranov, A. Generalov, S. Li, and A. Alù, "Coherently enhanced wireless power transfer," *Phys. Rev. Lett.*, vol. 120, no. 14, 143901, 2018.

[130] K. X. Wang, Z. Yu, S. Sandhu, and S. Fan, "Fundamental bounds on decay rates in asymmetric single-mode optical resonators," *Opt. Lett.*, vol. 38, no. 2, pp. 100–102, 2013.

[131] S. Fan, W. Suh, and J. D. Joannopoulos, "Temporal coupled-mode theory for the fano resonance in optical resonators," *J. Opt. Soc. Am. A*, vol. 20, no. 3, pp. 569–572, 2003.

[132] H. Haus, *Waves and Fields in Optoelectronics.* Englewood Cliffs: Prentice Hall, 1984.

Chapter 4

Time domain modeling with the generalized dispersive material model

Ludmila J. Prokopeva[1], William D. Henshaw[2], Donald W. Schwendeman[2] and Alexander V. Kildishev[1]

The generalized dispersive material (GDM) model, first introduced in 2010 [1,2], enables the universal modeling of the transient electromagnetic response of materials with an arbitrary linear dispersion. This approach to approximating arbitrary frequency-dependent dielectric functions was efficiently implemented in time-domain (TD) numerical schemes using the auxiliary differential equation (ADE) and recursive convolution (RC) techniques. The initial implementation in [2] was demonstrated with the classical finite-difference time domain (FDTD) scheme of Yee [3]. Later the GDM approach was extended to other TD methods, such as the finite volume (FVTD) [4] and the discontinuous Galerkin (DGTD) methods [5], along with the high-order accurate FDTD schemes for overset grids [6]. Overall, the GDM model provides a convenient and universal approach to implement an arbitrary linear dispersion in TD Maxwell codes, for example, see the Python FDTD package [7].

Initially, the primary application of the GDM model was the accurate modeling of plasmonic nanostructures that are highly dispersive at optical frequencies. Simulation examples of such structures built on the noble metals can be found in [8,9], including nano-slits, nano-nets, and semi-continuous composites. Later, the GDM model was applied to the efficient TD simulations of tunable multi-variate dispersion of graphene [10]. Recently, it has also been shown that the GDM approach handles multi-variate dispersion of phase change materials (PCM) [11] equally well and the synthetic dispersion curves retrieved from bianisotropic homogenization [12].

In all of the aforementioned cases, the GDM formulation is usually more compact than traditional approaches found in the literature. The fits with the GDM model can also provide intuition into the possible physical mechanisms that characterize the dispersive behavior of a material. Moreover, since the GDM model

[1]School of ECE and Birck Nanotechnology Center, Purdue University, West Lafayette, IN, USA
[2]Department of Mathematical Sciences and the Center for Modeling, Optimization and Computational Analysis, Rensselaer Polytechnic Institute, Troy, NY, USA

provides a TD formulation, it delivers a computational platform for multiphysics studies, where general linear dispersive effects are coupled to additional nonlinear equations. See, for example, our work on simulations of plasmon-enhanced gain, where our numerical models account for coupled linear dispersion of a plasmonic metallic nanostructure and nonlinear carrier kinetics of gain medium [13,14].

The goal of this chapter is to provide details of the GDM model and its properties, give examples of numerical implementations, and include critical derivations not available from [2] due to space restrictions. All of these crucial components should be helpful to readers developing TD Maxwell's solvers and performing multiphysics modeling of real-life devices with tunable and multivariate dielectric functions.

4.1 GDM model

The GDM model involves constitutive equations defining the electromagnetic response of a material as a function of frequency. The constitutive equations are then coupled to Maxwell's equations. We begin with a brief description of Maxwell's equations in the time and frequency domains, followed by analytical properties and representations of the GDM model. Physical interpretations of the parameters appearing in the model are illustrated by an explicit conversion of the formulas between the time and frequency domains, as well as the relationship to the classical models of optical dispersion. The formulae derived here provide the basis for the numerical schemes discussed in the next section.

4.1.1 Maxwell's equations

Maxwell's equations in the time domain are a system of partial differential equations connecting the inducing fields (the electric and magnetic field strength vectors, $\mathbf{E}(\mathbf{x},t)$ and $\mathbf{H}(\mathbf{x},t)$) with the induced fields (the magnetic flux density $\mathbf{B}(\mathbf{x},t)$ and electric displacement $\mathbf{D}(\mathbf{x},t)$), all functions of position $\mathbf{x}$ and time t, given by

$$\left.\begin{array}{ll}\partial_t\mathbf{D} = \nabla\times\mathbf{H}, & \mathbf{D} = \varepsilon_0\varepsilon * \mathbf{E},\\ \partial_t\mathbf{B} = -\nabla\times\mathbf{E}, & \mathbf{B} = \mu_0\mathbf{H},\end{array}\right\}\quad \mathbf{x}\in R,\quad t\geq 0, \tag{4.1}$$

where ε_0 and μ_0 are the electric permittivity and magnetic permeability in vacuum, respectively, and $R\subset\mathbb{R}^3$ denotes the spatial domain of interest. Also, we will use the notation $c_0 = 1/\sqrt{\mu_0\varepsilon_0}$ for speed of light in vacuum. The expression $\varepsilon * \mathbf{E}$ in (4.1) denotes a convolution in time, i.e.

$$\mathbf{D}(\mathbf{x},t) = \varepsilon_0\varepsilon * \mathbf{E} = \varepsilon_0\int_{-\infty}^{\infty}\varepsilon(\tau)\mathbf{E}(\mathbf{x},t-\tau)\mathrm{d}\tau, \tag{4.2}$$

where $\varepsilon(t)$ is the relative permittivity with causality requiring that $\varepsilon(t) = 0$ for $t < 0$. Here, for simplicity, we assume isotropic nonmagnetic optical materials, so that the relative magnetic permeability is $\mu = 1$.

Maxwell's equations (4.1) are augmented with appropriate initial conditions, boundary conditions, and interface conditions. Boundary and interface conditions for the dispersive case are of a similar form to the non-dispersive case since the spatial derivatives in the equations have not changed. The dispersive equations, however, require additional initial conditions since the time convolution (4.2) depends on past time values of **E**. The number of additional initial conditions required depends on the form of $\varepsilon(t)$ and these will be discussed in subsequent sections.

In the frequency domain, the system of equations in (4.1) become

$$\left.\begin{aligned} -\iota\omega\hat{\mathbf{D}} &= \nabla\times\hat{\mathbf{H}}, & \hat{\mathbf{D}} &= \varepsilon_0\hat{\varepsilon}\hat{\mathbf{E}}, \\ -\iota\omega\hat{\mathbf{B}} &= -\nabla\times\hat{\mathbf{E}}, & \hat{\mathbf{B}} &= \mu_0\hat{\mathbf{H}}, \end{aligned}\right\} \quad \mathbf{x}\in R, \quad \omega\in\mathbb{R}, \tag{4.3}$$

where the symbols with hats denote the Fourier transform of their corresponding quantities in the time domain. For example, $\hat{\mathbf{E}}(\mathbf{x},\omega)$ corresponds to $\mathbf{E}(\mathbf{x},t)$ via the transform pair

$$\hat{\mathbf{E}}(\mathbf{x},\omega) = \int_{-\infty}^{\infty} e^{\iota\omega t}\mathbf{E}(\mathbf{x},t)\mathrm{d}t, \quad \mathbf{E}(\mathbf{x},t) = \frac{1}{2\pi}\int_{-\infty}^{\infty} e^{-\iota\omega t}\hat{\mathbf{E}}(\mathbf{x},\omega)\mathrm{d}\omega.$$

The relative electric permittivity in the frequency domain, $\hat{\varepsilon} = \hat{\varepsilon}(\omega)$, defines the dispersion characteristics of the material, and it can take a very general form subject to physical realizability conditions. In the frequency domain, the time convolution (4.2) transforms to a multiplication condition

$$\hat{\mathbf{D}}(\mathbf{x},\omega) = \varepsilon_0\hat{\varepsilon}(\omega)\hat{\mathbf{E}}(\mathbf{x},\omega).$$

Constraints on the form of $\hat{\varepsilon}(\omega)$ to enforce causality and other properties will be described in the next subsection.

To separate the dielectric constant and finite conductivity from additional dispersive terms in the electric displacement field **D**, we introduce the electric polarization vector **P** as

$$\mathbf{D}(\mathbf{x},t) = \varepsilon_0\varepsilon_\infty\mathbf{E}(\mathbf{x},t) + \sigma\int_0^t \mathbf{E}(\mathbf{x},\tau)\mathrm{d}\tau + \mathbf{P}(\mathbf{x},t),$$

which transforms to the frequency domain as

$$\hat{\mathbf{D}} = \varepsilon_0\varepsilon_\infty\hat{\mathbf{E}} - \frac{\sigma}{\iota\omega}\hat{\mathbf{E}} + \hat{\mathbf{P}},$$

where the dielectric constant ε_∞ is the limiting value of $\hat{\varepsilon}(\omega)$ as $\omega\to\infty$, σ is a conductivity constant, and **P** is related to **E** through the electric susceptibility χ as

$$\mathbf{P} = \varepsilon_0\chi * \mathbf{E},$$

in the time domain, or

$$\hat{\mathbf{P}}(\mathbf{x},\omega) = \varepsilon_0 \hat{\chi}(\omega)\hat{\mathbf{E}}(\mathbf{x},\omega),$$

in the frequency domain. These formulas imply that $\hat{\varepsilon}$ and $\hat{\chi}$ are related by

$$\hat{\varepsilon}(\omega) = \varepsilon_\infty - \frac{\sigma}{\iota\omega\varepsilon_0} + \hat{\chi}(\omega).$$

4.1.2 The GDM model

According to Landau *et al.* [15], $\hat{\varepsilon}(\omega)$ should satisfy the two properties

$$\hat{\varepsilon}(\iota\omega) \in \mathbb{R}, \quad \text{for } \omega > 0, \tag{4.4a}$$

and

$$\hat{\varepsilon}(\omega) \to \varepsilon_\infty > 0 \quad \text{for } \omega \to \infty. \tag{4.4b}$$

Note also that since $\varepsilon(t)$ is real, then $\hat{\varepsilon}(-\omega) = \overline{\hat{\varepsilon}}(\omega)$ and thus $\hat{\varepsilon}(\omega)$ is fully defined for all $\omega \in \mathbb{R}$ knowing its form for $\omega \geq 0$. The conditions (4.4) imply that the dielectric function $\hat{\varepsilon}(\omega)$ can be approximated by a rational function of the form

$$\hat{\varepsilon}(\omega) = \varepsilon_\infty + \frac{\alpha_0 + \alpha_1 s + \cdots + \alpha_m s^m}{\beta_0 + \beta_1 s + \cdots + \beta_n s^n}, \quad s = -\iota\omega, \tag{4.5}$$

where $m < n$ and all coefficients $\{\alpha_i, \beta_i\}$ are real. We assume that α_m and β_n are nonzero and take $\beta_n = 1$ without loss of generality. Since the polynomial in the denominator of (4.5) has real coefficients, its roots are either real or appear as complex conjugate pairs. Thus, this polynomial can be factored as

$$\beta_0 + \beta_1 s + \cdots + \beta_{n-1}s^{n-1} + s^n = \prod_{j=1}^{n_1} \left(s + b_{1,j}\right)^{p_j} \prod_{j=1}^{n_2} \left(s^2 + b_{1,j}s + b_{0,j}\right)^{q_j},$$

where the linear factors $\left(s + b_{1,j}\right)$, $j = 1,2,\ldots,n_1$, correspond to (any) real roots with multiplicity p_j, and the quadratic factors $\left(s^2 + b_{1,j}s + b_{0,j}\right)$, $j = 1,2,\ldots,n_2$, correspond to the pairs of complex-conjugate roots (assuming $b_{1,j}^2 < 4b_{0,j}$) with multiplicity q_j. We assume that the approximation has no multiple roots, i.e., $p_j = q_j = 1$ for all j (which excludes polynomial in time factors in the solution), and as a result the rational function can be expressed as a partial fraction decomposition of the form

$$\hat{\varepsilon}(\omega) = \varepsilon_\infty - \frac{\sigma}{\iota\omega\varepsilon_0} + \sum_{i\in I_1} \underbrace{\frac{a_{1,i}}{b_{1,i} - \iota\omega}}_{\hat{\chi}_i(\omega)} + \sum_{i\in I_2} \underbrace{\frac{a_{0,i} - \iota\omega a_{1,i}}{b_{0,i} - \iota\omega b_{1,i} - \omega^2}}_{\hat{\chi}_i(\omega)}, \tag{4.6}$$

where again all parameters ε_∞, σ, $a_{k,i}$, and $b_{k,i}$ are real. Here, we have specifically identified the conductivity term with coefficient σ corresponding to a pole at the origin (i.e., if $\beta_0 = 0$). Note that by setting $a_{0,i} = b_{0,i} = 0$ in (4.6) a conjugate pole pair, $i \in I_2$, reduces to a single pole term and thus one could exclude the single pole terms, $i \in I_1$, in (4.6) without loss of generality. Here we keep both sets since this may lead to more efficient numerical implementations. Note that the special case of a non-dissipative media (i.e., lossless media) has $\hat{\varepsilon}(\omega) = \hat{\varepsilon}(-\omega)$ for time reversability, which implies $\sigma = a_{1,i} = b_{1,i} = 0$, i.e. we obtain multipole Sellmeir model.

As noted in equation (4.6), we introduce the notation for partial susceptibility terms $\hat{\chi}_i(\omega)$, and the corresponding partial polarizations, $\hat{\mathbf{P}}_i = \varepsilon_0 \hat{\chi}_i(\omega)\hat{\mathbf{E}}$, so that

$$\hat{\chi}(\omega) = \sum_{i \in I_i \cup I_2} \hat{\chi}_i(\omega).$$

The relations, $\hat{\mathbf{P}}_i = \varepsilon_0 \hat{\chi}_i(\omega)\hat{\mathbf{E}}$, for $i \in I_i \cup I_2$, transform to ordinary differential equations in the time domain (formally by the replacement $-\iota\omega \rightarrow \partial_t$) as

$$\partial_t \mathbf{P}_i + b_{1,i}\mathbf{P}_i = \varepsilon_0 \left(a_{1,i}\mathbf{E}\right), \quad i \in I_1, \tag{4.7a}$$

$$\partial_t^2 \mathbf{P}_i + b_{1,i}\partial_t \mathbf{P}_i + b_{0,i}\mathbf{P}_i = \varepsilon_0 \left(a_{0,i}\mathbf{E} + a_{1,i}\partial_t \mathbf{E}\right), \quad i \in I_2. \tag{4.7b}$$

These differential equations are then coupled to Maxwell's equations in (4.1) for **E** and **H** to give

$$\varepsilon_0 \varepsilon_\infty \partial_t \mathbf{E} = \nabla \times \mathbf{H} - \sigma \mathbf{E} - \partial_t \mathbf{P}, \tag{4.7c}$$

$$\mu_0 \partial_t \mathbf{H} = -\nabla \times \mathbf{E}, \tag{4.7d}$$

$$\mathbf{P} \stackrel{\text{def}}{=} \sum_{i \in I_1 \cup I_2} \mathbf{P}_i. \tag{4.7e}$$

The system of equations in (4.7) forms the basis for the ADE formulation of the GDM model.

Note that from the form of the time derivatives in (4.7) we require initial conditions for **E** and **H**, as well as initial conditions for $\mathbf{P}_i$ when $i \in I_1$ or initial conditions for both $\mathbf{P}_i$ and $\partial_t \mathbf{P}_i$ when $i \in I_2$. Furthermore, note that (4.7) defines a well-posed initial-boundary-value problem (with appropriate boundary conditions) when ε_∞ is positive, and for *any* real values of the GDM parameters σ, $a_{0,i}$, $a_{1,i}$, $b_{0,i}$ and $b_{1,i}$, although certain values of the GDM parameters lead to solutions with bounded growth in time (e.g., for a gain material). This is discussed further in Section 4.1.3.

The RC formulation is based on an inverse Fourier transform of (4.6) to the time domain, which gives

$$\varepsilon(t) = \varepsilon_\infty \delta(t) + \frac{\sigma}{\varepsilon_0} U(t) + \chi(t), \tag{4.8a}$$

$$\chi(t) \stackrel{\text{def}}{=} \sum_{i \in I_1 \cup I_2} \chi_i(t), \tag{4.8b}$$

where $\delta(t)$ and $U(t)$ denote the Dirac delta function and Heaviside function, respectively, and $\chi_i(t)$ is the inverse Fourier transform of $\hat{\chi}_i(\omega)$ (expressions for which are described later in Section 4.2.2). The dispersive Maxwell's equations are then given by

$$\varepsilon_0\varepsilon_\infty\partial_t\mathbf{E} = \nabla\times\mathbf{H} - \sigma\mathbf{E} - \varepsilon_0\partial_t(\chi * \mathbf{E}), \tag{4.8c}$$

$$\mu_0\partial_t\mathbf{H} = -\nabla\times\mathbf{E}. \tag{4.8d}$$

The system of equations in (4.8) is the basis for the RC formulation. Similiar to (4.7), system (4.8) requires initial conditions for **E** and **H**, as well as one additional initial condition for each χ_i, $i \in I_1$, and two additional initial conditions for each χ_i, $i \in I_2$.

The approximation (4.6) is called the GDM model and gives the mathematical (rational) approximation to a very broad class of dielectric functions in the frequency domain (omitting the case of multiple poles). The GDM model was first introduced by us at the conferences in 2010 [1,16] with details published in the extended proceedings [2]. The term *GDM model*, however, was first introduced in our later work in 2011 [17].

4.1.3 The dispersion relation for the GDM model

The dispersion relation for the GDM model is derived in the usual way by considering plane wave solutions of the form $\mathbf{E}(\mathbf{x},t) = e^{st}e^{\imath\mathbf{k}\cdot\mathbf{x}}\hat{\mathbf{E}}_0$ for constants $s \in \mathbb{C}$ and $\mathbf{k} \in \mathbb{C}^3$. Substituting this form into the governing equation for **E** written as a vector wave equation

$$\partial_t^2\mathbf{E} + \frac{1}{\varepsilon_0}\partial_t^2\mathbf{P} - c_0^2\Delta\mathbf{E} = 0,$$

where $c_0^2 = 1/(\varepsilon_0\mu_0)$ leads to the dispersion relation

$$s^2 + s^2\sum_{i=1}^{N_1}\frac{a_{1,i}}{b_{1,i}+s} + s^2\sum_{i=1}^{N_2}\frac{a_{0,i}+a_{1,i}s}{b_{0,i}+b_{1,i}s+s^2} + (c_0k)^2 = 0, \tag{4.9}$$

where $k^2 = \mathbf{k}^T\mathbf{k}$, $k \in \mathbb{C}$, and we have set $\varepsilon_\infty = 1$ for simplicity. Note also that we have included the (possible) conductivity term in (4.6) as one of the linear terms in (4.9). After multiplying through by the product of the denominators (4.9) becomes a polynomial of degree $M = 2 + N_1 + 2N_2$ in s. One can view (4.9) as an equation for s given k or an equation for k given s.

When $\mathbf{k} \in \mathbb{R}^3$, we can view the plane wave decomposition as a Fourier transform in space. Then, for a given $k \in \mathbb{R}$, there will be M roots, $s_n \in \mathbb{C}$, $n = 1, 2, \ldots, M$, that define M modes of propagation for these spatial Fourier modes. Growth or decay in time of these solutions depends on whether $\Re(s_n) > 0$ or $\Re(s_n) < 0$, respectively. To avoid growth in time one must enforce that all roots satisfy $\Re(s_n) \leq 0$. A necessary condition is that (4.9) has no poles with $\Re(s) > 0$, which requires $b_{0,i} \geq 0$ and $b_{1,i} \geq 0$. Routh–Hurwitz theory can be used to derive necessary and sufficient conditions to ensure that all roots satisfy $\Re(s_n) \leq 0$. For example, for the case of a complex-conjugate pair term

$$s^2 + (c_0 k)^2 + s^2 \frac{a_0 + a_1 s}{b_0 + b_1 s + s^2} = 0, \tag{4.10}$$

the Routh–Hurwitz theory leads to the conditions

$$a_0 \geq 0, \quad b_0 \geq 0, \quad a_1 \geq 0, \quad b_1 \geq 0, \quad a_1 b_0 \leq a_0 b_1, \tag{4.11}$$

for no growth in time for all $k \in \mathbb{R}$, see [6] for further details. Note that these same conditions (4.11) also arise when the susceptibility,

$$\hat{\chi} = \frac{a_0 + a_1 s}{b_0 + b_1 s + s^2},$$

is required to have no poles with $\Re(s) > 0$ (which implies $b_0 \geq 0$ and $b_1 \geq 0$) and that $\Im(\chi) \geq 0$ for $\omega \geq 0$.

Further information about the dispersive nature of the plane-wave solutions can be obtained through an asymptotic analysis of the roots to the dispersion relation (4.10). The asymptotic forms of the roots for large $c_0 k$ are given by

$$s = \pm \iota c_0 k - \frac{a_1}{2} \mp \frac{\iota}{8 c_0 k}\left(a_1^2 + 4 a_1 b_1 - 4 a_0\right) + \mathcal{O}\left((c_0 k)^{-2}\right), \tag{4.12}$$

$$s = -\frac{b_1}{2} \pm \sqrt{(b_1/2)^2 - b_0} + \mathcal{O}\left((c_0 k)^{-2}\right). \tag{4.13}$$

The first root (4.12) represents plane waves that, to leading order, propagate with wave number k but that are damped in time due to the term $-a_1/2$. The second root (4.13) is a purely dispersive mode. When $b_1 = 0$ (no-damping) and $s \approx \pm \iota \sqrt{b_0}$, this mode represents a response of the material at frequency $\omega \approx \mp \sqrt{b_0}$ that arises from the location of the pole in the second-order GDM term. A positive value for b_1 leads to damping in the response. Thus for $(b_1/2)^2 - b_0 < 0$, each second-order GDM term can be viewed roughly as providing a material response at frequency $\omega \approx \sqrt{b_0 - (b_1/2)^2}$.

Another way to analyze dispersive properties of a GDM term is to review the linear second-order ODE for polarization vector in time domain $\mathbf{P}(t)$,

$$\partial_t^2 \mathbf{P} + b_1 \partial_t \mathbf{P} + b_0 \mathbf{P} = \varepsilon_0 (a_0 \mathbf{E} + a_1 \partial_t \mathbf{E}). \tag{4.14}$$

This equation represents a phase-relaxed damped harmonic oscillator, driven by a given external electric field $\mathbf{E}(t)$. We note that (4.14) describes a 0D local material response to optical excitation. The full-wave coupling of $\mathbf{E}(t)$ and $\mathbf{P}(t)$ to Maxwell's equations is neglected for simplicity in the present discussion.

For a zero scalar field $E = 0$, (4.14) is a classical equation of unforced damped oscillator, where the solution depends on $\mathrm{sgn}\left(b_0 - b_1^2/4\right)$. If $b_0 > b_1^2/4$, the solution is a *damped harmonic oscillator* of the form $P_0(t) = A_0 \mathrm{e}^{-\gamma t} \sin(\delta t - \varphi_0)$. Here the amplitude A_0 and the phase φ_0 depend on the initial displacement $P(0)$ and initial momentum $\dot{P}(0)$, while the decay rate $\gamma = b_1/2$ and the oscillation frequency

$\delta = \sqrt{b_0 - \gamma^2}$ are fixed characteristics of the system that do not depend on the initial conditions (this property is known as *isochronism*). On the other hand, if $b_0 \leq b_1^2/4$, then the unforced solutions decay exponentially without oscillation. The subcase $b_0 < b_1^2/4$ is called *over-damped oscillator*, and the solution is $P_0(t) = A_1 e^{-\gamma_1 t} + A_2 e^{-\gamma_2 t}$, where $\gamma_{1,2} = \gamma \pm \sqrt{\gamma^2 - b_0} > 0$. The subcase $b_0 = b_1^2/4$ is called *critically damped oscillator* with $P_0(t) = (A + Bt)e^{-t}$. Note that the solution of homogeneous equation $P_0(t)$ becomes negligible after the relaxation time $t > \tau = \gamma^{-1}$, unless $b_1 \leq 0$.

If we now impose a harmonic external field $E = \sin(\omega t)$, the solution $P(t)$ is a superposition of the previous undriven oscillator $P_0(t)$ (decaying if $b_1 > 0$) and steady-state oscillation driven by the external field $P_s(t) = A\cos(\omega t - \varphi)$, so that $P(t) = P_0(t) + P_s(t)$. The amplitude and phase of the steady state solution, A and φ, are given in terms of the complex-valued susceptibility $\hat{\chi}(\omega) = (a_0 + a_1 s)/(b_0 + b_1 s + s^2)$, $s = -\iota\omega$, as

$$A^2 = \varepsilon_0^2 |\hat{\chi}|^2 = \varepsilon_0^2 \frac{a_0^2 + a_1^2\omega^2}{\left(b_0^2 - \omega^2\right)^2 + b_1^2\omega^2},$$

$$\varphi = \mathrm{Arg}(\hat{\chi}).$$

By analyzing the susceptibility function $\hat{\chi}(\omega)$ we can identify special points of the spectrum. The *resonance frequency* ω_0 is defined as the frequency for which the amplitude $A(\omega)$ is maximized. This frequency is given by the formula $\omega_0^2 = \left(\sqrt{\left(a_0^2 + b_0 a_1^2\right)^2 - b_1^2 a_0^2 a_1^2} - a_0^2\right)/a_1^2$ for the general case, while for the Lorentz case $(a_1 = 0)$, we have $\omega_0^2 = b_0 - b_1^2/2$. The *crossing point* ω_c defined by $\Re(\hat{\chi}(\omega_c)) = 0$, is given by $\omega_c^2 = a_0 b_0/(a_0 - a_1 b_1)$, and it reduces to $\omega_c^2 = b_0$ for the Lorentz case. The *absorption peak* ω_a defined as the frequency for which $\Im(\hat{\chi})$ is maximized has a simple expression for the Lorentz case $\omega_a = b_0/3 - b_1^2/6 + \sqrt{16b_0^2 - 4b_0 b_1^2 + b_1^4}/6$, while for the general case its value can be found as a root of a cubic equation which we omit here. Typical lineshapes and special points of the susceptibility for different classical models are shown in Figure 4.1.

4.1.4 Special GDM cases: classical dispersion models

Analytical dispersion models for real materials are often found by fitting an approximating functional form to experimental ellipsometry data for the real and/or imaginary parts of the dielectric function $\hat{\varepsilon}(\omega)$ in frequency space. Some of these frequency domain models may have rather complicated forms in the time domain, for example, models containing fractional-order derivatives (such as the Cole–Cole [18], Cole–Davidson [19], and Havriliak–Negami [20] models), and thus, are not easily implemented in TD codes, see e.g., [21,22]. Another approach is rooted in a physical understanding of a given material and the nature of its optical properties, as opposed to a formal mathematical fit. With this approach, the dispersion models are derived from theoretical arguments based on classical or quantum mechanical descriptions of the material response due to light-matter interaction.

In this section, we present the main classical dispersion relationships and demonstrate that all of them are the special cases of the GDM model. This means that all of these models can be implemented within the GDM framework in the time domain using either the ADE or RC formulations. Table 4.1 summarizes the correspondence between the coefficients in the classical dispersion models with their GDM coefficients.

Figure 4.1 shows typical lineshapes of the susceptibility, $\hat{\chi}(\omega)$, for the Debye, Lorentz, Drude, and CP models as functions of the photon energy ($\hbar\omega$) in eV. Here we identify some physical characteristics of the models. We take $\varepsilon_\infty = 0$, and as a result $\hat{\chi}(\omega) \to 0$ as $\omega \to \infty$ for all models. The constants $\Delta\varepsilon$ and Drude's Ω_p are adjusted so that the permittivity jump is unity, $|\Re(\chi(0))| = 1$. The constants Γ and Ω are chosen so that $b_1 = 3.1$ eV and $\sqrt{b_0} = 4$ eV for all models. With this adjustments, the Lorentz model becomes equivalent to the CP model with zero phase,

Table 4.1 GDM representations of classical dispersion models.

Model	Susceptibility $\hat{\chi}(\omega)$	GDM coefficients
Debye	$\frac{\Delta\varepsilon\Gamma}{\Gamma - \iota\omega}$	$a_0 = 0,\ b_0 = 0,\ a_1 = \Delta\varepsilon\Gamma,\ b_1 = \Gamma$
Lorentz	$\frac{\Delta\varepsilon\Omega^2}{\Omega^2 - \iota\Gamma\omega - \omega^2}$	$a_0 = \Delta\varepsilon\Omega^2,\ b_0 = \Omega^2,\ a_1 = 0,\ b_1 = \Gamma$
Drude	$\frac{\Omega^2}{-\iota\Gamma\omega - \omega^2}$	$a_0 = \Omega^2,\ b_0 = 0,\ a_1 = 0,\ b_1 = \Gamma$
Sellmeier	$\frac{\Delta\varepsilon\Omega^2}{\Omega^2 - \omega^2}$	$a_0 = \Delta\varepsilon\Omega^2,\ b_0 = \Omega^2,\ a_1 = 0,\ b_1 = 0$
CP	$\frac{\Delta\varepsilon\Omega}{2}\left[\frac{e^{\iota\phi}}{\Omega - \iota\Gamma - \omega} + \frac{e^{-\iota\phi}}{\Omega + \iota\Gamma + \omega}\right]$	$a_0 = \Delta\varepsilon\Omega(\Omega \cos\phi - \Gamma \sin\phi), b_0 = \Omega^2 + \Gamma^2$ $a_1 = -\Delta\varepsilon\Omega \sin\phi, b_1 = 2\Gamma$

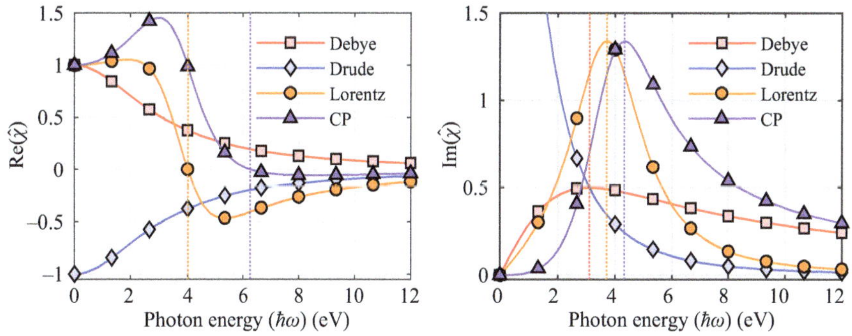

Figure 4.1 Typical form of the susceptibility, $\hat{\chi}(\omega)$, for classical dispersion models as a function of photon energy $\hbar\omega$. The model parameters are chosen as follows. Debye: $\Delta\varepsilon = 1, \Gamma = 3.1$ eV; Drude: $\Omega = \Gamma = 3.1$ eV; Lorentz: $\Delta\varepsilon = 1, \Omega = 4$ eV$, \Gamma = 3.1$ eV; and CP: $\Delta\varepsilon = 1.1716$ eV, $\Omega = 3.6815$ eV, $\Gamma = 1.55$ eV$, \phi = -\pi/4$ (see Table 4.1 for the definitions of the models and parameters).

so we can compare the lineshapes of two oscillators with $\phi = 0$ and $\phi = -\pi/4$. We can see that nonzero phase breaks the symmetry of a generic Lorentz oscillator.

In reference to the curves in Figure 4.1, we observe that the transition from unity static value to zero is different for all models. The Lorentz and CP curves exhibit the resonance located roughly at $\omega \approx \sqrt{b_0} = 4$ eV, and their real parts change sign at the crossing point. On the other hand, real part is positive and decreases monotonically from 1 to 0 for the Debye model. For the Drude model, its real part resembles a flipped Debye model, while its imaginary part tends to infinity as $\omega \to 0$ due to the conductivity term (see (4.20)). The crossing point frequencies, ω_c, are marked by the vertical dashed lines in the plots of $\Re(\hat{\chi})$ on the left, while the absorption peak frequencies, ω_a, are marked by the vertical dashed lines in the plots of $\Im(\hat{\chi})$ on the right.

Note that, in agreement with the results of the previous section, since the conditions $a_i, b_i, a_0 b_1 - a_1 b_0 \geq 0$ are satisfied for all models, the imaginary part of $\hat{\chi}$ is non-negative, corresponding to lossy or lossless materials (with no gain). The functional forms shown in Figure 4.1 are the basis functions, which, when added together can provide an approximate fit to the data of an arbitrary dielectric function $\hat{\varepsilon}(\omega)$. It is typical, but not necessary, to require the individual GDM terms to have non-negative imaginary parts so that the imaginary part of the sum is non-negative. For example, terms with negative imaginary part arise in Drude decomposition (4.20) and approximations based on several CP terms. In such cases, special care should be taken to analyze stability and convergence of the numerical schemes in time domain.

Debye model. For a dielectric medium and high optical frequencies, the electronic polarization is a dominant effect in the presence of the external electric field. The material response can be explained by (negatively charged) electrons and (positively charged) nuclei pulled in different direction by the electric field, and therefore, a dipole moment is induced in the dielectric. For static fields, the displacement is characterized by the static permittivity ε_s. In the absence of the external field, the simplest model suggests that the dielectric media restores exponentially with a constant relaxation time τ. This simple TD model for the dielectric response is called *Debye relaxation* [23]. In the frequency domain, the corresponding dielectric function is given by

$$\hat{\varepsilon}_{\text{Debye}}(\omega) = \varepsilon_\infty + \frac{\Delta\varepsilon}{1 - \iota\omega\tau} = \varepsilon_\infty + \frac{\Delta\varepsilon\Gamma}{\Gamma - \iota\omega}, \tag{4.15}$$

where ε_∞ is the high-frequency dielectric constant, $\Delta\varepsilon = \varepsilon_\infty - \varepsilon_s$. If we introduce the decay rate as $\Gamma = \tau^{-1}$ then in the time domain, (4.15) implies

$$\varepsilon_{\text{Debye}}(t) = \varepsilon_\infty \delta(t) + \Delta\varepsilon\Gamma e^{-\Gamma t} U(t). \tag{4.16}$$

The measured dispersion of dielectric materials is typically more complicated than a single relaxation system. In fact, the dielectric relaxation is a many particle phenomenon, connected to re-adjustments of atoms, molecules, ions, defects, etc. This observation suggests a generalization of the simple Debye relaxation model leading to a *multipole Debye model* that is a superposition of several Debye terms, with an added DC conductivity term that reflects transport properties,

$$\hat{\varepsilon}_{\text{MDebye}}(\omega) = \varepsilon_\infty - \frac{\sigma}{\iota\varepsilon_0\omega} + \sum_{i\in I}\frac{\Delta\varepsilon_i}{1-\iota\omega\tau_i}. \tag{4.17}$$

The quantum mechanical mechanism behind DC conductivity and its connection to the relaxation constant (Debye–Stokes–Einstein equation) is discussed, for example, in [24] and the references therein. In terms of the GDM model, conductivity is an independent parameter that can be obtained from complex permittivity data and the Kramers–Kronig relation. The multipole Debye model can be viewed as a first-order approximation of the GDM model under the assumption of small frequencies compared to the relaxation times, i.e. $\omega^2 \ll \omega\tau^{-1}$. In this approximation, second-order terms reduce to a sum of first-order ones leading to the model in (4.17).

Lorentz model. For dielectric and metallic media, the dispersion at optical frequencies, assumed in the *Lorentz model*, is based on an oscillatory motion of an electron in the presence of external field. The derivation starts with the equation of motion for a single electron of charge $-q_\text{e}$ and mass m_e, bound to a nucleus of charge $+q_\text{e}$ fixed in position inside the material. Then, the oscillatory behavior of the electron is described by Newton's second law [25],

$$\partial_t^2\mathbf{x}_{\text{e},i} + \Gamma\partial_t\mathbf{x}_{\text{e},i} + \Omega^2\mathbf{x}_{\text{e},i} = -\frac{q_\text{e}}{m_\text{e}}\mathbf{E}\ , \tag{4.18}$$

where $\mathbf{x}_{\text{e},i}$ is the position of the ith electron, Γ is the rate of damping, and Ω gives the frequency of the (undamped) oscillation associated with the natural restoring force. With a Fourier transform, we solve this equation for $\mathbf{x}_{\text{e},i}$, and then obtain the electric dipole moment of the charge q_e displaced at $\mathbf{x}_{\text{e},i}$ as $\mathbf{p}_{\text{e},i} = q_\text{e}\mathbf{x}_{\text{e},i}$. The collective oscillations of N electrons in a material volume V described by (4.18) induce a macroscopic polarization density $\mathbf{P} = N\mathbf{p}_\text{e}$, where $\mathbf{p}_\text{e} = \sum_i\mathbf{p}_{\text{e},i}/V$ is the volume averaged dipole moment, and the *plasma frequency*, Ω_p, is defined by $\Omega_\text{p}^2 = Nq_\text{e}^2/(m_\text{e}\varepsilon_0)$. The corresponding macroscopic relative permittivity is thus given by

$$\hat{\varepsilon}_{\text{Lorentz}}(\omega) = \varepsilon_\infty + \frac{\Omega_\text{p}^2}{\Omega^2 - \iota\omega\Gamma - \omega^2} = \varepsilon_\infty + \frac{\Delta\varepsilon\Omega^2}{\Omega^2 - \iota\omega\Gamma - \omega^2}. \tag{4.19}$$

As in the Debye model, we can introduce a static jump in the permittivity (as $\omega \to 0$) given here by $\Delta\varepsilon = \Omega_\text{p}^2/\Omega^2$. The corresponding GDM coefficients are listed in Table 4.1.

Drude model. The *Drude model* describes the motion of free electrons, and it can be viewed as the degenerate case of the Lorentz model with negligible restoring force (i.e. $\Omega = 0$). In this case the motion is not oscillatory, but rather can be viewed as a superposition of terms representing conductivity and a Debye relaxation, i.e.

$$\hat{\varepsilon}_{\mathrm{Drude}}(\omega) = \varepsilon_\infty - \frac{\Omega_{\mathrm{p}}^2}{\iota\omega\Gamma + \omega^2} = \varepsilon_\infty - \frac{\Omega_{\mathrm{p}}^2}{\iota\omega\Gamma} - \frac{\Omega_{\mathrm{p}}^2\Gamma^{-2}}{1 - \iota\omega\Gamma^{-1}}. \tag{4.20}$$

In the Drude case, we also use the notation $\Omega = \Omega_{\mathrm{p}}$ since there is no ambiguity.

Sellmeier model. The *Sellmeier model* [26] is commonly used to approximate the response of transparent media (e.g., glasses), and again can be viewed as a special case of the Lorentz model but now with negligible damping ($\Gamma = 0$). The resulting model in the frequency domain is given by

$$\hat{\varepsilon}_{\mathrm{Sellmeier}}(\omega) = \varepsilon_\infty + \frac{\Delta\varepsilon\Omega^2}{\Omega^2 - \omega^2}. \tag{4.21}$$

Note that the dielectric function (4.21) is dispersive but it has no imaginary part contradicting analytical properties (4.4) and representation (4.5). The correct analytical form of the Sellmeier model, derived as Fourier transform of $\varepsilon(t) = \varepsilon_\infty\delta(t) + \Delta\varepsilon\Omega\sin(\Omega t)U(t)$, includes a pair of imaginary delta function terms $\pm\frac{\pi\iota}{2}\Delta\varepsilon\Omega\delta(\omega \mp \Omega)$ corresponding to a pair of real poles $\omega = \pm\Omega$. The parameters of the Sellmeier approximations in the experiment-based dispersion models are typically retrieved from the absorption bands outside the transparent region with the Kramers-Kronig formula. However, in the simplified analysis, the remote absorption terms are often neglected, and the Sellmeier models are obtained by fitting the real part of permittivity only in the transparent region.

In addition to the first-order models, *multipole* second-order models are used to approximate actual macroscopic optical dispersion of some materials. The response of metals, for example, is often described by a Drude term and several Lorentz terms [27]. Dispersion of transparent media, on the other hand, is usually approximated as a superposition of several Sellmeier terms with all poles being outside the target range [26]. In both cases, for metallic and transparent media, the 'theoretical' high-frequency limit should be $\varepsilon_\infty = 1$. However, the models are usually valid within a limited wavelength bandwidth, so this asymptotic value is commonly adjusted to improve the fit to the ellipsometry data.

CP model. The CP model arises in semiconductor modeling and describes the excitation of an optical CP in terms of the amplitude $\Delta\varepsilon$, energy gap Ω, broadening Γ, and phase ϕ, see [28]. Unlike higher-order models, the first-order CP model ($\mu_i = -1$ in [28]) is a rational function well suited for time domain modeling. It takes the form

$$\hat{\varepsilon}_{\mathrm{CriticalPoint}}(\omega) = \varepsilon_\infty + \frac{\Delta\varepsilon\Omega}{2}\left(\frac{\mathrm{e}^{\iota\phi}}{\Omega - \omega - \iota\Gamma} + \frac{\mathrm{e}^{-\iota\phi}}{\Omega + \omega + \iota\Gamma}\right). \tag{4.22}$$

A factor of 1/2 is included for convenience to match the scale used in the Lorentz model. Collecting the two first-order terms into a single second-order term gives

$$\hat{\varepsilon}_{\text{CriticalPoint}}(\omega) = \varepsilon_\infty + \Delta\varepsilon\Omega\frac{\Omega\cos\phi + (\iota\omega - \Gamma)\sin\phi}{(\Omega^2 + \Gamma^2) - 2\iota\omega\Gamma - \omega^2}. \tag{4.23}$$

This form indicates that the Lorentz model (4.19) is just a special case of the CP model when the phase is zero, $\phi = 0$. A minor difference is that the poles of the Lorentz model can be real (if $\Omega^2 < \Gamma^2/4$) or complex conjugates (if $\Omega^2 > \Gamma^2/4$), while the poles of the CP model always appear as complex conjugates (assuming $\Omega > 0$ and $\Gamma \in \mathbb{R}$).

Table 4.1 defines the GDM parameters a_0, a_1, b_0, and b_1 in terms of the physical parameters $\Delta\varepsilon$, Ω, ϕ, and Γ. The reverse conversion from GDM to physical parameters is

$$\begin{aligned}
a &= \frac{\sqrt{a_0^2 - a_0 a_1 b_1 + a_1^2 b_0}}{\delta} \quad \text{(amplitude)},\\
\varphi &= \iota \ln \frac{a_0 + a_1(-\gamma + \iota\delta)}{a\delta} \quad \text{(phase)},\\
\gamma &= \frac{b_1}{2} \quad \text{(damping constant)},\\
\delta &= \sqrt{b_0 - \gamma^2} \neq 0 \quad \text{(oscillation frequency)}.
\end{aligned}$$

These parameters define properties of a damped phase-relaxed oscillator in the time domain. The corresponding Fourier pair for the susceptibility reads

$$\hat{\chi}(\omega) = \frac{a_0 - \iota\omega a_1}{b_0 - \iota\omega b_1 - \omega^2} \rightarrow \chi(t) = a e^{-\gamma t}\sin(\delta t - \varphi)U(t). \tag{4.24}$$

It has been noted in recent papers [29–31] that the dispersion in the noble metals can be described more accurately and with half the number of oscillators by using a multipole Drude-CP model instead of a classical Drude–Lorentz model. The reason for this is that while the intraband transitions in metals are modeled classically by the Drude term, the interband transition lines add an asymmetric shape to the dispersion curve that can only be fitted efficiently after adding a nonzero phase to the Lorentz model. Therefore, if only Lorentz terms are used, then artificial oscillators are needed, one for each (left and right) slope of the resonance, resulting in twice the number of oscillator terms. Later, accurate fits of Drude-CP models were reported for the ellipsometry data of other noble metals, graphene, and PCM [11,31,32]. These results provide some experimental justification for adding non-zero phase to the classical Lorentz model.

The aforementioned classical models incorporate some simplified physics of light-matter interactions taken from the first principles using idealized quantum mechanical and semi-classical models. In reality, there are multiple dispersion mechanisms taking place in the material on an atomic level (dipoles, electrons, lattice, etc.), so that there are other resonances not necessarily caused by electron transitions. Also, each mechanism in real materials is usually described with multiple terms, e.g., atoms have multiple electron levels.

At the macroscopic level, major linear dispersion mechanisms can be characterized effectively through a superposition of simpler terms. However, each term should be physically justified by material properties, otherwise mathematical approximations with artificial terms may result in an ill-defined dispersion model. Using a Lorentz model instead of a CP model for an approximation of interband transitions in some metals, for example, can lead to an ill-defined model. As described in the literature and from the derivation of the GDM model, adding a phase parameter can be essential to approximate the optical response of a given material effectively with a minimal number of terms. As noted previously, Table 4.1 provides a list of the classical models discussed above and the corresponding coefficients of the GDM model.

4.1.5 GDM fits

The coefficients of different classic models can be chosen to fit given data for the real and imaginary parts of the dielectric function $\hat{\varepsilon}(\omega)$. The plots in Figure 4.2 show selected approximations known from the literature [27,29,30,33] to the experimental data of the refractive index of gold taken from Johnson and Christy [34]. Here, the real $\Re(n)$ and imaginary $\Im(n)$ parts of the refractive index are plotted as a function of the frequency in eV (ω [eV] $= \hbar\omega$ [rad/s]).

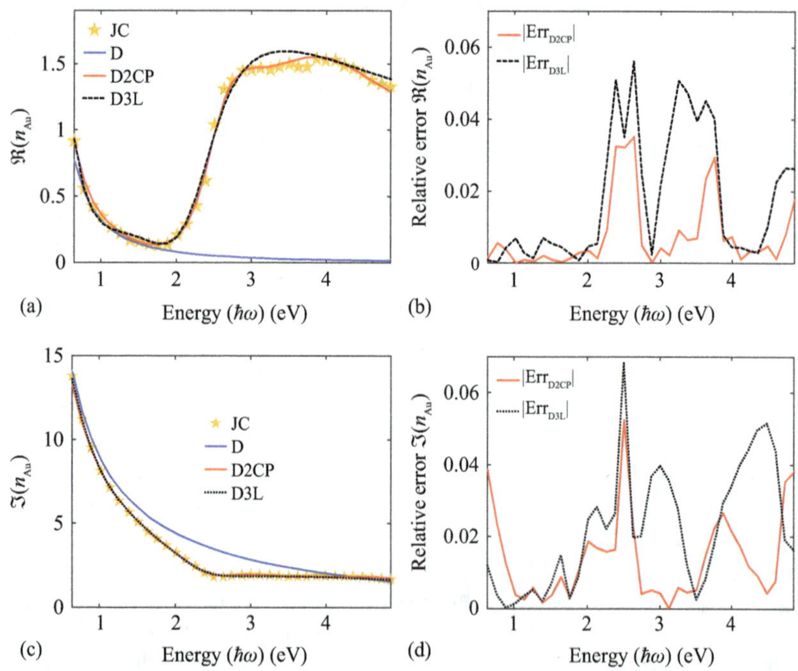

Figure 4.2 Comparison of models fitting Johnson and Christy data [34] for Au: the Drude model [33], the D3L (Drude + 3 Lorentz) model [27], and the D2CP (Drude + 2 CP) model [29,30].

Table 4.2 Parameters of the Drude—two critical points (D2CP) model of gold used in the simulations.

	ε_∞	$\Delta\varepsilon_i$	$\lambda_i = \frac{2\pi c_0}{\Omega_i}$, **nm**	$\nu_i = \frac{2\pi c_0}{\Gamma_i}$, **nm**	φ_i, **rad**
D	1.54	–	143	14,500	–
CP_1		2.54	470	1,900	$-\pi/4$
CP_2		2.20	325	1,060	$-\pi/4$

The Drude (D) model mostly recovers the trend in the $\Im(n)$, but completely misses the interband transitions in $\Re(n)$. The Drude–Lorentz (D3L) model does a better job of matching both $\Re(n)$ and $\Im(n)$, but does not capture the two interband transitions between 3 and 4 eV, despite having three oscillators in the model. The Drude + 2CP terms (D2CP) model fits both peaks in the interband region with only two oscillators. Note that the D2CP approximation of the Au dielectric function was obtained for the truncated wavelength range (below 1 μm, or equivalently above 1.24 eV), while the D3L model was fitted across the entire Johnson and Christy range, this explains the larger error below 1.2 eV. In general, the D3L model performs worse and misses important dispersion features in the interband region despite having more fitting parameters than D2CP (12 vs. 10). Furthermore, the D3L model is more computationally expensive than D2CP due to the additional oscillator. A detailed comparison of these two models is given in [31].

This comparison demonstrates why simpler models (e.g., multipole Debye, multiple Lorentz) should generally not be used for a modular approach in the general dispersive case, as they may lead to an unnecessarily large number of terms and artificial behaviors. GDM terms, on the other hand, are derived from general algebraic assumptions, thus with proper fitting techniques the GDM model usually fits the experimental data more accurately and with a shorter list of physically meaningful terms. In the case of gold, for example, the two CP terms stand for two asymmetric interband transitions, while a Drude term represents intraband features. Table 4.2 shows the parameters of the D2CP model that were used for the numerical simulations presented in Section 4.3.

4.2 Numerical implementation of the GDM model

The literature on the implementation of TD schemes for the dispersive Maxwell's equations, and for the FDTD method in particular, is quite extensive. The key papers that address different numerical schemes for different classical models can be found in [35–43], and the references therein. Most of the schemes in the time domain fall into the following two categories: (i) ADE schemes based on finite-difference approximations of the differential equations for the partial polarizations, and (ii) RC schemes based on recursive numerical integration of the convolution integral.

In this section, we describe three numerical approximations of dispersive Maxwell's equations in the time domain with a single universal GDM model. All three approximations are extensions of the classical Yee finite-difference TD (FDTD) scheme [3]. In the first of the three approximations, the Yee FDTD scheme is coupled to a second-order accurate ADE formulation of the GDM model. The second approach couples the Yee FDTD scheme to a second-order accurate approximation of the RC formulation. Finally, we show how the first two approximations can be described by a universal GDM-Yee scheme that treats either the ADE or RC formulations. Although we restrict attention to the Yee scheme, it is also possible to implement the GDM model within other FDTD schemes, as well as finite-element and finite-volume approximations.

4.2.1 Yee-based ADE GDM scheme

Maxwell's equations in first-order form with the ADE formulation of the GDM model are described by the system of equations in (4.7). Here, we present a second-order accurate approximation to these equations based on Yee's FDTD scheme for Maxwell's equations. We note that second and fourth-order accurate approximations of the corresponding system of equations for $\mathbf{E}(\mathbf{x}, t)$ in second-order form are also available, see [6].

Let $\mathbf{E}^n$ and $\mathbf{H}^{n+1/2}$ denote discrete approximations in space and time to the electric and magnetic fields at times $t^n = n\Delta t$ and $t^{n+1/2} = t^n + \Delta t/2$, respectively. On a Cartesian grid, $\mathbf{E}^n$ is located at cell edges, while $\mathbf{H}^{n+1/2}$ is positioned at cell faces following the usual construction of the Yee FDTD scheme on a staggered grid. This configuration of the grid functions is assumed with spatial indices suppressed for notational convenience. The second-order accurate Yee-ADE-GDM scheme can then be written compactly as

$$\varepsilon_0\varepsilon_\infty D_{+t}\mathbf{E}^n = \nabla_h^{(f)} \times \mathbf{H}^{n+1/2} - \sigma A_{+t}\mathbf{E}^n - D_{+t}\mathbf{P}^n, \tag{4.25a}$$

$$\mu_0 D_{+t}\mathbf{H}^{n+1/2} = -\nabla_h^{(e)} \times \mathbf{E}^n, \tag{4.25b}$$

$$\left[D_{+t} + b_{1,i}A_{+t}\right]\mathbf{P}_i^n = \varepsilon_0 a_{1,i}A_{+t}\mathbf{E}^n, \quad i \in I_1, \tag{4.25c}$$

$$\left[D_{+t}D_{-t} + b_{1,i}D_{0t} + b_{0,i}A_{+t}A_{-t}\right]\mathbf{P}_i^n = \varepsilon_0\left[a_{1,i}D_{0t} + a_{0,i}A_{+t}A_{-t}\right]\mathbf{E}^n, \quad i \in I_2. \tag{4.25d}$$

Here, D_{+t}, D_{-t} and D_{0t} are the standard forward, backward, and central difference operators [44], respectively, while A_{+t} and A_{-t} are the forward and backward averaging operators, all defined by

$$D_{+t}f^n \stackrel{\text{def}}{=} \frac{f^{n+1} - f^n}{\Delta t}, \quad D_{-t}f^n \stackrel{\text{def}}{=} \frac{f^n - f^{n-1}}{\Delta t}, \quad D_{0t}f^n \stackrel{\text{def}}{=} \frac{f^{n+1} - f^{n-1}}{2\Delta t},$$

$$A_{+t}f^n \stackrel{\text{def}}{=} \frac{f^{n+1} - f^n}{2}, \quad A_{-t}f^n \stackrel{\text{def}}{=} \frac{f^n + f^{n-1}}{2}, \quad A_{0t}f^n \stackrel{\text{def}}{=} \frac{f^{n+1} + f^{n-1}}{2}.$$

It follows that the composite operators $D_{+t}D_{-t}$ and $A_{+t}A_{-t}$ appearing in (4.25) are

$$D_{+t}D_{-t}f = \frac{f^{n+1} - 2f^n + f^{n-1}}{\Delta t^2}, \quad A_{+t}A_{-t}f^n = \frac{f^{n+1} + 2f^n + f^{n-1}}{4}.$$

The spatial operators, $\nabla_h^{(f)}\times$ and $\nabla_h^{(e)}\times$, in (4.25a) and (4.25b) denote discrete face-based and edge-based approximations to the curl operator, respectively, employed in the standard Yee scheme.

Equations (4.25a) and (4.25c) are centered in time at $t^{n+1/2}$, while the approximations in (4.25b) and (4.25d) are centered at t^n. This follows the staggered space-time construction used in the Yee FDTD scheme and leads to a second-order accurate approximation in time. The averaging operator $A_{+t}A_{-t}$ is used in (4.25d) so that the time-step restriction is determined by the usual CFL condition for the non-dispersive Yee scheme [45].

Note that an alternative scheme formed by replacing $A_{+t}A_{-t}$ with the central average A_{0t} in (4.25d) leads to a more restrictive time-step. For details on a comparison of different ADE approximations, and a stability analysis for classical models see [45]. Finally, we note that the approximations in (4.25) are point-wise implicit in the variables $\mathbf{E}^{n+1}$ and $\mathbf{P}_i^{n+1}$, $i \in I_1 \cup I_2$, and thus require the solution of a relatively small linear system at each grid point.

4.2.2 Yee-based recursive convolution GDM scheme

The system of equations in (4.8) describes the RC formulation of the GDM model. Here we present a second-order accurate approximation to these equations based on the Yee FDTD scheme for Maxwell's equations.

The form of the single-pole and double-pole GDM susceptibility terms in the frequency and time domains are given in Table 4.3, where $s = -\iota\omega$. For the double-pole term we assume that $b_0 > (b_1/2)^2$ so that there are complex conjugate poles. If $b_0 < (b_1/2)^2$, then there are two real poles (the case of an over-damped oscillator), and the double-pole term can be written as two single-pole terms. In the special case when $b_0 = (b_1/2)^2$ there is a double real pole, and we exclude this case in the discussion below.

A second-order accurate Yee-RC-GDM scheme takes the form

$$\varepsilon_0\varepsilon_\infty D_{+t}\mathbf{E}^n = \nabla_h^{(f)} \times \mathbf{H}^{n+1/2} - \sigma A_{+t}\mathbf{E}^n - D_{+t}\mathbf{P}^n, \tag{4.26a}$$

$$\mu_0 D_{+t}\mathbf{H}^{n+/2} = -\nabla_h^{(e)} \times \mathbf{E}^n, \tag{4.26b}$$

Table 4.3 GDM susceptibilities in the frequency and time domains.

$\widehat{\chi}(\omega)$	$\chi(t)$	**Notes**
$\frac{a_1}{b_1+s}$	$\alpha e^{\beta t}U(t) = a_1 e^{-b_1 t}U(t)$	$U(t)$: Heaviside function; $\alpha, \beta \in \mathbb{R}$
$\frac{a_0 + a_1 s}{b_0 + b_1 s + s^2}$	$\left(\alpha e^{\beta t} + \bar{\alpha} e^{\bar{\beta} t}\right) U(t)$	$b_0 > (b_1/2)^2, \delta \stackrel{\text{def}}{=} \sqrt{b_0 - (b_1/2)^2} \in \mathbb{R}$; $\alpha = \frac{a_0 + a_1\beta}{2\iota\delta}, \beta = -\frac{b_1}{2} + \iota\delta \in \mathbb{C}$

$$\mathbf{P}_i^{n+1} = \mathrm{e}^{\beta_i \Delta t} \mathbf{P}_i^n + \alpha_i \varepsilon_0 \Delta t \left(\theta_{0,i} \mathbf{E}^{n+1} + \theta_{1,i} \mathrm{e}^{\beta_i \Delta t} \mathbf{E}^n \right), \quad i \in I_1, \tag{4.26c}$$

$$\mathbf{Q}_i^{n+1} = \mathrm{e}^{\beta_i \Delta t} \mathbf{Q}_i^n + \alpha_i \varepsilon_0 \Delta t \left(\theta_{0,i} \mathbf{E}^{n+1} + \theta_{1,i} \mathrm{e}^{\beta_i \Delta t} \mathbf{E}^n \right), \quad i \in I_2, \tag{4.26d}$$

$$\mathbf{P}_i^{n+1} \stackrel{\text{def}}{=} 2\Re\left[\mathbf{Q}_i^{n+1}\right], \quad i \in I_2, \tag{4.26e}$$

$$\mathbf{P}^n \stackrel{\text{def}}{=} \sum_{i \in I_1 \cup I_2} \mathbf{P}_i^n. \tag{4.26f}$$

Here we have chosen to handle the conductivity term, $\sigma \mathbf{E}$, as in the Yee-ADE-GDM scheme (4.25); this term could also be treated with a RC-type approximation if desired. There are two free coefficients in the scheme, $\theta_{0,i}$ and $\theta_{1,i}$, and these are determined for accuracy and stability as discussed below.

First-order GDM term ($i \in I_1$). In this case, the susceptibility function $\hat{\chi}_i(\omega)$ is a single real pole which transforms to exponential decay in the time domain, $\chi_i(t) = \alpha_i \mathrm{e}^{\beta_i t}$ with real coefficients α_i and β_i. At first sight, evaluation of the convolution $\mathbf{P}_i(\mathbf{x}, t) = \varepsilon_0 \chi_i * \mathbf{E}(\mathbf{x}, t)$ would seem to imply that we need to know $\mathbf{E}(\mathbf{x}, t)$ for all past time values of $\mathbf{E}(\mathbf{x}, t)$. However, when $\chi_i(t)$ is of the form of an exponential, the convolution can be written in terms of a recurrence formula. This can be seen from the following manipulation of the convolution formula for $\mathbf{P}_i(\mathbf{x}, t + \Delta t)$. Starting from

$$\mathbf{P}_i(\mathbf{x}, t + \Delta t) = \varepsilon_0 \int_0^\infty \chi_i(\tau) \mathbf{E}(\mathbf{x}, t + \Delta t - \tau) \mathrm{d}\tau,$$

then replace τ by $\tau + \Delta t$,

$$\mathbf{P}_i(\mathbf{x}, t + \Delta t) = \varepsilon_0 \int_{-\Delta t}^\infty \chi_i(\tau + \Delta t) \mathbf{E}(\mathbf{x}, t - \tau) \mathrm{d}\tau,$$

then split the integral into two parts,

$$\mathbf{P}_i(\mathbf{x}, t + \Delta t) = \varepsilon_0 \int_0^\infty \chi_i(\tau + \Delta t) \mathbf{E}(\mathbf{x}, t - \tau) \mathrm{d}\tau + \varepsilon_0 \int_{-\Delta t}^\infty \chi_i(\tau + \Delta t) \mathbf{E}(\mathbf{x}, t - \tau) \mathrm{d}\tau.$$

Now, use $\chi_i(\tau + \Delta t) = \mathrm{e}^{\beta_i \Delta t} \chi_i(t)$ in the first integral and replace τ with $-\tau$ in the second to give the exact two-level recursion

$$\mathbf{P}_i(\mathbf{x}, t + \Delta t) = \mathrm{e}^{\beta_i \Delta t} \mathbf{P}_i(\mathbf{x}, t) + \varepsilon_0 \int_0^{\Delta t} \chi_i(\Delta t - \tau) \mathbf{E}(\mathbf{x}, t + \tau) \mathrm{d}\tau. \tag{4.27}$$

Note that identity (4.27) is derived for continuous fields (prior to any approximation). It indicates that $\mathbf{P}_i(\mathbf{x}, t + \Delta t)$ is determined from $\mathbf{P}_i(\mathbf{x}, t)$ and the finite time history of $\mathbf{E}(\mathbf{x}, t)$ for $t \in [t, t + \Delta t]$.

Table 4.4 Weights for first- and second-order RC schemes (4.28). Note that if $\beta = 0$ then the expressions should be treated as limit $\beta \to 0$.

θ_0	θ_1	Order	Notes
1	0	1	Forward-Euler
0	1	1	RRC (Rectangular [37]), or backward-Euler
0	$\frac{e^{\beta\Delta t} - 1}{\beta\Delta t} \approx 1$	1	PCRC (Piecewise-constant [39])
$\frac{1}{2}$	$\frac{1}{2}$	2	TRC (Trapezoidal [38])
$\frac{1-e^{-\beta\Delta t/2}}{\beta\Delta t} \approx \frac{1}{2}$	$\frac{e^{\beta\Delta t/2} - 1}{\beta\Delta t} \approx \frac{1}{2}$	2	PCRC2 (Piecewise-constant [40])
$\frac{e^{-\beta\Delta t} - 1 + \beta\Delta t}{(\beta\Delta t)^2} \approx \frac{1}{2}$	$\frac{e^{\beta\Delta t} - 1 - \beta\Delta t}{(\beta\Delta t)^2} \approx \frac{1}{2}$	2	PLRC (Piecewise-linear [41])

Next, we assume that the integral in (4.27) is approximated with a two-point quadrature, leading to a recursive update formula on $\mathbf{P}_i^{n+1}$

$$\mathbf{P}_i^{n+1} = e^{\beta_i\Delta t}\mathbf{P}_i^n + \varepsilon_0\Delta t a_i\left[\theta_{0,i}\mathbf{E}^{n+1} + \theta_{1,i}e^{\beta_i\Delta t}\mathbf{E}^n\right], \tag{4.28}$$

where the weights $\theta_{0,i}$ and $\theta_{1,i}$ are determined by the quadrature rule used and may depend on β_i and Δt. For example, using the trapezoidal approximation to the integral in (4.27) implies $\theta_{0,i} = \theta_{1,i} = 1/2$. Table 4.4 provides the list of weights for different quadrature rules including those conventionally used in the literature for RC-FDTD schemes: rectangular (RRC) [37], trapezoidal (TRC) [38], piecewise-constant of the first and second order (PCRC, PCRC2) [39,40], and piecewise-linear (PLRC) [41].

Assuming a two-point approximation to the integral, based only on values of $\mathbf{E}^n$ and $\mathbf{E}^{n+1}$, the highest order of accuracy is two. In this chapter, the polarization equation (4.27) is coupled to the *second-order* Yee's FDTD scheme. To maintain the order of accuracy it is best to use second-order RC schemes, for example, TRC, PCRC2, and PLRC. Then the coupled RC-FDTD Yee's scheme will have global second order of accuracy. The same approach may be applied to derive higher-order RC schemes in time domain, e.g., see GDM implementation for DGTD and high-order FDTD solvers [5,6].

Second-order GDM term ($i \in I_2$). In this case, the frequency domain susceptibility $\hat{\chi}_i(\omega)$ is a sum of two complex-conjugate poles

$$\chi_i(t) = a_i e^{\beta_i t} + \bar{a} e^{\bar{\beta}_i t} \stackrel{\text{def}}{=} \chi_i^{(\text{cp})} + \bar{\chi}_i^{(\text{cp})},$$

where, as opposed to the case of a single real pole ($i \in I_1$), the parameters a_i and β_i are complex. The complex coefficients a_i, β_i characterize the harmonic behavior of $\chi_i(t)$, which is essentially a phase-relaxed damped oscillator. This is to be compared to the non-harmonic case of a real pole representing an exponential decay in time domain, see Table 4.3.

Letting $\mathbf{Q}_i = \varepsilon_0 \chi_i^{(\mathrm{cp})} * \mathbf{E}$ (note that $\mathbf{Q}_i$ is a complex-valued function) and following the steps above for the susceptibility in the single-pole case gives

$$\mathbf{Q}_i(\mathbf{x}, t+\Delta t) = \mathrm{e}^{\beta_i \Delta t}\mathbf{Q}_i(\mathbf{x}, t) + \varepsilon_0 \int_0^{\Delta t} \chi_i^{(\mathrm{cp})}(\Delta t - \tau)\mathbf{E}(\mathbf{x}, t+\tau)\mathrm{d}\tau, \tag{4.29}$$

which leads to the approximation

$$\begin{aligned} \mathbf{Q}_i^{n+1} &= \mathrm{e}^{\beta_i \Delta t}\mathbf{Q}_i^n + \varepsilon_0 \Delta t \alpha_i \left[\theta_{0,i}\mathbf{E}^{n+1} + \theta_{1,i}\mathrm{e}^{\beta_i \Delta t}\mathbf{E}^n\right], \\ \mathbf{P}_i^{n+1} &\stackrel{\mathrm{def}}{=} 2\Re\left[\mathbf{Q}_i^{n+1}\right]. \end{aligned} \tag{4.30}$$

The formula for $\mathbf{Q}_i^{n+1}$ in (4.30) can be implemented without complex arithmetic, if desired, by separating its real and imaginary parts, and then using these two real-valued formulas. Furthermore, these two formulas involving time levels $n+1$ and n could be combined into a single three-level update for $\mathbf{P}^{n+1}$ involving time-levels $n+1$, n, and $n-1$ of $\mathbf{E}$ and $\mathbf{P}$; see Section 4.2.3 and the high-order RC-FDTD implementation in [2] for details.

4.2.3 Yee-based universal GDM scheme

Finally, we note that the second-order accurate Yee-based scheme for the GDM model using the ADE formulation (4.25) and RC formulation (4.26) can both be written in the universal form

$$\varepsilon_0 \varepsilon_\infty D_{+t}\mathbf{E}^n = \nabla_h^{(f)} \times \mathbf{H}^{n+1/2} - \sigma A_{+t}\mathbf{E}^n - D_{+t}\mathbf{P}^n, \tag{4.31a}$$

$$\mu_0 D_{+t}\mathbf{H}^{n+1/2} = -\nabla_h^{(e)} \times \mathbf{E}^n, \tag{4.31b}$$

$$\mathbf{P}_i^{n+1} = d_{1,i}\mathbf{P}_i^n + \varepsilon_0\left(f_{0,i}\mathbf{E}^{n+1} + f_{1,i}\mathbf{E}^n\right), \quad i \in I_1, \tag{4.31c}$$

$$\mathbf{P}_i^{n+1} = d_{1,i}\mathbf{P}_i^n + d_{2,i}P_i^{n-1} + \varepsilon_0\left(f_{0,i}\mathbf{E}^{n+1} + f_{1,i}\mathbf{E}^n + f_{2,i}\mathbf{E}^{n-1}\right), \quad i \in I_2, \tag{4.31d}$$

$$\mathbf{P}^n \stackrel{\mathrm{def}}{=} \sum_{i \in I_1 \cup I_2} \mathbf{P}_i^n, \tag{4.31e}$$

where the coefficients $d_{m,i}, f_{m,i} \in \mathbb{R}$ depend on the particular choice of an ADE or RC approximation, and are shown in Table 4.5.

The generalization (4.31) makes it convenient to implement many different ADE and RC schemes with the same code, where a conditional operator sets the values of $d_{m,i}$ and $f_{m,i}$ depending on the specific scheme. This is contrary to conventional approaches where each scheme requires a separate implementation, thus multiplying the amount of code by the number of schemes.

Also, while the RC implementation (4.26) relies on complex conjugation and does not support the over-damped case with real poles $\left(b_1 \le b_0^2/4\right)$, the universal implementation (4.31) operates with real coefficients and functions and works for any generic second-order GDM term without this restriction. To make it work for an over-damped case, one has to change the f_i-formulas in Table 4.5 to a more

Table 4.5 Coefficients $d_{m,i}$ and $f_{m,i}$ for the universal implementation of ADE and RC schemes (4.31).

	$\chi(\omega) = \frac{a_1}{b_1 - \iota\omega}$		$\chi(\omega) = \frac{a_0 - \iota\omega a_1}{b_0 - \iota\omega b_1 - \omega^2}$	
	ADE	**RC**	**ADE**	**RC**
d_0	$2 + b_1\Delta t$	–	$4 + 2b_1\Delta t + b_0(\Delta t)^2$	–
d_1	$\frac{2 - b_1\Delta t}{d_0}$	$e^{\beta\Delta t}$	$\frac{8 - 2b_0(\Delta t)^2}{d_0}$	$2e^{-\gamma\Delta t}\cos\delta\Delta t$
d_2	0	0	$\frac{-4 + 2b_1\Delta t - b_0(\Delta t)^2}{d_0}$	$-e^{-2\gamma\Delta t}$
f_0	$\frac{a_1\Delta t}{d_0}$	$\alpha\Delta t\theta_1$	$\frac{2a_1\Delta t + a_0(\Delta t)^2}{d_0}$	$2\Delta t\Re(\alpha\theta_1)$
f_1	$\frac{a_1\Delta t}{d_0}$	$\alpha\Delta t e^{\beta t}\theta_0$	$\frac{2a_0(\Delta t)^2}{d_0}$	$2\Delta t\Re\left(\alpha e^{\beta\Delta t}\theta_0 - \alpha e^{\bar{\beta}\Delta t}\theta_1\right)$
f_2	0	0	$\frac{-2a_1\Delta t + a_0(\Delta t)^2}{d_0}$	$2d_2\Delta t\Re(\alpha\theta_0)$

general form, assuming that two exponential poles $\alpha e^{\beta t}$ and $\widetilde{\alpha} e^{\widetilde{\beta} t}$ are not necessarily complex conjugates, but may be both real. To do this, we substitute the real part operator with a sum $2\Re(f(\alpha,\beta)) = f(\alpha,\beta) + f\left(\widetilde{\alpha},\widetilde{\beta}\right)$, where explicit formulas for both poles are

$$\alpha = \frac{a_0 + a_1\beta}{2\iota\delta}, \quad \beta = \frac{b_1}{2} + \iota\delta, \quad \delta = \sqrt{b_0 - b_1^2/4}, \tag{4.32}$$

$$\widetilde{\alpha} = \frac{a_0 + a_1\widetilde{\beta}}{-2\iota\delta}, \quad \widetilde{\beta} = \frac{b_1}{2} - \iota\delta. \tag{4.33}$$

If $\left(b_1 \geq b_0^2/4\right)$, then δ is real and pairs of $\alpha, \widetilde{\alpha}$ and $\beta, \widetilde{\beta}$ are complex conjugate, following the standard assumption for the RC scheme and Table 4.3. If $\left(b_1 < b_0^2/4\right)$, then δ is pure imaginary and all four coefficients $\alpha, \widetilde{\alpha}, \beta, \widetilde{\beta}$ are real. In both cases, Table 4.5 will produce correct real coefficients for a universal scheme without restriction on b_0, b_1.

Furthermore, as shown in [2], the universal implementation (4.31) can be written in a compact form that has less floating-point operations (FLOPs) and uses two real recursive accumulators Ψ_i and Φ_i.

The comparison of the three implementations – complex RC (4.26), universal (4.31), and compact universal [2] – is summarized in Table 4.6. First, we count the number of elementary operations, assuming that complex multiplication and addition take 6 and 2 FLOPs, respectively, with the universal scheme taking 30% more FLOPs and the RC formulation taking twice more FLOPs than the compact version. In terms of storage, all three implementations require storing two numerical arrays of polarization terms (either the real and imaginary parts of $\mathbf{Q}_i^n$, or two time levels of a real-valued polarization $\mathbf{P}_i^n$ and $\mathbf{P}_i^{n-1}$, or two real-valued accumulators Ψ_i^n and

Table 4.6 Comparison of the GDM implementations (complex numbers are in boldface).

Scheme	FLOPs	Storage	Notes
Complex RC (4.26): $\mathbf{Q}^{n+1} = \mathbf{a}\mathbf{Q}^n + \mathbf{b}E^{n+1} + \mathbf{c}E^n$	14	$\Re(\mathbf{Q}^n)$, $\Im(\mathbf{Q}^n)$ E^{n+1}, E^n	RC methods only Must be $b_0 > b_1^2/4$
Universal (4.31): $P^{n+1} = aP^n + bP^{n-1} + cE^{n+1} + dE^n + eE^{n-1}$	9	P^n, P^{n-1} E^{n+1}, E^n, E^{n-1}	ADE and RC Any b_0, b_1
Compact universal eq. (17) in [2]: $\Psi^{n+1} = a\Psi^n + bE^n + \Phi^n$ $\Phi^{n+1} = c\Psi^n + dE^n$	7	Ψ^n, Φ^n E^n	ADE and RC Any b_0, b_1

Φ_i^n). Also, organizing computation without additional storage is more convenient in the compact universal formulation since the update formulas involve only one time-step of the electric field, $\mathbf{E}^n$. Note that on the contrary to both universal schemes, the RC formulation does not support ADE schemes and over-damped case of the oscillator $\left(b_0 < b_1^2/4\right)$.

4.3 Numerical results

In this section, we present some examples of numerical results that illustrate second-order accurate convergence of the ADE and RC schemes. The problem of interest concerns the one-dimensional propagation of light normally incident on a thin 50 nm gold slab, ($h \stackrel{\text{def}}{=} 50$ nm). The dispersion properties of gold are modeled with the D2CP model, see Table 4.2. For illumination we use a short Gaussian pulse with normal incidence whose spectrum is broad enough to cover the frequency range of interest, 0.6 eV to 6 eV (or 200 nm to 2,000 nm).

The modulated Gaussian pulse soft source term is taken of the form

$$E_{\text{src}}(t) = \exp\left[-\frac{(t-t_0)^2}{2\sigma_0^2}\right]\sin\omega_0 t$$

with $\omega_0 = 2\pi c_0/(366\text{ nm})$, $\sigma_0 = 1.3$ fs, $t_0 = 11$ fs. This source generates the incident field $E_{\text{inc}}(t,x) = E_{\text{src}}(t-(x-x_{\text{src}})/c_0)$, where x_{src} indicates the location of the source in front of the film.

Let $\hat{E}(x_m,\omega)$ denote the Fourier transform of the computed solution (total field) at a probe location $x = x_m$. Similarly, let $\hat{E}_{\text{inc}}(x_m,\omega)$ denote the Fourier transform of the computed incident field. The reflection $r(\omega)$ and transmission $t(\omega)$

coefficients are then determined from the computed solution sampled at two probe locations using

$$r_{\text{num}}(\omega) = \frac{\hat{E}(x_a, \omega) - \hat{E}_{\text{inc}}(x_a, \omega)}{\hat{E}_{\text{inc}}(x_a, \omega)}, \quad t_{\text{num}}(\omega) = \frac{\hat{E}(x_b, \omega)}{\hat{E}_{\text{inc}}(x_a, \omega)},$$

where x_a and x_b are the coordinates of the film edges. The numerically computed reflection and transmission coefficients are verified with the analytical formulas

$$r(\omega) = \frac{(\hat{\varepsilon} - 1)\sinh\gamma}{2\hat{\varepsilon}\cosh\gamma - (\hat{\varepsilon} + 1)\sinh\gamma}, \quad t(\omega) = \frac{2}{2\hat{\varepsilon}\cosh\gamma - (\hat{\varepsilon} + 1)\sinh\gamma},$$

where $\gamma = \imath kh$, $k = \omega\sqrt{\hat{\varepsilon}}/c_0$, $h = x_b - x_a$ is film thickness, and $\hat{\varepsilon} = \hat{\varepsilon}(\omega)$ is the relative permittivity of the film taken from the D2CP model.

In Figure 4.3, we compare exact and numerical values for the spectra of the reflectance and transmittance, $R(\omega) = |r(\omega)|^2$ and $T(\omega) = |t(\omega)|^2$ (Figure 4.3(a)), as well as complex coefficients of reflection and transmission, r and t (Figure 4.3(b), (c)). Acceptable agreement (less than 1% error) in magnitude and phases is achieved with 10 numerical cells per film thickness h, i.e., $\Delta x = h/10$ (in these simulations we used the ADE scheme).

Finally, we verify the theoretically predicted second-order convergence of one ADE scheme and three RC schemes. In Figure 4.4, we plot the relative numerical error in the ratio t/r

$$\text{Err} = \left|\frac{t/r - t_{\text{num}}/r_{\text{num}}}{t/r}\right| \times 100\%$$

computed using mesh spacings Δx equal to $h/10$, $h/20$, $h/40$, and $h/80$, Figure 4.4(a). When the mesh spacing decreases by a factor of two, it is seen that the relative error decreases by approximately a factor four, as expected for a second-order accurate scheme where the error is proportional to Δx^2. The actual computed ratio is shown in Figure 4.4(b) for the mesh size going from $h/10$ to

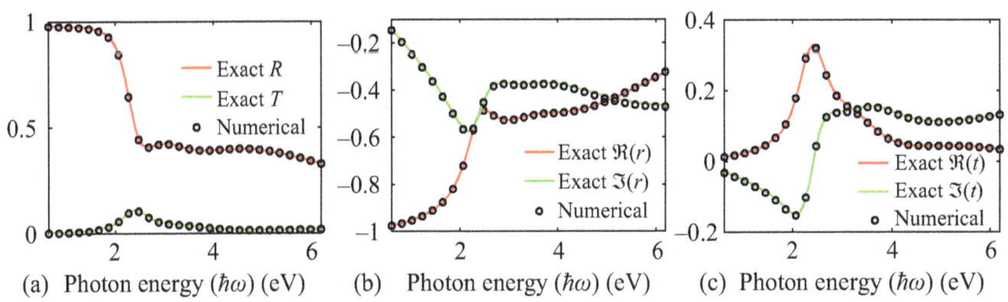

Figure 4.3 Comparison of exact and numerical spectra for scattering from a 50-nm gold slab: (a) reflectance and transmittance, R and T, (b) complex reflection coefficient, r, (c) complex transmission coefficient, t.

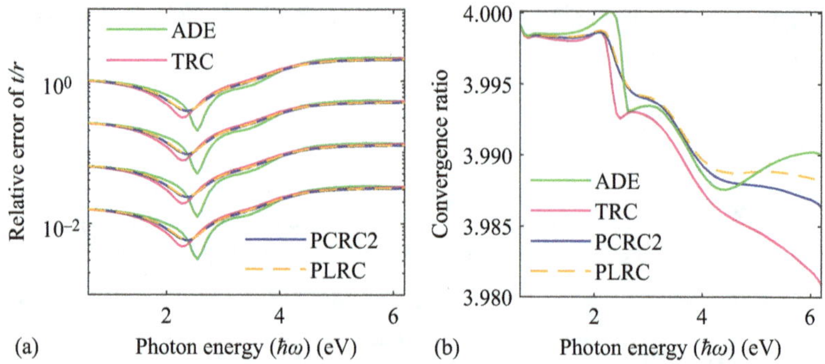

Figure 4.4 (a) Convergence of the relative error of the ratio t/r for spatial mesh spacings Δx equal to $h/10$, $h/20$, $h/40$, and $h/80$, where h is the film thickness, (b) ratio of the errors for $\Delta x = h/10$ to $\Delta x = h/20$ (a factor of 4 implies second-order accuracy).

$h/20$. The second-order accurate convergence of the Yee-based GDM approach is thus confirmed for all shown ADE and RC schemes.

4.4 Conclusions

This chapter presented the critical derivations and detailed analysis of the recently proposed GDM model. The model aims at efficient and universal numerical modeling of arbitrary dispersion in engineered optical materials using TD solvers. The proposed modular approach does not depend on the choice of a numerical TD scheme (FDTD, FVTD, or FETD) and incorporates the major classical dispersion models (Debye, Drude, Lorentz, Sellmeier) along with more complex general terms. Using the standard Yee scheme, this chapter presented a detailed mathematical survey on the vital components of the GDM implementation, validation, numerical properties, and parameters constraints that has not been published before in full detail.

The most recent progress in this direction is focused on the advanced implementation of the GDM model beyond the standard FDTD methods, such as the higher-order FDTD and DGTD schemes [5,6]. The main advantage is that the GDM formalism allows for arbitrary dispersion and can be interfaced directly with existing material databases that are maintained along with the recent research in nanophotonics and optical materials science.

References

[1] Prokopeva L, Borneman J, Kildishev A. Time-domain modeling of metal-dielectric nanostructures characterized by a set of single-pole dispersion

terms. In: Digests of the 2010 14th Biennial IEEE Conference on Electromagnetic Field Computation. IEEE; 2010. p. 1.

[2] Prokopeva L, Borneman J, Kildishev A. Optical dispersion models for time-domain modeling of metal-dielectric nanostructures. IEEE Transactions on Magnetics. 2011;47(5):1150–1153.

[3] Yee K. Numerical solution of initial boundary value problems involving Maxwell's equations in isotropic media. IEEE Transactions on Antennas and Propagation. 1966;14(3):302–307.

[4] Prokopeva L, Lebedev A, Fedoruk M, *et al.* A Generalized Dispersive Material Model for FVTD Method. In: 27th Annual Review of Progress in Applied Computational Electromagnetics. ACES; 2011. pp. 1–6.

[5] Ren Q, Bao H, Campbell SD, *et al.* Continuous-discontinuous Galerkin time domain (CDGTD) method with generalized dispersive material (GDM) model for computational photonics. Optics Express. 2018;26(22):29005–29016.

[6] Angel JB, Banks JW, Henshaw WD, *et al.* A high-order accurate scheme for Maxwell's equations with a generalized dispersive material model. Journal of Computational Physics. 2019;378:411–444.

[7] Chun K, Kim H, Kim H, *et al.* GMES: a Python package for solving Maxwell's equations using the FDTD method. Computer Physics Communications. 2013;184(4):1272–1279.

[8] Fang J, Prokopeva L, Alam M, *et al.* Numerical studies on randomly distributed gold nano-net structures. In: 28th Annual Review of Progress in Applied Computational Electromagnetics. ACES; 2012. pp. 1028–1033.

[9] Thoreson M, Fang J, Kildishev A, *et al.* Fabrication and realistic modeling of three-dimensional metal-dielectric composites. Journal of Nanophotonics. 2011;5(1):051513.

[10] Prokopeva L, Emani N, Boltasseva A, *et al.* Time domain modeling of tunable response of graphene. In: CLEO: QELS_Fundamental Science. Optical Society of America; 2013. p. QTh1A–8.

[11] Prokopeva LJ, Liberman V, Chou J, *et al.* Time domain modeling of bi-anisotropic media and phase change materials with generalized dispersion (Conference Presentation). In: Metamaterials, Metadevices, and Metasystems. International Society for Optics and Photonics (Vol. 11080); 2019. p. 1108006.

[12] Prokopeva LJ, Kudyshev ZA, Kildishev AV. Time Domain Modeling of Bi-Anisotropic Homogenized Media with Analytical Dispersion. In: 2018 Conference on Lasers and Electro-Optics (CLEO). IEEE; 2018. pp. 1–2.

[13] Prokopeva L, Trieschmann J, Klar T, *et al.* Numerical modeling of active plasmonic metamaterials. In: Optical Complex Systems: OCS11. vol. 8172. International Society for Optics and Photonics; 2011. p. 81720B.

[14] Azzam SI, Fang J, Liu J, *et al.* Exploring time-resolved multiphysics of active plasmonic systems with experiment-based gain models. Laser & Photonics Reviews. 2019;13(1):1800071.

[15] Landau LD, Bell J, Kearsley M, *et al.* Electrodynamics of Continuous Media. vol. 8. Elsevier; 2013.

[16] Prokopeva L, Borneman J, Kildishev A. Time-domain modeling of metal-dielectric nanostructures. In: Quantum Electronics and Laser Science Conference. Optical Society of America; 2010. p. JWA14.
[17] Prokopeva L, Kildishev A, Fang J, *et al.* Nanoplasmonics FDTD Simulations Using a Generalized Dispersive Material Model. In: 27th Annual Review of Progress in Applied Computational Electromagnetics. ACES; 2011. pp. 963–968.
[18] Cole KS, Cole RH. Dispersion and absorption in dielectrics I. alternating current characteristics. The Journal of Chemical Physics. 1941;9(4): 341–351.
[19] Davidson DW, Cole RH. Dielectric relaxation in glycerol, propylene glycol, and *n*-propanol. The Journal of Chemical Physics. 1951;19(12): 1484–1490.
[20] Havriliak S, Negami S. A complex plane analysis of α-dispersions in some polymer systems. Journal of Polymer Science Part C: Polymer Symposia. 1951;14(1):99–117.
[21] Rekanos IT, Yioultsis TV. Approximation of Grünwald-Letnikov fractional derivative for FDTD modeling of cole-cole media. IEEE Transactions on Magnetics. 2014;50(2):181–184.
[22] Mescia L, Bia P, Caratelli D. Fractional derivative based FDTD modeling of transient wave propagation in Havriliak-Negami media. IEEE Transactions on Microwave Theory and Techniques. 2014;62(9):1920–1929.
[23] Debye P. Berichte der deutschen physikalischen Gesellschaft. 1913. New York: Interscience Publishers; 1954.
[24] Power G, Vij J, Johari G. Structure-dependent conductivity and relaxation time in the Debye–Stokes–Einstein equation. The Journal of Physical Chemistry B. 2007;111(38):11201–11208.
[25] Wooten F. Optical properties of solids. American Journal of Physics. 1973;41(7):939–940.
[26] Sellmeier W. Zur Erklärung der abnormen Farbenfolge im Spectrum einiger Substanzen. Annalen der Physik und Chemie. 1871;219:272–282.
[27] Hao F, Nordlander P. Efficient dielectric function for FDTD simulation of the optical properties of silver and gold nanoparticles. Chemical Physics Letters. 2007;446(1–3):115–118.
[28] Etchegoin P, Kircher J, Cardona M. Elasto-optical constants of Si. Physical Review B. 1993;47(16):10292.
[29] Etchegoin PG, Le Ru E, Meyer M. An analytic model for the optical properties of gold. The Journal of Chemical Physics. 2006;125(16):164705.
[30] Etchegoin PG, Le Ru E, Meyer M. Erratum: "An analytic model for the optical properties of gold". The Journal of Chemical Physics. 2007;127(18):189901.
[31] Vial A, Laroche T. Comparison of gold and silver dispersion laws suitable for FDTD simulations. Applied Physics B. 2008;93(1):139–143.
[32] Prokopeva L, Kildishev A. Efficient time-domain model of the graphene dielectric function. In: Metamaterials: Fundamentals and Applications VI. vol. 8806. International Society for Optics and Photonics; 2013. p. 88060H.

[33] Ni X, Liu Z, Kildishev AV. PhotonicsDB: Optical Constants. 2007.

[34] Johnson PB, Christy RW. Optical constants of the noble metals. Phys Rev B. 1972;6:4370–4379.

[35] Taflove A, Hagness SC. Computational electrodynamics: the finite-difference time-domain method. Artech House; 2005.

[36] Gandi O. Advances in computational electrodynamics–the finite-difference time-domain method. Taflove, Ed. Boston: Artech House; 1998.

[37] Hawkins RJ, Kallman JS. Linear electronic dispersion and finite-difference time-domain calculations: a simple approach (integrated optics). Journal of Lightwave Technology. 1993;11(11):1872–1874.

[38] Siushansian R, LoVetri J. A comparison of numerical techniques for modeling electromagnetic dispersive media. IEEE Microwave and Guided Wave Letters. 1995;5(12):426–428.

[39] Luebbers RJ, Hunsberger F. FDTD for *N*th-order dispersive media. IEEE Transactions on Antennas and Propagation. 1992;40(11):1297–1301.

[40] Schuster JW, Luebbers RJ. An accurate FDTD algorithm for dispersive media using a piecewise constant recursive convolution technique. In: IEEE Antennas and Propagation Society International Symposium. 1998 Digest. Antennas: Gateways to the Global Network. Held in conjunction with: USNC/URSI National Radio Science Meeting (Cat. No. 98CH36), vol. 4, IEEE; 1998. pp. 2018–2021.

[41] Kelley DF, Luebbers RJ. Piecewise linear recursive convolution for dispersive media using FDTD. IEEE Transactions on Antennas and Propagation. 1996;44(6):792–797.

[42] Vial A. Implementation of the critical points model in the recursive convolution method for modelling dispersive media with the finite-difference time domain method. Journal of Optics A: Pure and Applied Optics. 2007;9(7):745.

[43] Lu J, Chang Y. Implementation of an efficient dielectric function into the finite difference time domain method for simulating the coupling between localized surface plasmons of nanostructures. Superlattices and Microstructures. 2010;47(1):60–65.

[44] Gustafsson B, Kreiss HO, Oliger J. Time Dependent Problems and Difference Methods. New York: John Wiley and Sons Inc.; 1995.

[45] Hulse C, Knoesen A. Dispersive models for the finite-difference time-domain method: design, analysis, and implementation. Journal of the Optical Society of America A. 1994;11(6):1802–1811.

Chapter 5

Inverse-design of plasmonic and dielectric optical nanoantennas

Sawyer. D. Campbell[1], Eric B. Whiting[1], Danny Z. Zhu[2] and Douglas H. Werner[1]

5.1 Introduction

Antennas are devices that transmit or receive electromagnetic waves and have been of significant interest for many decades. Just as Maxwell's equations are valid all the way from radio frequencies (RF) to the visible spectrum, antennas can be designed to operate across various portions of these regimes depending on their geometrical size and material composition. In fact, antenna design at optical and infrared wavelengths has seen a tremendous amount of research focus over the past decade [1–6]. Unsurprisingly, antennas at these wavelengths are met with a number of potential challenges including a limited selection of materials and fabrication techniques compared to their more conventional RF counterparts. Fortunately, there has been a lot of work in the development of new manufacturing technologies for micro- and nano-scale fabrication including 3D printing and additive-manufacturing. Still, one of the most common challenges that arises for optical antenna design is the limited availability conductivity of metals at visible wavelengths.

Noble metals such as gold, silver, and copper are usually thought of more as lossy dielectrics rather than metals in optical and infrared nanoantenna and meta-device design. Interestingly, the real parts of the refractive indices of these metals become quite small in different parts of the optical regime. This corresponds to the plasma frequency transition from the classical Drude model where the permittivity passes from negative to positive as the incident electromagnetic wave frequency increases. More succinctly, on either side of their plasma frequency ω_p, metals behave as epsilon positive (EPS) or epsilon negative (ENG) materials (i.e., $\varepsilon > 0$ and $\varepsilon < 0$, respectively). Fascinatingly, when certain pairs of EPS and ENG materials are intelligently arranged and illuminated with the appropriate wavelength

[1]School of Electrical Engineering and Computer Science, Department of Electrical Engineering, The Pennsylvania State University, University Park, PA, USA
[2]Department of Electrical Engineering and Computer Science, U.S. Military Academy, West Point, NY, USA

source, the incident light can couple into localized longitudinal surface plasmon-polariton (SPP) wave modes. Quantized excitations of these modes are known as plasmons and they are quite intriguing due to how their dispersion relation differs from that of photons. Specifically, surface plasmons can be created with very large longitudinal wave numbers k_x when the frequency of the incident light is near the surface plasmon frequency ω_{sp} [7]. Consequently, plasmonic devices are able to "shrink" light since the wavenumber is inversely proportional to the effective wavelength in the material [8]. This phenomenon is the enabling-mechanism for the realization of highly subwavelength plasmonic nanoantenna-based devices [9–14].

Unfortunately, metals at infrared and optical frequencies are typically quite lossy, which limits the performance of plasmonic devices in applications such as transmissive metasurfaces or when long distance surface wave propagation is desired. On the other hand, dielectrics in these regimes can have little loss, but their permittivity range is somewhat limited at optical frequencies. Like plasmonic materials, dielectric materials can also be used to design subwavelength resonators albeit when their dielectric constants are large. Furthermore, dielectric-based nanoantennas can enable a wide range of beam shaping applications [15] that are typically realized with metasurfaces: patterned surfaces which are engineered to alter the reflection and transmission of electromagnetic waves in a desired way [16,17]. Metasurfaces are commonly built from periodic or aperiodic arrangements of nanoantennas which may alter the polarization, magnitude, or phase of reflected and transmitted light. Dielectric nanoantenna-based metasurfaces are highly desirable alternatives to their plasmonics counterparts due to their ability to achieve very high transmission efficiencies. Moreover, dielectric materials can be used to design conformal [18] and tunable [19] metasurfaces.

Regardless of the materials used to fabricate them or their frequency of operation, all nanoantennas can be improved through optimization. To this end, RF engineers routinely employ numerical techniques to optimize antenna and metamaterial structures; however, use of these techniques is much less common in the optical and infrared regimes [20]. Fortunately, optimization methods are becoming more popular in optical nanoantenna design as the community gains more knowledge on the subject and this chapter seeks to further these gains. To do so, the chapter is organized as follows. First, a brief introduction to the various methods that exist for nanoantenna and meta-device optimization is provided. This is followed by in-depth discussions of three different optical nanoantenna examples based on plasmonic and dielectric materials. Finally, a set of conclusions and outlooks is provided.

5.2 Optimization methods for plasmonic and dielectric optical nanoantennas

While no single optimization algorithm is perfectly suited for all problems one may encounter in nanoantenna design [21], several optimizers have proven quite successful in finding optimal solutions for a variety of nanoantenna and meta-device design problems [20]. These algorithms are broadly categorized as either "local" or "global" optimizers depending on the ability to find function minima: local algorithms generally quickly converge on local minima while global algorithms attempt to find the true global minimum. Regardless of their construction, both local and global optimizers seek to find optimal solutions to user-constructed fitness or cost functions depending on the goal of maximizing or minimizing the function, respectively. Generally, these functions are hyper-dimensional and their input values are limited to finite ranges in order to incorporate constraints that may arise from the chosen fabrication method, materials available, and other factors such as size, weight, power, and monetary cost (SWaP-C). Optimization algorithms are commonly employed in inverse-design problems where the goal is to find a device that achieves a specific functionality while meeting a set of input constraints. This process is formally referred to as system-by-design [22] and is a powerful design methodology for discovering unintuitive designs that achieve extremely high-performances not possible with traditional techniques. Furthermore, while conventional local and global optimizers generally only seek to optimize a single cost function, multi-objective optimizers (MOO) are able to simultaneously optimize problems with multiple competing design goals [23].

In this section, brief introductions to local, global, and multi-objective optimization algorithms and their application to plasmonic and optical nanoantenna and metasurface design are given. Additionally, a flowchart which discusses how to select the appropriate optimization algorithm for a given inverse-design problem is provided.

5.2.1 Local optimization algorithms

Optimization has a history dating back nearly three hundred years. Newton's [24] and Raphson's [25] works from the late seventeenth century have been adapted into the well-known Newton–Raphson method which is an algorithm for finding the roots of a given function. The Newton–Raphson method is one of the earliest known optimization algorithms. In 1847, Cauchy introduced the gradient method which attempts to find a function's minima by moving in the direction of a locally evaluated gradient [26]. Since then, advanced gradient-based algorithms such as the method of conjugate gradients [27] and damped least squares [28] have improved upon the stability and convergence rates of earlier algorithms while seeing success in a wide range of problems. In particular, the damped least squares algorithm has seen wide adoptance in the lens design community for its success in optical system optimization [29,30].

A limitation of local optimizers is their ability to find optimal solutions to multi-modal problems. Unsurprisingly, since these algorithms generally follow cost gradients towards lower function values, they often get stuck in local minima (see Figure 5.1). Generally, local optimizers are ill-suited for problems whose response surfaces (the hyperdimensional surface which maps the input dimensions to the function output) contain many local minima. Additionally, some cost functions may not be differentiable which precludes the use of gradient-based local algorithms in the optimization process. Moreover, gradient calculations are usually

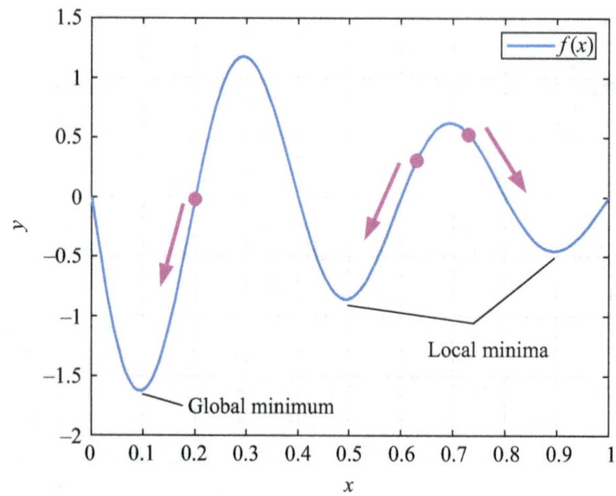

Figure 5.1 Conceptual figure of local optimization. The cost function $f(x)$ and its gradients are evaluated at three locations given by the magenta dots. Of course, since the algorithm follows the gradient towards its basin, it is possible to end up in local minima rather than at the global minimum.

based on finite-differencing which requires additional function evaluations for each optimization variable and can be time-prohibitive for computationally costly function evaluations such as those that are typically encountered in full-wave analysis.

Recently, local algorithms have found success in the application of topology optimization to optical nanoantenna and metasurface design [31]. Topology optimization exploits gradients obtained from adjoint analysis in which the physical electric field $\mathbf{E}$ and the adjoint field $\mathbf{E}^{\mathrm{adj}}$ (i.e., $\overline{\mathbf{E}^{\top}}$) can be used to calculate gradients of a nearly unlimited number of variables with minimal computational cost. In the case of nanophotonic devices, these variables are typically the values of a spatially varying refractive index distribution within a pre-defined mesh. The gradients generated from the adjoint analysis are used by algorithms such as the gradient descent method to alter the index distribution (usually binary) in order to realize a structure which minimizes a desired cost function. Numerous examples of dielectric-based metagratings [32] and beam steering metasurfaces [33] published in the literature have successfully been realized with topology-optimization and local optimizers. Still, topology optimization suffers from the same drawbacks that limit local optimizers. Namely, the tendency to get stuck in local minima and the inability to optimize highly multi-modal problems. Fortunately, global optimization algorithms are largely able to overcome these issues.

5.2.2 Global optimization algorithms

Perhaps the most notable difference between local and global optimization algorithms is the use of gradient information. As discussed above, local optimizers generally exploit locally calculated gradients to move solutions towards function minima. Meanwhile, this is typically done in serial with only a single agent whose position in the input space is iteratively updated by the optimizer. On the contrary, global optimizers typically employ a set of agents, called a population, which explore the solution in some sort of intelligent manner. To this end, global optimizers generally do not use gradient information in the aid of finding function minima. Rather, they employ some type of metaheuristic which is applied to the data gathered from the evaluated cost function and used to update the population and move them towards the global minimum.

Perhaps the most well-known global optimization algorithm is the genetic algorithm (GA) [34,35] which is a nature-inspired algorithm that mimics the process of natural selection in order to find optimal solutions. The GA represents population members as binary codes of 1's and 0's and after the population has been evaluated by a predetermined cost function the best performing population members are used to produce the next generation through a procedure known as crossover. Additionally, random evolution seen in nature is achieved through an additional process known as mutation. Together these methods enforce a Darwinistic survival of the fittest type of evolutionary process. The GA has seen tremendous success in RF antenna and metamaterial design [36,37] where it has been utilized to produce a variety of devices with engineered electromagnetic properties. In nanoantenna and metasurface design, the GA is commonly employed to alter the topology of a unit cell that is described by a pixelated grid in order to achieve a desired functionality. Figure 5.2 shows an example unit cell with four-fold symmetry that is to be optimized by the GA. Here, the black and white pixels are assigned binary values of 1 and 0 and represent the inclusion or absence of some dielectric or metallic material, respectively. The GA seeks to minimize a chosen cost function by intelligently arranging the 1's and 0's within the unit cell to produce the desired behavior (e.g., complex transmission or reflection values or effective material parameters). One drawback of the original incarnation of the GA is that it was limited to combinatorial problems (i.e., those which possess a finite number of solutions). That being said, even the simple example shown in Figure 5.2 contains 2^{36} solutions which would require nearly 70 billion function evaluations to analyze each solution: an obviously prohibitive task. Fortunately, global optimizers such as the GA can find optimal solutions to such problems with a relatively limited numner of function evaluations. Still, real-valued algorithms are desirable since they are applicable to a larger variety of problems with continuous input spaces.

Following the GA, the next generation of global optimizers brought forth the capability to optimize continuous problems. Like the GA, these algorithms were inspired by natural and evolutionary processes. Perhaps the best known nature-inspired real-valued algorithm to emerge from this era is the particle swarm

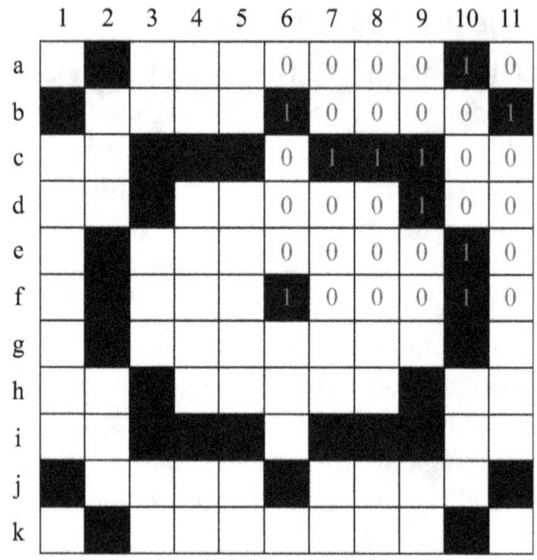

Figure 5.2 Example unit cell topology explored with the GA. The black pixels (1's) indicated the inclusion of some material (e.g., a dielectric or metal) whereas the white pixels (0's) indicate an absence of the inclusion and are assumed to be of some background material (usually air). Here, the GA explores only a quarter of the unit cell geometry due to the imposed symmetry constraints.

optimization (PSO) algorithm [38,39]. PSO is an optimizer within the class of swarm intelligence algorithms that mimics the movements and behaviors of communal species such as flocks of brids or schools of fish in order to solve global optimization problems. PSO has seen tremendous success in applied electromagnetics optimization in the RF community [40,41]. Moreover, PSO's success led to the development of a number of other nature-inspired algorithms including those based on the behaviors of bats [42], wolves [43], and bees [44], among others. Interestingly, PSO has been shown to be a special case of the wind-driven optimization (WDO) algorithm [45], which mimics the behavior of atmospheric forces to steer solutions towards the global minimum. Since then, evolutionary algorithms such as the covariance matrix adaptation evolutionary strategy (CMA-ES) [46] have been shown to be significantly more robust and efficient over a wide range of problems compared to older algorithms such as the GA and PSO. Since its introduction to the electromagnetics community in [47], CMA-ES has seen a wide range of success is antenna, metamaterial, and optical design problems [48–51]. Figure 5.3 shows CMA-ES applied to a difficult test problem known as the "dropwave" function. While this is a highly multi-modal problem with many local minima, CMA-ES is able to find the global minimum with a relatively small population of 12 members in approximately 30 generations. A gradient-based local optimization algorithm would routinely fail to find the global minimum unless its starting point was chosen to be in the very narrow center well: a highly unlikely case without relying on *a priori* knowledge.

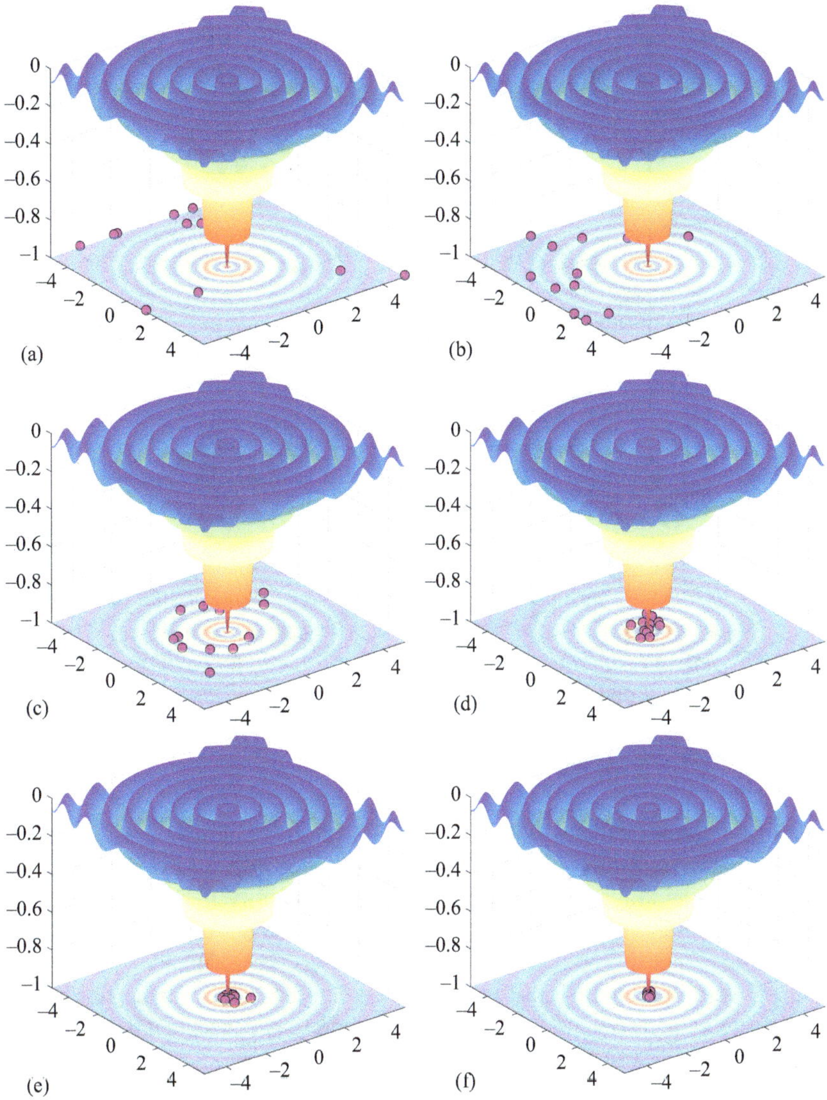

Figure 5.3 Optimization of the dropwave test function by CMA-ES. The pink dots represent members of the population which intelligently explore the solution space and their locations are shown as they evolve after (a) 0, (b) 10, (c) 15, (d) 20, (e) 25, and (f) 30 generations.

5.2.3 Multi-objective optimization algorithms

MOO algorithms represent an emerging class of algorithms for nanoantenna inverse-design and optimization [52]. MOO differs from single objective algorithms in that they present the user with not a single solution, but rather a collection

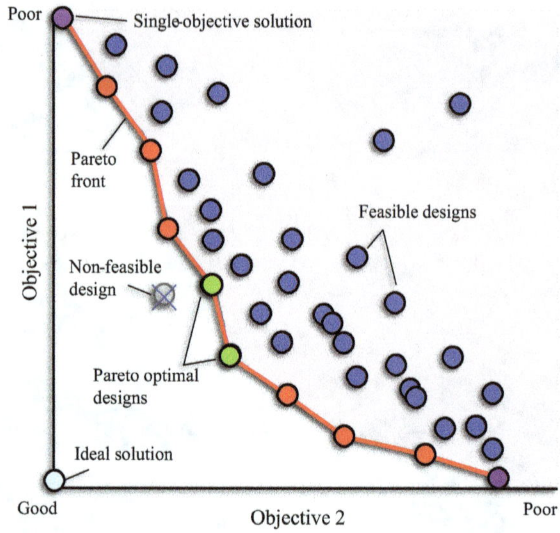

Figure 5.4 Hypothetical Pareto front resulting from a two-objective optimization problem. The Pareto set of optimal designs (red dots) lie on the Pareto front (red line) and dominate other designs (blue dots) which are feasible, but outperformed by the Pareto optimal solutions. The Pareto front shows the inherent tradeoffs between objectives "1" and "2" and can be used a posteriori to select a final design that achieves the best compromise between the different goals.

of optimal solutions called the Pareto set [53]. These solutions represent those at which no improvement in a single objective can come without the degradation in another. This condition is also known as Pareto optimality [54]. Members of the Pareto set lie on a virtual and potentially hyperdimensional surface known as the Pareto front. Analysis of the Pareto front is a very powerful tool and can aid designers in understanding the tradeoffs that exist between competing design objectives (e.g., performance versus SWaP-reduction). A conceptual diagram of the Pareto set is shown in Figure 5.4.

While local and global single-objective algorithms are judged according to their ability to quickly converge towards optimal solutions, MOO algorithms are, unsurprisingly, judged by a different set of criteria; concepts like crowding distance and breadth are used to compare different MOO algorithms to one another [55]. Crowding distance is a measure of how closely spaced solutions on the Pareto front are to one another; a good algorithm is able to produce a Pareto set with little to no crowding. Meanwhile, breadth is a measure of the range of solutions found along the orthogonal objective axes; more breadth is generally better than little breadth. Furthermore, hypervolume, which measures the volume of objective space that the Pareto front occupies, has been found to be a good all-around metric to measure the performance of different MOO algorithms [55]. Interestingly, while the population-based nature of global evolutionary algorithms lends itself well to MOO, gradient-based local optimizers including the multiple gradient descent algorithm [56] and multi-objective Newton's method [57] have been extended to support multiple objectives. Still, MOO algorithms such as the NSGA-III [58], multi-objective particle swarm optimization [59], multi-objective covariance matrix evolution strategy [60], and BORG [61] are among the most popular choices for multi-objective meta-device optimization.

5.2.4 Applied nanoantenna and meta-device optimization

While applying optimization to nanoantenna and meta-device problems is growing increasingly easy with the number of internal functions, libraries, and toolboxes available for high level scripting languages like MATLAB® and Python, it is not always clear how to choose the appropriate optimization algorithm for a given problem. Figure 5.5 provides a flowchart which presents a simplified view of how to select the appropriate optimization algorithm for a chosen nanoantenna and/or meta-device inverse-design problem.

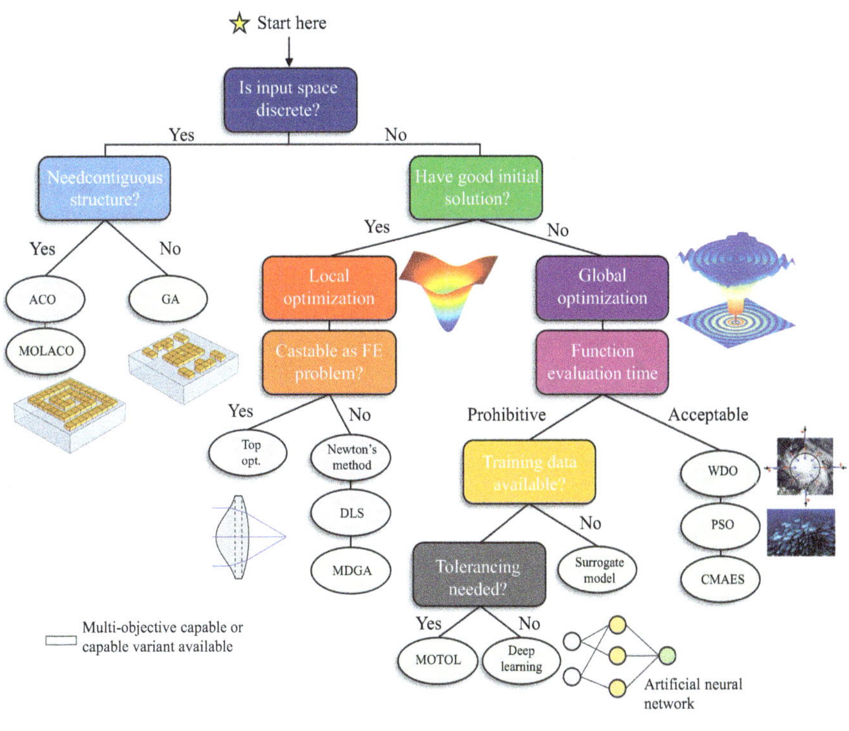

Figure 5.5 Flowchart for general meta-device design. Designers can follow the chart to determine the optimization strategy best suited for their problems. [Reprinted/Adapted] with permission from ref [20], [OSA].

The choice of optimizer can generally be narrowed down not by the desired performance goal (e.g., tailored resonance, field enhancement, or transmission phase or magnitude), but by the dimensionality of and constraints placed on the input space. Starting from the top of the flowchart (see "Start Here") the first question one must answer is with regards to the nature of the input space. If the input space is discrete, then the cost function is a combinatorial problem (i.e., possessing a finite number of possible solutions). For such problems, the proper choice of algorithm is generally a combinatorial optimizer such as the GA [35,37], ant colony optimization (ACO) [62], or multi-objective lazy ant colony optimization (MOLACO) [63] algorithms. Then, one must decide if contiguous structures are needed. If so, then algorithms such as ACO or MOLACO should be chosen *in lieu* of the GA. On the other hand, real-valued algorithms should be chosen when the input space is continuous. However, the choice of whether to employ local or global algorithms depends on the presence of a good starting solution (or lack thereof). If a good initial solution (i.e., guess) exists, then a range of local optimization algorithms can be chosen from to solve the problem. Moreover, if the designer is able to cast their problem in such a form that it can be incorporated into a topology optimization framework (as discussed above), then a combination of adjoint analysis and gradient-based algorithms can be used to optimize the problem. However, most often a good initial solution or starting point is not known and thus the recommended approach is to employ a global optimizer. Moreover, many problems are multi-modal and contain many local minima which necessitates the use of global algorithms. The next factor to consider in determining which algorithm to employ is the computational cost or evaluation time for the chosen problem. When these times are acceptable to the user, a range of real-valued global optimization algorithms (e.g., WDO, PSO, CMA-ES) can be directly applied to the problem.

Many problems in electromagnetics and optical design require tremendous computational resources and are intractable to the direct application of global optimization. In this case, surrogate models [64] can be employed to help accelerate the optimization process of expensive cost functions. This is achieved through either the use of an existing analytical model (e.g., equivalent circuit models [8,65]) or a lower fidelity representation of the computationally expensive model [66] which can help to guide the algorithm into good regions by correlating the input and output spaces of the expensive cost function and the computationally cheap surrogate model. Otherwise, a trained model based on an intelligent hyper-dimensional spatial mapping procedure such as polynomial regressions [67] or the Kriging method [68] can be generated with a limited number of full function evaluations and queried by the optimizer in pursuit of the global minimum. Additionally, when enough training data is available machine learning techniques such as deep neural networks [20] can be used to provide extremely high fidelity replacements to the full cost functions. Moreover, specific algorithms such as the multi-objective optimization with tolerance (MOTOL) algorithm [69] employ machine-learning to provide *in situ* tolerance estimation and robustness optimization which is intractable for most problems without the aid of such surrogate modeling techniques. Finally, most of these global optimization techniques support multiple objectives and this capability should be explored by the user when their problems are best defined by multiple distinct cost functions.

Finally, while the no free lunch theorem [21] states that all optimization algorithms perform equally well when averaged over all possible problems, the authors want to reassure the reader that this condition is never truly realized in meta-device optimization. Of course, some algorithms are more applicable to certain problem types than others and the flowchart provided in Figure 5.5 was constructed in order to address that issue. The readers should note that most problems are best addressed through the use of evolutionary single- or multi-objective global algorithms and that these algorithms (e.g., NSGA-III, PSO, CMA-ES, BORG) can essentially be thought of as "black-box" optimizers for most problems they can expect to encounter. The following sections provide several examples of these algorithms applied to plasmonic and dielectric nanoantenna design problems.

5.3 Optimized plasmonic nanoantennas for large field enhancement

There has been tremendous interest in plasmonic (i.e., metallic) optical nanoantennas due to their ability to confine light into highly subwavelength dimensions by coupling incident radiation into SPP modes. One of the most common examples of plasmonic optical nanoantennas are spherical core-shell nanoparticles [70–72] which can achieve strong resonance behaviors while still being highly-subwavelength in nature. These nanoparticles have a number of applications including surface-enhanced Raman scattering [73] and cancer therapy [74]. Moreover, applied optimization can help designers to quickly tune these nanoparticle geometries to achieve a desired resonance behavior [75]. Plasmonic nanoantennas have also been designed to realize a variety of interesting optical phenomena including the superscattering of light [76], sensors [77], focusing metasurfaces [78], and broadband absorbers [79]. Interestingly, their ability to localize the electric field at the subwavelength scale makes them ideal for achieving gigantic field enhancement which can be exploited to enhance nonlinear optical phenomena [80] or for sensing applications where enhancing the field can make it easier to detect a small signal. Moreover, the plasmonic nanoantenna geometry can be optimized in order to maximize the field enhancement in a desired area [81–83] where a quantum dot or some other nonlinear material may be placed to enhance its behavior. The maximum field enhancement region commonly occurs in a narrow space between two nanoantenna elements. Of course, the chosen design and optimization strategy can have a huge impact on the achieved field enhancement and convergence rate in finding the optimal solution. To this end, a field-enhancing nanoantenna is optimized with two different strategies and their resulting geometries and performances are compared.

Both designs are inspired by the setup employed in [81], where a gold nanostructure is patterned in order to enhance the field at a particular location and a

GA was used to optimize the location of 10 nm×10 nm×11 nm gold nanobricks in a doubly periodic unit cell. For the examples presented in this section, these same geometrical constraints are imposed. A fast periodic finite element boundary integral (PFEBI) solver [84] was employed to model the electromagnetic behavior of candidate unit cells due to its computational efficiency for planar nanoantenna structures that can be well-represented by brick elements. All unit cells are illuminated by a y-polarized normally incident plane wave at a wavelength of 647 nm. Optimizers were employed to minimize a single cost function constructed in order to maximize the electric field value at a designer-specified position ($x = 0$, $y = 0$, $z = 5.5$ nm). For the first design, the unit cell was mapped to a grid of 21×21 pixels whose values were optimized by a GA. Figure 5.6 shows a top-down view of the optimized pattern and the corresponding field intensity spatial map on a dB scale. A value greater than 0 dB indicates field enhancement. Clearly, the GA successfully patterned the gold nanobricks to place the maximum field enhancement in the unit cell center as desired. However, it took a tremendous number of function evaluations (i.e., different nanoantenna configurations) in order to achieve the desired performance. This is due, in part, to the fact that $2^{21\times21} = 2^{441}$ designs exist in the solution space. Moreover, this number would only increase if smaller gold nanobricks were desired. Thus, it is highly desirable to limit the number of variables used to optimize the nanoantenna geometry.

Fortunately, there are design generation strategies that can produce a wide range of geometries with only a few input parameters. For example, Gielis'

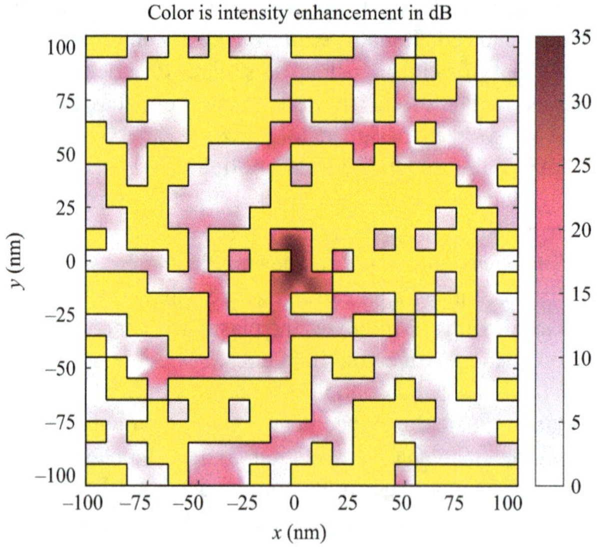

Figure 5.6 Field intensity plot for Au nanoantenna optimized by the GA. The black outline represents the nanoantenna border while the colorbar represents the electric field intensity enhancement on a dB scale.

superformula has been used to generate a highly diverse range of nanoparticle geometries in order to realize performances beyond what canoncial spheroidal shapes can achieve [85]. Another approach is to employ Bézier surfaces in conjunction with a level-set method to generate nearly arbitrary shapes with a limited number of control points [50]. Compared to the unit cell optimized by the GA with 441 binary variables, good success was found with the Bézier surface shape optimization method and only 25 continuous variables. Figure 5.7 shows a top-down view of the optimized unit cell geometry and corresponding field intensity map. Due to the reduction in variables needed to optimize the structure, a finer grid of 5 nm × 5 nm × 5.5 nm nanobricks was used in order to increase the nanoantenna resolutions. The same cost function used in the GA example was applied here, but this time the optimization was performed using CMA-ES. Unsurprisingly, the optimal design is a nearly continuous single nanoantenna shape which produces significant field enhancement at the unit cell center.

Figure 5.8 compares the convergence behavior observed for the two competing optimization methods. The maximum electric field enhancement at the unit cell center is plotted against the number of function evaluations required by the optimizer. It is easy to see that not only did the control-surface shape optimization method achieve a higher field enhancement level than found by the GA, but it did so in just a fraction of the number of evaluations needed by the GA to find its optimal design. Interestingly, in both designs there are likely large areas of

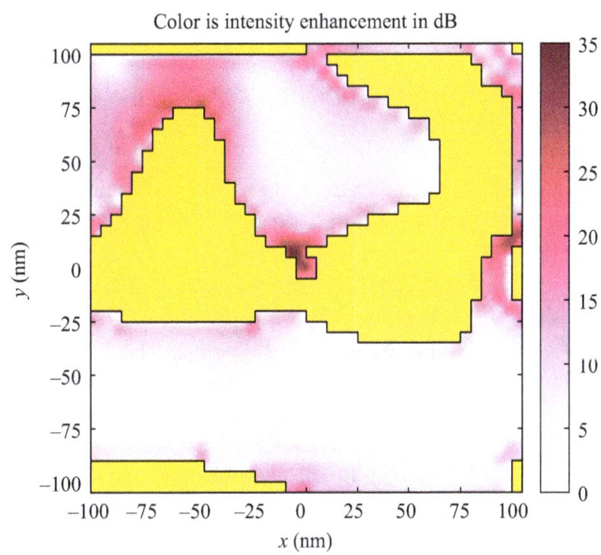

Figure 5.7 Field intensity plot for Au nanoantenna optimized by CMA-ES using the control surface method. The black outline represents the nanoantenna border while the colorbar represents the electric field intensity enhancement on a dB scale.

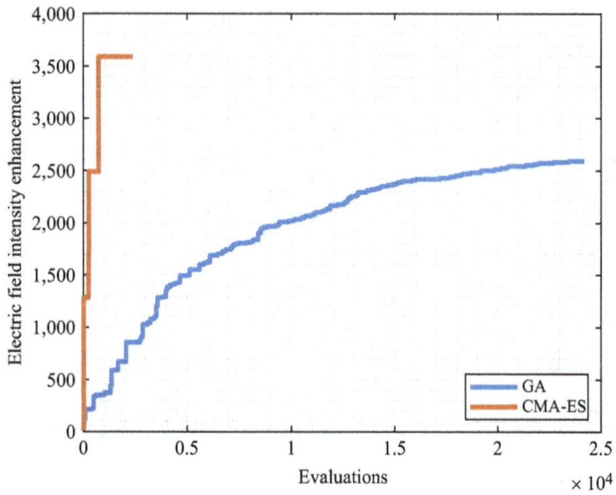

Figure 5.8 Maximum electric field intensity enhancement achieved as a function of the number of evaluations required using (blue) the GA and (red) CMA-ES.

unneeded gold nanobricks found at places with minimal field enhancement. To this end, MOO could be employed in a future study to simultaneously enhance the electric field at the unit cell center while minimizing the volume of gold needed in the nanoantenna structure.

5.4 Nanoantenna optimization for phase-gradient metasurface applications

Recently it has been shown that metasurfaces [17] can enable one to realize a generalized form of Snell's law for reflection and refraction at surfaces [86]. This is accomplished through the intelligent patterning of a metasurface to achieve a specific spatially varying reflection and/or transmission phase commonly referred to as a phase-gradient [87]. Properly choosing the phase-gradient affords one the ability to have nearly arbitrary control over the angles at which an incident wave reflects and refracts. Moreover, the phase-gradient can have polarization [88], wavelength [12], and angular dependencies which can be exploited to offer additional control over the transmission and reflection behaviors of the metasurface. This control has led to the realization of a number of metasurface-based planar optical components [89,90] including retro-reflectors [91] and so-called metalenses [92]. Metalenses have the potential to significantly reduce optical system SWaP, a limiting factor in a number of infrared applications [93], which makes them extremely desirable for disruptive optical systems. For example, metalenses can be paired with more conventional optical components to achieve broadband color correction [94] which would typically require three or more homogeneous lenses to yield the same behavior in traditional systems. Of course, the performance achievable by any phase-gradient metasurface is dependent on the underlying behavior of its constituent building blocks (i.e., unit cells). Therefore, one should synthesize phase-gradient metasurfaces from the best possible set of unit cells available and to do that, an efficient optimization framework should be employed in order to discover them. In this section, two example phase-gradient beam-steering metasurfaces based on 2D and 3D unit cells are presented along with discussions of their performance results and the design methodologies used to optimize them.

5.4.1 Two-dimensional dielectric nanoantennas

Dielectric nanoantennas are of considerable interest due to their low loss [95] which makes them desirable for a number of applications where high transmission or reflection efficiency is required [96,97]. Moreover, broadband dielectric nanoantenna-based metasurfaces have been demonstrated [98]. This functionality and their ultra-thin size makes them attractive options to disrupt the traditional design paradigm used in conventional optical device design [99]. While most commonly realized with positive inclusions, dielectric nanoantennas based on holey (i.e., negative) inclusions have also been demonstrated [100]. In fact, dielectric nanoantenna geometries are only limited by current micro- and nano-fabrication methods available to designers. That being said, two-dimensional (2D) (i.e., planar) dielectric nanoantenna geometries represent the vast majority of all designs presented in the literature to date. Moreover, just as the previous plasmonic nanoantenna example could be optimized to maximize its desired field enhancement performance, dielectric nanoantennas intended for phase-gradient applications can be optimized to achieve a desired complex transmission or reflection performance. To this end, numerous studies of optimized dielectric metasurfaces have been reported in the literature [101–103]. As discussed with the previous plasmonic example, a nanoantenna's performance is highly dependent on its shape and this remains true for dielectric designs [104].

Traditionally, nanoantenna design has been carried out by picking a canonical shape such as an ellipse or rectangle, and then changing its dimensions (e.g., thickness, width, etc.) to tune its electromagnetic response in order to achieve a desired behavior. An advantage to basing nanoantenna geometries on canonical shapes is that very few variables are needed to describe the nanoantenna, which allows designers to tune designs with simple parametric sweeps. Additionally, standard shapes can often be generated to couple to, interact with, or produce physical phenomena like electric ("I" type structure) or magnetic dipole ("ring" like structures) resonance, which leads to physical intuition of how the nanoantenna works. Nevertheless, it has been demonstrated that better performing devices can be created by exploring more complex nanoantenna geometries [104]. Of course, some challenges arise when expanding the design space to include near-arbitrary geometries. For example, it is often the case that more complex shapes require more variables to describe them. Consequently, this makes it more difficult to

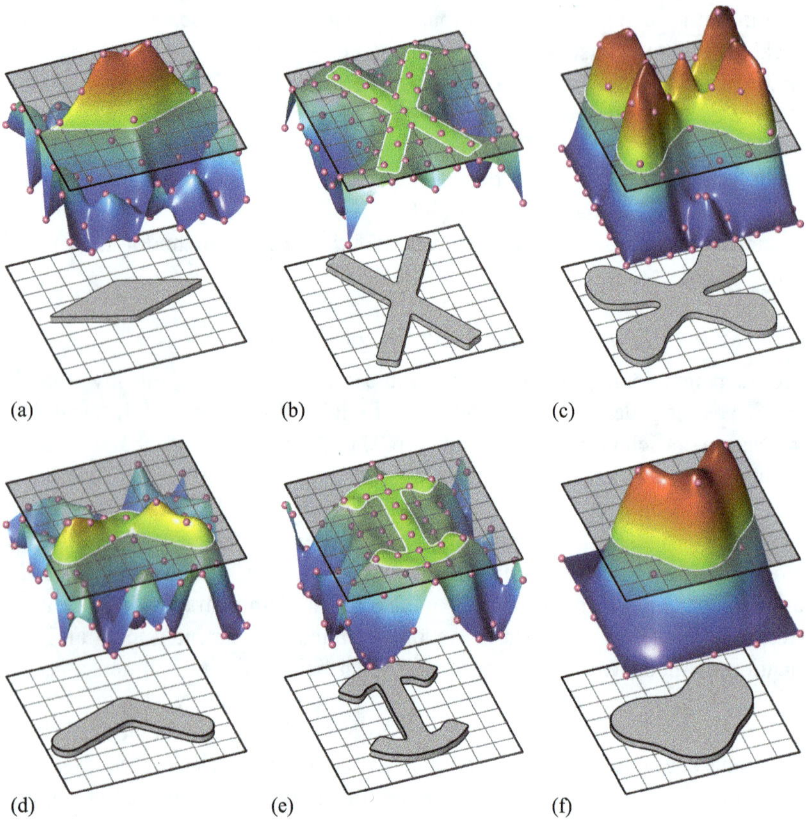

Figure 5.9 Plots showing how the control surface method is used to generate nanoantenna geometries. Standard shapes like (a) diamond, (b) cross, (d) V-antenna, and (e) end-loaded dipole as well as arbitrary shapes (c) with 4-fold symmetry and (f) with no symmetry can be generated through this procedure.

perform parameter sweep-based tuning both because of the increased time to run the simulations and difficulty in visualizing the multidimensional input space. Furthermore, it is unclear what shape will be optimal beforehand which makes seeding an optimization process nearly impossible. In general, one seeks a balance between the requisite geometrical complexity needed to achieve the desired functionality and number of design variables in order to make the inverse-design process tractable.

Figure 5.9 shows a nanoantenna shape generation technique based on a 3D control-surface method. Here, control points (pink dots) are used to alter the surface topology which, when intersected with a level-set plane, determines the 2D planar nanoantenna geometry. Essentially, a thresholding procedure is applied to the surface and everywhere the surface is greater than the threshold becomes the nanoantenna, and everywhere else is left as background material (e.g., air). This procedure is extremely flexible and allows the creation of both canonical and truly arbitrary shapes. As Figure 5.9 shows, the method can create traditional symmetric shapes such as a diamond or cross (Figure 5.9(a) and (b)) as well as asymmetric shapes like the popular V-antenna or end-loaded dipole (Figure 5.9(d) and (e)). On the other hand, one can make arbitrary shapes like in Figure 5.9 (c) and (f). Employing symmetries helps reduce the number of variables required to control the surface and, thus, determine the nanoantenna design. To further reduce the number of variables needed to describe the surface, the control points can be fixed on an xy grid, and then the surface is changed by simply moving each control point in the z-direction. For most cases, it is sufficient to describe the shape with four-fold symmetry using either an arrangement of 3×3 or 5×5 control points for a total of 9 or 25 variables, respectively.

Combining this shape generation strategy with an optimizer enables the inverse-design of high performing nanoantenna arrays. With such a large search space of designs, it is practically impossible to perform a parametric sweep of all variable variations and tune the design by hand to achieve the desired performance. A more efficient approach is to use a MOO algorithm to search for optimal designs. The first step in this process is to decide what the cost functions should be. As seen in Figure 5.10(a), one could set the first objective to be the negative transmission magnitude. Note that, minimizing a cost function based on this objective will maximize the transmission. The problem with this strategy is that one wants not only to maximize transmission (or reflection) magnitude for each nanoantenna, but also to achieve a diverse set of phase options which can be used to form a phase-gradient metasurface. However, if one chooses a second objective to be the transmission phase, then the optimizer will try to minimize the phase (see Figure 5.10(b)). A better strategy is to perform a coordinate rotation of these objectives so that the two objectives are a linear combination of the magnitude and phase (see Figure 5.10(c), (d)). Then by minimizing both objectives simultaneously, the optimizer is able to maximize the transmission magnitude while simultaneously producing a range of phase options. This is well suited for MOO algorithms that seek to map out the Pareto front which is shown as magenta lines in Figure 5.10. Additionally, this approach is more efficient than optimizing nanoantennas to yield a specific phase value, because one does not know what phase options achieve the best performance beforehand. Furthermore, since changing the shape of nanoantennas will modify the phase and magnitude simultaneously, it will be challenging to hold the phase constant and only improve the magnitude of the reflection (transmission) coefficient. Therefore, it is a better strategy to optimize for all phase options simultaneously.

To showcase the efficacy of this design approach, the solution space of an example nanoantenna unit cell configuration is found using an efficient MOO algorithm and is shown in Figure 5.11. The nanoantenna designs are made of silicon ($\varepsilon_r = 11.71$) on a infinite halfspace silica substrate ($\varepsilon_r = 1.797$) with air above. The wavelength of operation is 5 μm. The unit cells are simulated in a doubly periodic environment with a periodicity of 2.25 μm, while the nanoantenna

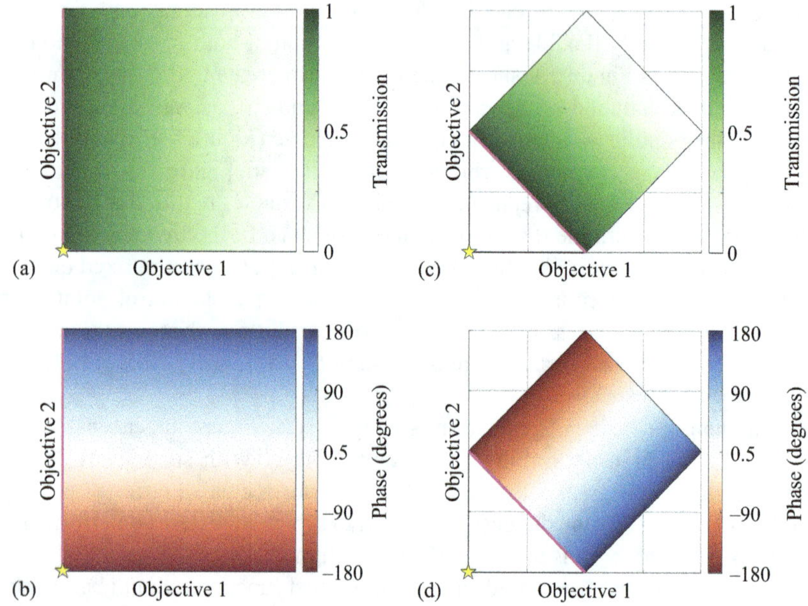

Figure 5.10 Demonstration of how to use MOO to create a diverse set of meta-atoms with varying phase values. Subplots (a-b) show an ineffective way of using the magnitude and phase of the transmission coefficient to create two objectives. Subplots (c-d) show a more efficient way of using the magnitude and phase of the transmission coefficient to create two objectives that when optimized lead to maximized transmission for a diverse range of phase.

height is 1 μm. In this case, the goal is to maximize nanoantenna transmittance and phase diversity and then pick designs from the solution space to form a beam-steering supercell. It can be seen that the optimization found a large number of designs with greater than 90% transmittance across a complete 2π phase range. Using these nanoantenna solutions, a supercell was formed by selecting unit cells from a subset of only the best performing designs found within a narrow range around the target phase options. The supercell was arranged to achieve the desired phase gradient needed to steer an incident planewave into the -1 transmitted diffraction order with very high diffraction efficiency. Figure 5.12(a) shows the design of the optimal supercell. It should be noted that the middle element was chosen to be an empty unit cell. This choice has two interesting benefits; (1) the empty unit cell intrinsically has a large transmission magnitude and (2) the lack of a nanoantenna decreases the chances for inter-cell coupling which can degrade metasurface performance. Looking more in-depth at the supercell, one notes that since the transmission phase is correlated to nanoantenna size, it makes sense that the supercell is created from small (empty unit cell), medium, and large nanoantennas. This is similar to traditional designs where a beamsteering supercell is created from

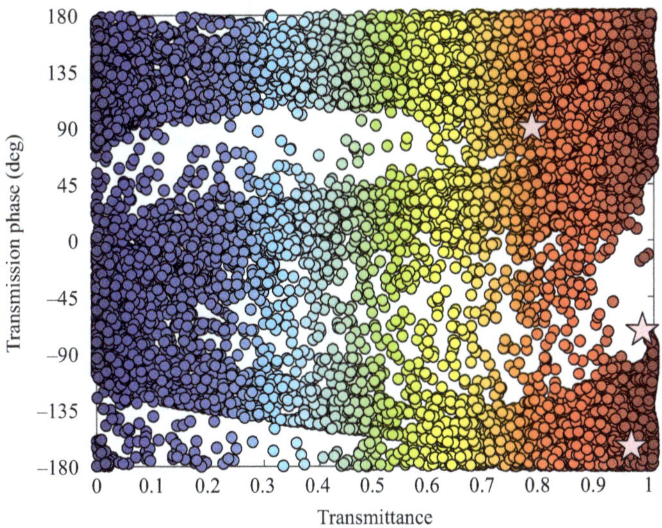

Figure 5.11 *The transmittance and transmission phase for a library of silicon on silica nanoantenna designs found using MOO with the coordinate transformation approach. The pink stars indicate the locations of the chosen unit cells in the transmittance and transmission phase solution space.*

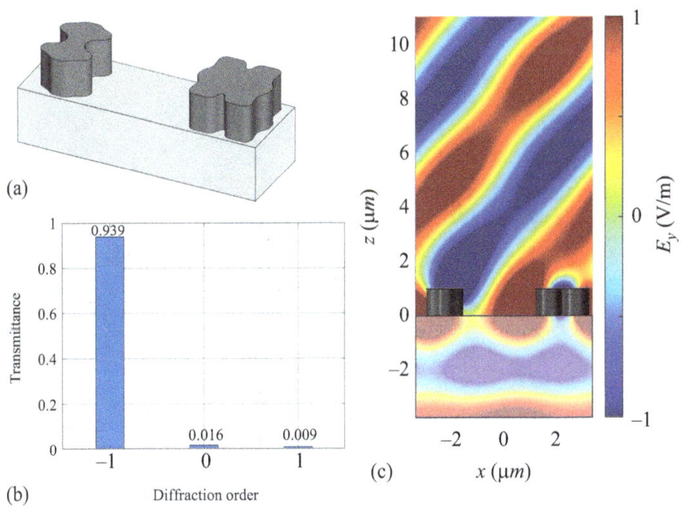

Figure 5.12 *(a) The finalized supercell design of the silicon on silica meta-atom. The middle meta-atom has no element. (b) The diffraction efficiencies for the supercell design. (c) The field plot showing y-polarized light normally incident from the bottom.*

a cylinder with increasing radii. The synthesized supercell was simulated in COMSOL Multiphysics v. 5.3 [105]. One can see from Figure 5.12(b) that the supercell achieves a diffraction efficiency of 93.9% while steering very little into the other unwanted diffraction orders. Figure 5.12(c) shows the near field for a y-polarized wave incident from the bottom which is steered by the metasurface into the intended -1 diffraction order. The near field range is quite short in this design and one can see that the transmitted field quickly takes on the desired plane wave behavior.

The proposed nanoantenna design approach is easily extended to multiple design objectives including polarization-, angular-, and wavelength-dependent responses. Here, a polarization-dependent response is chosen as an additional objective which can be straightforwardly included in the coordinate-transformation based augmented MOO approach. Furthermore, by having a diverse set of amplitude and phase options for multiple polarizations a multifunctional metasurfaces can be readily designed. As an example of how this can be performed, a beam-steering device is designed to steer x- and y-polarized light into opposite directions. The design is based on silicon nanoantennas ($\varepsilon_r = 11.71$) on top of an infinite halfspace calcium fluoride substrate ($\varepsilon_r = 1.957$) with air above the metasurface. Each nanoantenna is simulated at a wavelength of 5 μm in a doubly periodic environment with both x- and y-polarized normally incident plane wave illumination traveling in the $+z$ direction. The periodicity of the unit cells is 2.25 µm, and the nanoantenna height is 2 µm. Figure 5.13 shows the resulting transmission parameters for both polarizations of light generated by the proposed MOO design technique. As shown before, each circle represents a different nanoantenna design with the x-axis showing the transmission phase for y-polarized light, and the y-axis showing the phase for x-polarized light. For all designs, only co-pol light

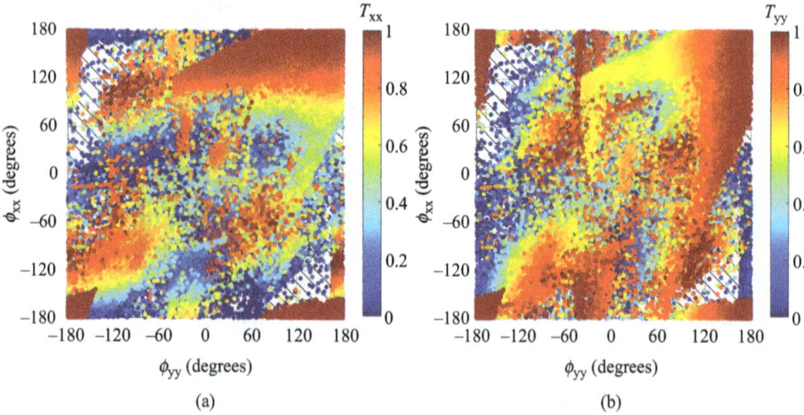

Figure 5.13 Unit cell transmission phase obtained through MOO for y- (x-axis) and x-polarized incident light (y-axis). The dot color indicates the transmittance for (a) x- and (b) y-polarized incident light.

transmission was considered although a design could easily be generated to achieve large cross-pol transmission. The colorbars show the transmittance for x- and y-polarized light, Figure 5.13(a) and (b), respectively. To synthesize a beamsteering device that steers different polarizations into different diffraction orders, one must carefully select the unit cells. For both polarizations, one wants to maximize transmission, but one has to choose the phase differences carefully. For the supercell design shown in Figure 5.14 the neighboring nanoantennas must have an increasing phase difference for x-polarized light and a decreasing phase difference for y-polarized light. Figure 5.14(a) shows the synthesized optimal design. No symmetry was imposed to define the nanoantenna geometries in order to achieve the necessary large phase differences for both polarizations. Figure 5.14(b) shows that there is high transmittance for the x-polarized light of about 85.4%, and moderate performance for y-polarized light (60.6%). In Figure 5.14(c), one can see the multifunctional nature of the synthesized supercell. TE (y-polarized) and TM (x-polarized) incident light scatter into the -1 and $+1$ diffraction orders, respectively. Some residual zero order transmission is present in both configurations, but is quite low overall.

Planar metasurfaces are extremely desirable for manipulating the transmission and reflection characteristics of surfaces in an arbitrary manner. By coupling powerful optimization algorithms with intelligent design procedures, one can realize phase-gradient metasurface-based devices that achieve significantly higher performances than unoptimized designs. The examples discussed in this section

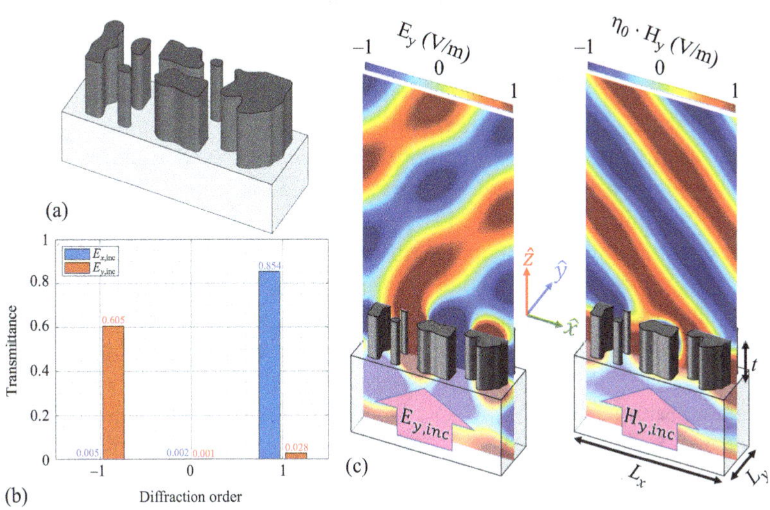

Figure 5.14 Polarization-dependent multifunctional beam-steering supercell design summary: (a) the optimized supercell design, (b) the diffraction efficiencies for x- and y-polarized incident light, and (c) the fields for (left) y- and (right) x-polarized incident light. The light is normally incident from the bottom.

represent only a fraction of the potential design and application spaces that are available for 2D metasurfaces. Naturally, these spaces are even larger when expanding to 3D metasurfaces.

5.4.2 Three-dimensional metallodielectric nanoantennas

Enabling nanoantennas with three-dimensional (3D) geometrical variation significantly increases the degrees of freedom available to designers. Moreover, 3D unit cell geometries can have significantly higher performance in a number of objectives (e.g., polarization control, bandwidth, or field-of-view) compared to their 2D counterparts. While the performance gains by going to 3D are extremely exciting, moving into the third dimension brings with it a number of fabrication and computational challenges [106]. Fortunately, recent advancements in micro- and nano-fabrication techniques have opened the doors for the realization of high-performance 3D optical nanoantenna-based devices. Methods such as direct-write [107], e-beam lithography [108], and nano-imprint lithography [109] have all been used to successfully demonstrate 3D metamaterial structures. On the computational side, expanding the design space from 2D to 3D does increase the time and resources needed to evaluate candidate designs. Both frequency- and time-domain methods (e.g., finite element method (FEM) and finite difference time domain (FDTD), respectively) require meshing of the computational domain in order to solve Maxwell's equations and analyze the performance of a given nanoantenna design. Consequently, if the unit cell's geometric complexity and volume increases then the computational time and memory required to solve the problem correspondingly grows. Furthermore, an expanded solution space also increases the difficulty of finding good solutions and taxes the capabilities of any chosen optimization algorithm. Nevertheless 3D nanoantenna inverse-design is tractable when efficient full-wave solvers are paired with powerful MOO algorithms.

In this section, the MOLACO algorithm [63,110] is used to optimize 3D unit cell geometries for phase-gradient metasurface applications. MOLACO is an extension of traditional ACO [62,111] which exploits stigmergy in ant colonies to solve combinatorial graph problems. Artifical "ants" (henceforth referred to simply as ants) march along the graph and deposit pheromones in proportion to the quality of the solution found. In electromagnetics problems, the trails marched by the ants are converted into metallic meander-lines and evaluated via an appropriate full-wave solver (e.g., FEM or FDTD) to assess their ability to achieve a desired electromagnetic performance. MOLACO's key difference from traditional ACO algorithms is the inclusion of "lazy" ants [112] which assigns each ant a finite probability to stop marching forward at any given point on the graph. This feature expands the variety of trails produced by enabling variable-length meander-lines. MOLACO is able to solve arbitrary graph problems such as those seen in Figure 5.15 so long as the user is able to map the graph to a chosen EM problem.

To implement MOLACO for phase-gradient metasurface optimization, a corresponding unit cell configuration must be chosen. To this end, a cuboid unit cell based on the novel membrane projection lithography (MPL) is selected as the

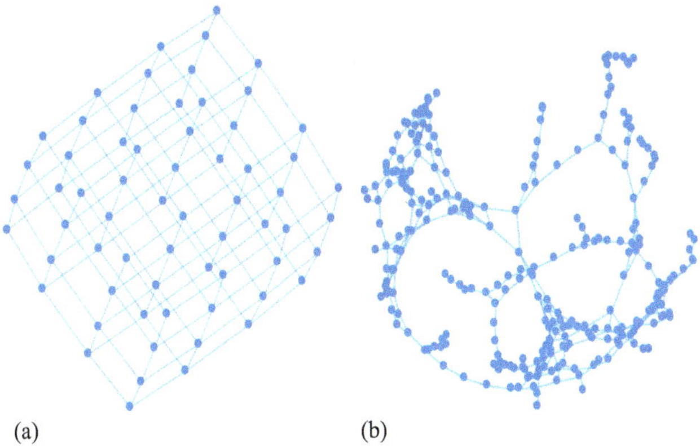

Figure 5.15 *Example graph representations of problems that MOLACO can explore: (a) 5 × 5 × 5 and (b) arbitrary graphs.*

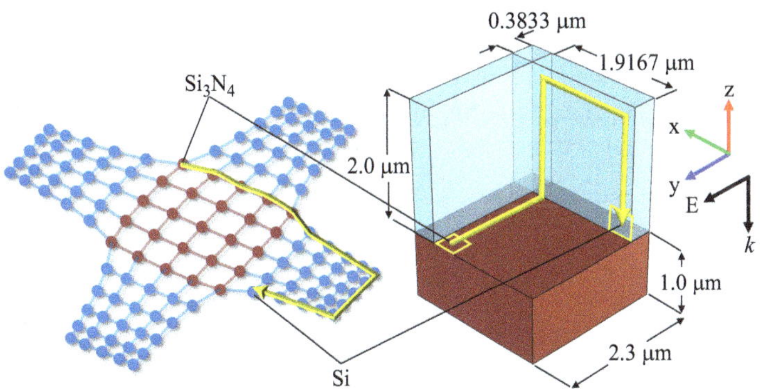

Figure 5.16 *Translating the 5 × 5 × 5 folded-net graph to the MPL-based unit cell. The yellow trail corresponds to the ant path and represents where gold will be deposited.*

fabrication method of choice due to its ability to place multiple metamaterial inclusions within a single unit cell [113–115]. Figure 5.16 shows a 5 × 5 × 5 folded-net graph applied to an empty MPL cavity. The walls and substrate are comprised of Si_3N_4 and Si, respectively, while the nanoantennas are deposited with gold. One can see from Figure 5.16 how the folded-net graph is mapped to the unit cell.

MOLACO deploys virtual "ant" agents who explore the graph and leave behind contiguous trails (see the yellow line) which become the gold inclusions on the realized unit cells. Fabrication constraints are easily employed to eliminate closed loops, single pixel islands, and "hanging" pixels from the final designs. Loops and hanging pixels are eliminated by simply not allowing the ants to cross over their existing trail. Consequently, the ants are also not permitted to move backwards at any point. Nevertheless, there exists a tremendous variety of possible trails in the system and it would be intractable to evaluate all of them in pursuit of optimal solutions. Fortunately, optimizers like MOLACO are able to find high performance designs in only a limited number of function evaluations making them a valuable tool in the nanoantenna design process.

Each unit cell is evaluated with doubly periodic boundary conditions under normally incident y-polarized plane-wave illumination at a wavelength of 5.5 μm. A periodic finite element boundary integral solver [84] is chosen to evaluate the complex transmission and reflection coefficients of the infinite unit cell array due to its extreme efficiency for problems that can be easily represented with rectangular prismatic elements. For the intended application of transmissive phase-gradient metasurfaces, the ideal unit cells would possess unity magnitude and span a complete 2π transmission phase range. Of course, in the presence of dielectrics and metals, it is very challenging to achieve extremely high transmission magnitude and a diverse phase range simultaneously. Specifically, the empty cavity has an intrinsic transmission magnitude and phase that the nanoantennas must alter in order to move the phase across the 2π range. To achieve a wide range of phase options MOLACO employs a Pareto Locus method which simultaneously maximizes transmission while promoting phase diversity. Figure 5.17 showcases a variety of ant-trails generated during the optimization process with each achieving its own unique complex transmission characteristics.

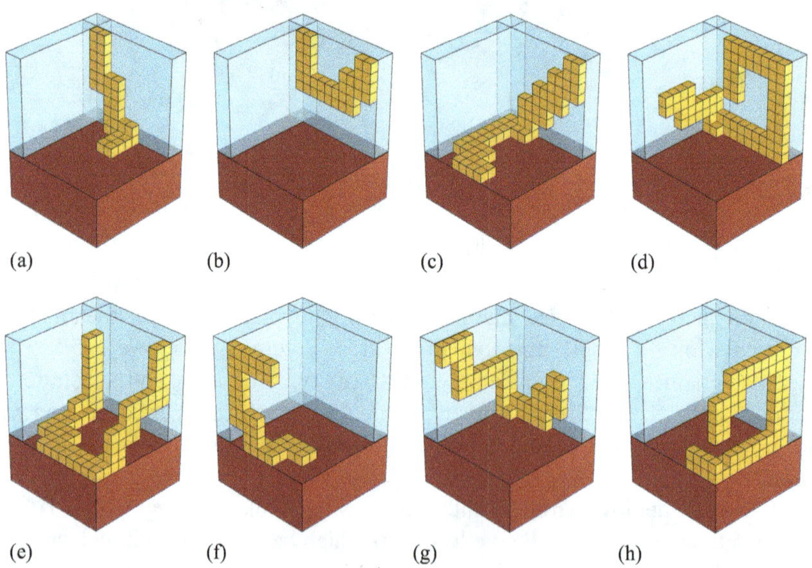

Figure 5.17 Example unit cells generated by MOLACO showcasing the diversity of gold trails produced by the ants. The bottom two vertical walls are removed to aid in viewing the gold trails.

Figure 5.17 provides an insight into the variety of unit cell options generated by MOLACO during the optimization process. While these designs represent a small fraction of the entire design space that MOLACO is able to explore, they nevertheless provide insight into how the algorithm works. Each trail can begin at a random node location and meander a different length based on the chosen ant's laziness and the presence of any design constraints (e.g., intersection trails). Over 100,000 unique unit cell geometries were evaluated during the optimization process. These solutions are collected into a sort of "periodic table" which stores information about the various unit cell geometries and their transmission phase and amplitude options. Figure 5.18 provides a scatter plot of transmittance $\left(|t|^2\right)$ as a function of the transmission phase.

One can see that MOLACO was able to discover unit cells that produce a full 360° of transmission phase range while simultaneously achieving high transmittance. Solutions near the empty cavity solution (indicated by the star) generally have very little gold in them or possess trails that are completely perpendicular to the *y*-polarized incident field. Furthermore, since gold is lossy in the MWIR and optical regimes, adding gold generally comes at the expense of lower transmittance and increased absorption. This is why solutions near the "empty cavity" case have the highest transmittances and solutions further away have lower values. Moreover, shifting the transmission phase 180° away from the "natural" behavior of the empty cavity unit is generally the most difficult phase option to achieve due to it being the furthest phase value away.

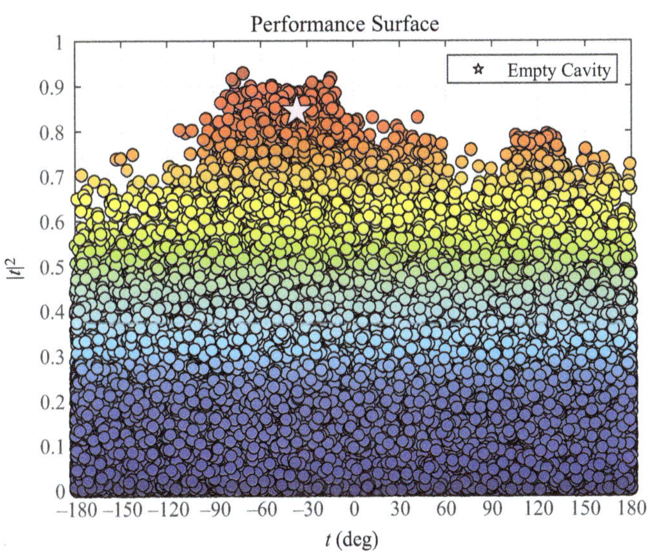

Figure 5.18 Transmission solution space generated by MOLACO during optimization. The star represents the empty cavity (i.e., no gold trail) solution.

With the periodic table generated, one can next intelligently pick unit cells with different transmission behaviors to serve as building blocks for devices based on a periodic array of supercells or completely aperiodic tilings. The efficacy of the single unit cell optimization paradigm requires that any supercell synthesized from the periodic table generated during the optimization process achieves behavior similar to what would be predicted by a linear array of sources with individual unit cell complex transmission and reflection values. To test this assumption, a beam-steering metasurface is again chosen as a proof-of-concept device since, while being relatively simple in construction, it can provide insight into inter-cell coupling and be easily extended to metalens-based devices. From scalar diffraction theory, the minimum supercell width Λ needed to achieve beam-steering must be greater than the incident wavelength λ in the surrounding medium. Also, the supercell needs to possess a minimum of three equally spaced phase options in order to steer the beam. Given the unit cell spacing and incident wavelength of 2.3 μm and 5.5 μm, respectively, both conditions will be satisfied when three or more equally spaced unit cells are chosen to form a supercell. With these requirements in place, three unit cells with 120° relative phase shifts from one another were selected from the periodic table of elements to form the supercell. A view of the chosen supercell is given in Figure 5.19.

The supercell periodicity is small enough that only the -1, 0 , and $+1$ diffraction orders are available to steer the beam into. Next, full-wave simulations were employed to measure the supercell's performance as a beam steerer. The same excitation setup used for the unit cell simulations was applied to illuminate the supercell: a normally incident *y*-polarized plane wave traveling in the $-z$ direction. In choosing the unit cells, special attention was paid to the orientation of the gold trails within in order to avoid inter-cell coupling when pairing them in the supercell. Unit cells with gold trails on or close to a single wall were chosen so as to separate the trails in adjacent cells as far as possible. This strategy proved fruitful as a minimal amount of coupling-related performance degradation was observed in the evaluated supercell. COMSOL Multiphysics v. 5.3 was employed to validate the supercell results. A 3D view of the supercell steering the normally incident

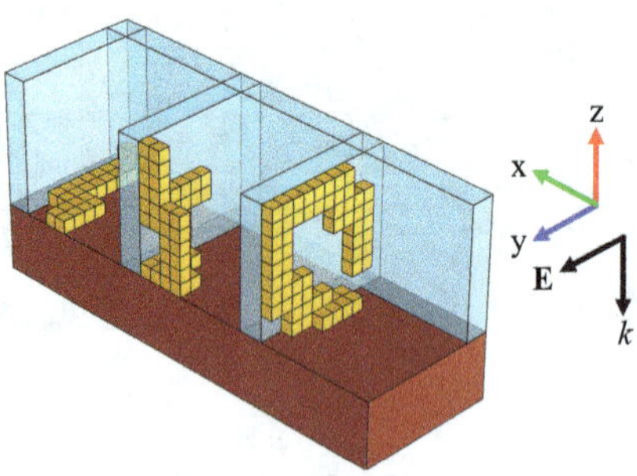

Figure 5.19 Isometric view of the assembled three-element supercell which is illuminated by a y-polarized planewave traveling in the z-direction.

planewave into the +1 order is shown in Figure 5.20(a). The reflected field is omitted from the plot to aid in visualizing how the supercell steers the incident beam.

To visualize the presence of coupling within the supercell, the electric field distribution is mapped and compared to that measured for each unit cell when evaluated in the single unit cell simulation environment. These results are captured in Figure 5.20(b) which provides a top-down view of the difference in the electric fields between the two simulation environments. It should be noted that wherever blue is observed in this plot one can safely assume that the presence of coupling is weak or nonexistent. In fact, there are only a few small areas with noticeable changes in the electric field distribution which suggests that the assumption of using individually evaluated unit cells from the periodic table to form phase-gradient based devices was valid. Finally, Figure 5.20(c) summarizes the diffraction efficiencies of the transmitted field. One observes that there is a small presence ($\approx 5\%$) of the zeroth order in the transmitted field which is the reason for the slightly wavy nature of the steered beam in Figure 5.20(a). The supercell is able to achieve over 64% diffraction efficiency into the +1 order which compares favorably to other designs from the literature which operate in the MWIR [97].

Supercells that achieve high diffraction efficiencies have previously relied on all-dielectric materials to realize such high performances levels, but the 3D nature of the MPL-based cavities enabled MOLACO to find similar performing designs

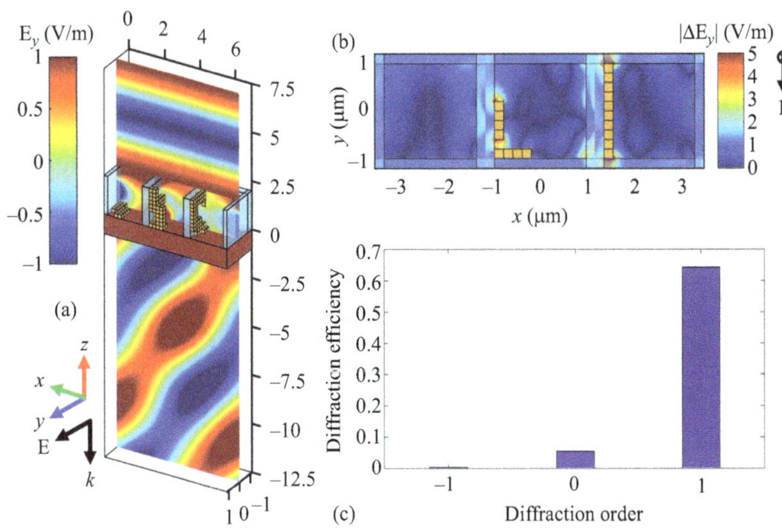

Figure 5.20 Summary of the supercell beam-steerer performance. (a) The incident and transmitted fields in the supercell vicinity. (b) A top-down view of the difference in the electric field distributions in the unit cell and supercell simulation environments. (c) Diffraction efficiency of the transmitted plane-waves as a function of the diffraction order.

with metals which significantly outperform other metal-based examples from the literature [116]. Moreover, the MPL fabrication technique opens the door to optimizing unit cells that achieve performances beyond what is possible with planar nanoantenna and metasurface designs. For example, 3D unit cells have previously shown extremely wide field-of-view performance [110] at RF frequencies and this behavior should extend to designs in the MWIR and optical regimes. While these unit cells have only been optimized for normal incidence it is straightforward to analyze them over a desired angular range (i.e., field-of-view) and incorporate their performance at other angles into one or more cost functions.

5.5 Conclusions

Optimization is a powerful tool for plasmonic and dielectric nanoantenna design. In fact, a variety of high performance designs could not be realized without the aid of an optimizer. This chapter served as an overview of applied optimization for nanoantenna design. Optimization does not need to make a problem intractable. Furthermore, MOO algorithms are extremely desirable for their ability to simultaneously optimize multiple performance goals. These algorithms can also optimize pseudo objectives like "phase diversity" through intelligent transformations of the objective space. Finally, readers inexperienced with optimization are strongly encouraged to try out the algorithms and design methodologies discussed in this chapter to help to enable the discovery of high-performance solutions for the meta-device problems of interest to them.

Acknowledgments

The opinions in this work are solely of the authors and do not necessarily reflect those of the US Military Academy, the US Army, or the Department of Defense. The authors thank Dr. Bruce Burckel of Sandia National Laboratories for providing fabrication expertise on MPL and material properties for silicon, silicon nitride, and gold used in modeling the proposed unit cell structures. Partial support for this work was provided by the Defense Advanced Research Projects Agency (DARPA) (HR00111720032).

References

[1] Crozier KB, Sundaramurthy A, Kino GS, *et al.* Optical antennas: resonators for local field enhancement. Journal of Applied Physics. 2003;94(7):4632–4642.

[2] Taminiau TH, Stefani FD, Segerink FB, *et al.* Optical antennas direct single-molecule emission. Nature Photonics. 2008;2(4):234–237.

[3] Merlein J, Kahl M, Zuschlag A, *et al.* Nanomechanical control of an optical antenna. Nature Photonics. 2008;2(4):230–233.

[4] Schuller JA, Taubner T, Brongersma ML. Optical antenna thermal emitters. Nature Photonics. 2009;3(11):658–661.
[5] Gallo M, Mescia L, Losito O, *et al.* Design of optical antenna for solar energy collection. Energy. 2012;39(1):27–32.
[6] Yu N, Genevet P, Aieta F, *et al.* Flat optics: controlling wavefronts with optical antenna metasurfaces. IEEE Journal of Selected Topics in Quantum Electronics. 2013;19(3):4700423.
[7] Maier SA. Plasmonics: fundamentals and applications. Springer Science & Business Media; 2007.
[8] Duan H, Fernández-Domínguez AI, Bosman M, *et al.* Nanoplasmonics: classical down to the nanometer scale. Nano Letters. 2012;12(3):1683–1689.
[9] Muskens OL, Giannini V, Sánchez-Gil JA, *et al.* Optical scattering resonances of single and coupled dimer plasmonic nanoantennas. Optics Express. 2007;15(26):17736–17746.
[10] Muskens OL, Giannini V, Sánchez-Gil JA, *et al.* Strong enhancement of the radiative decay rate of emitters by single plasmonic nanoantennas. Nano Letters. 2007;7(9):2871–2875.
[11] Boriskina SV, Negro LD. Multiple-wavelength plasmonic nanoantennas. Optics Letters. 2010;35(4):538–540.
[12] Ni X, Emani NK, Kildishev AV, *et al.* Broadband light bending with plasmonic nanoantennas. Science. 2012;335(6067):427.
[13] Giannini V, Fernández-Domínguez AI, Heck SC, *et al.* Plasmonic nanoantennas: fundamentals and their use in controlling the radiative properties of nanoemitters. Chemical Reviews. 2011;111(6):3888–3912.
[14] Akselrod GM, Argyropoulos C, Hoang TB, *et al.* Probing the mechanisms of large Purcell enhancement in plasmonic nanoantennas. Nature Photonics. 2014;8(11):835–840.
[15] Stellinga DP. Shaping light beams with dielectric metasurfaces [Phd]. University of York; 2016.
[16] Holloway CL, Kuester EF, Gordon JA, *et al.* An overview of the theory and applications of metasurfaces: the two-dimensional equivalents of metamaterials. IEEE Antennas and Propagation Magazine. 2012;54(2):10–35.
[17] Chen HT, Taylor AJ, Yu N. A review of metasurfaces: physics and applications. Reports on Progress in Physics. 2016;79(7):076401.
[18] Kamali SM, Arbabi A, Arbabi E, *et al.* Decoupling optical function and geometrical form using conformal flexible dielectric metasurfaces. Nature Communications. 2016;7:11618.
[19] Kamali SM, Arbabi E, Arbabi A, *et al.* Highly tunable elastic dielectric metasurface lenses. Laser & Photonics Reviews. 2016;10(6):1002–1008.
[20] Campbell SD, Sell D, Jenkins RP, *et al.* Review of numerical optimization techniques for meta-device design [Invited]. Optical Materials Express. 2019;9(4):1842–1863.
[21] Wolpert DH, Macready WG. No free lunch theorems for optimization. IEEE Transactions on Evolutionary Computation. 1997;1(1):67–82.

[22] Massa A, Oliveri G, Rocca P, *et al.* System-by-design: a new paradigm for handling design complexity. In: The 8th European Conference on Antennas and Propagation (EuCAP 2014). The Hague, Netherlands: IEEE; 2014. p. 1180–1183.

[23] Deb K, Pratap A, Agarwal S, *et al.* A fast and elitist multiobjective genetic algorithm: NSGA-II. IEEE Transactions on Evolutionary Computation. 2002;6(2):182–197.

[24] Newton I. The method of fluxions and infinite series: with its application to geometry of curve-lines. London; 1736.

[25] Raphson J. Analysis Aequationum Universalis, seu ad Aequationes Algebraicas resolvendas Methodus generalis, & expedita, ex nova infinitarum serierum methodo, deducta ac demonstrata. Editio secunda cui accessit Appendix de Infinito Infinitarum Serierum progressu ad Equationum Algebraicarum Radices eliciendas. Cui etiam Annexum est; De Spatio reali, seu Ente Infinito Conamen Mathematico-Metaphysicum. Typis T. Braddyll, Prostant Venales Apud Johannem Taylor, Ad Insigne Navis in Cœeterio D. Pauli; 1697.

[26] M Augustine Cauchy. Méthode Générale Pour La Résolution Des Systemes d'équations Simultanées. Comp Rend Sci Paris. 1847;25:536–538.

[27] Hestenes MR, Stiefel E. Methods of conjugate gradients for solving linear systems. Journal of Research of the National Bureau of Standards. 1952; 49(6):409.

[28] Marquardt DW. An algorithm for least-squares estimation of nonlinear parameters. Journal of the Society for Industrial and Applied Mathematics. 1963;11(2):431–441.

[29] Jamieson TH. Optimization techniques in lens design. No. 5 in Monographs on Applied Optics. New York: American Elsevier Pub. Co; 1971.

[30] Yabe A, Society of Photo-optical Instrumentation Engineers. Optimization in lens design. 2018. OCLC: 1035446873.

[31] Jensen JS, Sigmund O. Topology optimization for nano-photonics. Laser & Photonics Reviews. 2011;5(2):308–321.

[32] Sell D, Yang J, Doshay S, *et al.* Large-angle, multifunctional metagratings based on freeform multimode geometries. Nano Letters. 2017;17(6):3752–3757.

[33] Sell D, Yang J, Doshay S, *et al.* Periodic dielectric metasurfaces with high-efficiency, multiwavelength functionalities. Advanced Optical Materials. 2017;5(23):1700645.

[34] Holland JH. Natural and artificial systems: an introductory analysis with applications to biology, control, and artificial intelligence. 1st ed. Complex Adaptive Systems. Cambridge, MA: MIT Press; 1992.

[35] Holland JH. Genetic algorithms. Scientific American. 1992;267(1): 66–72.

[36] Weile DS, Michielssen E. Genetic algorithm optimization applied to electromagnetics: a review. IEEE Transactions on Antennas and Propagation. 1997;45(3):343–353.

[37] Haupt RL, Werner DH. Genetic algorithms in electromagnetics. Hoboken, NJ: IEEE Press: Wiley-Interscience; 2007. OCLC: ocm70673291.
[38] Jin N, Rahmat-Samii Y. Advances in particle swarm optimization for antenna designs: real-number, binary, single-objective and multiobjective implementations. IEEE Transactions on Antennas and Propagation. 2007; 55(3):556–567.
[39] Kennedy J. Swarm intelligence. In: Zomaya AY, editor. Handbook of nature-inspired and innovative computing: integrating classical models with emerging technologies. Boston, MA: Springer US; 2006. p. 187–219.
[40] Boeringer DW, Werner DH. Particle swarm optimization versus genetic algorithms for phased array synthesis. IEEE Transactions on Antennas and Propagation. 2004;52(3):771–779.
[41] Massa A, Oliveri G, Salucci M, *et al.* Dealing with EM functional optimization through new generation evolutionary-based methods. In: 2014 International Conference on Numerical Electromagnetic Modeling and Optimization for RF, Microwave, and Terahertz Applications (NEMO); 2014. p. 1–4.
[42] Yang XS. A new metaheuristic bat-inspired algorithm. In: González JR, Pelta DA, Cruz C, *et al.* editors. Nature inspired cooperative strategies for optimization (NICSO 2010). Studies in Computational Intelligence. Berlin, Heidelberg: Springer Berlin Heidelberg; 2010. p. 65–74.
[43] Mirjalili S, Mirjalili SM, Lewis A. Grey wolf optimizer. Advances in Engineering Software. 2014;69:46–61.
[44] Karaboga D, Akay B. A comparative study of artificial bee colony algorithm. Applied Mathematics and Computation. 2009;214(1):108–132.
[45] Bayraktar Z, Komurcu M, Bossard JA, *et al.* The wind driven optimization technique and its application in electromagnetics. IEEE Transactions on Antennas and Propagation. 2013;61(5):2745–2757.
[46] Hansen N, Müller SD, Koumoutsakos P. Reducing the time complexity of the derandomized evolution strategy with covariance matrix adaptation (CMA-ES). Evolutionary Computation. 2003;11(1):1–18.
[47] Gregory MD, Bayraktar Z, Werner DH. Fast optimization of electromagnetic design problems using the covariance matrix adaptation evolutionary strategy. IEEE Transactions on Antennas and Propagation. 2011;59(4):1275–1285.
[48] Gregory MD, Martin SV, Werner DH. Improved electromagnetics optimization: the covariance matrix adaptation evolutionary strategy. IEEE Antennas and Propagation Magazine. 2015;57(3):48–59.
[49] Martin SH, Martinez I, Turpin JP, *et al.* The synthesis of wide- and multi-bandgap electromagnetic surfaces with finite size and nonuniform capacitive loading. IEEE Transactions on Microwave Theory and Techniques. 2014; 62(9):1962–1972.
[50] Sieber PE, Werner DH. Infrared broadband quarter-wave and half-wave plates synthesized from anisotropic Bézier metasurfaces. Optics Express. 2014;22(26):32371.

[51] Campbell SD, Nagar J, Werner DH. Multi-element, multi-frequency lens transformations enabled by optical wavefront matching. Optics Express. 2017;25(15):17258–17270.

[52] Nagar J, Campbell SD, Ren Q, *et al.* Multiobjective optimization-aided metamaterials-by-design with application to highly directive nanodevices. IEEE Journal on Multiscale and Multiphysics Computational Techniques. 2017;2:147–158.

[53] Deb K. Multi-objective optimization. Springer US; 2014. p. 403–449.

[54] Censor Y. Pareto optimality in multiobjective problems. Applied Mathematics and Optimization. 1977;4(1):41–59.

[55] Nagar J, Werner DH. A comparison of three uniquely different state of the art and two classical multiobjective optimization algorithms as applied to electromagnetics. IEEE Transactions on Antennas and Propagation. 2017;65(3):1267–1280.

[56] Désidéri JA. Multiple-gradient descent algorithm (MGDA) for multi-objective optimization. Comptes Rendus Mathematique. 2012;350(5):313–318.

[57] Fliege J, Drummond LMG, Svaiter BF. Newton's method for multiobjective optimization. SIAM Journal on Optimization. 2009;20(2):602–626.

[58] Seada H, Deb K. U-NSGA-III: a unified evolutionary optimization procedure for single, multiple, and many objectives: proof-of-principle results. In: Gaspar-Cunha A, Henggeler Antunes C, Coello CC, editors. Evolutionary Multi-Criterion Optimization. Lecture Notes in Computer Science. Springer International Publishing; 2015. p. 34–49.

[59] Alvarez-Benitez JE, Everson RM, Fieldsend JE. A MOPSO algorithm based exclusively on Pareto dominance concepts. In: Hutchison D, Kanade T, Kittler J, *et al.*, editors. Evolutionary Multi-Criterion Optimization. vol. 3410. Berlin, Heidelberg: Springer; 2005. p. 459–473.

[60] Igel C, Hansen N, Roth S. Covariance matrix adaptation for multi-objective optimization. Evolutionary Computation. 2007;15(1):1–28.

[61] Hadka D, Reed P. Borg: an auto-adaptive many-objective evolutionary computing framework. Evolutionary Computation. 2013;21(2):231–259.

[62] Dorigo M, Maniezzo V, Colorni A. Ant system: optimization by a colony of cooperating agents. IEEE Transactions on Systems, Man, and Cybernetics, Part B (Cybernetics). 1996;26(1):29–41.

[63] Zhu DZ, Werner PL, Werner DH. Design and optimization of 3-D frequency-selective surfaces based on a multiobjective lazy ant colony optimization algorithm. IEEE Transactions on Antennas and Propagation. 2017;65(12):7137–7149.

[64] Queipo NV, Haftka RT, Shyy W, *et al.* Surrogate-based analysis and optimization. Progress in Aerospace Sciences. 2005;41(1):1–28.

[65] Kim M, Wong AMH, Eleftheriades GV. Optical Huygens' metasurfaces with independent control of the magnitude and phase of the local reflection coefficients. Physical Review X. 2014;4(4).

[66] Koziel S. Surrogate-based optimization of microwave structures using space mapping and Kriging. In: 2009 European Microwave Conference (EuMC); 2009. p. 1062–1065.

[67] Easum JA, Campbell SD, Nagar J, *et al.* Analytical surrogate model for the aberrations of an arbitrary GRIN lens. Optics Express. 2016;24(16):17805–17818.

[68] Kleijnen JPC. Kriging metamodeling in simulation: a review. European Journal of Operational Research. 2009;192(3):707–716.

[69] Easum JA, Nagar J, Werner PL, *et al.* Efficient multiobjective antenna optimization with tolerance analysis through the use of surrogate models. IEEE Transactions on Antennas and Propagation. 2018;66(12):6706–6715.

[70] Campbell SD, Ziolkowski RW. Simultaneous excitation of electric and magnetic dipole modes in a resonant core-shell particle at infrared frequencies to achieve minimal backscattering. IEEE Journal of Selected Topics in Quantum Electronics. 2012;19(3):4700209.

[71] Campbell SD, Ziolkowski RW. Impact of strong localization of the incident power density on the nano-amplifier characteristics of active coated nanoparticles. Optics Communications. 2012;285(16):3341–3352.

[72] Campbell SD, Ziolkowski RW. Near-field directive beams from passive and active asymmetric optical nanoantennas. IEEE Journal of Selected Topics in Quantum Electronics. 2014;21(4):312–323.

[73] Li JF, Huang YF, Ding Y, *et al.* Shell-isolated nanoparticle-enhanced Raman spectroscopy. Nature. 2010;464(7287):392–395.

[74] Loo C, Lin A, Hirsch L, *et al.* Nanoshell-enabled photonics-based imaging and therapy of cancer. Technology in Cancer Research & Treatment. 2004; 3(1):33–40.

[75] Ginzburg P, Berkovitch N, Nevet A, *et al.* Resonances on-demand for plasmonic nano-particles. Nano Letters. 2011;11(6):2329–2333.

[76] Mirzaei A, Miroshnichenko AE, Shadrivov IV, *et al.*Superscattering of light optimized by a genetic algorithm. Applied Physics Letters. 2014; 105(1):011109.

[77] Kim KY, Jung J. Multiobjective optimization for a plasmonic nanoslit array sensor using kriging models. Applied Optics. 2017;56(21):5838.

[78] Pors A, Nielsen MG, Eriksen RL, *et al.* Broadband focusing flat mirrors based on plasmonic gradient metasurfaces. Nano Letters. 2013;13(2): 829–834.

[79] Bossard JA, Lin L, Yun S, *et al.* Near-ideal optical metamaterial absorbers with super-octave bandwidth. ACS Nano. 2014;8(2):1517–1524.

[80] Kauranen M, Zayats AV. Nonlinear plasmonics. Nature Photonics. 2012; 6(11):737–748.

[81] Feichtner T, Selig O, Kiunke M, *et al.* Evolutionary optimization of optical antennas. Physical Review Letters. 2012;109(12).

[82] Forestiere C, Pasquale AJ, Capretti A, *et al.* Genetically engineered plasmonic nanoarrays. Nano Letters. 2012;12(4):2037–2044.

[83] Feichtner T, Selig O, Hecht B. Plasmonic nanoantenna design and fabrication based on evolutionary optimization. Optics Express. 2017;25(10): 10828–10842.

[84] Eibert TF, Volakis JL, Wilton DR, *et al.* Hybrid FE/BI modeling of 3-D doubly periodic structures utilizing triangular prismatic elements and an MPIE formulation accelerated by the Ewald transformation. IEEE Transactions on Antennas and Propagation. 1999;47(5):843–850.

[85] Babaei F, Javidnasab M, Rezaei A. Supershape nanoparticle plasmons. Plasmonics. 2018;13(4):1491–1497.

[86] Yu N, Genevet P, Kats MA, *et al.* Light propagation with phase discontinuities: generalized laws of reflection and refraction. Science, New Series. 2011;334(6054):333–337.

[87] Ding F, Pors A, Bozhevolnyi SI. Gradient metasurfaces: a review of fundamentals and applications. Reports on Progress in Physics. 2018;81 (2):026401.

[88] Arbabi A, Horie Y, Bagheri M, *et al.* Dielectric metasurfaces for complete control of phase and polarization with subwavelength spatial resolution and high transmission. Nature Nanotechnology. 2015;10(11):937–943.

[89] Yu N, Capasso F. Flat optics with designer metasurfaces. Nature Materials. 2014;13(2):139–150.

[90] Genevet P, Capasso F, Aieta F, *et al.* Recent advances in planar optics: from plasmonic to dielectric metasurfaces. Optica. 2017;4(1):139.

[91] Arbabi A, Arbabi E, Horie Y, *et al.* Planar metasurface retroreflector. Nature Photonics. 2017;11(7):415–420.

[92] Zuo H, Choi DY, Gai X, *et al.* High-efficiency all-dielectric metalenses for mid-infrared imaging. Advanced Optical Materials. 2017;5(23):1700585.

[93] Hogan H. For IR imaging, better SWaP beckons. Photonics Spectra. 2015; 49(8):48–51.

[94] Nagar J, Campbell SD, Werner DH. Apochromatic singlets enabled by metasurface-augmented GRIN lenses. Optica. 2018;5(2):99–102.

[95] Krasnok AE, Miroshnichenko AE, Belov PA, *et al.* All-dielectric optical nanoantennas. Optics Express. 2012;20(18):20599–20604.

[96] Shalaev MI, Sun J, Tsukernik A, *et al.* High-efficiency all-dielectric metasurfaces for ultracompact beam manipulation in transmission mode. Nano Letters. 2015;15(9):6261–6266.

[97] Zhang L, Ding J, Zheng H, *et al.* Ultra-thin high-efficiency mid-infrared transmissive Huygens meta-optics. Nature Communications. 2018;9(1):1481.

[98] Devlin RC, Khorasaninejad M, Chen WT, *et al.* Broadband high-efficiency dielectric metasurfaces for the visible spectrum. Proceedings of the National Academy of Sciences. 2016;113(38):10473–10478.

[99] Kamali SM, Arbabi E, Arbabi A, *et al.* A review of dielectric optical metasurfaces for wavefront control. Nanophotonics. 2018;7(6):1041–1068.

[100] Ong JR, Chu HS, Chen VH, *et al.* Freestanding dielectric nanohole array metasurface for mid-infrared wavelength applications. Optics Letters. 2017;42(13):2639.

[101] Egorov V, Eitan M, Scheuer J. Genetically optimized all-dielectric metasurfaces. Optics Express. 2017;25(3):2583.

[102] Sui S, Ma H, Lv Y, *et al.* Fast optimization method of designing a wideband metasurface without using the Pancharatnam–Berry phase. Optics Express. 2018;26(2):1443.

[103] Jafar-Zanjani S, Inampudi S, Mosallaei H. Adaptive genetic algorithm for optical metasurfaces design. Scientific Reports. 2018;8(1).

[104] Campbell SD, Zhu DZ, Nagar J, *et al.* Inverse design of engineered materials for extreme optical devices. In: 2018 International Applied Computational Electromagnetics Society Symposium (ACES); 2018. p. 1–2.

[105] COMSOL Multiphysics v. 5.3. www.comsol.com. COMSOL AB, Stockholm, Sweden.

[106] Soukoulis CM, Wegener M. Past achievements and future challenges in the development of three-dimensional photonic metamaterials. Nature Photonics. 2011;5(9):523–530.

[107] Winkler R, Schmidt FP, Haselmann U, *et al.* Direct-write 3D nanoprinting of plasmonic structures. ACS Applied Materials & Interfaces. 2017;9(9): 8233–8240.

[108] Leong ESP, Deng J, Khoo EH, *et al.* Fabrication of suspended, three-dimensional chiral plasmonic nanostructures with single-step electron-beam lithography. RSC Advances. 2015;5(117):96366–96371.

[109] Wu W, Kim E, Ponizovskaya E, *et al.* Optical metamaterials at near and mid-IR range fabricated by nanoimprint lithography. Applied Physics A. 2007;87(2):143–150.

[110] Zhu DZ, Gregory MD, Werner PL, *et al.* Fabrication and characterization of multiband polarization independent 3-D-printed frequency selective structures with ultrawide fields of view. IEEE Transactions on Antennas and Propagation. 2018;66(11):6096–6105.

[111] Stützle T, Hoos HH. MAX–MIN ant system. Future Generation Computer Systems. 2000;16(8):889–914.

[112] Charbonneau D, Poff C, Nguyen H, *et al.* Who are the "lazy" ants? the function of inactivity in social insects and a possible role of constraint: inactive ants are corpulent and may be young and/or selfish. Integrative and Comparative Biology. 2017;57(3):649–667.

[113] Burckel DB, Wendt JR, Eyck GAT, *et al.* Fabrication of 3D metamaterial resonators using self-aligned membrane projection lithography. Advanced Materials. 2010;22(29):3171–3175.

[114] Burckel DB, Wendt JR, Eyck GAT, *et al.* Micrometer-scale cubic unit cell 3D metamaterial layers. Advanced Materials. 2010;22(44):5053–5057.

[115] Burckel BD, Finnegan PS, David Henry M, *et al.* Oblique patterned etching of vertical silicon sidewalls. Applied Physics Letters. 2016;108(14): 142103.

[116] Hu X, Wei X. Metallic metasurface for high efficiency optical phase control in transmission mode. Optics Express. 2017;25(13):15208–15215.

Chapter 6

Multi-level carrier kinetics models for computational nanophotonics

Shaimaa I. Azzam[1] *and Alexander V. Kildishev*[1]

Multi-level rate equations (MREs) have been extremely useful for describing active laser media (gain media) in response to an external pulse [1]. This approach has also been critical for the analysis of lasing dynamics including pumping, population inversion, and saturation. Due to the limited computational power, approximate nonlinear material models were employed to mimic the material response in the time and frequency domains [2]. Relying on perturbation theory, power series expansions of the nonlinear susceptibility have been frequently used to model the nonlinear interactions of light waves with materials [3]. This approach, however, has many limitations including the inability to capture transient response, failure to converge in some crucial real-life cases, and incapacity to account for the quantum effects. Moreover, it is proved to be inadequate for modeling nonlinearities in some singular classes of materials, e.g., epsilon-near-zero materials [4]. Currently, with the advancement and the availability of high-performance computing (HPC) machinery, more full-wave analysis techniques are being adopted to enable better predictive power of more advanced numerical models at a sufficiently low computational cost. Some of these methods, including those described in this chapter, are critical components of the multiphysics computational frameworks that provide a more realistic physical picture of light-matter interaction dynamics, opening up pathways for design and optimization of nonlinear photonic devices and systems.

This chapter describes the interaction of the electromagnetic fields with nonlinear media by using a semi-classical approach in which the carriers of the nonlinear medium are treated quantum-mechanically using a multi-level quantum system based on the type of the material while the electromagnetic waves are treated classically with Maxwell's equations [5–8]. Our discussion in the chapter focuses mainly on the work done using the finite difference time domain method (FDTD) [9] with the auxiliary differential equation (ADE) technique [10]. The ADE-FDTD approach is one of the most versatile techniques as it allows direct integration of other material properties such as nonlinearity, dispersion,

[1]School of Electrical and Computer Engineering and Birck Nanotechnology Center, Purdue University, West Lafayette, IN, USA

nonlocality, and others. In Section 6.1, we address the use of various MRE for the numerical modeling of several essential cases discussing the interaction of light with gain media; selected examples include spontaneous emission [11] and plasmonic [6,8,12] and random lasers [13]. The numerical simulations of saturable absorption [14–16] and reverse saturable absorption [7] are discussed in Section 6.2. In Section 6.3, we discuss the application of MRE to simulate two-photon absorption [17] that could be directly generalized to other more sophisticated multiphoton absorption processes.

6.1 Gain media models

The ADE-FDTD method has been used to model the gain material response in laser systems using the frequency-dependent negative conductivity term [18]. This approximation holds when the intensity of the input light is small, and the pulses are slowly varying. For rapidly changing pulses and high light intensities, the optical material responses may become significantly nonlinear, and their complete physics-based dependence on the input intensity should be incorporated. Rate equations have been used to describe the excitation and relaxation dynamics in gain media for a long time [1]. However, it was not until 1998 that they were adopted as a material model in the time domain [19]. It has been shown that such a model is fully explicit, eliminating the need for iterative schemes. This physics-based approach can capture the rich dynamics of gain media such as dephasing, saturation, etc. However, it did not include the Pauli's exclusion principle which was later accounted for [6]. More recently, the four-level atom model has been extended to six- and ten-level models to account for more complex atomic transition topologies [8] and even for the photobleaching effect in organic dyes [12].

Figure 6.1 shows the Jablonski diagram of a four-level atomic system, Figure 6.1(a), and a six-level atomic system, Figure 6.1(b), that are used to model the excitation and relaxation dynamics in a gain medium. The system of MRE

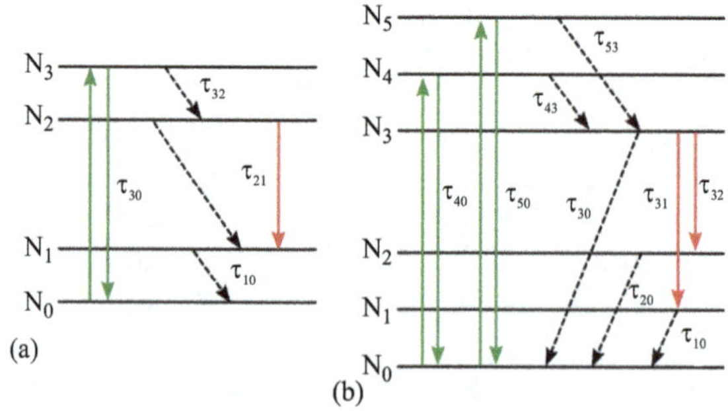

Figure 6.1 Jablonski diagrams of (a) four-level and (b) six-level atomic systems showing the allowed transitions and the corresponding lifetimes of a gain medium.

describing the different transitions and pumping dynamics in the four-level case is given by,

$$\frac{\partial N_3}{\partial t} = -\frac{N_3}{\tau_{32}} + \frac{1}{\hbar\omega_{30}}\mathbf{E}\cdot\frac{\partial \mathbf{P}_{30}}{\partial t} \tag{6.1a}$$

$$\frac{\partial N_2}{\partial t} = -\frac{N_2}{\tau_{21}} - \frac{N_3}{\tau_{32}} + \frac{1}{\hbar\omega_{21}}\mathbf{E}\cdot\frac{\partial \mathbf{P}_{21}}{\partial t} \tag{6.1b}$$

$$\frac{\partial N_1}{\partial t} = -\frac{N_1}{\tau_{10}} + \frac{N_2}{\tau_{21}} + \frac{1}{\hbar\omega_{21}}\mathbf{E}\cdot\frac{\partial \mathbf{P}_{21}}{\partial t} \tag{6.1c}$$

where N_3, N_2, and N_1 are the population density of carriers in the different energy levels. The total carrier density, N_d, is conserved such that $N_d = N_3 + N_2 + N_1 + N_0$ at any time and space point. The lifetimes of the transition between energy level $|i\rangle$ to level $|j\rangle$ are given by τ_{ij} as depicted in Figure 6.1. The energy between levels $|3\rangle$ and $|0\rangle$ corresponds to the angular frequency ω_{30} and that between $|1\rangle$ and $|2\rangle$ corresponds to ω_{21}. The term $\frac{1}{\hbar\omega_{ij}}\mathbf{E}\cdot\frac{\partial \mathbf{P}_{ij}}{\partial t}$ represents the coherent emission and absorption between levels $|i\rangle$ and $|j\rangle$.

The induced macroscopic polarizations due to the transitions satisfy the Lorentz ordinary differential equations,

$$\frac{\partial^2 \mathbf{P}_{ij}}{\partial t^2} + \gamma_{ij}\frac{\partial \mathbf{P}_{ij}}{\partial t} + \omega_{ij}^2 \mathbf{P}_{ij} = \boldsymbol{\kappa}_{ij}\left(N_j - N_i\right)\mathbf{E} \tag{6.2}$$

where γ_{ij} is the dephasing constant between levels $|i\rangle$ and $|j\rangle$. The source term in (6.2) is proportional to a product of the local electric field and the difference in populations between the upper and the lower levels of this transition with a proportionality constant κ_{ij}, where $ij \in \{30, 21\}$. The coupling coefficient κ_{ij} represents the dipole matrix element between levels $|i\rangle$ and $|j\rangle$ and is given by $\kappa_{ij} = 2\omega_{ij}|\mu_{ij}|^2/\hbar = 6\pi\varepsilon_0 c^3/\omega_{ij}^2\tau_{ij}$, where ε_0 and c are the free space permittivity and speed of light, respectively, and τ_{ij} is the decay time constant between levels $|i\rangle$ and $|j\rangle$ [19,20].

The polarization vectors are then coupled to Maxwell's curl equations,

$$\nabla \times \mathbf{H}(t, \mathbf{r}) = \varepsilon_0\varepsilon_{\mathrm{h}}\frac{\partial \mathbf{E}(t, \mathbf{r})}{\partial t} + \frac{\partial \mathbf{P}(t, \mathbf{r})}{\partial t}, \tag{6.3}$$

$$\nabla \times \mathbf{E}(t, \mathbf{r}) = -\mu_0\frac{\partial \mathbf{H}(t, \mathbf{r})}{\partial t}, \tag{6.4}$$

where μ_0 is the permeability of free space. Equations (6.1)–(6.4) are then solved simultaneously with a full-wave solver on a staggered Yee grid by applying the classical finite-difference approximation to the spatial and time derivatives [9,20]. The time-dependent electromagnetic fields are recorded and converted to the frequency domain using the fast Fourier transform to obtain the transmission and reflection coefficients.

Here, we show an application example of using a four-level system coupled to Maxwell's equations to model lasing dynamics in a silver nanohole array [8,21]. The structure, depicted in Figure 6.2(a), consists of an array of silver nanoholes with a periodicity of 565 nm, 175 nm diameter, and 100 nm thickness [21]. The array is covered with a layer of Rhodamine-101 (R101) dye in PVA and exhibits a resonance around 617 nm [21], close to the emission line (605 nm) of R101. The 2-μm-thick R101-PVA sample is illuminated by two beams, a 595-nm narrowband pump beam and a femtosecond probe beam delayed by 2 ps. Total R101 carrier density is 10 mM $\left(N_{\mathrm{d}} = 6.0221 \times 10^{24}\mathrm{m}^{-3}\right)$. The four-level parameters retrieved from the experiment are as follows: transition wavelengths $\lambda_{30} = 575$ nm and $\lambda_{21} = 605$ nm; dephasing times $T_{2,30} = 9$ fs and $T_{2,21} = 25.5$ fs; decay times $\tau_{\mathrm{r},30} = 5$ ns, $\tau_{\mathrm{nr},32} = 0.3$ ps, $\tau_{\mathrm{r},21} = 6$ ns, $\tau_{\mathrm{nr},21} = 7.33$ ns, and $\tau_{\mathrm{nr},10} = 0.35$ ps. The lasing peak emerges at 617 nm with increasing normalized intensity as we increase the pump power. The simulation results for the lasing spectrum are shown in Figure 6.2(c) and are in perfect agreement with the results from the experiment depicted in Figure 6.2(b) [21].

A crucial advantage of the full-wave numerical framework built on MRE is that it allows access to the spatiotemporal internal system dynamics such as the population inversion distribution as a function of time and space similar to examples depicted in Figure 6.3. Hence, the simulations uncover the critical spatiotemporal interplay in the interaction between population inversion and electromagnetic fields within the lasing action. For input powers above the threshold, 3.5 mW, the stimulated emission appears to be several orders of magnitude higher than the spontaneously emitted electric fields for 0.5 mW pumping power, which is below the threshold. These findings confirm that the silver nanoholes array supports SPP-Bloch waves which exhibit coherent, constructive interference of SPP waves, and they are amplified by the gain medium. Figure 6.3 also shows that population inversion is achieved in the system both below the threshold, Figure 6.3(c), and above the threshold, Figure 6.3(d). Below the lasing threshold, the population of the upper lasing level $|2\rangle$ stays at ~45% (the lower level $|1\rangle$ is almost empty).

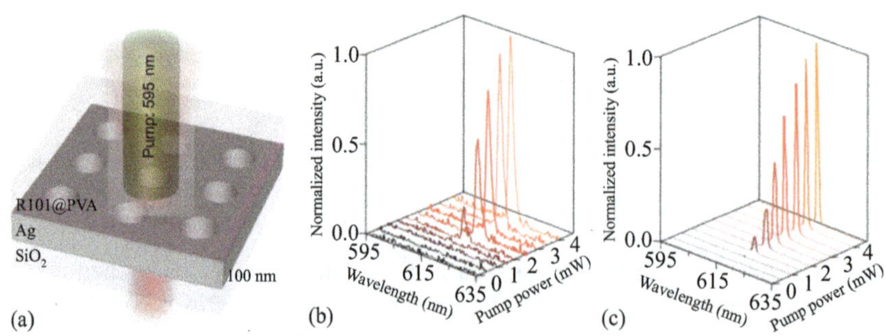

Figure 6.2 Evolution of the emission spectra vs. increasing pump power; (b) experimental and (c) simulation results. © 2018. Reprinted, with permission, from [8].

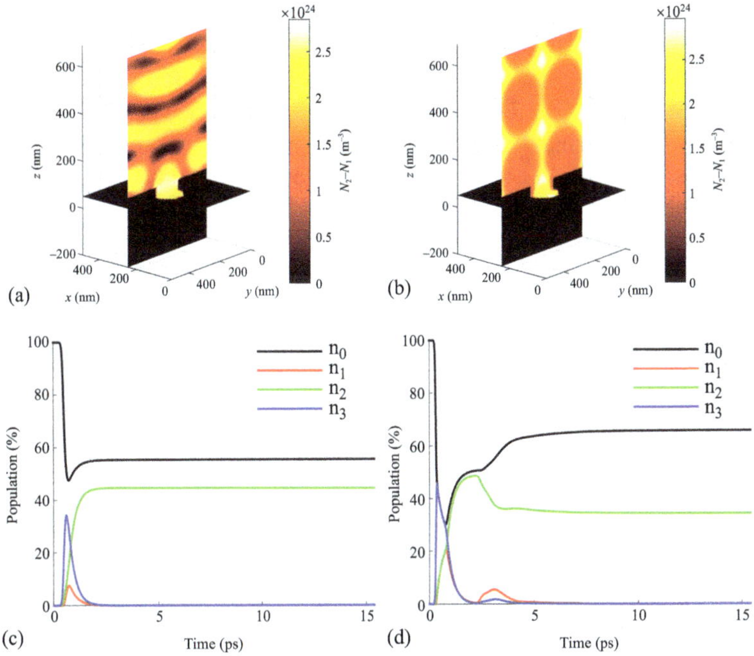

Figure 6.3 The simulated results of the steady state population inversion distribution $(N_2 - N_1)$ *for the pumping powers of (a) 0.05 mW and (b) 3.5 mW. The time domain population evolution at the position* ($x = 250$ nm, $y = 250$ nm, $z = 50$ nm) *for pumping power of (c) 0.5 mW and (d) 3.5 mW.*

However, above the threshold, stimulated transitions in the strong localized electromagnetic fields produced by lasing deplete the population of the excited-state molecules from ~48% to ~35% of the total population. More interestingly, when the system is above the threshold, the spatial profile of the population density $(N_2(\vec{r}, t) - N_1(\vec{r}, t))$ inversely correlates with the electromagnetic mode profile because of the inversion depletion by strong lasing fields.

This framework and its application-specific variations have been successfully used to design, simulate, and optimize numerous active photonics systems including nanolasers and spasers [8,22,23], organic dye nanolasers [12,24], plasmonic nanolasers with stopped-light [25,26], and all-optical computing [27]. The four-level rate equations coupled to Maxwell's equations have also been employed to simulate disordered systems such as random lasers [13]. The MRE-based numerical framework in [13] has been used to simulate the field patterns and spectra of localized lasing modes inside a random laser and to capture the mode competition and saturation as a function of the pumping rate and system length.

6.2 Saturable absorbing media

Materials exhibiting saturation of absorption have a broad scope of applications. Saturable absorbing media find their applications in passive mode-locking [28,29] and passive Q-switching [30–32] to generate ultrashort laser pulses. Reverse saturable absorbers (RSA) are commonly used for optical limiters (OL) to protect optical devices and human eyes from intense lasers [33–38]. Material models for saturable absorbing media have generally been limited to phenomenological approaches where the absorption coefficient is linked to some saturation intensity value taking the form:

$$\alpha(I) = \frac{\alpha_0}{1 + I/I_s}, \tag{6.5}$$

where $I = 2n\varepsilon_0 c|E|^2$ is the light intensity, n is the material linear refractive index, I_s is the saturation intensity, and α_0 is the linear absorption coefficient.

Classically, the modeling of a nonlinear optical material response has been accomplished by expanding the susceptibility in a power series to connect the material response to the electric field [2],

$$P(t) = \varepsilon_0\chi^1 E(t) + \varepsilon_0\chi^2 E(t)^2 + \varepsilon_0\chi^3 E(t)^3 + \ldots. \tag{6.6}$$

However, this model puts forward many limitations. First, the power series expansion of the absorption coefficient does not converge for $I > I_s$ [2]. Second, it assumes that the material response to light is instantaneous, which is nonphysical. Also, it gives no direct access to the electric or magnetic fields inside the medium and provides no means to predict the spatiotemporal dynamics inside the absorbing medium. That is why more advanced models employing carrier kinetics and based on the MREs have been developed to model the excitation and relaxation dynamics inside the absorbing materials accurately. However, until recently, the majority of these models relied on linking the MRE systems to the absorption coefficient and the field intensity instead of the field itself [39–42]. Using the MREs, those models mimicked a wide range of nonlinear absorption phenomena such as saturable absorption [39,40], reverse saturable absorption [41], two- and multiple photon absorption, and excited-state absorption [42]. However, that modeling approach did not provide full-wave solutions and hence also suffered from significant limitations including the inability to accurately simulate illumination with tightly focused or tightly structured beams, along with the incapacity to handle resonant nano-structured composite elements.

In the following section, we focus on the ADE-FDTD approaches developed to model saturable absorbing media. Section 6.2.1 deals with the saturable absorbers (SAs) while Section 6.2.2 presents the time domain models for RSAs.

6.2.1 Saturable absorption models with MRE

An SA is a material with a nonlinear absorption coefficient that decreases as the incident light intensity increases. Most materials exhibit saturation of their absorption. However, this typically occurs at very high intensities close to the damage threshold of the material. When the light of weak intensity is incident on the SA, a photon is absorbed, and a molecule is excited from the ground to a higher energy atomic level causing the absorption of light. Due to the abundance of absorbent molecules in the ground state, the material can absorb more photons. However, when the lifetime of the higher energy state is high enough that the atoms cannot decay to participate in the absorption process again, saturation occurs. A broad range of materials exhibit saturable absorption including 2D materials such as graphene [43], MoS_2 [44,45], black phosphorus [46], carbon nanotubes [47,48], topological insulators [49], along with dielectric [50,51] and metallic [52,53] nanoparticles.

To overcome the shortcomings of the modeling techniques mentioned earlier, more physics-based models using the population kinetics with the MRE have been introduced [14,16,54,55]. Initially, only steady-state solutions were considered in those studies [54,55]. An important nonlinear generalization of the Lorentz model has been applied by Varin *et al.* to model second- and third-order optical nonlinearities [56,57] with FDTD. Recently, two- and four-level atom models have been used to simulate SA in the time domain using the ADE-FDTD framework [14,16]. The MRE system describing the transition kinetics in the four-level atom (see the Jablonski diagram in Figure 6.4(b)) is given by,

$$\frac{\partial N_3}{\partial t} = -\frac{N_3}{\tau_{30}} - \frac{N_3}{\tau_{31}} - \frac{N_3}{\tau_{32}} + \frac{1}{\hbar\omega_{30}}\mathbf{E}\cdot\frac{\partial \mathbf{P}_{30}}{\partial t} \tag{6.7a}$$

$$\frac{\partial N_2}{\partial t} = -\frac{N_2}{\tau_{20}} - \frac{N_2}{\tau_{21}} + \frac{N_3}{\tau_{32}} + \frac{1}{\hbar\omega_{20}}\mathbf{E}\cdot\frac{\partial \mathbf{P}_{20}}{\partial t} \tag{6.7b}$$

$$\frac{\partial N_1}{\partial t} = -\frac{N_1}{\tau_{10}} + \frac{N_2}{\tau_{21}} + \frac{N_3}{\tau_{31}} + \frac{1}{\hbar\omega_{10}}\mathbf{E}\cdot\frac{\partial \mathbf{P}_{10}}{\partial t} \tag{6.7c}$$

where N_1, N_2, N_3, and N_4 are the population density of carriers at the different energy levels. The system describes the absorption of the incident light at three distinct frequencies, ω_{30}, ω_{20}, and ω_{10}, as well as the rapid decay of the excited-state population from higher levels to lower energy levels due to non-radiative

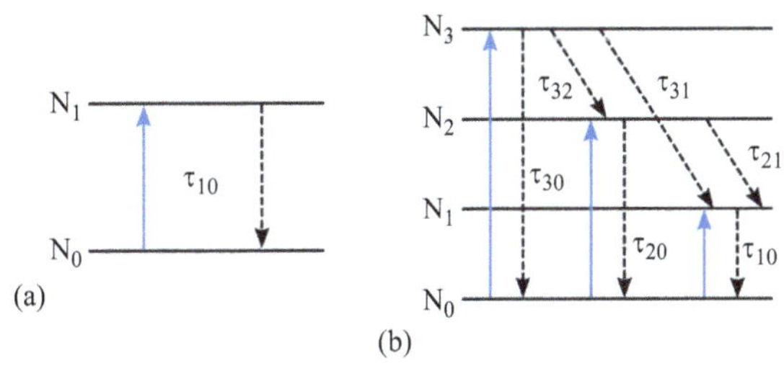

Figure 6.4 Jablonski diagrams of (a) a two-level and (b) a four-level atomic systems showing the allowed transitions and the corresponding lifetimes of a saturable absorbing medium.

relaxations. The transition lifetimes between levels i and j are τ_{ij}, as shown in Figure 6.4(b).

6.2.2 Modeling reverse saturable absorption with MRE

Reverse saturable absorption is a property of some nonlinear optical materials which exhibit an increased absorption when the light intensity incident on the material increases. A wide range of materials can exhibit reverse saturable absorption such as metals [34], organic dyes [58,59], clustered metal particles [60], synthetic compounds [61], carbon nanotubes [62], and two-dimensional materials [63]. RSAs are generally used for OL to protect optical devices from intense laser light and in protective goggles.

Recently, a full wave time-domain framework for RSAs has been implemented using the ADE-FDTD method based on carrier kinetics with a system of MRE [7]. This framework provides a physics-based numerical modeling environment for optimizing RSAs with the ability to capture the reverse saturation dynamics inside the material. As a result, it makes possible the real-life experimental-based design and optimization of the devices and systems utilizing RSAs including OL [37,38] and all-optical computing [27].

Unlike other nonlinear phenomena, there are surprisingly few efforts towards the developments of material models for RSAs. A material model compatible with full-wave solutions would help the development and optimization of OL devices with an overall objective of reducing their intensity threshold and hence drastically expanding their areas of applicability. Recently, the systems of MRE have been used to model RSAs in the time domain using the ADE-FDTD framework [7]. The Jablonski diagrams in Figure 6.5 show two possible RSA topologies. In Figure 6.5(a), the RSA is associated with the extra absorption by the triplet pair $(T_2 - T_1)$. The condition for proper RSA operation, in this case, is that the absorption cross-section of the triplet state (σ_{T}) is larger than that of the ground singlet state (σ_{S}). In the topology shown in Figure 6.5(b), there are three ways how the RSA can function: due to absorption of the triplet state, or absorption on the higher singlet state, or a combination of both.

The following set of rate equations describe the transitions shown in Figure 6.5(a):

$$\frac{\partial N_{\mathrm{S0}}}{\partial t} = \frac{N_{\mathrm{S1}}}{\tau_{\mathrm{S1}}} + \frac{N_{\mathrm{T1}}}{\tau_{\mathrm{T1}}} - \frac{1}{\hbar\omega_0}\mathbf{E}\cdot\frac{\partial \mathbf{P}_{\mathrm{G}}}{\partial t}, \tag{6.8a}$$

$$\frac{\partial N_{\mathrm{S1}}}{\partial t} = -\frac{N_{\mathrm{S1}}}{\tau_{\mathrm{S1}}} - \frac{N_{\mathrm{S1}}}{\tau_{\mathrm{ISC}}} + \frac{1}{\hbar\omega_0}\mathbf{E}\cdot\frac{\partial \mathbf{P}_{\mathrm{G}}}{\partial t}, \tag{6.8b}$$

$$\frac{\partial N_{\mathrm{T1}}}{\partial t} = -\frac{N_{\mathrm{T1}}}{\tau_{\mathrm{T1}}} + \frac{N_{\mathrm{S1}}}{\tau_{\mathrm{ISC}}} + \frac{N_{\mathrm{T2}}}{\tau_{\mathrm{T2}}} - \frac{1}{\hbar\omega_0}\mathbf{E}\cdot\frac{\partial \mathbf{P}_{\mathrm{T}}}{\partial t}, \tag{6.8c}$$

$$\frac{\partial N_{\mathrm{T2}}}{\partial t} = -\frac{N_{\mathrm{T2}}}{\tau_{\mathrm{T2}}} + \frac{1}{\hbar\omega_0}\mathbf{E}\cdot\frac{\partial \mathbf{P}_{\mathrm{T}}}{\partial t}, \tag{6.8d}$$

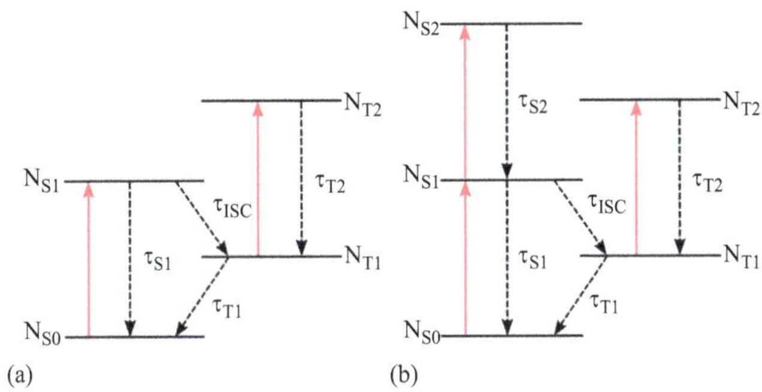

Figure 6.5 The Jablonski diagrams of an RSA represented by (a) four-level and (b) five-level atomic systems with the allowed transitions and the sets of corresponding lifetimes.

where ω_0 is the frequency of the absorbed light. The population densities in the lower and upper singlet and triplet states, respectively, are N_{S0}, N_{S1}, N_{T1}, and N_{T2}. The total population, $N_d = N_{S0} + N_{S1} + N_{T1} + N_{T2}$, is conserved at all times, where N_d is the density of absorbing atoms. The lifetime of state S_1 is $1/(1/\tau_{S1} + 1/\tau_{ISC})$, where τ_{ISC} is the intersystem crossing lifetime. Lifetimes of the upper and lower triplet states are τ_{T2} and τ_{T1} (Figure 6.5(a)). The term $\frac{1}{\hbar\omega_0}\mathbf{E}\cdot\frac{\partial \mathbf{P}_{ij}}{\partial t}$ represents the coherent absorption between levels $|i\rangle$ and $|j\rangle$. The averaged polarization vectors are governed by (6.2) and they are solved in the same way as previously discussed in Section 6.1.

The dynamics of a four-level atomic level for a 1-μm thin film RSA material are shown in Figure 6.6. The lifetimes and scattering rate parameters of the model are: $\tau_{T2} = 1$ ps, $\tau_{T1} = 300$ ns, $\tau_{S1} = 1$ ns, $\tau_{ISC} = 10$ ps, and $\gamma_G = \gamma_T = 10^{14}$ Hz and the host medium's refractive index is assumed to be 1. The density of the carriers is 0.3 mM $\left(1.806 \times 10^{17}\text{cm}^{-3}\right)$ and $\omega_0 = 2\pi/\lambda_0$, where λ_0 is 532 nm. The RSA layer is illuminated with two pulses (a strong 1-ps pump pulse followed by a weak 50-fs probe) separated by a short delay of 5 ps. The wavelength of the pump and the probe is 532 nm. The nonlinear transmission and absorption of the RSA layer are studied at a fixed probe fluence of 100 nJ/cm^2 and a pump fluence varying from 10^{-3} μJ/cm^2 to 10^2 μJ/cm^2. We observe the expected RSA behavior in Figure 6.6(a); the transmittance is constant at low input fluences, but it becomes nonlinear at higher fluence values. The saturation fluence for the RSA layer is around 2.5 μJ/cm^2 in this case. At a pump fluence of 10^{-2} μJ/cm^2, well into the linear regime, the normalized ground state population density is almost unity during all times, and consequently, the carrier build-up in the upper states is negligible. Figure 6.6(b) shows the time evolution of the macroscopic polarization densities at the ground (P_G), and excited triplet (P_T) states. As expected in this case, P_T is almost negligible. For clarity, we normalized all polarizations to the maximum value of P_G. Further, in the nonlinear regime with a pump fluence of 50 μJ/cm^2,

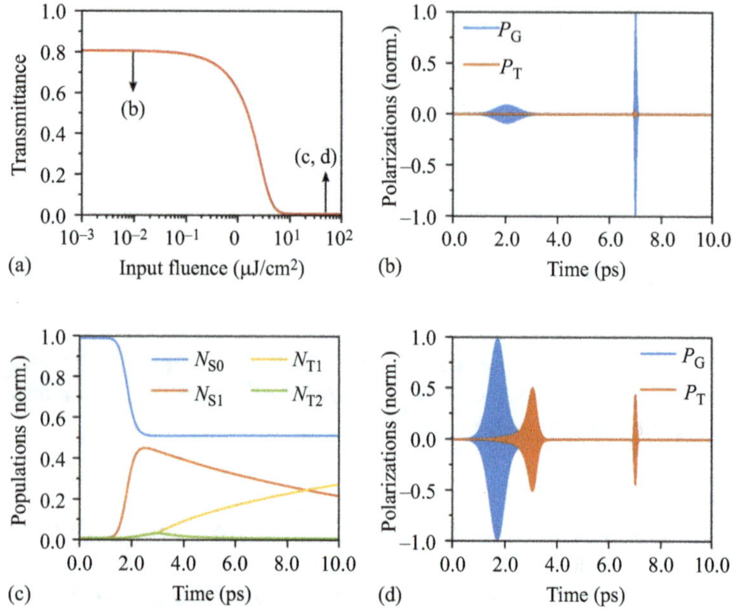

Figure 6.6 Simulated behavior and time dynamics of a four-level RSA system: (a) transmittance vs. input fluence of a 1-μm film and (b) induced polarizations densities inside the RSA for a pump fluence of 0.01 μJ/cm^2. The polarizations are governed by the ground state, while the triplet state contribution in the linear regime is negligible (not shown). Nonlinear time dynamics for a pump fluence of 50 μJ/cm^2: (c) population densities and (d) polarizations. Reproduced from [7].

the carrier dynamics depicted in Figure 6.6(c) start showing depopulation of the ground state and increased build-up of triplet state density. Hence, P_T starts contributing to the material response, see Figure 6.6(d). We may notice that in this case P_T entirely dominates the response of the system to the probe (around 7 ps), which explains the enhanced absorption.

Reducing the light intensity required for activating an OL is crucial for expanding their areas of application. For example, in protective goggles, OL devices should be activated at normal levels of the incoherent light illumination. From the reverse saturable absorption materials standpoint, this translates to a very long triplet lifetime. Thus far, most of the development in this direction has been focused on developing materials with these favorable properties. However, here we use the multi-level material models developed for RSAs to optimize OL devices where, due to the local field enhancement from plasmonic structures, the nonlinear response from RSAs is further intensified. This way, the light intensity required for

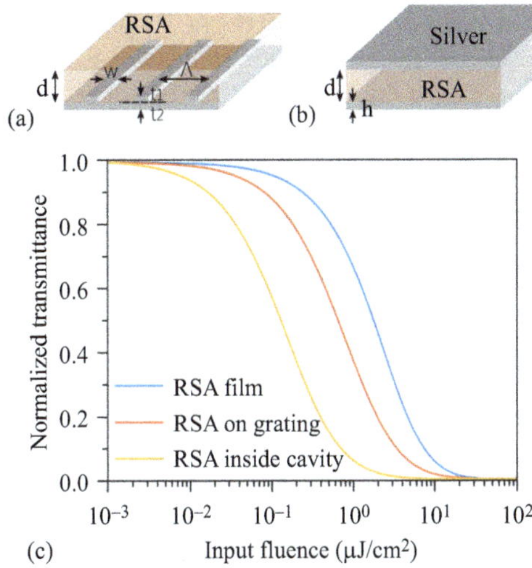

Figure 6.7 Enhancement of OL with plasmonic structures. 200-nm-thick RSA film deposited: (a) on top of a silver grating; or (b) between two thin silver layers; (c) normalized transmittance vs. pump fluence for the following cases: a bare 200 nm RSA layer, the same RSA layer on top of a silver grating (panel (a)), and the same RSA sandwiched between two thin silver films (panel (b)). Reproduced from [7].

activating a given OL is significantly reduced. As a proof of concept, we design two plasmonic structures with an embedded RSA to study the reduction in the saturation energy. Figure 6.7(a) shows the same RSA, with a thickness d of 200 nm on top of a silver grating. The bottom silver film is 20-nm thin (t_2) and the periodic ridges of the grating are 10-nm high (t_1) and 50-nm wide (w) . The pitch of the grating structure (Λ) is 500 nm, and the optical properties of silver are obtained from [64]. Figure 6.7(b) illustrates the second case, where the same 200-nm RSA layer is sandwiched between two silver films with a thickness of $h = 10$ nm. We optimized the geometrical parameters of the grating and the cavity to maximize the linear transmission and reduce the saturation intensity. Spatial grid steps were set to 10 nm and 2.5 nm inside the RSA layer and silver films, respectively. The linear transmittance is 0.93 in the case of bare RSA film, 0.46 in the case of the RSA layer on the plasmonic grating, and 0.68 in the case of the RSA layer inside the cavity-like structure. Figure 6.7(c) shows the normalized transmittance as a function of the input fluence for all three structures under test: a bare 200-nm RSA, the 200-nm RSA film on the grating, and the 200-nm RSA inside the cavity. The transmittance in each case is normalized to its maximum for a more straightforward comparison. As observed in Figure 6.7(c), the saturation intensity for the bare RSA layer is $2.651\ \mu J/cm^2$, for the RSA layer on a silver grating is $1.019\ \mu J/cm^2$, and for the

RSA layer embedded in a cavity-like structure is 0.2006 μJ/cm^2. This example demonstrates a 2.6-fold reduction in the saturation intensity for the structure with the silver grating and 13.2-fold decrease in the case of plasmonic cavity configuration.

The availability of material models for the RSAs allows full-wave numerical simulations aimed at optimizing the nonlinear light-matter interaction metrics. These models enable the integration of other important multiphysics models that lay a necessary foundation for HPC machinery required for the experimental-based design, optimization, and realization of new OL devices, all-optical switching systems, and optical computing logic.

6.3 Multiphoton absorption models

The language of the time-domain MRE can be also straightforwardly applied to multiphoton absorption. A 1D FDTD model based on a system of MRE has been already used to model two-photon absorption (TPA) in organic dyes [17]. A six-level scheme can be extended to a 3D framework to model complex photonic devices with embedded TPA materials. Two-photon absorption ignores the absorption from the triplet states due to a long intersystem crossing lifetime. A virtual state (N_ν) is used in the model to simulate a delay between the absorption of the first photon (between N_0 and N_ν) and the second one (between N_ν and N_4). The MRE system for the band diagram shown in Figure 6.8 is defined as,

$$\frac{\partial N_4}{\partial t} = -\frac{N_4}{\tau_{42}} + \frac{1}{\hbar\omega_{\nu 4}} \mathbf{E} \cdot \frac{\partial \mathbf{P}_{\nu 4}}{\partial t}, \tag{6.9a}$$

$$\frac{\partial N_\nu}{\partial t} = -\frac{N_\nu}{\tau_{\nu 0}} + \frac{1}{\hbar\omega_{\nu 0}} \mathbf{E} \cdot \frac{\partial \mathbf{P}_{\nu 0}}{\partial t} - \frac{1}{\hbar\omega_{\nu 4}} \mathbf{E} \cdot \frac{\partial \mathbf{P}_{\nu 4}}{\partial t}, \tag{6.9b}$$

$$\frac{\partial N_3}{\partial t} = -\frac{N_3}{\tau_{32}} + \frac{1}{\hbar\omega_{30}} \mathbf{E} \cdot \frac{\partial \mathbf{P}_{30}}{\partial t}, \tag{6.9c}$$

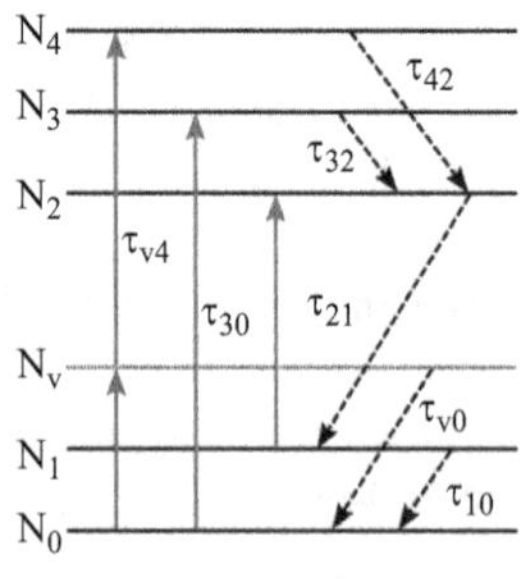

Figure 6.8 The Jablonski diagram of a TPA medium defined by its six-level topology, allowed transitions and corresponding lifetimes.

$$\frac{\partial N_2}{\partial t} = -\frac{N_2}{\tau_{21}} + \frac{N_3}{\tau_{32}} + \frac{N_4}{\tau_{42}} + \frac{1}{\hbar\omega_{12}}\mathbf{E}\cdot\frac{\partial \mathbf{P}_{12}}{\partial t}, \tag{6.9d}$$

$$\frac{\partial N_1}{\partial t} = -\frac{N_1}{\tau_{10}} + \frac{N_2}{\tau_{21}} - \frac{1}{\hbar\omega_{21}}\mathbf{E}\cdot\frac{\partial \mathbf{P}_{21}}{\partial t}, \tag{6.9e}$$

$$\frac{\partial N_0}{\partial t} = \frac{N_1}{\tau_{10}} + \frac{N_\nu}{\tau_{\nu 0}} - \frac{1}{\hbar\omega_{30}}\mathbf{E}\cdot\frac{\partial \mathbf{P}_{30}}{\partial t} - \frac{1}{\hbar\omega_{\nu 0}}\mathbf{E}\cdot\frac{\partial \mathbf{P}_{\nu 0}}{\partial t}, \tag{6.9f}$$

where N_0, N_1, N_ν, N_2, N_3, and N_4 are the population densities at the different energy levels. The above MRE are coupled to Maxwell's equations and solved as described in Section 6.1.

This model has successfully been used to match the spectral response of the organic 4,4′-bis(diphenylamino) stilbene (BDPAS) dye [65] at low and high excitation energies. The approach can be further generalized to two and three dimensions to allow for accurate full-wave analysis of structured nanophotonic devices with multiphoton absorption media.

To summarize, this chapter discussed an accurate and powerful semi-classical multiphysics approach to numerical modeling of several critical nonlinear phenomena employing carrier kinetics. We conclude that coupling a system of MRE with Maxwell's equations provides a versatile full-wave framework with significantly enhanced predictive power that is capable of simulating sophisticated nanostructured photonic and optoelectronic devices with embedded nonlinearities and vectorial input beams of arbitrary spatiotemporal complexity.

References

[1] Siegman A. Lasers. University Science Books, Mill Valley, CA; 1986.

[2] Boyd RW. Nonlinear optics. Cambridge, MA, USA: Academic Press; 2003.

[3] Bravo-Abad J, Fan S, Johnson SG, *et al.* Modeling nonlinear optical phenomena in nanophotonics. Journal of Lightwave Technology. 2007;25(9):2539–2546.

[4] Reshef O, Giese E, Alam MZ, *et al.* Beyond the perturbative description of the nonlinear optical response of low-index materials. Optics Letters. 2017;42(16):3225–3228.

[5] Nagra AS, York RA. FDTD analysis of wave propagation in nonlinear absorbing and gain media. IEEE Transactions on Antennas and Propagation. 1998;46(3):334–340.

[6] Chang SH, Taflove A. Finite-difference time-domain model of lasing action in a four-level two-electron atomic system. Optics Express. 2004;12 (16):3827–3833.

[7] Azzam SI, Kildishev AV Time-domain dynamics of reverse saturable absorbers with application to plasmon-enhanced optical limiters. Nanophotonics. 2018;8(1):145–151.

[8] Azzam SI, Fang J, Liu J, *et al.* Exploring time-resolved multiphysics of active plasmonic systems with experiment-based gain models. Laser & Photonics Reviews. 2019;13(1):1800071.

[9] Yee K. Numerical solution of initial boundary value problems involving Maxwell's equations in isotropic media. IEEE Transactions on Antennas and Propagation. 1966;14(3):302–307.
[10] Goorjian PM, Taflove A. Direct time integration of Maxwell's equations in nonlinear dispersive media for propagation and scattering of femtosecond electromagnetic solitons. Optics Letters. 1992;17(3):180–182.
[11] Vuckovic J, Painter O, Xu Y, *et al.* Finite-difference time-domain calculation of the spontaneous emission coupling factor in optical microcavities. IEEE Journal of Quantum Electronics. 1999;35(8):1168–1175.
[12] Chua SL, Zhen B, Lee J, *et al.* Modeling of threshold and dynamics behavior of organic nanostructured lasers. Journal of Materials Chemistry C. 2014;2 (8):1463–1473.
[13] Jiang X, Soukoulis C. Time dependent theory for random lasers. Physical Review Letters. 2000;85(1):70.
[14] Mock A. Modeling passive mode-locking via saturable absorption in graphene using the finite-difference time-domain method. IEEE Journal of Quantum Electronics. 2017;53(5):1–10.
[15] Hill S C, Videen G, Sun W. Scattering and internal fields of a microsphere that contains a saturable absorber: finite-difference time-domain simulations. Applied Optics. 2001;40(30):5487–5494.
[16] Azzam SI, Kildishev AV. Time-domain dynamics of saturation of absorption using multilevel atomic systems. Optical Materials Express. 2018;8 (12):3829–3834.
[17] Borneman JD. Optical metamaterials: Linear and nonlinear characterization and application to optical limiting. Purdue University; 2010.
[18] Hawkins R, Kallman J. Lasing in tilted-waveguide semiconductor laser amplifiers. Optical and Quantum Electronics. 1994;26(3):S207–S217.
[19] Nagra AS, York RA. FDTD analysis of wave propagation in nonlinear absorbing and gain media. IEEE Transactions on Antennas and Propagation. 1998;46(3):334–340.
[20] Chang S, Taflove A. Finite-difference time-domain model of lasing action in a four-level two-electron atomic system. Optics express. 2004;12(16):3827–3833.
[21] Meng X, Liu J, Kildishev AV, *et al.* Highly directional spaser array for the red wavelength region. Laser & Photonics Reviews. 2014;8(6):896–903.
[22] Yang A, Hoang TB, Dridi M, *et al.* Real-time tunable lasing from plasmonic nanocavity arrays. Nature Communications. 2015;6:6939.
[23] Wang D, Yang A, Wang W, *et al.* Band-edge engineering for controlled multi-modal nanolasing in plasmonic superlattices. Nature Nanotechnology. 2017;12(9):889.
[24] Coles DM, Trichet AA, Dolan PR, *et al.* Diffusion-driven continuous-wave-pumped organic dye lasers. Laser & Photonics Reviews. 2015;9(5):538–544.
[25] Wuestner S, Pickering T, Hamm JM, *et al.* Ultrafast dynamics of nanoplasmonic stopped-light lasing. Faraday Discussions. 2015;178:307–324.

[26] Pickering T, Hamm JM, Page AF, *et al.* Cavity-free plasmonic nanolasing enabled by dispersionless stopped light. Nature Communications. 2014;5:4972.
[27] Miscuglio M, Mehrabian A, Hu Z, *et al.* All-optical nonlinear activation function for photonic neural networks. arXiv preprint arXiv:181001216. 2018.
[28] Ippen E, Shank C, Dienes A. Passive mode locking of the cw dye laser. Applied Physics Letters. 1972;21(8):348–350.
[29] Paschotta R, Keller U. Passive mode locking with slow saturable absorbers. Applied Physics B. 2001;73(7):653–662.
[30] Xiao G, Bass M. A generalized model for passively Q-switched lasers including excited state absorption in the saturable absorber. IEEE Journal of Quantum Electronics. 1997;33(1):41–44.
[31] Zhang X, Zhao S, Wang Q, *et al.* Optimization of Cr-doped saturable-absorber Q-switched lasers. IEEE Journal of Quantum Electronics. 1997;33 (12):2286–2294.
[32] Morris JA, Pollock CR. Passive Q switching of a diode-pumped Nd: YAG laser with a saturable absorber. Optics Letters. 1990;15(8):440–442.
[33] Hollins RC. Materials for optical limiters. Current Opinion in Solid State and Materials Science. 1999;4(2):189–196.
[34] Prem Kiran P, Shivakiran Bhaktha BN, Narayana Rao D. Nonlinear optical properties and surface-plasmon enhanced optical limiting in Ag–Cu nanoclusters co-doped in SiO_2 sol-gel films. Journal of Applied Physics. 2004;96(11):6717–6723.
[35] Xia T, Hagan DJ, Dogariu A, Said AA, van Stryland EW. Optimization of optical limiting devices based on excited-state absorption. Applied Optics. 1997;36(18):4110–4122.
[36] Lepkowicz R, Hagan DJ, Kobyakov A, van Stryland EW. Picosecond optical limiting in reverse saturable absorbers: a theoretical and experimental study. Journal of the Optical Society of America B. 2002;19(1):94–101.
[37] Hirata S, Totani K, Yamashita T, Adachi C, Vacha M. Large reverse saturable absorption under weak continuous incoherent light. Nature Materials. 2014;13(10):938.
[38] Hirata S, Vacha M. Large reverse saturable absorption at the sunlight power level using the ultralong lifetime of triplet excitons. The Journal of Physical Chemistry Letters. 2017;8(15):3683–3689.
[39] Huff L, DeShazer L. Saturation of optical transitions in organic compounds by laser flux. Journal of the Optical Society of America. 1970;60(2):157–165.
[40] Sharan A, Sharma RC, Sandhya S, *et al.* Modeling absorption in saturable absorbers. Optics Communications. 2001;199(1–4):267–275.
[41] McKerns MM, Sun W, Lawson CM, *et al.* Higher-order triplet interaction in energy-level modeling of excited-state absorption for an expanded porphyrin cadmium complex. Journal of the Optical Society of America B. 2005;22 (4):852–861.

[42] Parilov E, Potasek M. Generalized theoretical treatment and numerical method of time-resolved radially dependent laser pulses interacting with multiphoton absorbers. Journal of the Optical Society of America B. 2006;23 (9):1894–1910.

[43] Bao Q, Zhang H, Wang Y, *et al.* Atomic-layer graphene as a saturable absorber for ultrafast pulsed lasers. Advanced Functional Materials. 2009;19 (19):3077–3083.

[44] Du J, Wang Q, Jiang G, *et al.* Ytterbium-doped fiber laser passively mode locked by few-layer molybdenum disulfide (MoS_2) saturable absorber functioned with evanescent field interaction. Scientific Reports. 2014;4:6346.

[45] Zhang H, Lu S, Zheng J, *et al.* Molybdenum disulfide (MoS_2) as a broadband saturable absorber for ultra-fast photonics. Optics Express. 2014;22(6): 7249–7260.

[46] Chen Y, Jiang G, Chen S, *et al.* Mechanically exfoliated black phosphorus as a new saturable absorber for both Q-switching and mode-locking laser operation. Optics Express. 2015;23(10):12823–12833.

[47] Yamashita S, Inoue Y, Maruyama S, *et al.* Saturable absorbers incorporating carbon nanotubes directly synthesized onto substrates and fibers and their application to mode-locked fiber lasers. Optics Letters. 2004;29(14):1581–1583.

[48] Set SY, Yaguchi H, Tanaka Y, *et al.* Laser mode locking using a saturable absorber incorporating carbon nanotubes. Journal of Lightwave Technology. 2004;22(1):51.

[49] Zhao C, Zhang H, Qi X, *et al.* Ultra-short pulse generation by a topological insulator based saturable absorber. Applied Physics Letters. 2012;101 (21):211106.

[50] Ahmad H, Lee C, Ismail M, *et al.* Zinc oxide (ZnO) nanoparticles as saturable absorber in passively Q-switched fiber laser. Optics Communications. 2016;381:72–76.

[51] Bai X, Mou C, Xu L, *et al.* Passively Q-switched erbium-doped fiber laser using Fe3O4-nanoparticle saturable absorber. Applied Physics Express. 2016;9(4):042701.

[52] Gao Y, Zhang X, Li Y, *et al.* Saturable absorption and reverse saturable absorption in platinum nanoparticles. Optics Communications. 2005;251 (4–6):429–433.

[53] Wu D, Peng J, Cai Z, *et al.* Gold nanoparticles as a saturable absorber for visible 635 nm Q-switched pulse generation. Optics Express. 2015;23 (18):24071–24076.

[54] Hill SC, Videen G, Sun W, *et al.* Scattering and internal fields of a microsphere that contains a saturable absorber: finite-difference time-domain simulations. Applied Optics. 2001;40(30):5487–5494.

[55] Feng Y, Dong N, Wang G, *et al.* Saturable absorption behavior of freestanding graphene polymer composite films over broad wavelength and time ranges. Optics Express. 2015;23(1):559–569.

[56] Varin C, Bart G, Emms R, *et al.* Saturable Lorentz model for fully explicit three-dimensional modeling of nonlinear optics. Optics Express. 2015;23 (3):2686–2695.
[57] Varin C, Emms R, Bart G, *et al.* Explicit formulation of second and third order optical nonlinearity in the FDTD framework. Computer Physics Communications. 2018;222:70–83.
[58] Li C, Zhang L, Wang R, Song Y, Wang Y. Dynamics of reverse saturable absorption and all-optical switching in C60. Journal of the Optical Society of America B. 1994;11(8):1356–1360.
[59] Naga Srinivas NKM, Venugopal Rao S, Narayana Rao D. Saturable and reverse saturable absorption of Rhodamine B in methanol and water. Journal of the Optical Society of America B. 2003;20(12):2470–2479.
[60] Tutt LW, McCahon SW. Reverse saturable absorption in metal cluster compounds. Optics Letters. 1990;15(12):700–702.
[61] Blau W, Byrne H, Dennis W, *et al.* Reverse saturable absorption in tetraphenylporphyrins. Optics Communications. 1985;56(1):25–29.
[62] Sun X, Xiong Y, Chen P, *et al.* Investigation of an optical limiting mechanism in multiwalled carbon nanotubes. Applied Optics. 2000;39 (12):1998–2001.
[63] Ouyang Q, Yu H, Wu H, Lei Z, Qi L, Chen Y. Graphene/MoS_2 organic glasses: fabrication and enhanced reverse saturable absorption properties. Optical Materials. 2013;35(12):2352–2356.
[64] Johnson PB, Christy RW. Optical constants of the noble metals. Physical Review B. 1972;6(12):4370.
[65] Drobizhev M, Rebane A, Suo Z, *et al.* One-, two-and three-photon spectroscopy of π-conjugated dendrimers: cooperative enhancement and coherent domains. Journal of Luminescence. 2005;111(4):291–305.

Chapter 7

Nonlinear multipolar interference: from nonreciprocal directionality to one-way nonlinear mirror

Ekaterina Poutrina[1,2] and Augustine Urbas[1]

Light manipulation at the nanoscale is at the heart of an immense variety of potential applications, providing a possible path for controlling optical response at an almost atomic scale. In this vein, engineered optical nanomaterials and structured nanocomposite media bring unprecedented capabilities for tailoring such nanoscale interactions. In this chapter, we delve into the unique capabilities enabled by engineering the electromagnetic response of nonlinear nanomaterials and develop a framework for describing and balancing the inherently nanoscale properties of these optical systems through multipolar interference. We discuss how nonlinear multipolar response differs qualitatively both from its linear counterpart and from the response of conventional nonlinear materials. In particular, we show that a special design of the effective nonlinear polarizabilities of a nanoelement can ensure a nonreciprocal directionality of nonlinear generation from that element, with the generation direction decoupled from the direction of the excitation beam. Such nonreciprocal directionality is desirable in a number of applications, including optical isolation or enhanced harmonic generation, but is fundamentally unachievable in linear, time-independent, and nonmagnetic media. Alternatively, a tailored design of a nanoelement can ensure directionally selective inhibition of a selected nonlinear process. Such inhibition can be of interest in radio-physics applications, where strong and spurious nonlinear response is often present and undesirable. Both nonreciprocal directionality and inhibition of the nonlinear response require engineering of (a) the relative strengths of nonlinear electric and magnetic multipolar modes of a nanoelement and (b) the relative strengths of various terms within each of these modes. This balance of magnitudes and the resulting phenomena are not expected to readily occur in the nonlinear response of natural materials, but, as we reveal, can be realistically achieved by designing the nonlinear multipolar response of nanostructures. The proposed concept offers the flexibility of achieving all the explored phenomena via the response of

[1]Air Force Research Laboratory, Wright-Patterson, OH, USA
[2]UES, Inc., Dayton, OH, USA

subwavelength elements. These elements can become building blocks which can then be used in developing an effective nonlinear medium with unique features inherently incorporated in the material structure and design.

7.1 Introduction

The ability to control directionality and efficiency of light scattering is imperative in any application involving the use and manipulation of light at the nanoscale. Many of these applications rely on the inherent reciprocity of light scattering. By reciprocity, we refer to the fundamental theorem stating that electromagnetic wave transmission through a linear, time-independent medium characterized by a symmetric permittivity tensor remains unchanged if one interchanges the source and the detector [1–4]. Breaking the reciprocity of light transmission requires overcoming at least one of the above-listed properties of the scattering medium. This fundamental limitation renders nonreciprocal light propagation to be rather uncommon. It is, however, required in any application or arrangement where optical isolation or directional absorption are needed. Examples range from single devices such as lasers and optical amplifiers, to large-scale applications such as lightwave communication systems. With the current drive of miniaturization in optical and photonic technology to subwavelength devices, control of light scattering becomes even more challenging. Not only is nonreciprocal transmission unachievable on a subwavelength scale with current state-of-the-art optical isolators, Faraday rotators being the most standard example, but even reciprocal and directional light scattering requires qualitatively new approaches. Conventional optical elements and approaches are in many cases not suitable for operating at the nanoscale.

The reason for such a reduced control of scattering directionality is equally fundamental: it stems from the reduction of the degree of coherence occurring with the interaction of even a highly coherent light beam with a nanoscale object [5,6]. The ability to tailor scattering characteristics of nanoscale elements, is, however, of critical importance in a variety of applications, including scattered light spectroscopy in biomedicine [7,8], optical nanoantennas [9–22], solar energy technologies, nanoparticle sensors, photovoltaic systems, perfect absorbers, or single-photon emitters. Achieving a nonreciprocal transmission on the nanoscale would revolutionize such technologies, amounting to the development of an optical diode with all the functionalities an electrical diode offers, and benefiting a variety of miniaturized all-optical and optoelectronic devices.

In order to control the directionality of scattering from subwavelength elements, it is natural to employ processes inherent to nanoscale interactions, and multipolar interference is a convenient and clear choice. It can be shown that light scattered off a nanoelement (or a group of nanoelements) in an otherwise uniform and source-free medium can be presented as a linear combination of two sets of spherical harmonics, with a given set representing either electric or magnetic multipolar modes [23–25]. All harmonics in the expansion have the same origin at the center of coordinates, hence scattering directionality strongly depends on the interference effect between various multipolar partial waves. Control of particle response via multipolar interference relies on engineering mode characteristics whose combined scattering yields the desired profile, e.g., to yield directional scattering by selecting modes of opposite parity. As an example, the x-component of the electric field radiated by several electric and magnetic multipoles is shown in Figure 7.1. The even parity of electric field is seen for electric dipolar and magnetic quadrupolar modes, while the magnetic dipole and electric quadrupole exhibit odd parity of their electric field. Similar parity alternation exists for the magnetic field (not shown), with the parity being opposite to that of the electric field in each case. Due to the opposite field parity, one can expect the field radiated by, e.g., electric and magnetic dipolar modes to be interferometrically suppressed in one hemisphere, while doubled in the other hemisphere, if one is able to engineer comparable amplitudes of these spherical waves. An example of such suppression is shown in Figure 7.2(a). Note that when obtained by scattering of the incident field off a nanoelement, the electric and magnetic harmonics shown in Figure 7.1 would represent the *scattered*, rather than the total, field. The *total* field outside the nanoelement can then be found as a sum of the incident and the scattered fields.

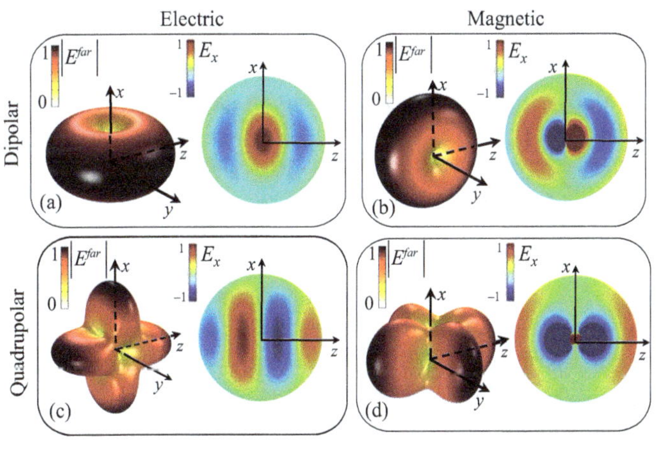

Figure 7.1 Illustration of the alternating field parity with increasing multipolar order. (a) Electric dipole oriented along x-axis; (b) magnetic dipole oriented along y-axis; (c) Q_{xz+zx} component of the electric quadrupole; and (d) Q_{yz+zy} component of magnetic quadrupole. In each figure, (left) shows angular distribution of the far field norm and (right) provides the xz cross section of E_x component of the electric field. The field parity is seen to alternate both with the increasing multipolar order and with a change from electric to magnetic multipolar type.

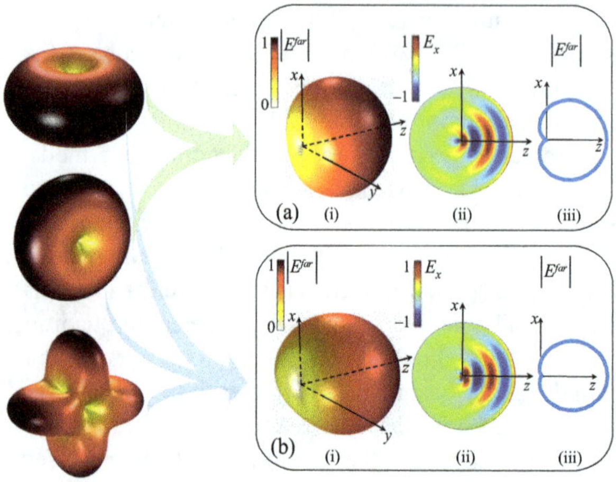

Figure 7.2 Illustration of field suppression in one hemisphere via multipolar interference. (a) Field suppression in the negative z hemisphere due to the interference of electric and magnetic dipolar modes from Figure 7.1, when the two dipoles are superimposed at the origin, matched in amplitudes, and oscillate out of phase. Suppression would occur in the positive z hemisphere for in-phase electric and magnetic dipolar modes (not shown). (b) Similar field suppression as in (a), achieved by matching the amplitude of electric dipolar mode with the combined amplitude of magnetic dipolar and electric quadrupolar modes from Figure 7.1, all superimposed at the origin. In both (a) and (b), (i)–(iii) indicate (i) the resulting angular distribution of the far field norm, (ii) xz cross section of the resulting amplitude of the E_x field component, and (iii) xz cross section of the norm of the far field angular distribution.

Suppression of scattering in one direction via interference between electric and magnetic dipolar modes produced by scattering off of a small sphere is known as the Kerker condition [26–29] and has been demonstrated in several experiments [30–32]. A modified condition for unidirectionality can be developed based on the fact that parities of the electric and magnetic modes interchange with increasing multipolar order [33,34]. Suppression of radiation in either the forward or backward direction can then be achieved whenever the *combined* strength of multipolar modes of a *certain parity*, radiating along the propagation direction, matches that of an opposite parity. As an example, in Figure 7.2(b), the field scattered into the electric dipolar mode is assumed to be counterweighted by that scattered into a sum of magnetic dipolar and electric quadrupolar modes (both having parity opposite to the one of the electric dipolar mode), with all higher multipolar modes assumed to be negligible. The resulting far field is suppressed in the negative z hemisphere, just like in Figure 7.2(a), where such suppression was achieved by balancing the amplitudes of the two dipolar modes. There have been several discussions on the improved directionality when higher-order multipoles are involved in achieving scattering suppression in one of the hemispheres [35–37].

Note that, in the case of linear scattering, the directions of field doubling and suppression will always switch when reversing the direction of the excitation beam. This switch in direction is a consequence of the reciprocity theorem mentioned earlier. The predominant "forward" or "backward" scattering direction can be defined with respect to the direction of the k-vector of the incident field, rather than to some fixed coordinate system. We term such scattering "unidirectional." This scenario is different from nonreciprocal directionality, where the scattering direction remains unchanged with respect to some fixed laboratory coordinate system when reversing the direction of the excitation [38].

The deeply subwavelength size of an element employed in multipolar interference-controlled scattering allows for arranging such elements within a plane with a subwavelength periodicity, forming a metasurface [39–41]. Assuming the spacing between the elements is optimized to minimize coupling [42] while avoiding a significant spatial dispersion [43], the scattering characteristics of an individual element can be expected to transfer to those of the metasurface, allowing for a nanoscale engineering of its scattering response. Light scattered along the surface by each individual element will then be interferometrically redistributed between the symmetric directions on the two sides of the metasurface, following the momentum conservation requirements. In this manner, one can achieve, e.g., a zero reflection of the incident field from the metasurface, in the case of a perfect forward scattering by each individual element. More complicated designs, which employ space-variant nanoelement arrangements and account for coupling between the elements within the array, can also be developed, bringing a variety of new functionalities to the response [44–53].

We emphasize upfront that while there is no limit on the level of forward-scattering directionality, there is a fundamental limitation, imposed by the optical theorem [24], preventing perfect backscattering (or complete *suppression* of scattering in the forward direction) [29,54]. The theorem states that the extinction cross section (sum of absorption and scattering cross sections) is a function of the portion of light scattered in the forward direction. Hence, a complete cancelation of forward scattering would assume no scattering overall due to power conservation considerations. The theorem, however, does not prevent minimizing the scattering in the forward direction by interferometrically suppressing the field quadrature oscillating in phase with the incident field that does not contribute to extinction [27]. As a result, a pronounced backscattering directionality can be achieved while maintaining a small but nonzero forward scattering amplitude, even when accounting for realistic losses in a nanoelement. One can understand the reason for the unattainability of a perfect backscattering in a more intuitive way when considering the complex electric and magnetic polarizabilities of a nanoelement. Matching the strengths of, e.g., electric and magnetic dipolar modes assumes matching simultaneously the values of both real and imaginary parts of electric and magnetic polarizabilities of that nanoelement [34,42]. As will be seen in more detail in Section 7.2.1, perfect backscattering requires matching these values with the opposite signs. For a passive element, however, the imaginary parts of both electric and magnetic polarizabilities are always positive, preventing them being matched with the opposite sign, and thus preventing a perfect suppression of scattering in the forward direction. There are approaches that can be utilized to alleviate or entirely overcome the restriction on forward scattering suppression imposed by the optical theorem. The aforementioned design strategy of minimizing forward scattering by suppressing one of the field quadratures can be pursued by minimizing the imaginary parts of the polarizabilities to the extent possible, while matching their real parts, for example, by utilizing the off-resonant magnetic and electric responses. Such an approach has been used to design deeply subwavelength plasmonic dimer structures with over 20 dB backscattering directionality [34]. The requirement of matching both the real and the imaginary parts of polarizabilities with the opposite signs can also be fulfilled by adding gain to the scattering medium, as has been done in [29], utilizing a sphere with a negative imaginary part of either its permittivity or permeability. In that vein, the use of nonlinear response brings a variety of qualitatively new possibilities, including the opportunity of a zero nonlinear generation in the forward direction [38,55]. Such generation can be regarded as nonlinear scattering, where the scattered field emerges at a different frequency. One can see this as an extension of implementing gain in the scattering response—in this case, a nonlinear gain.

The nonlinear response of a medium to an incident electromagnetic wave plays a crucial role in a variety of applications. The demand for its use is manyfold: in some areas, such as radiophysics or strong field interactions, the nonlinear response can be strong and hence simply needs to be taken into account [56,57]. In other areas, the use of nonlinear response offers unique capabilities and is not substitutable, allowing phenomena unattainable via linear interactions. Examples of such phenomena include sum-frequency generation—the process employed in two-photon spectroscopy [58], imaging of drug delivery [59], a variety of pharmaceutical studies [60], biological and chemical sensing [61] and potentially in infrared imaging; Raman scattering, which serves as a basis of low-noise optical amplifiers [62], Raman sensors, and multiple biological applications such as tissue imaging [63]; entangled photon production—a key element in quantum information technology and quantum imaging [64–66]; electro-optical response, which allows the realization of optical modulators [67]; super-continuum generation, producing broadband light sources [68]; optical parametric amplification [62]; two-photon absorption [69], serving as a basis for optical limiters, microfabrication, of biological imaging; and the list can be continued.

Perhaps one of the most striking examples of a phenomenon uniquely achievable via nonlinear response is the ability of breaking the reciprocity of light transmission. Nonlinear response is not subject to the reciprocity theorem noted earlier, and many works exploit this fact to predict or demonstrate nonreciprocal light propagation based on nonlinear interactions [70–81]. The majority of these approaches rely on either geometric [76–81] or material [70–75] asymmetry in the geometry along the propagation direction, with some works exploiting a similar asymmetry in chiral structures [82,83]. The distribution of the local field inside such an asymmetric structure does not remain the same, nor does it necessarily produce its mirror image, when reversing the direction of the incident field. Placing the nonlinear element within the region of the highest contrast between the local field patterns induced by forward versus backward-propagating incident beams results in the variation of the efficiency of the transmitted light for the excitation intensities strong enough to invoke the nonlinear response. While helping to reduce significantly the footprint of optical isolators compared to other mechanisms, e.g., regular Faraday rotation-based devices, this approach still assumes the geometric dimensions comparable to a wavelength or larger, relying, for example, on some type of standing wave pattern inside the structure, or on a particular location of a nonlinear element within the decaying evanescent field [84–86]. As a result, the suggested designs are "device-type," which prevents their use as "artificial molecules" in building a medium exhibiting a similar nonreciprocal transmission. Limitations of this approach have also been discussed in terms of a necessary trade-off between the strength of the forward to backward local field contrast and the level of absorption [87]. A significant reduction of the footprint of a nonreciprocal element could be achieved if one could combine possibilities offered by the nonlinear response with those of multipolar interference, as the latter necessarily assumes operating at subwavelength-scales. In this chapter, we will see that nonreciprocal directionality of nonlinear generation is inherently incorporated into multipolar interference when electric and magnetic multipolar modes are produced via a nonlinear interaction.

Due to the small interaction volumes involved, nonlinear generation from a nanoelement is not subject to phase matching constraints [88,89]. As a result, typically, the generation direction is almost isotropic, for all nonlinear processes involved. Similar to linear response, however, multipolar interference can take place between partial waves in the nonlinearly produced field, thereby helping to restore control of scattering directionality. Indeed, the interference between the nonlinearly produced electric and magnetic partial waves was shown to induce some asymmetry between the "transmission" and "reflection" sides in the far field pattern of the second harmonic generated from plasmonic nanostructures, which was subsequently used for the analysis of multipolar contributions to the nonlinear response [90–92]. Furthermore, a complete suppression of the nonlinearly generated field in one of the hemispheres has been predicted theoretically [88] and demonstrated experimentally [93] in the microwave range for deeply subwavelength metamaterial layers. Nonlinear multipolar response from nanoelements has been attracting increasing attention recently [94–97]. It has been shown to enhance the efficiency of the second [94,98–103], and third [104–108] order nonlinear response in dielectric nanoelements, to produce circular dichroism of four wave mixing in microwave metamaterials [109], to achieve resonant Raman scattering from silicon nanoelements [110], and to allow ultrafast optical switching [111]. More recently, shaping of angular distributions and enhancement of directionality of nonlinear generation via multipolar interference from dielectric nanoelements have been demonstrated [55,112–116].

Once the nonlinearly generated field has been produced by a nanoelement, the interference process leading to the asymmetry of that field in the far field zone, reported in the aforementioned references, is similar to the interference leading to one-side suppression of linear scattering, depicted in Figure 7.2. It similarly requires matching the combined strengths of all even-parity multipoles with that of the odd parity. The only difference is that it is the field at the frequency generated by the nonlinear interactions that is now decomposed into sets of interfering electric and magnetic spherical harmonics depicted in Figures 7.1 and 7.2, rather than the linearly scattered field. In that sense, directional nonlinear generation is not different from unidirectional linear scattering where the suppressed and doubled sides switch when the direction of the excitation is reversed. For example, the achieved dominant direction of the second [55] or the third harmonic [115] is expected to reverse together with the reversal of the fundamental excitation beam, remaining "forward" with respect to the direction of that beam.

There is, however, a qualitative difference between the process by which a given multipole is produced when utilizing a nonlinear, versus linear, response; this difference becomes even more significant when a magnetic dipolar (or higher-order multipolar) mode is nonnegligible in the linear response of the nanoelement. In particular, the majority of nanoelements made of nonmagnetic materials are not bianisotropic or biisotropic [117–119] in their linear response, unless a specific geometric design is employed [120–125]. There is no cross-coupling of the incident electric and magnetic fields through linear response in typical, non-bianisotropic nanoelements, i.e., it is solely the electric (magnetic) vector of the incident field that controls the amplitude and phase of the linearly induced electric (magnetic) dipolar mode. Due to the multiplicative nature of nonlinear polarization and magnetization, however, the absence of *nonlinear* magneto-electric cross-coupling is almost never the case if a magnetic dipolar or other order multipolar mode is present in the *linear* scattering response.

Cross-coupling of electric and magnetic fields, manifesting as magneto-electric terms in the *intrinsic* nonlinear susceptibility, is certainly expected to happen for a bianisotropic material, as has been shown in the case of chromium oxide [126,127].

Cross-coupling of electric and magnetic fields through *effective nonlinear* polarizability (electric and/or magnetic one) will, however, equally be present, in the majority of cases, even for an element which has no biisotropy or bianisotropy in its linear response, but possesses a linear, geometric shape-based, magnetic dipolar (or higher-order multipolar) moment. The simplest example can be an entirely symmetric dielectric nanosphere made of non-centrosymmetric material with a purely electric nonlinear second-order susceptibility (assuming the sphere diameter is large enough to produce a pronounced linear magnetic dipolar response). Second-order nonlinear magneto-electric polarizabilities have been analyzed in more complex geometries, such as L-shaped plasmonic particles [92,128], nanostrip-dimer geometries [38], and in a series of microwave structures [88,93,109,129]. The presence of magneto-electric terms in the nonlinear polarizability significantly increases the variety of pathways that can be utilized to induce electric and magnetic multipolar modes via a nonlinear interaction. Moreover, it creates shared pathways, where a single (electric *or* magnetic) vector of the fundamental field can control the phase of *both* electric *and* magnetic, nonlinearly induced, multipolar modes. It is this overall variety of pathways, as well as the existence of shared pathways, that brings a qualitative difference to the control of directionality and efficiency of nonlinear generation via multipolar interference, compared to a similar control of linear scattering.

In particular, as we will see, a careful engineering of the respective strengths of various pathways *within* each type of the *nonlinearly* produced multipolar partial wave can result in the nonlinear generation from a nanoelement occurring predominantly in a single direction which *remains unchanged* when reversing the direction of the fundamental beam [38]. Furthermore, the presence of magneto-electric terms leads to interference between various terms *within* the (nonlinearly produced) electric and magnetic multipolar modes themselves. As a result of this interference, a slight change in the efficiency of nonlinear generation occurs when the direction of any subset of the fundamental beams is reversed. Consequently, nonreciprocity in terms of ensuring just a change in the efficiency of nonlinear generation when reversing the direction of any subset of the fundamental beams is inherent to and expected in the nonlinear response of most nanoelements, even symmetric ones, and for most nonlinear processes. Engineering a certain relation between various interfering terms within each (electric and magnetic) nonlinearly generated partial wave can then allow an inhibition of a specified nonlinear process.

As in the case with linear response, the subwavelength nature of the interaction volumes involved with this approach allows for a relatively straightforward arrangement of such nonreciprocal nonlinear elements forming a metasurface with similar unique features in its nonlinear response. In this work, we present an operation of a one-way nonlinear mirror based on such an approach, with an image of the source produced by difference frequency generation (DFG) on one and the same side of a metasurface independent of the source location. Combined with the recent advances demonstrating orders of magnitude enhancement of the efficiency of nonlinear response with metasurfaces, the presented example, and the approach in general, can find numerous applications in integrated photonic circuits, image upconversion designs, electro-optical devices, or quantum optical systems.

7.2 Single-element scattering response: An overview

In this section, we provide a framework for describing nonlinear multipolar response in terms of effective nonlinear magneto-electric polarizabilities and discuss new features in the nonlinear response of nanoelements brought by the presence of such magneto-electric terms. While, generally, intrinsic magnetization can be included in the analysis, the study presented in this chapter assumes nonmagnetic materials for nanoelement geometries. In Section 7.2.1, we start from expressions for the far field of the interfering electric and magnetic multipolar modes, where such modes can be created by either linear scattering or nonlinear generation. We use the developed expressions for a brief review of linear magneto-electric interference, providing a mathematical description of one-side field suppression shown earlier in Figure 7.2. Without reducing the generality and for clarity of further discussion, we truncate multipolar series in this section at the dipolar terms. (We include electric quadrupolar response later in Sections 7.3 and 7.4, when developing the retrieval procedure for nonlinear magneto-electric polarizabilities and applying it to physical geometries. Similar expressions, accounting for up to a magnetic quadrupolar mode, can be also found in [37]. The latter reference also provides a detailed analysis of the balance of various terms required to maximize interferometric suppression of power scattered in one of the hemispheres. Higher-order multipoles, however, do not bring a qualitative difference to the phenomena discussed in this section; to avoid complicating the notation, we omit them, therefore, for now.) The analysis presented here also assumes no bianisotropy (or biisotropy) in the linear response of a nanoelement. We next use, in Section 7.2.2, the example of interferometric scattering suppression to illustrate the reciprocal directionality of linear scattering. In Section 7.2.3, we present expressions for nonlinearly generated electric and magnetic dipolar modes, written in terms of effective nonlinear magneto-electric polarizabilities. Throughout the chapter, we use the second-order nonlinear response as an example. We review the conditions when effective nonlinear magneto-electric polarizabilities can become significant, emphasizing that no bianisotropy or optical activity in the linear response is required for such polarizabilities to develop. We discuss the qualitative differences brought by the presence of such magneto-electric terms in the nonlinear response of nanoelements, compared to purely electric nonlinear response of natural materials. We also show that the outcome of multipolar interference can be fundamentally different when multipolar modes are produced by a nonlinear, versus linear, interaction. We discuss several interesting possible phenomena resulting from interference of nonlinearly generated multipolar terms. In the remainder of the section, we discuss two of these phenomena in more detail: (i) the possibility of nonreciprocal directionality of nonlinear generation (Section 7.2.3.2) and (ii) directionally selective interferometric suppression of a certain nonlinear process (Section 7.2.3.3).

7.2.1 Expressions for the scattered field

Consider a nanoelement that has a set of electric and magnetic dipolar modes induced along the x- and y-axes, respectively, in response to an electromagnetic wave incident along the z direction, as shown in Figure 7.3. We do not specify, for now, the particular process responsible for the formation of these modes—they can be induced by ether linear or nonlinear response. Assume further that no higher-order multipoles acquire significant strength within the same spectral range. The field radiated via a combination of such orthogonal, spatially superimposed electric and magnetic dipoles can be found by considering the standard expressions for the far field radiated by linear electric and magnetic dipoles of

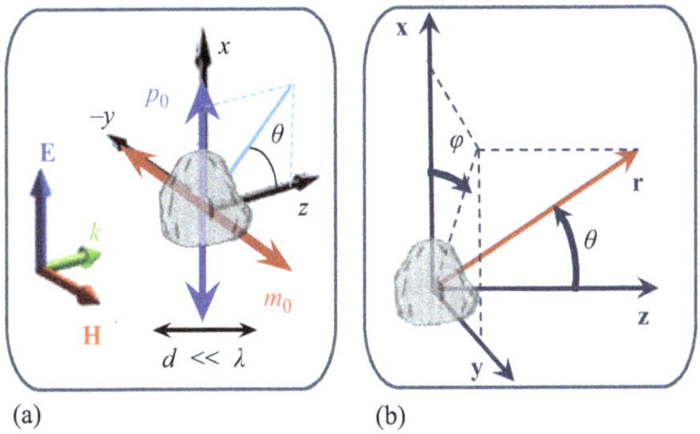

Figure 7.3 Visualization of the orientation of the induced dipolar moments and of the coordinate systems. (a) Orientation of the induced dipolar moments, assumed to be along the incident field vectors. (b) Cartesian and spherical coordinate systems notation used in the text.

moments $\mathbf{p}e^{-i\omega t}$ and $\mathbf{m}e^{-i\omega t}$ in vacuum, given by (here and further, bold font indicated vector quantities) [5,34]:

$$\mathbf{E}^{el} = -\frac{k^2 e^{ikr}}{4\pi\varepsilon_0}\frac{1}{r^3}\mathbf{r}\times\mathbf{r}\times\mathbf{p}, \tag{7.1a}$$

$$\mathbf{E}^{m} = -\frac{k^2 e^{ikr}}{4\pi}\sqrt{\frac{\mu_0}{\varepsilon_0}}\frac{1}{r^2}\mathbf{r}\times\mathbf{m} \tag{7.1b}$$

where the superscripts indicate the field scattered by electric and magnetic dipoles, ε_0 and μ_0 are the permittivity and the permeability of vacuum, r is the radius-vector directed from the scatterer (located at the origin) to the observation point in the far field, and harmonic time dependence factor has been omitted. Assuming x-oriented electric ($\mathbf{p} = (p_{x0}, 0, 0)$) and y-oriented magnetic ($\mathbf{m} = (0, m_{y0}, 0)$) dipolar modes in the above expressions, the Cartesian components of the electric far field produced by such a combination of dipoles are as following:

$$E_x^{el+m} = \frac{k^2}{4\pi}\frac{e^{ikr}}{r}\left[\frac{1}{\varepsilon_0}p_{x0}(\sin^2\theta\,\sin^2\varphi + \cos^2\theta) + \eta m_{y0}\cos\,\theta\right], \tag{7.2a}$$

$$E_y^{el+m} = -\frac{k^2}{4\pi}\frac{e^{ikr}}{r}\frac{1}{\varepsilon_0}p_{x0}\sin^2\theta\,\sin\,\varphi\,\cos\,\varphi, \tag{7.2b}$$

$$E_z^{el+m} = \frac{k^2}{4\pi}\frac{e^{ikr}}{r}\left[-\frac{1}{\varepsilon_0}p_{x0}\,\sin\,\theta\,\cos\,\theta\,\cos\,\varphi - \eta m_{y0}\,\sin\,\theta\,\cos\,\varphi\right]. \tag{7.2c}$$

Here, p_{x0} (m_{y0}) denotes the amplitude of the x (y) component of the induced electric (magnetic) dipole, $\eta \equiv \sqrt{\mu_0/\varepsilon_0}$ is the vacuum impedance, and $k \equiv \omega/c$ is the wavenumber. To simplify further the illustration, we consider a single (xz) cross section of the above field pattern. Making the transformation from Cartesian to spherical coordinates and considering the xz $(\varphi = 0)$ scattering plane, one arrives at the following angular distribution of the E_θ far field component of the electric field within this plane [130]:

$$E_\theta\big|_{\varphi=0} = \frac{k^2}{4\pi}\frac{e^{ikr}}{r}\left[\frac{1}{\varepsilon_0}p_{x0}\cos\theta + \eta m_{y0}\right] \tag{7.3}$$

with angles θ and φ in a standard spherical coordinate system (Figure 7.3). One can show $E_\varphi = 0$ in the $\varphi = 0$ scattering plane, and the radial component E_r falls off in the far field, so that the scattered far field, denoted as $\mathbf{E}^{sc}_{far}$, is completely represented by its E_θ component in the xz plane: $\mathbf{E}^{sc}_{far}\big|_{\varphi=0} = E_\theta\big|_{\varphi=0}$.

We emphasize that, in the case of linear response, (7.1)–(7.3) represent the *scattered*, rather than the total, field. The total field outside the scatterer is then found as the sum of the incident field and that of (7.1)–(7.3).

As seen from (7.3), the radiation from the above set of dipoles can be interferometrically suppressed in either the backward $(\theta = 180^\circ)$ or forward $(\theta = 0^\circ)$ direction when the strengths of the electric and magnetic dipolar modes are being matched with either the same $(1/\varepsilon_0 p_x = \eta m_y)$ or opposite $(1/\varepsilon_0 p_x = -\eta m_y)$ signs, respectively. These scenarios are known as the first and the second Kerker [26] conditions mentioned earlier, with the latter one (suppression in the forward direction) being subject to the optical theorem limitations discussed in the introduction. The field distribution resulting from the first of these scenarios was illustrated earlier in Figure 7.2(a). The treatment can be extended to higher-order multipoles—see, e.g., [34,37]. Note, however, that, as can be expected from Figure 7.1, all even (in electric field) and odd higher-order multipoles will contribute to (7.3) in a manner similar to that of electric and magnetic dipolar modes, respectively (e.g., the sum of electric quadrupolar and magnetic dipolar modes would need to be matched in strength with the sum of magnetic quadrupolar and electric dipolar modes, if quadrupolar modes become significant, etc.). Without reducing the generality, for clarity of the illustration given below, and to simplify the notation, we retain here only the dipolar modes. We will include quadrupolar terms when retrieving the polarizabilities in Section 7.4.

One can see from (7.3) that, while the Kerker conditions are often associated with matching the values of electric (α^{el}) and magnetic (α^m) polarizabilities of a nanoelement, it is the strengths of the dipolar modes themselves that regulate the resulting interference pattern. Consequently, the particular directions of enhancement and suppression of the scattered field can be controlled by regulating the phase of the induced dipolar modes. The latter can be achieved by both changing the sign of one of the polarizabilities (e.g., in the case of linear response, by moving the spectral position from below to above the resonance in the resonant region), as well as by switching the phase of the inducing electric or magnetic field. Both these scenarios can result in switching the phase of one or both dipolar modes, thus changing the sign before one or both of the terms in (7.3). In fact, this dependence of scattering directionality on the phase of the excitation fields allows for a simple illustration of the reciprocity of linear scattering, even when a unidirectional scattering is achieved, as explained below.

7.2.2 Linear multipolar interference

In the case of linear response, assuming no bianisotropy exhibited by a nanoelement, the dipolar modes entering (7.3) are related to the electric ($\mathbf{E}_0$) and magnetic ($\mathbf{H}_0$) vectors of the incident field as:

$$\mathbf{p} = \varepsilon_0 \overline{\overline{\alpha^{el}}} \mathbf{E}_0 \quad (7.4a)$$

$$\mathbf{m} = \overline{\overline{\alpha^{m}}} \mathbf{H}_0 \quad (7.4b)$$

Figure 7.4(b) provides examples of interferometric scattering suppression given by (7.3) when varying the phase and the direction of the incident field according to the four configurations indicated in Figure 7.4(a). In each case, a match in strength between the two dipolar amplitudes entering (7.3) is assumed: $(|1/\varepsilon_0 p_x| = |\eta m_y|)$. With arrangement (i), such a balance of the dipolar amplitudes leads to scattering suppression in the negative z hemisphere (with a complete suppression for $\theta = 180°$) and in doubling of the scattered field for $\theta = 0°$. Reversing the direction and/or the phase of the incident beam using the arrangements shown in sets (ii)–(iv) leads to phase switching of either the electric or magnetic vector of the excitation field (or of both vectors, as in arrangement (iii)), with respect to their direction and phase in set (i). According to (7.4), this switch results in sign reversal of the corresponding induced dipolar mode(s). In particular, denoting electric and magnetic dipolar modes induced with the arrangement (i) as p_0 and m_0, the modes induced with the rest of the excitation arrangements in

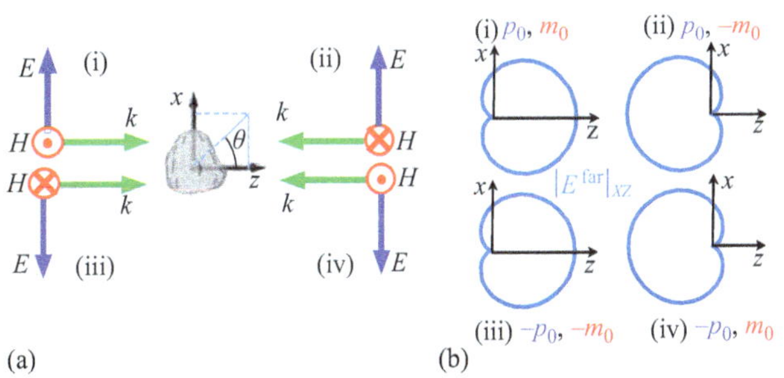

Figure 7.4 Illustration of the reciprocal directionality of linear scattering. (a) The four orientation of the incident field used for inducing electric and magnetic dipolar modes in a nanoelement located at the origin. (b) Angular distribution of the norm of the linearly scattered far field in the xz plane for the four excitation arrangements shown in (a), calculated according to (7.3). The amplitudes of electric and magnetic dipolar modes are assumed to match in all cases. p_0 and m_0 denote electric and magnetic dipolar modes induced with arrangement (i). Forward unidirectional linear scattering is seen staying "forward" with respect to the direction of incidence of the excitation beam, rather than to a fixed laboratory coordinate system.

Figure 7.4(b) are given by (ii) p_0 and $-m_0$; (iii) $-p_0$ and $-m_0$; and (iv) $-p_0$ and m_0, as shown on the figure. Phase change of a given dipolar mode will thus transfer to a sign change in front of the corresponding dipolar term in (7.3). One can see further that, whenever the sign change occurs in front of *a single* term in (7.3), such as in cases (ii) and (iv), it results in switching of the sides with the suppressed and doubled scattered field, compared to arrangement (i) (e.g., in both sets (ii) and (iv), field doubling will occur for $\theta = 180°$, versus doubling for $\theta = 0°$ in set (i)). Since both configurations (ii) and (iv) represent the reversed direction of the incident beam, one can see that forward (backward) unidirectional linear scattering always stays "forward" ("backward") with respect to the direction of incidence of the excitation beam, rather than to a fixed laboratory coordinate system. (The case of backward scattering is not shown in Figure 7.4, since matching of the dipolar amplitudes with the same sign was assumed; if the amplitudes are matched with the opposite sign, one can track the backscattering direction to vary similarly between the sets (i)–(iv).) This behavior makes the linear scattering unidirectional with respect to the incident beam, but this directionality is not nonreciprocal.

In that vein, it is instructive to consider the arrangement in set (iii) in Figure 7.4(b). In this case, the phase of oscillations of *both* electric and magnetic excitation field vectors is opposite to that in (i). This results in a *simultaneous* sign change before both electric and magnetic terms in (7.3) and, as follows from that equation, in a preserved direction of the scattered far field. The situation is somewhat similar to the "negative index" media scenario, in which the direction of energy propagation is preserved when the phase velocities of both electric and magnetic partial waves are simultaneously reversed [131,132]. However, in the present example, such a reversal occurs due to the simultaneous phase switching of the electric and magnetic vectors of the incident field, rather than the simultaneously negative values of electric and magnetic polarizabilities (the latter scenario occurs in the negative index media).

Inclusion of higher-order multipoles in (7.3) will follow similar guidelines, with the phase of linearly induced odd (even) in the electric field modes effectively following the phase of the magnetic (electric) incident field vector, as, for example, has been shown for up to a magnetic quadrupolar mode in [37] and as discussed in more detail in Section 7.3.1.

7.2.3 *Nonlinear multipolar interference*

Once the dipolar modes entering (7.3) happen to be induced in a nanoelement, radiation produced by such a combination of multipoles can clearly be suppressed in one hemisphere, whether these modes are induced through a linear or a nonlinear interaction. In the latter case, the interference between the two modes leads to an asymmetry in the far field pattern of light at the new, nonlinearly produced, frequency. This phenomenon has been discussed in a series of works [55,90–93]. In contrast with the linear case, however, interference between multipolar modes allows for a much broader palette of phenomena when such modes have been originally produced via a nonlinear response. One qualitative difference with the linear case is that, as we show here, nonlinear multipolar interference allows for a nonreciprocal directionality of nonlinear generation, where the sides with the interferometrically suppressed and doubled fields remain unchanged with respect to a fixed laboratory coordinate system when reversing the direction of an excitation beam. Another interesting outcome is the possibility of an overall suppression of a certain nonlinear process via the interference of the nonlinearly produced multipolar modes. Several other interesting phenomena have been predicted based on nonlinear multipolar interference, such as an extension of the electro-optic effect to the electro-magnetic case, where the effective nonlinear magneto-electric polarizabilities provided field-induced contributions to both linear permittivity and permeability; or field-controlled electro-magnetic coupling, which vanished in the absence of the applied electric field [88]. In the remainder of this section, we present the description of nonlinear multipolar modes in terms of the effective nonlinear magneto-electric polarizabilities and discuss in more details the possibility of nonreciprocal directionality and suppression of a nonlinear process mentioned above.

7.2.3.1 Nonlinear multipolar modes as a result of interference between various nonlinear magnetoelectric polarizability terms

The presence of a magnetic dipolar mode in the linear response results in the fact that both electric and magnetic, as well as magneto-electric terms can contribute to nonlinear polarizabilities of a nanoelement. For example, nonlinearly induced electric and magnetic dipolar modes, which are quadratic in the fundamental fields in the case of second-order nonlinear response, can be expected to have terms containing products of electric and magnetic vectors of the fundamental fields, in addition to terms quadratic in the electric fundamental fields and those quadratic in the magnetic fundamental fields. We denote the presence of such terms containing products of electric and magnetic field vectors as "nonlinear magneto-electric coupling." In the case of magnetic media, the presence of tensor components of intrinsic nonlinear susceptibility responsible for nonlinear magneto-eclectic coupling has been pointed out since the original works on nonlinear optical material response [133]. Microscopic analysis of magneto-electric terms in nonlinear susceptibility of Cr_2O_3 has been performed in [127], and the interference between these terms was used to identify antiferromagnetic domains through second harmonic generation in this material. More recently, a formalism deriving a complete set of such terms via density matrix analysis has been suggested, and the symmetry properties of such magneto-electric nonlinear susceptibility tensor has been analyzed [134]. Nonlinear magneto-optic effects enabled by magnetic media have been attracting attention over the years, with numerous applications being proposed [135–143].

Few natural materials, however, have nonlinear magneto-electric susceptibilities strong enough to produce a significant magneto-electric interference, with Cr_2O_3 being one of the few examples [127]. Furthermore, the particular nonzero magneto-electric terms are fixed by the atomic structure in the case of natural materials, preventing configurations required by many of the phenomena described in this chapter. Here, we focus on nanoelements that do not possess intrinsic magnetization, but do exhibit geometric shape-based magnetic dipolar (and possibly higher-order multipolar) moments. In this case, nonlinearly induced multipolar modes can similarly be represented in terms of magneto-electric nonlinear polarizabilities, with, however, a much greater flexibility over the control of their respective amplitudes.

As discussed above, in the case of a nonlinear response, each of the dipolar moments induced in a nanoelement can result from a sum of several terms [88,126]. The number of terms depends on the order of the nonlinear process. In the present work, we limit consideration to the second-order response, while the treatment can be extended to higher-order processes. We consider a nonlinear mixing of two monochromatic fundamental waves at frequencies ω_1 and ω_2, producing a wave at frequency $\omega_3 = \omega_1 + \omega_2$. Either of the fundamental frequencies can take negative values [69], in which case a process of DFG would occur. If a magnetic dipolar mode is present in the linear response of a nanoelement, complete expressions for the second-order effective nonlinear polarizabilities, in general, can include up to eight terms, resulting from eight possible combinations of the mixed electric and magnetic fundamental modes [38,88]. The relations can be written as follows:

$$\begin{aligned}\mathbf{p}_0(\omega_3) &= \frac{\varepsilon_0}{\sqrt{I}}\left[\eta^2\alpha_{emm}(\omega_3;\omega_1,\omega_2):\mathbf{H}_0(\omega_1)\mathbf{H}_0(\omega_2)+\alpha_{eee}(\omega_3;\omega_1,\omega_2):\mathbf{E}_0(\omega_1)\mathbf{E}_0(\omega_2)\right]\\ &\quad+\frac{\varepsilon_0}{\sqrt{I}}\left[\eta\alpha_{eem}(\omega_3;\omega_1,\omega_2):\mathbf{E}_0(\omega_1)\mathbf{H}_0(\omega_2)+\eta\alpha_{eme}(\omega_3;\omega_1,\omega_2):\mathbf{H}_0(\omega_1)\mathbf{E}_0(\omega_2)\right],\end{aligned} \tag{7.5}$$

$$\begin{aligned}\mathbf{m}_0(\omega_3) &= \frac{1}{\sqrt{I}}\left[\alpha_{mem}(\omega_3;\omega_1,\omega_2):\mathbf{E}_0(\omega_1)\mathbf{H}_0(\omega_2)+\alpha_{mme}(\omega_3;\omega_1,\omega_2):\mathbf{H}_0(\omega_1)\mathbf{E}_0(\omega_2)\right]\\ &\quad+\frac{1}{\sqrt{I}}\left[\frac{1}{\eta}\alpha_{mee}(\omega_3;\omega_1,\omega_2):\mathbf{E}_0(\omega_1)\mathbf{E}_0(\omega_2)+\eta\alpha_{mmm}(\omega_3;\omega_1,\omega_2):\mathbf{H}_0(\omega_1)\mathbf{H}_0(\omega_1)\right];\end{aligned} \tag{7.6}$$

where $\mathbf{I} \equiv |\mathbf{E}_0(\omega_1)||\mathbf{E}_0(\omega_2)|$, and $\mathbf{p}_0$ and $\mathbf{m}_0$ are vectors of the *nonlinearly* induced electric and magnetic dipole moments, respectively. The time-harmonic factor has been omitted, and the double-dot product denotes the summation over all Cartesian components. The α_{abc} coefficients (*a*, *b*, and *c* standing for either "*e*" or "*m*") represent the effective nonlinear polarizabilties of various types. Subscripts "*e*" and "*m*" indicate the type of dipolar modes that are participating in the interaction: the first subscript denotes the type of the dipolar mode produced at frequency ω_3, while the second and the third subscripts denote the types of the fundamental modes at ω_1 and ω_2, respectively, contributing to this particular magneto-electric polarizability term. Since only the second-order nonlinear response will be presented, we will use the term "hyperpolarizability" throughout the chapter to denote nonlinear polarizability terms α_{abc}, following the standard name for the second-order nonlinear polarizability [69]. Normalization has been chosen in order to keep the single units of [m^3] for all hyperpolarizability types.

We emphasize that, noting a *pure electric intrinsic* response in the case of non-magnetic materials assumed here, the coefficients on the right-hand sides of (7.5) and (7.6) represent the effective, rather than the intrinsic, hyperpolarizabilities. That assumes that the field vectors entering these equations denote the incident fundamental fields, not the local fields inside the nanoelement. The origin of such effective nonlinear response is different for plasmonic or other nanoelements made of centrosymmetric materials [133,144–146], as opposed to those possessing an intrinsic second-order nonlinear response. For example, the nonlocal free-electron response at the surface provides the majority of contribution to the second-order nonlinear response in the case of plasmonic nanoelements [146,147]. The particular set of nonzero tensor components of the effective hyperpolarizability will, therefore, largely depend on the geometric shape of the nanoelement [148,149]. In this chapter, we employ nanoelements that have inclusions of noncentrosymmetric media in their geometry, hence the main contribution to the nonlinear response results from the intrinsic, electric second order susceptibility of these inclusions. Extending the analysis to the case of surface response would be an interesting and important step to pursue in the future.

One can think of the various terms entering the right-hand sides of (7.5) and (7.6) as of different pathways leading to the certain type (electric or magnetic) of the nonlinear response. Note that, in either (7.5) or (7.6), the terms in the first and second brackets comprise the sets of effective hyperpolarizabilities of two different types, polar and axial [88,150], respectively. Also note that unless one considers a degenerate nonlinear process where $\omega_1 = \omega_2$, there is no restriction on the respective values of various terms: for example, there is no requirement that $\alpha_{mem} = \alpha_{mme}$ unless $\omega_1 = \omega_2$. One can design the amplitudes of various terms, therefore, relatively independently. (That said, there are restrictions on the respective values of various Cartesian tensor components for a given polarizability type, imposed by the symmetry group it belongs to (polar or axial); the restrictions are similar to those for nonlinear magnetic media discussed in [134].)

As seen from (7.5) and (7.6), in addition to the interference between electric $(\mathbf{p}_0)$ and magnetic $(\mathbf{m}_0)$ modes, the interference can now also occur between various hyperpolarizability terms *within* each of these modes. As discussed later in Sections 7.2.3.3 and 7.4.2, it is this interference within a multipolar mode itself that allows for a suppression of a given nonlinear process for certain respective directions of the fundamental beams, if a required balance between the amplitudes of various terms is met.

One can also encounter a situation where electric and magnetic hyperpolarizabilities within a given symmetry group share the same (either electric or magnetic) "pathway" through the fundamental field of a given frequency. For example, the "emm" and "mme" terms, both belonging to the polar group, share a "pathway" through the magnetic field at frequency ω_1. The existence of such shared pathways clearly allows simultaneous control of the phase of *both* electric *and* magnetic terms sharing this pathway by varying the phase of the *single* (electric *or* magnetic) vector of the fundamental field. As we discuss in more detail in Section 7.2.3.2, such simultaneous control may allow for a nonreciprocity in the direction of nonlinear generation.

Note also that each of the effective polarizabilities is a tensor where certain relations exist between the amplitudes of various Cartesian tensor components, imposed by the symmetry group. If a nanoelement also possesses a nonzero intrinsic second-order susceptibility, interesting possibilities may arise due to the interplay of the effective and the intrinsic nonzero tensor components of the nonlinear polarizability tensors. Some study of the impact of the intrinsic hyperpolarizability tensor structure on second harmonic generation in a nanoelement has been done in [151]. This analysis is, however, beyond the scope of this chapter. In the remainder of the section, we discuss in more detail the aforementioned phenomena of nonreciprocal directionality and of the suppression of a nonlinear process.

7.2.3.2 Nonreciprocal directionality of nonlinear generation

Unless the nonlinear process is degenerate with $\omega_1 = \omega_2$, various terms entering the right hand sides of (7.5) and (7.6) are relatively decoupled and can be designed nearly independently. Their relative amplitudes can be controlled, for example, by the choice spectral positions of the fundamental fields with respect to linear electric and magnetic multipolar resonances. Consider now a situation where only the first term is non-zero in each of the (7.5) and (7.6), and assume that $\alpha^{(2)}_{emm,xyy} = \pm\alpha^{(2)}_{mem,yxy}$, with the rest of the Cartesian components being negligible. According to (7.5) and (7.6), these terms share the same pathway through the magnetic field of the fundamental beam at ω_2. As a result, switching the phase of *just the magnetic* vector of this fundamental wave—e.g., following arrangement (ii) in Figure 7.5(a)—while keeping the beam at ω_1 in place, will produce a *simultaneous* phase change in *both* (nonlinearly generated) electric and magnetic dipolar modes. This results in a *simultaneous* sign change for both terms in (7.3)—a scenario discussed earlier with respect to the case (iii) in Figure 7.4. As a result, the directions of the interferometric

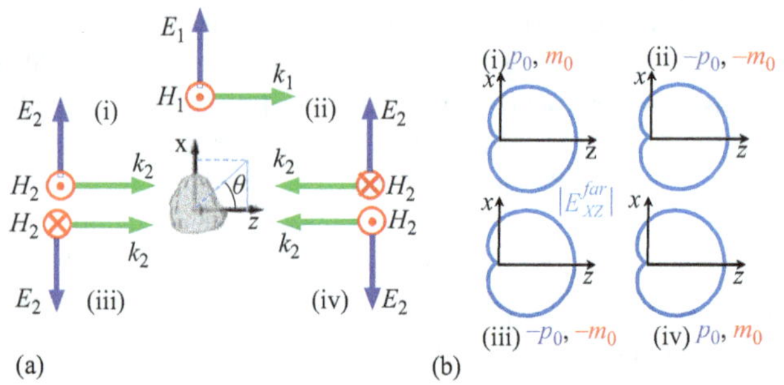

Figure 7.5 Illustration of a potential nonreciprocal directionality of nonlinear generation. (a) Excitation arrangements. The direction and phase of the beam at frequency ω_2 is varied between the same four arrangements as in Figure 7.4. The direction of the beam at frequency ω_1 is assumed unchanged in all four cases (i)–(iv). (b) The expected angular distribution of the norm of the nonlinearly generated far field in the xz-plane when the terms $\alpha_{emm,xyy} = \alpha_{mem,yxy}$ are dominant and the rest of the terms are negligible. Nonlinear generation is expected to be suppressed in negative z hemisphere and doubled in the positive z hemisphere in all four excitation arrangements, independently of the direction of the fundamental beam at ω_2.

suppression and doubling of the nonlinearly generated field will remain unchanged, compared to arrangement (i).

On the other hand, if one directs the pump wave at ω_2 according to set (iv) in Figure 7.5(a), none of the terms in (7.3) will change their sign: since both "emm" and "mem" terms only depend on the phase of the *magnetic* vector of the ω_2 beam, reversing the phase of just the electric vector of that beam, as it happens in case (iv), does not produce any impact. Thus, the generation direction again will be preserved as in arrangement (i). There will be, similarly, no change in the generation direction with arrangement (iii). In this instance, the opposite (with respect to set (i)) phase of the magnetic vector of the fundamental wave at frequency ω_2 will induce a switch of phase in both the electric and magnetic nonlinearly produced dipolar modes. Meanwhile, the opposite phase of the electric vector will induce no change at all, due to no dependence of the "emm" and "mem" terms on the electric field of the ω_2 beam. The result again is a preservation of the direction of nonlinear generation. The angular distribution of the nonlinearly generated field, expected to be produced with these four arrangements, is illustrated in Figure 7.5(b).

According to the above discussion, the direction of nonlinear generation is expected to be preserved with respect to a fixed laboratory coordinate system in all four arrangements shown in Figure 7.5(a). Thus, the generation is expected to be nonreciprocal with respect to the direction of the fundamental wave at ω_2. This behavior is qualitatively different from linear scattering illustrated in Figure 7.4(b) [34,130], where the "pathways" are different for electric and magnetic dipolar modes.

A similar nonreciprocity in the nonlinear response will occur for any other combination of effective nonlinear polarizabilities of a similar strength, as long as they share the same "pathway," whether electric or magnetic, while the rest of the effective hyperpolarizability terms are negligible. For example, the condition $\alpha^{(2)}_{eee,xxy} = \alpha^{(2)}_{mem,yxx}$, with the rest of hyperpolarizabilities negligible, ensures the preservation of the generation direction when reversing the fundamental wave at ω_1. Nonreciprocity with respect to *all* fundamental waves at once, however, requires the equality of hyperpolarizabilities belonging to different symmetry groups (e.g., $\alpha^{(2)}_{emm} = \alpha^{(2)}_{mmm}$), and thus would be forbidden in elements of a single group symmetry. This principle can be extended to higher-order nonlinear processes, as well as to higher-order multipolar modes.

As discussed, the nonreciprocity phenomena described here is not limited to a particular nonlinear process or even to a particular order of the nonlinear response. However, the *unidirectional* and *nonreciprocal* nonlinear scattering based on the principle described above is not expected to occur in elements of a given symmetry group for processes that are degenerate in frequency, such as harmonic generation. For example, for the second harmonic process, the condition $\alpha_{mme} = \alpha_{mem}$ will be always met, since $\omega_1 = \omega_2$. As a result, it would be impossible to isolate an effective magnetic hyperpolarizability of a single kind (either "mem" or "mme"), as required in the above example. Consequently, even if a condition of matching the values of electric and magnetic hyperpolarizabilities sharing the same "pathway" is met—e.g., $\alpha^{(2)}_{emm,xxy} = \alpha^{(2)}_{mem,yxx}$—the presence of the third comparable term (the "mme" one) that does not share that pathway will prevent the directionality of the second harmonic from being preserved on reversal of either of the excitation beams.

7.2.3.3 Inhibition of a nonlinear process

While an option of such nonreciprocity in the direction of nonlinear generation is unique by itself and can be employed in a variety of applications, magneto-electric interference between various effective hyperpolarizability terms can lead to other phenomena that are unachievable with natural nonlinear media. As seen from (7.5) and (7.6), each mode (electric or magnetic) can, by itself, result from the interference effects between various terms entering the right-hand sides of these equations. Note that, in either the electric or magnetic mode, the terms within a *single symmetry group* (the terms in each square bracket) depend on a different type of fundamental field vector of a given frequency. Consequently, reversing the direction of one of the fundamental waves will produce a sign change for one of the two interfering terms in each bracket, but not for the other one. As a result, the nonlinear dipolar responses will be different for the co- and counter-propagating pumps.

In general, all the terms within a single symmetry group on the right-hand sides of (7.5) and (7.6) are expected to be nonzero, at least to some extent, in an arbitrary nanoelement with a nonzero linear magnetic dipolar response (unless a special effort is made to design the nanoelement with only specific dominant effective hyperpolarizability terms). Therefore, when the direction of one or more (but not all) of the excitation beams is reversed, a majority of nanoelements should exhibit a change in the efficiency in the nonlinear response, due to the change in the interference pattern of various terms *within* each multipolar mode. This expectation is valid even for nanoelements that are entirely symmetric, both in terms of their geometry and material composition. Thus, nonreciprocity with respect to the direction of a subset of fundamental beams, in terms of the efficiency of the nonlinear response, is inherent to the nonlinear response of most nanoelements, even symmetric ones, and for most nonlinear processes.

Furthermore, if a geometry/spectral position can be found that ensures the equality between the two electric terms and between the two magnetic terms within the symmetry group (i.e., $\alpha_{eee} = \alpha_{emm}$ and $\alpha_{mem} = \alpha_{mme}$, for a polar nanoelement), the nonlinear process for which such a relation holds can be overall suppressed when reversing one of the beams. The suppression will happen with contra-directional excitation if the effective hyperpolarizabilities are matched with the same sign, and with co-directional excitation if they are matched with an opposite sign. If, additionally, all terms are equal within the symmetry group ($\alpha_{eee} = \alpha_{emm} = \alpha_{mem} = \alpha_{mme}$, for polar geometry), nonlinear generation is expected to be unidirectional for co-propagating excitation (suppressed in one hemisphere and doubled on the other one) but suppressed overall with contra-directional excitation.

In order to analyze the feasibility of achieving the described phenomena with physical elements, a procedure is needed allowing one to identify the nonlinear magnetoelectric polarizabilities of a nanoelement. In Section 7.3, we develop an approach allowing the retrieval of these polarizabilities from the angular distribution of the nonlinearly generated far field.

7.3 Retrieval of the effective nonlinear multipolar polarizabilities

While several methods have been developed allowing to both predict [56,152] and retrieve [153–155] purely electric and magnetic effective nonlinear polarizabilities, only a few approaches are available so far to determine a complete set of magneto-electric ones. First identification of several magneto-electric terms in the effective nonlinear polarizabilities of L-shaped nanoelements has been made in [92,128]. A comprehensive approach deriving a complete set of effective nonlinear magnetoelectric susceptibilities of a unit cell incorporating nonlinear elements has been done in [88]. The approach employed coupled-mode theory [156] adapted to Blokh modes propagating in the periodic cell arrangement, and used field averaging over a unit cell to express the effective nonlinear magneto-electric susceptibilities in terms of the intrinsic, purely electric ones. The method showed an excellent agreement with the experiment [93], as well as with the results obtained by previously developed methods for pure electric and pure magnetic effective nonlinear susceptibilities [56,152,157–160]. Being based on the coupled-mode theory, however, this approach relies on the analysis of the propagating fundamental waves within the medium. As such, it cannot be used when propagating solutions do not exist—for example, in such spectral regions of interest as those near a resonance, where the real part of either electric or magnetic linear effective susceptibility is negative. It also has a limitation of assuming a negligible intrinsic loss within the unit cell. An alternative approach has recently been proposed [161,162], extending the previously developed nonlinear transfer matrix method [153–155] to the magneto-electric case. Each of these methods allows to identify effective nonlinear susceptibilities of a periodic arrangement of unit cells within a layer as a whole, rather than to determine the nonlinear polarizability of a single, isolated unit cell or a single nanoelement. Here, we focus on the nonlinear response of a single nanoelement, which can then be utilized if employing the nanoelement alone, operating as a nanoscale nonlinear optical antenna, or if using it as a building block within an effective medium. The procedure has been developed in [38], and we describe its main steps below.

The retrieval approach implemented here consists of two steps. First, we develop a procedure allowing to identify the amplitudes of various multipolar modes in the nonlinearly generated field. Second, this procedure is repeated for several combinations of the respective directions and phases of the fundamental excitation beams. The number of arrangements depends on the order of the nonlinear process, with four combinations required for the second-order response considered here. The resulting sets of the retrieved multipolar amplitudes are then used to identify the eight magneto-electric polarizabilities entering (7.5) and (7.6). Since the first step only deals with the field that has been already produced, without tracking processes leading to the field generation, it is equivalent to a similar decomposition in the case of linear scattering (in the latter case, identifying multipolar contributions in the linearly scattered far field). The only difference is that the multipolar decomposition is now performed for the field at the nonlinearly generated frequency. This field can be found, e.g., from numerical analysis. As such, any of the procedures developed for the multipolar decomposition of the linearly scattered far field can, generally, be used as a first step of the retrieval. In Section 7.3.1, we provide a brief overview of the developed methods and describe a simplified approach employed here.

7.3.1 Retrieval of multipolar partial waves

A variety of methods exists for identifying the amplitudes of multipolar modes scattered linearly by a nanoelement or by a group of nanoelements. The most standard approach projects the scattered far field, detected on the surface of a sphere surrounding the group of scattering elements, onto spherical harmonics, directly indicating the multipolar moments in terms of electric and magnetic Mie coefficients [25,163–165]. The far field used in the analysis is found by simulations (e.g., by integrating the scattered near field using the Stratton-Chu [166] expression) or experimentally. A set of expressions can then be used [23,25,163,164] to relate these coefficients to the Cartesian components of various multipoles. While this method is the most general, it requires a relatively fine sampling of the far field on the surface of the imaginary sphere surrounding the scatterers, although some techniques can be used for reducing the number of sampling points [167]. The method is also not entirely suitable for periodic structures, as it assumes a group of nanoelements of finite dimensions. Alternatively, multipolar coefficients can be expressed directly in terms of volume integrals of the source current density [23], the latter found numerically. Several alternative approaches have been developed, enabling determination of multipolar coefficients from the knowledge of either surface or volume currents within the elements. One of these techniques [165] calculates effective electric and magnetic surface currents that reproduce the scattering response of the nanoelement. These currents can then be used to predict the far field and proceed with the standard approach, projecting the far field onto the spherical harmonics. Alternately, one can proceed to find multipolar moments directly from surface currents using the algorithm developed by the authors. The method has advantages over the standard projection on spherical harmonics, as it allows finding individual responses of scatterers in close proximity to other scatterers, or scatterers placed inside a lossy medium [165]. Another approach [164] maps multipolar expansions in terms of spherical harmonics to a multipolar decomposition of current density in terms of sets of unit currents, the so-called "current multipoles." A nanoelement is then treated as a collection of these unit currents.

Generally, any of the above-mentioned procedures can be used as a first step of the nonlinear retrieval, identifying multipolar amplitudes present in the nonlinearly generated far field in the same manner as it would have been done for the linearly scattered field (with the currents-based approaches, currents at the nonlinearly produced frequency, rather than those induced by a linear response, would of course have to be used; the former would have to be first found numerically, same as the nonlinearly generated far field). In the approach employed here, a simplified procedure is used [34,38], where the multipolar modes up to the electric quadrupolar term are retrieved by probing the far field in three orthogonal directions. The accuracy of the prediction is verified by reconstructing the whole angular distribution analytically and comparing it with the numerically found one. The method helps us to reduce the computational effort by probing the far field in only few points, while fully visualizing the angular distribution of each retrieved mode and ensuring its agreement with the numerically produced shape. It also allows for a rather intuitive visualization of even and odd partial waves constituting the scattered field. The method has been tested by comparing the results with Mie theory predictions [34] and has shown an excellent agreement with numerical results for most nanoelements that do not exhibit optical activity.

7.3.1.1 Retrieval procedure

In the current subsection, we use the notation assuming the retrieval of a linearly scattered field. The same procedure, with the corrected notation, will be used as a first step for nonlinear retrieval. We indicate the necessary corrections in the next subsection when describing the nonlinear retrieval.

From the expressions for Cartesian components of the dipolar modes given in Section 7.2.1 (Equation (7.2)), one can note that the far field amplitudes of these modes are decoupled in the xy $(\theta = 90^\circ)$ plane: the x- (z-)component of the scattered field taken in the xy plane represent purely electric (magnetic) dipolar mode. Furthermore, each of these components represents the full scattering amplitude in the y and x direction, respectively (since the remaining two Cartesian components vanish along the y- and x-axes). Therefore, if no higher-order multipoles contributed to scattering into the xy plane, the electric and magnetic dipolar amplitudes could have been retrieved by probing the scattered far field in the directions along the y- and x-axes, respectively. The contribution of higher-order multipoles into the far field grows, however, with frequency at a faster rate than that of the dipolar modes, getting an additional frequency factor for each increased multipolar order [25,168]. Consequently, at optical frequencies, higher-order multipoles are likely to make a noticeable addition into the scattered far field and need to be accounted for [168]. This comment is especially relevant to the electric quadrupolar mode, which results from the symmetric part of the same field gradient tensor responsible for the excitation of the magnetic dipolar mode [23,25,169]. For the geometries implemented in the present chapter, we find no multipoles higher than the electric quadrupolar mode being significant within the frequency range of interest. Assuming also no optical activity exhibited by the nanoelement, multipolar contributions can then still be found relatively easily by probing the scattered far field in as little as three directions, as described below.

Assuming the x-polarized field incident along z direction, and no optical activity in the nanoelement, the only tensor component of the electric quadrupolar mode that can be excited with such illumination is the $Q_{xz} = Q_{zx}$ one [168], where Q_{xz} denotes the amplitude of the xz tensor component of the quadrupolar moment. The angular distribution of the electric field produced by this mode can be obtained relatively easily via Taylor expansion of field potentials [23]. It is visualized best when written in spherical coordinate system (Figure 7.3), with its components given by [23]:

$$E_\theta^{el.quad.} = -\frac{ik^3}{8\pi}\frac{e^{ikr}}{r}Q_{xz}\left[\cos^2\theta\cos\varphi - \sin^2\theta\cos\varphi\right], \tag{7.7a}$$

$$E_\varphi^{el.quad.} = \frac{ik^3}{8\pi}\frac{e^{ikr}}{r}Q_{xz}\cos\theta\sin\varphi. \tag{7.7b}$$

The E_r component vanishes in the far field.

The field described by (7.7) has the 4-lobe pattern shown earlier in Figure 7.1(c). Furthermore, the shape of the quadrupolar lobes in the xy ($\theta = 90°$) plane is identical to that of the magnetic dipolar mode (Equation (7.2)), thus preventing the retrieval of the magnetic dipolar amplitude by probing the direction along the x-axis (or any other direction in the xy plane) alone. One can track the origin of the first and the second terms in brackets in (7.7a) to xz and zx components of the quadrupolar tensor [23], which have the same amplitude but orthogonal angular distributions, forming the figure 8-type patterns along the z- and x-axes, respectively. (Note that one can arrive to the identical angular distribution for each of these components by considering the field of a pair of two closely spaced, out-of-phase electric dipoles, with each dipole having an angular distribution given by (7.1a), as has been done for the xz term in [34].) As seen from (7.7a), these two terms contribute with the opposite phase to the scattered field, cancelling interferometrically in the direction along 45° in each quadrant in the xz ($\varphi = 0$) plane. As a result, one of the terms will always contribute constructively, and the other destructively, to the magnetic dipolar mode (which has the same parity as the electric quadrupolar one), elongating its (otherwise circular in the xz plane) shape either along z- or along x-axis (the latter depending on the overall relative phase of the dipolar and quadrupolar modes). This fact allows resolving the magnetic dipolar and electric quadrupolar amplitudes by probing the scattered field at two specific points in the xz plane, as soon as the electric dipolar mode is known.

Assuming the electric dipolar amplitude is known from probing the scattered field along the y direction, one option to resolve the magnetic dipolar and electric quadrupolar modes is to use the set of points at $\theta = 45°$ and at $\theta = 0°$ in the xz ($\varphi = 0$) plane. The first of these directions does not have a contribution from the electric quadrupolar mode, thus allowing to find the magnetic dipolar amplitude as the difference between the total scattering amplitude in the $\theta = 45°$ direction and the electric dipolar amplitude found earlier. The quadrupolar amplitude can then be found by subtracting the sum of the dipolar modes from the full scattering amplitude in the $\theta = 0°$ direction. Another option is to probe the scattering amplitudes along the x- and z-axes, noting that the quadrupolar contribution adds to the

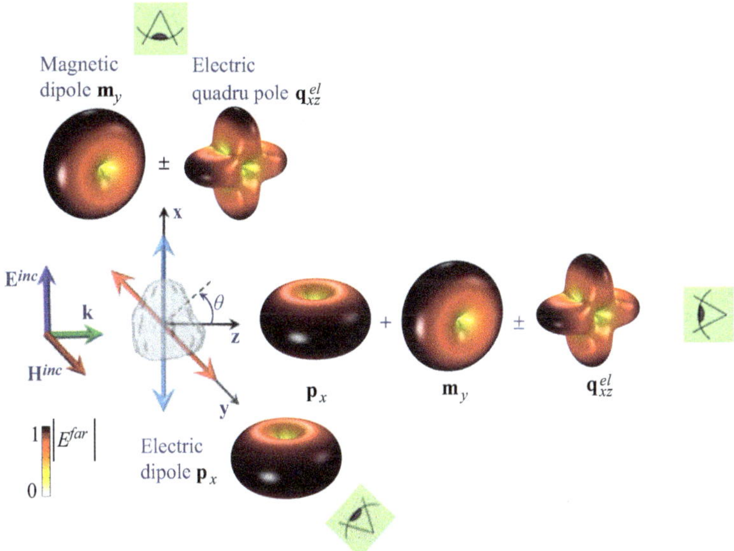

Figure 7.6 *Schematic illustration of the retrieval procedure described in the text. Assuming no multipoles higher than the electric quadrupolar mode are significant and no optical activity in a nanoelement, electric dipolar mode is retrieved by probing the scattering amplitude in the "y" direction. Magnetic dipolar and electric quadrupolar modes are then resolved by probing the scattering amplitude in two additional directions, along "x"- and "z"-axes. The "±" and "∓" signs indicate the fact that the XZ and ZX tensor components of the quadrupolar mode (the only ones possibly excited with the depicted arrangement) scatter with the opposite phases in the z and y directions.*

magnetic dipolar mode with the same amplitude but opposite phases along these two directions. This approach is depicted in Figure 7.6. The two approaches produce the same result. We proceed with the second approach below, providing the explicit expressions used in the retrieval.

Accounting for the electric quadrupolar mode given by (7.7), the E_θ component of the scattered far field in the $xz(\varphi = 0)$ plane, presented earlier in (7.3) for dipolar modes, is given by

$$E_\theta\big|_{\varphi=0} = \frac{k^2}{4\pi}\frac{e^{ikr}}{r}\left[\frac{1}{\varepsilon_0}p_{x0}\cos\theta + \eta m_{y0} - \frac{ik}{2\varepsilon_0}Q_{xz}\cos 2\theta\right] \tag{7.8}$$

The E_φ spherical component again vanishes in the xz plane, and E_r falls off overall in the far field. The scattered far field $\mathbf{E}_{far}^{sc}$, taken in the xz plane, is thus again completely represented by its E_θ component, the latter given by (7.8): $\mathbf{E}^{sc}_{far}\big|_{\varphi=0} = E_\theta\big|_{\varphi=0}$.

In general, assuming excitation by the external field that does not have sources within the volume of the nanoelement, quadrupolar moment is related to the symmetric part of the electric field gradient as following [23,25,169]:

$$\mathbf{Q} = \alpha^q \varepsilon_0 \frac{\nabla \mathbf{E}_0 + (\nabla \mathbf{E}_0)^T}{2}, \tag{7.9}$$

where α^q is the quadrupolar polarizability and superscript T indicates the transpose operation. For the x-polarized plane wave excitation along the z direction, $\mathbf{E}_0 = E_0 e^{ikx}\hat{\mathbf{x}}$, the Q_{xz} tensor component can be written from (7.9) as following [168]:

$$Q_{xz} = \frac{1}{2}\alpha^q \varepsilon_0 \frac{\partial E_x}{\partial z} = \frac{ik}{2}\alpha^q \varepsilon_0 E_0. \tag{7.10}$$

With the use of (7.10), the last term in the brackets in (7.8) transforms into $-\frac{k^2}{4\varepsilon_0}\alpha_q \varepsilon_0 E_0 \cos 2\theta$. Note that α^q in (7.9) and (7.10) has units of m^5, ensuring the units of the quadrupolar moment Q_{xz} to be m^2C and the overall units of the last term to agree with the rest of the terms in (7.8). Since the units of the dipolar polarizabilities α^{el} and α^m in (7.4) are m^3, for convenience of being able to present all the retrieved parameters on the same plot, we introduce a normalized quadrupolar tensor component according to

$$\widetilde{Q}_{xz} \equiv \widetilde{\alpha}^q \varepsilon_0 E_0, \tag{7.11}$$

where $\widetilde{\alpha}^q$ is the normalized, frequency-dependent quadrupolar polarizability which has units of m^3 and is related to α^q as

$$\widetilde{\alpha}^q \equiv \frac{k^2}{4}\alpha^q. \tag{7.12}$$

With the newly introduced normalized quadrupolar tensor component, (7.8) becomes

$$E^{sc}_{far}\Big|_{\varphi=0} = \frac{k^2}{4\pi}\frac{e^{ikr}}{r}\left[\frac{1}{\varepsilon_0}p_{x0}\cos\theta + \eta m_{y0} + \frac{1}{\varepsilon_0}\widetilde{Q}_{xz}\cos 2\theta\right]. \tag{7.13}$$

In writing (7.13), we also used the fact that all spherical components, except E_θ, vanish in the xz ($\varphi = 0$) plane, so that $E^{sc}_{far}\big|_{\varphi=0} = E_\theta\big|_{\varphi=0}$. We now introduce the scattering amplitude $\mathbf{A}$ according to $\mathbf{E}^{\mathbf{sc}}_{\mathbf{far}} \equiv \frac{\mathbf{e}^{\mathbf{ikr}}}{\mathbf{r}}|\mathbf{E}_0|\mathbf{A}$ (where both $\mathbf{E}^{sc}_{far}$ and $\mathbf{A}$ are assumed to be in the far field zone), which helps us express multipolar amplitudes in terms of the far field scattered in a given direction, rather than dealing with the distance-dependent scattered field incorporating the $exp(ikr)/r$ factor. The modulus of E_0 is used to emphasize the fact that any phase change in the scattering field is incorporated in the corresponding phase change of the scattering amplitude, rather than in the E_0 coefficient. Using this definition on the left-hand side of (7.1a), the electric dipolar amplitude is found in terms of scattering amplitude probed in the y $(\varphi = 90^\circ, \theta = 90^\circ,)$ direction as following:

$$p_{x0} = \frac{4\pi\varepsilon_0|E_0|}{k^2} A\Big|_{\substack{\theta = 90^\circ \\ \varphi = 90^\circ}} \tag{7.14}$$

In writing (7.14), we used the fact that the E_x^{el+m} field component given by (7.1a) is the only nonzero component contributing to scattering along the y-axis (assuming no multipolar contributions above the electric quadrupolar mode), hence the full scattering amplitude A, rather than its A_x Cartesian component, is used in (7.14). The result is identical if one probes the A_x component in that direction, which might be preferable if retrieving the amplitudes from numerical simulations.

Using the result of (7.14) in (7.13), and resolving between the m_{y0} and $\widetilde{Q}_{xz}$ terms by evaluating (7.13) in the $\theta = 90^\circ$ and $\theta = 0^\circ$ directions, one arrives at the following expressions for the magnetic dipolar and the (normalized) electric quadrupolar amplitudes in terms of the scattered far field amplitude probed along the x, y, and z directions:

$$m_{y0} = \frac{1}{2}\frac{4\pi|E_0|}{\eta k^2}\left[A\Big|_{\substack{\theta = 0^\circ \\ \varphi = 0^\circ}} + A\Big|_{\substack{\theta = 90^\circ \\ \varphi = 0^\circ}} - A\Big|_{\substack{\theta = 90^\circ \\ \varphi = 90^\circ}}\right] \tag{7.15a}$$

$$\widetilde{Q}_{xz} = \frac{1}{2}\frac{4\pi\varepsilon_0|E_0|}{k^2}\left[A\Big|_{\substack{\theta = 0^\circ \\ \varphi = 0^\circ}} - A\Big|_{\substack{\theta = 90^\circ \\ \varphi = 0^\circ}} - A\Big|_{\substack{\theta = 90^\circ \\ \varphi = 90^\circ}}\right] \tag{7.15b}$$

Linear polarizabilities follow directly from (7.14) and (7.15) if one uses (7.4) (with $\mathbf{p} = (p_{x0}, 0, 0)$ and $m = (0, m_{y0}, 0)$) and (7.11) to relate polarizabilities to the corresponding multipolar amplitudes. Note that, in contrast with polarizabilities, which are fixed numbers that do not change when reversing the direction of the incident field, the sign of the multipolar modes on the left-hand side of (7.14) and (7.15) may change upon reversal of the excitation beam. In the case of linear scattering, as discussed in Section 7.2.2 and according to (7.4), the sign of each dipolar mode follows the phase of the corresponding excitation field vector on beam reversal, repeating the pattern shown earlier in Figure 7.4(a). The values of the scattering amplitude A (still probed in the same directions) will also change on beam reversal, resulting in a preserved value of the retrieved polarizability. For example, denoting the multipolar modes and scattering amplitudes induced with arrangement (i) from Figure 7.4(a) with superscript "+," and using (7.4) and (7.11) on the left-hand side of (7.14) and (7.15), the retrieved polarizabilities are as following:

$$\alpha_{xx}^{el} = \frac{4\pi}{k^2} A^+ \Big|_{\substack{\theta = 90^\circ \\ \varphi = 90^\circ}} \tag{7.16a}$$

$$\alpha_{yy}^{m} = \frac{1}{2}\frac{4\pi}{k^2} \left[A^+ \Big|_{\substack{\theta = 0^\circ \\ \varphi = 0^\circ}} + A^+ \Big|_{\substack{\theta = 90^\circ \\ \varphi = 0^\circ}} - A^+ \Big|_{\substack{\theta = 90^\circ \\ \varphi = 90^\circ}} \right] \tag{7.16b}$$

$$\widetilde{\alpha}^{q} = \frac{1}{2}\frac{4\pi}{k^2} \left[A^+ \Big|_{\substack{\theta = 0^\circ \\ \varphi = 0^\circ}} - A^+ \Big|_{\substack{\theta = 90^\circ \\ \varphi = 0^\circ}} - A^+ \Big|_{\substack{\theta = 90^\circ \\ \varphi = 90^\circ}} \right], \tag{7.16c}$$

where $\widetilde{\alpha}^{q}$ is the normalized quadrupolar polarizability of the Q_{xz} tensor component defined in (7.12).

If the incident beam is reversed according to, e.g., set (ii) from Figure 7.4(a), the sign on the left-hand side of (7.15a), and hence on the left-hand side of (7.16b), flips (since $m_y^- = -m_y^+$, according to (7.15a)). The same sign change would occur for the quadrupolar amplitude $\widetilde{Q}_{xz}$ in (7.15b) with the excitation according set (ii) in Figure 7.4(a), while the sign of the dipolar mode remains the same. In the same time, all A^+ amplitudes on the right-hand sides of (7.16) will change to A^- (still probed in the same directions with respect to a fixed coordinate system), resulting in the preserved values of the retrieved polarizabilities α^{el}, α^{m}, and $\widetilde{\alpha}^{q}$. We note that linearly induced quadrupolar amplitude Q_{xz} will overall have its phase changing in the same manner as the magnetic dipolar mode m_{y0} when varying the direction and phase of the excitation beam between sets (i)–(iv) from Figure 7.4(a). One can see that by considering the amplitude of the linearly polarized incident wave for the four excitation arrangements of Figure 7.4(a), which are as following: (i) $\mathbf{E}_0 = E_0\widehat{x}\exp(ikz)$; (ii) $\mathbf{E}_0 = E_0\widehat{x}\exp(-ikz)$; (iii) $\mathbf{E}_0 = -E_0\widehat{x}\exp(ikz)$; and (iv) $\mathbf{E}_0 = -E_0\widehat{x}\exp(-ikz)$. Using these expressions in (7.10) and (7.9), one can see that the sign of the quadrupolar mode Q_{xz} in the sets (ii) and (iii) is expected to be opposite to that in (i) and (iv), thus varying in a similar fashion as the phase of the induced magnetic dipolar mode. An alternative derivation of the quadrupolar mode changing its phase similarly to the magnetic dipolar one on beam reversal can be found in [37]. This fact allows us later to use a single effective hyperpolarizability for each of the "emm" and "emq" and similar terms ("q" standing for the quadrupolar contribution), rather than accounting separately for the magnetic dipolar and quadrupolar contributions into the nonlinearly generated dipolar modes, as explained in more detail in Section 7.4.2.

7.3.1.2 Relation with numerical procedures and validation of the linear retrieval

The scattering amplitude **A** can be found from simulations or measured experimentally (as soon as a technique to identify both its amplitude and phase is implemented). In the present work, we use finite elements method (FEM) incorporated within COMSOL Multiphysics package [170] for the numerical analysis. The far field scattering amplitude, $A(\widehat{\mathbf{r}}_0)$, in a given direction $\widehat{\mathbf{r}}_0$, is found numerically by integrating the scattered near fields **E** and **H** over an arbitrary surface S enclosing the scattering objects using the Stratton-Chu formula [166]:

$$\mathbf{A}(\widehat{\mathbf{r}}_0) = -\frac{1}{E_0}\frac{ik}{4\pi}\widehat{\mathbf{r}}_0\times\int_S\left[\mathbf{n}\times\mathbf{E}-\eta\widehat{\mathbf{r}}_0\times(\mathbf{n}\times\mathbf{H})\right]e^{ik\mathbf{r}'\cdot\widehat{\mathbf{r}}_0}dS', \tag{7.17}$$

where $\mathbf{n}$ is a normal to surface S and $\widehat{\mathbf{r}}_0$ is a unit vector along the direction of interest.

The validity of the assumptions of a negligible amplitude of higher-order modes and of no optical activity exhibited by the nanoelement, made in deriving the retrieval procedure in the previous subsection, can be verified by comparing the numerically (or experimentally) found angular distribution of the scattered field with the analytical one; the latter distribution can be obtained by using the retrieved values of multipolar amplitudes in the theoretical expressions for the angular dependence of various modes. An example of such a comparison for the dimer geometry used later in Section 7.4, is presented in Figure 7.7, showing exact agreement between the theoretical and the numerical results. For convenience, we provide below the expressions for the Cartesian components of the quadrupolar amplitudes used in figure construction and obtained by coordinate transformation from (7.7):

$$E_x^{el.quad.}=\frac{ik^3}{8\pi}\frac{e^{ikr}}{r}Q_{xz}\left\{\left[\sin^2\theta\sin^2\varphi+\cos^2\theta\right]\cos\theta+\sin^2\theta\cos^2\varphi\cos\theta\right\} \tag{7.18a}$$

$$E_y^{el.quad.}=2\frac{ik^3}{8\pi}\frac{e^{ikr}}{r}Q_{xz}\sin^2\theta\cos\theta\cos\varphi\sin\varphi \tag{7.18b}$$

$$E_z^{el.quad.}=-\frac{ik^3}{8\pi}\frac{e^{ikr}}{r}Q_{xz}\left\{\cos^2\theta\cos\varphi\sin\theta-\sin^2\theta\cos\varphi\sin\theta\right\} \tag{7.18c}$$

Some trigonometric identities were used to arrive at (7.18a) in the form given. Note that the two terms in curly brackets in both (7.18a) and (7.18c) correspond to the xz and zx portions of the quadrupolar mode, while the E_y component is identical

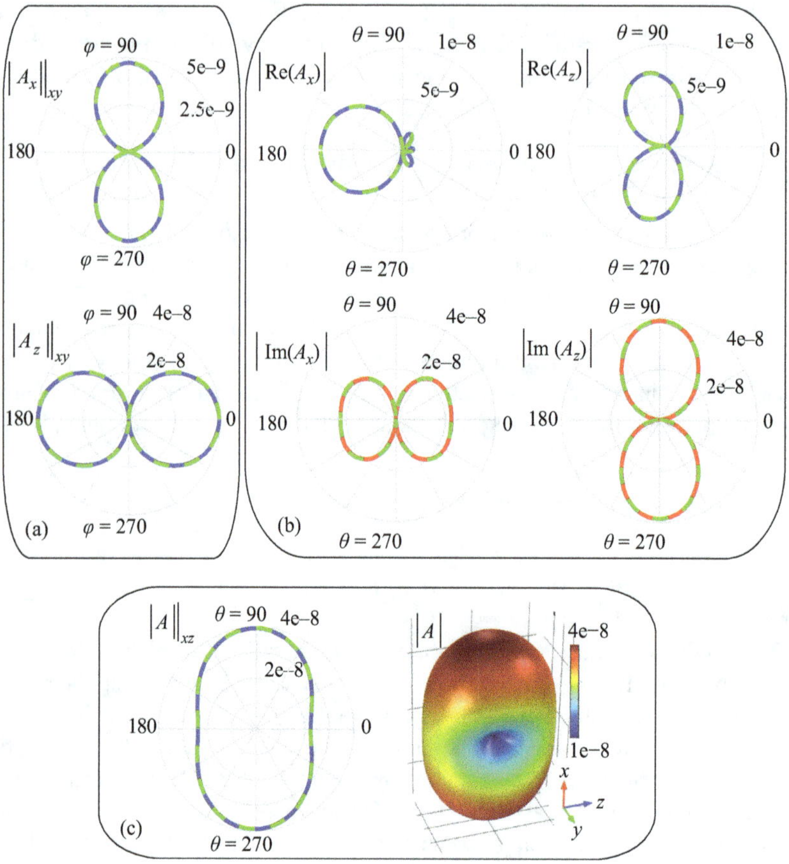

Figure 7.7 Comparison of the numerical and analytical far field distributions, validating the linear retrieval procedure. (a) xy ($\theta = 90°$) plane. (b) xz ($\varphi = 0°$) plane. (c) Left: norm of the scattering amplitude in the xz($\varphi = 0°$) plane. Right: 3D angular distribution of the norm of the far field scattering amplitude. In all figures, solid green lines represent numerical distribution and dashed lines indicate analytical distribution, with dashed green (dashed red) in (b) indicating the real (the imaginary) part. The comparison is shown for the magnetic dipolar resonance spectral position from Figure 7.11 at 300 THz.

for both *xz* and *zx* contributions: one can arrive to the first (the second) term by combining the corresponding Cartesian components of the two dipolar modes oscillating with an opposite phase along *x* (along *z*) and separated by a small distance along *z* (along *x*) direction. Such derivation was made in [34] for the *xz* part of the quadrupolar mode. In order to produce the numerical curves in Figure 7.17, polarizabilities retrieved using (7.16) were used in (7.2) and (7.18); the sum of the resulting amplitudes was then presented in Figure 7.7. A detailed analysis of the

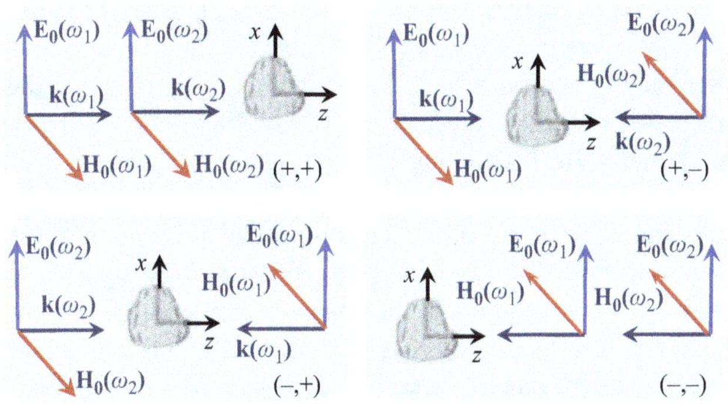

Figure 7.8 *The four pump arrangements used for the effective hyperpolarizabilities retrieval. The "+" ("–") sign denotes the fundamental beam incident in the positive (negative) "z" direction, with the signs in the first and the second positions denoting the direction of the fundamental waves at ω_1 and ω_2, respectively.*

accuracy of the similar retrieval for various geometries, as well as a comparison with Mie coefficients up to the dipolar terms for analytically treatable shapes, has been also done in [34].

In addition to polarizabilities retrieval, the knowledge of the scattering amplitude allows one to calculate the numerical values of the scattering cross sections [24]: the extinction can be found as a function of the imaginary part of the scattering amplitude in the forward direction; the absorption cross section can be calculated by integrating the heat losses over the volume of the particle and normalizing it by the intensity of the incident field; finally, the scattering cross section is obtained as the difference between the extinction and absorption. As an illustration of the accuracy of the simulations performed, such a calculation has been done for spherical geometries in [34], showing an exact agreement of the scattering amplitudes obtained from simulations with those found analytically from Mie theory.

7.3.2 Retrieval of nonlinear magnetoelectric polarizabilities

The approach developed in Section 7.3.1 can equally be used for identifying multipolar components in the nonlinearly generated field: one just has to probe the *nonlinearly produced* field amplitude along the x, y, and z directions using (7.14) and (7.15). In order to retrieve all eight terms of the effective hyperpolarizabilities entering (7.5) and (7.6), one can repeat the retrieval procedure using four combinations of the respective directions of the fundamental beams, with one possible set of arrangements shown in Figure 7.8. As seen in the figure, each of the fundamental beams is alternated between the positive and the negative z directions, with sets (i) and (ii) from Figure 7.4(a) being chosen in Figure 7.8 example for positive and negative excitations. A combination of any other two sets with opposite excitation directions can be used, e.g., sets (i) and (iv), accounting for a corresponding change of phase of the fundamental field vectors. Using (7.14) and (7.15) with each of the arrangements shown in Figure 7.8, one arrives at four sets of the nonlinearly produced multipolar amplitudes, which we denote as $p_3^{+,+}$, $m_3^{+,+}$, $Q_3^{+,+}$, etc. Here, the subscript "3" denotes the "(ω_3)" frequency dependence; and the first (the second) sign in the superscript indicates the direction of the fundamental beam at frequency ω_1 (ω_2) used to produce that particular mode, with $+$ denoting propagation in the positive z direction. For example, the $p_3^{+,+}$ amplitude denotes the $\mathbf{p}(\omega_3)^{+,+}$ mode, and so on.

Each set of the nonlinearly produced multipolar modes can be related to the effective nonlinear polarizabilities using (7.5) and (7.6). The signs of various terms on the right sides in these equations will change between the four arrangements shown in Figure 7.8, following the product of signs of the fundamental vectors for each term in a given arrangement. Denoting by "$\mathbf{E}_{01}$" and "$\mathbf{H}_{01}$" ("$\mathbf{E}_{02}$" and "$\mathbf{H}_{02}$") the electric and magnetic vectors of the fundamental field at frequency ω_1 (ω_2) in the arrangement "(+,+)," the fundamental fields in the rest of the arrangements in Figure 7.8, written in terms of those in the arrangement "(+,+)," appear as following: ($\mathbf{E}_{01}$, $\mathbf{H}_{01}$, $\mathbf{E}_{02}$, $-\mathbf{H}_{02}$) in arrangement "(+,–)"; ($\mathbf{E}_{01}$, $-\mathbf{H}_{01}$, $\mathbf{E}_{02}$, $\mathbf{H}_{02}$) in arrangement "(–,+)"; and ($\mathbf{E}_{01}$, $-\mathbf{H}_{01}$, $\mathbf{E}_{02}$, $-\mathbf{H}_{02}$) in arrangement "(–,–)." Using this notation in (7.5) and (7.6), one arrives at the following sets of equations for the electric and magnetic dipolar modes:

$$\mathbf{p}_3^{+,+} = \frac{\varepsilon_0}{\sqrt{I}}\left[\alpha_{eee}^{(2)}:\mathbf{E}_{01}\mathbf{E}_{02} + \eta^2\alpha_{emm}^{(2)}:\mathbf{H}_{01}\mathbf{H}_{02} + \eta\alpha_{eem}^{(2)}:\mathbf{E}_{01}\mathbf{H}_{02} + \eta\alpha_{eme}^{(2)}:\mathbf{H}_{01}\mathbf{E}_{02}\right], \tag{7.19a}$$

$$\mathbf{p}_3^{+,-} = \frac{\varepsilon_0}{\sqrt{I}}\left[\alpha_{eee}^{(2)}:\mathbf{E}_{01}\mathbf{E}_{02} - \eta^2\alpha_{emm}^{(2)}:\mathbf{H}_{01}\mathbf{H}_{02} - \eta\alpha_{eem}^{(2)}:\mathbf{E}_{01}\mathbf{H}_{02} + \eta\alpha_{eme}^{(2)}:\mathbf{H}_{01}\mathbf{E}_{02}\right], \tag{7.19b}$$

$$\mathbf{p}_3^{-,+} = \frac{\varepsilon_0}{\sqrt{I}}\left[\alpha_{eee}^{(2)}:\mathbf{E}_{01}\mathbf{E}_{02} - \eta^2\alpha_{emm}^{(2)}:\mathbf{H}_{01}\mathbf{H}_{02} + \eta\alpha_{eem}^{(2)}:\mathbf{E}_{01}\mathbf{H}_{02} - \eta\alpha_{eme}^{(2)}:\mathbf{H}_{01}\mathbf{E}_{02}\right], \tag{7.19c}$$

$$\mathbf{p}_3^{-,-} = \frac{\varepsilon_0}{\sqrt{I}}\left[\alpha_{eee}^{(2)}:\mathbf{E}_{01}\mathbf{E}_{02} + \eta^2\alpha_{emm}^{(2)}:\mathbf{H}_{01}\mathbf{H}_{02} - \eta\alpha_{eem}^{(2)}:\mathbf{E}_{01}\mathbf{H}_{02} - \eta\alpha_{eme}^{(2)}:\mathbf{H}_{01}\mathbf{E}_{02}\right], \tag{7.19d}$$

$$\mathbf{m}_3^{+,+} = \frac{1}{\sqrt{I}}\left[\frac{1}{\eta}\alpha_{mee}^{(2)}:\mathbf{E}_{01}\mathbf{E}_{02} + \eta\alpha_{mmm}^{(2)}:\mathbf{H}_{01}\mathbf{H}_{02} + \alpha_{mem}^{(2)}:\mathbf{E}_{01}\mathbf{H}_{02} + \alpha_{mme}^{(2)}:\mathbf{H}_{01}\mathbf{E}_{02}\right], \tag{7.20a}$$

$$\mathbf{m}_3^{+,-} = \frac{1}{\sqrt{I}}\left[\frac{1}{\eta}\alpha_{mee}^{(2)}:\mathbf{E}_{01}\mathbf{E}_{02} - \eta\alpha_{mmm}^{(2)}:\mathbf{H}_{01}\mathbf{H}_{02} - \alpha_{mem}^{(2)}:\mathbf{E}_{01}\mathbf{H}_{02} + \alpha_{mme}^{(2)}:\mathbf{H}_{01}\mathbf{E}_{02}\right], \tag{7.20b}$$

$$\mathbf{m}_3^{-,+} = \frac{1}{\sqrt{I}}\left[\frac{1}{\eta}\alpha_{mee}^{(2)}:\mathbf{E}_{01}\mathbf{E}_{02} - \eta\alpha_{mmm}^{(2)}:\mathbf{H}_{01}\mathbf{H}_{02} + \alpha_{mem}^{(2)}:\mathbf{E}_{01}\mathbf{H}_{02} - \alpha_{mme}^{(2)}:\mathbf{H}_{01}\mathbf{E}_{02}\right], \tag{7.20c}$$

$$\mathbf{m}_3^{-,-} = \frac{1}{\sqrt{I}}\left[\frac{1}{\eta}\alpha_{mee}^{(2)}:\mathbf{E}_{01}\mathbf{E}_{02} + \eta\alpha_{mmm}^{(2)}:\mathbf{H}_{01}\mathbf{H}_{02} - \alpha_{mem}^{(2)}:\mathbf{E}_{01}\mathbf{H}_{02} - \alpha_{mme}^{(2)}:\mathbf{H}_{01}\mathbf{E}_{02}\right]. \tag{7.20d}$$

Noting the fundamental fields are linearly polarized along x, the only nonzero vector components entering (7.19) and (7.20) are E_x and H_y, for both fundamental fields at frequencies ω_1 and ω_2. We additionally assume for now that the intrinsic electric nonlinear susceptibility of the nanoelement material has a dominant $\chi_{xxx}^{(2)}$ tensor component. The dominance of the $\chi_{iii}^{(2)}$ component is the case in many second-order nonlinear materials from the 3m group, such as $LiNbO_3$ or BBO. This condition will also be precisely the scenario with the geometry implemented later in Section 7.4.2, both due to the dominance of the intrinsic $\chi_{xxx}^{(2)}$ tensor component of the nonlinear material and the additional enhancement of that component by the resonant local field. With such assumptions, the tensor components of the effective nonlinear magneto-electric polarizabilities probed with the present approach are those along the vectors of the incident fundamental fields—e.g., $\alpha_{eee,xxx}$ and $\alpha_{emm,xyy}$ for the "eee" and "emm" terms, respectively, etc. We follow the standard notation for the Cartesian components of the effective nonlinear polarizabilities [69], with the first letter in the Cartesian subscript indicating the polarization of the field produced by that term at frequency ω_3, and the second (the third) denoting the polarization of the corresponding fundamental field vector at frequency $\omega_1(\omega_2)$ responsible for the excitation of that term. With that in mind, the effective hyper-polarizabilities can be then resolved from (7.19) and (7.20) in terms of the electric and magnetic dipolar amplitudes as following:

$$\alpha_{eee,xxx}^{(2)} = \frac{0.25}{\varepsilon_0\sqrt{I}}\left(p_{3x}^{+,+} + p_{3x}^{+,-} + p_{3x}^{-,+} + p_{3x}^{-,-}\right), \tag{7.21a}$$

$$\alpha_{emm,xyy}^{(2)} = \frac{0.25}{\varepsilon_0\sqrt{I}}\left(p_{3x}^{+,+} - p_{3x}^{+,-} - p_{3x}^{-,+} + p_{3x}^{-,-}\right), \tag{7.21b}$$

$$\alpha_{eem,xxy}^{(2)} = \frac{0.25}{\varepsilon_0\sqrt{I}}\left(p_{3x}^{+,+} - p_{3x}^{+,-} + p_{3x}^{-,+} - p_{3x}^{-,-}\right), \tag{7.21c}$$

$$\alpha^{(2)}_{eme,xyx} = \frac{0.25}{\varepsilon_0\sqrt{I}}\left(p_{3x}^{+,+} + p_{3x}^{+,-} - p_{3x}^{-,+} - p_{3x}^{-,-}\right), \tag{7.21d}$$

$$\alpha^{(2)}_{mee,yxx} = \frac{\eta}{4\sqrt{I}}\left(m_{3y}^{+,+} + m_{3y}^{+,-} + m_{3y}^{-,+} + m_{3y}^{-,-}\right), \tag{7.22a}$$

$$\alpha^{(2)}_{mmm,yyy} = \frac{\eta}{4\sqrt{I}}\left(m_{3y}^{+,+} - m_{3y}^{+,-} - m_{3y}^{-,+} + m_{3y}^{-,-}\right), \tag{7.22b}$$

$$\alpha^{(2)}_{mem,yxy} = \frac{\eta}{4\sqrt{I}}\left(m_{3y}^{+,+} - m_{3y}^{+,-} + m_{3y}^{-,+} - m_{3y}^{-,-}\right), \tag{7.22c}$$

$$\alpha^{(2)}_{mme,yyx} = \frac{\eta}{4\sqrt{I}}\left(m_{3y}^{+,+} + m_{3y}^{+,-} - m_{3y}^{-,+} - m_{3y}^{-,-}\right). \tag{7.22d}$$

The amplitude of each of the dipolar modes entering the right-hand sides of (7.21) and (7.22) is retrieved using (7.14) and (7.15), with the amplitude A entering these equations being, in the case, the nonlinearly produced field at frequency ω_3 and the E_0 factor being effectively taken as $\sqrt{I}$. Note that this factor cancels with the $\frac{1}{\sqrt{I}}$ factor in (7.21) and (7.22), so that there is *no* dependence on the intensity I in the retrieved hyperpolarizabilities—e.g., in terms of scattering amplitude the $\alpha_{eee,xxx}$ expression is as following: $\alpha^{(2)}_{eee,xxx} = \frac{\pi}{k^2}\left(A_3^{+,+} + A_3^{+,-} + A_3^{-,+} + A_3^{-,-}\right)\Big|_{\substack{\theta = 90^\circ,\\ \phi = 90^\circ}}$ with corresponding expressions holding for the rest of the terms. A complete set of expressions for the retrieved effective hyperpolarizabilities written in terms of the far-field amplitude A_3 can be found in the supplementary material to [38].

A similar set of expressions holds for quadrupolar nonlinear polarizabilities, with the $m^{+,+}$ and similar terms on the right-hand side of (7.22) being replaced with $\widetilde{Q}_{xz}^{+,+}$, and so on (the tilde sign here denotes the normalized units of the quadrupolar amplitude, similar to the linear case, defined in (7.11) and (7.12)). The terms on the left-hand side would transform to $\widetilde{\alpha}_{qee}$ and so on, with the first subscript "m" being replaced by "q" in each of (7.22a)–(7.22d). The resulting expressions look similar to those for the dipolar modes, and we omit these expressions here for compactness. In Section 7.4, we proceed with numerical demonstration of some of the phenomena predicted earlier in Section 7.2.3 using physical nanoelement geometry.

7.4 Single-element scattering response: implementation with physical geometry

To illustrate the feasibility of observing the phenomena described in Section 7.2.3 in physical nanoelements, we consider the geometry of plasmonic

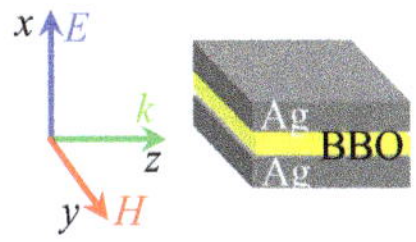

Figure 7.9 *Geometry of the dimer structure used in the analysis and orientation of the incident field. Dimer is made of two silver plates with the dimensions of 120 nm along y and z directions and 20 nm along x direction. The plates are separated by a 16 nm thick layer of BBO. The incident field is assumed to be plane-polarized along the x direction and propagating along the spacer layer, as shown.*

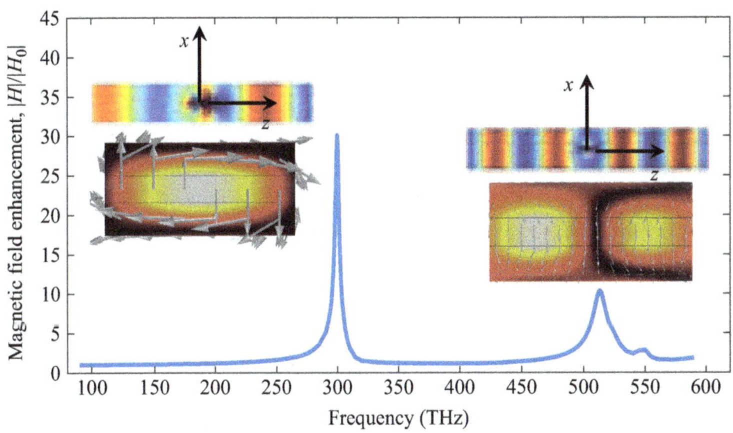

Figure 7.10 *Spectral dependence of the peak enhancement of the magnetic field within the dimer volume. Top color insets: xz plane cross section of the E_x component of the scattered field for the two resonances, showing odd (even) parity of the lower (higher) in frequency resonance, resembling that of the magnetic (electric) dipolar mode. The dimer is located at the origin, in the center of each color image. Lower color insets: xz cross section of the norm of the total magnetic field within the dimer volume at the two resonant spectral positions. Arrows indicate the direction of the total displacement field vector.*

dimers formed by two silver strips separated by a thin layer of BBO (Figure 7.9), mentioned earlier in Section 7.3.1.2. The spectral dependence of linear field enhancement within the dimer volume, under the excitation arrangement of Figure 7.9, is shown in Figure 7.10. As seen from the insets in the figure, the two resonances exhibited by the nanostructure have opposite parity of their scattered field, with the parity of the lower (higher) in frequency resonance resembling that of a magnetic (electric) dipolar mode. (As will be seen shortly in this section, the first resonance has, in fact, contributions from both magnetic dipolar and electric quadrupolar modes, which have the same parity and, in this case, the same resonant frequency.) As an aside, for the geometry and the excitation orientation employed, the parity of a particular resonance can intuitively be traced to the number of nodes of the magnetic field (lower insets in Figure 7.10), which define the direction of the x-component of the electric field at the opposite facets of the dimer—the major field component that couples to the far field [130].

Noting a comparable strength of the two resonant modes, one can expect a noticeable multipolar interference to occur both in the linear and in the nonlinear response of the nanoelement. For example, unidirectional linear scattering can be expected at some spectral position between the resonances where the odd and the even parity modes seen in Figure 7.10 match in strength. Indeed, strongly unidirectional linear scattering, achieving directionality over 20 dB (17 dB) in the forward (backward) directions, was demonstrated numerically with a similar geometry of just 69 nm in size [34]. Some analysis of dimer extinction and absorption cross sections and their dependence on the spacer thickness and material composition has also been performed in the above-mentioned work. In the present chapter, we focus on the nonlinear response of the dimer structure. In the remainder of this section, we retrieve linear and nonlinear polarizabilities of the dimer using the approach described in Section 7.3. We use the retrieved nonlinear parameters to estimate the optimal spectral positions of the fundamental beams where some of the phenomena predicted earlier in Section 7.2.3 are most likely to occur. We then demonstrate numerical examples of these phenomena using dimer geometry. In the course of the discussion, we also emphasize the existence of the analogy to Miller's rule [69] in the case of effective nonlinear magneto-electric polarizabilities, and the possibility of using permutation symmetries in magneto-electric terms to predict spectral positions where the effective nonlinear magneto-electric polarizabilities are expected to match in amplitude.

7.4.1 Linear multipolar polarizabilities of a dimer structure

Linear polarizabilities for the dimer structure from Figure 7.9, retrieved using (7.16), are shown in Figure 7.11. As seen from the figure, quadrupole response contributes about 15% of the total odd-parity resonant response near 300 THz. The comparison of the theoretical and numerical angular distribution of the scattered field amplitude for multipolar modes up to electric quadrupolar term, taken at 300 THz (the "magnetic" resonant spectral position), has been shown earlier in Figure 7.7, demonstrating an excellent agreement between the numerical and the analytical results, with less than 0.1% discrepancy between the two data. We observed a similar excellent agreement for all frequency range below and slightly above the magnetic resonant position. At frequencies above around 400 THz, multipoles higher than the electric quadrupolar mode gradually become more noticeable. The dipolar modes, however, still dominate in this range, with the total contribution of higher-order terms staying below 10% in the range up the electric dipolar resonance around 513 THz. Since all the examples presented here will involve spectral positions of the fundamental beams not higher than 450 THz, we

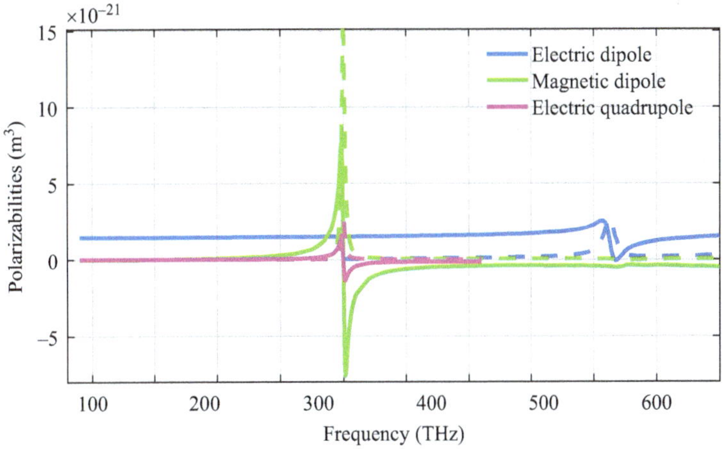

Figure 7.11 *Retrieved linear polarizabilities of the dimer structure depicted in Figure 7.9. The α_{xx}^{el} and α_{yy}^{m} tensor components of electric and magnetic dipolar polarizabilities, and the normalized xz tensor component of the quadrupolar polarizability are found following (7.16). Solid (dashed) lines represent real (imaginary) parts.*

do not aim at retrieving multipolar modes higher than the electric quadrupolar term. We next consider the process of DFG in dimer structure.

7.4.2 Nonlinear magnetoelectric polarizabilities of a dimer structure

We assume a simultaneous excitation of the dimer with two fundamental beams at frequencies ω_1 and ω_2, producing a DFG signal at frequency $\omega_3 = \omega_1 - \omega_2$. The excitation beams are linearly polarized along x direction, and the dimer orientation is the same as in Figure 7.9. The choice of the DFG process allows us to ensure that the nanoelement remains small compared to the nonlinearly generated wavelength, thus minimizing the manifestation of higher-order multipoles in the nonlinearly generated field.

Frequency domain analysis within the COMSOL Multiphysics software package [170] is used to simulate the nonlinear response of the dimer, using the procedure developed in [171]. We first perform a hydrodynamic model [146,147,172] analysis, extended to the DFG process, to verify the contribution of the metallic strips to the second-order nonlinear response of the dimer. The analysis has shown that the nonlinear response from metallic parts (predominantly surface-based) is about two orders of magnitude weaker than that from the BBO spacer, even at the resonant spectral position. The main contribution to the second-order nonlinear response in such a geometry thus results from the nonlinearity of the spacer material. This determination is in line with the similar conclusion regarding the third-order response in similar structures [173]. The nonlinear response due to metal elements has been neglected, therefore, in further analysis. The simulations have also shown that the major contribution to the nonlinear response comes from the $\chi^{(2)}_{xxx}$ component of the intrinsic second-order nonlinear susceptibility tensor; other tensor components produce a small impact, even though accounted for. This result is expected, due to the dominance (up to two orders of magnitude) of the x component of the local fundamental electric field within the spacer volume [34,130]. We thus chose the $\chi^{(2)}_{iii}$ tensor component of the BBO spacer to be along the x-axis in the simulations. The values for the nonzero $\chi^{(2)}$ tensor components follow those in [174], with $\chi^{(2)}_{xxx} = 2.3\,\text{pm}/V$.

We consider two sets of the fundamental excitation frequencies. In the first case, we fix the fundamental frequency f_2 at 100 THz and sweep f_1 over the "magnetic" resonance region, between 250 and 350 THz (here and further, $\omega_{1,2} \equiv 2\pi f_{1,2}$). In the second case, fundamental frequency f_1 is fixed at the "magnetic" resonance position at 300 THz, while f_2 is swept between 80 and 220 THz. We then employ the four arrangements of the respective directions of the fundamental beams, shown earlier in Figure 7.8, to retrieve the nonlinear effective polarizabilities. Note that, with the choice of the polarization of the excitation beams implemented here, and with the dominant $\chi^{(2)}_{xxx}$ tensor component of the intrinsic nonlinearity, the probed tensor components of the effective second-order polarizabilities are similar to those given in Section 7.3.2—e.g., the "emm" term is "emm,xyy," etc. Keeping this in mind, we omit the subscripts indicating Cartesian components further on.

The result of the retrieval for the first set of the excitation frequencies is shown in Figure 7.12. To illustrate the accuracy of the retrieval, we show in Figure 7.13 a comparison between the numerical and the analytical angular distributions for the nonlinearly generated dipolar and quadrupolar terms. An exact agreement is seen between the two datasets (the discrepancy is less than 0.1%). We observe similar excellent agreement for other combinations of the fundamental frequencies within the spectral range considered. As seen from Figure 7.12, all axial polarizabilities vanish identically, indicating the polar nature of the dimer geometry.

We note here a small modification to the notation implemented in Figure 7.12 and further on in the text: normally , the "m" letter in the second (third) position in the subscripts of the type "emm," "mme," etc., would indicate the *magnetic dipolar* contribution from the fundamental wave at ω_1 (ω_2) into the nonlinear response. Similar terms would then be required to describe the contribution of the fundamental quadrupolar terms to the nonlinear response—e.g., one should also have α_{eqq}, α_{meq}, etc., terms, where "q" stands for the contribution of the fundamental electric quadrupolar mode into the nonlinearly generated electric or magnetic dipolar modes. Noting that linear electric quadrupolar mode contributes about 15% of the total linear scattering response in the "magnetic" resonant region, such terms might not be negligible. As discussed earlier in Section 7.3.1, however, not only does the electric quadrupolar mode have the same parity as the magnetic dipolar one, but also it switches its phase together with the magnetic dipolar mode on beam reversal. As such, linear electric quadrupolar and linear magnetic dipolar modes are expected to always contribute cumulatively in terms of the phenomena

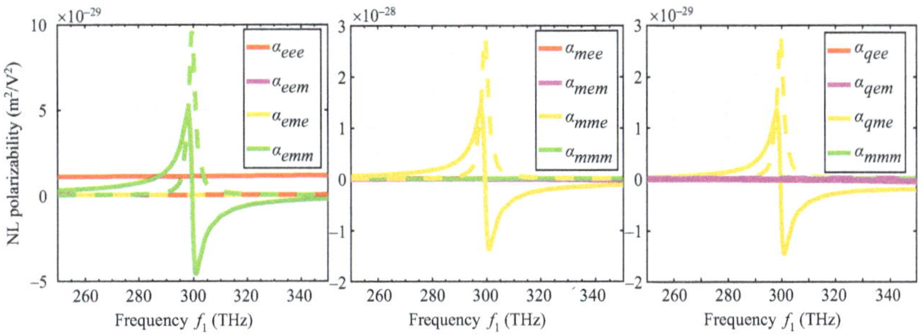

Figure 7.12 Retrieved effective nonlinear polarizabilities for the DFG process in the dimer when f_2 is fixed at 100 THz and f_1 is swept over the fundamental "magnetic" resonance region, from 250 to 350 THz, shown as a function of f_1. (a) Electric dipolar hyperpolarizabilities. (b) Magnetic dipolar hyperpolarizabilities. (c) Electric quadrupolar hyperpolarizabilities (normalized according to (7.12)). Solid (dashed) lines represent real (imaginary) parts. Axial hyperpolarizabilities vanish identically. Note that among polar hyperpolarizabilties, only the terms that exploit a "pathway" through the magnetic field of the wave at ω_1, which sweeps over the fundamental magnetic resonance range, are resonant. This behavior can be seen as an extension of Miller's rule to the effective nonlinear magneto-electric polarizabilities.

discussed in this chapter. In order to simplify the notation, we denote, therefore, a *combined* contribution from *all odd-parity modes* (both magnetic dipolar *and electric quadrupolar* ones) of the *fundamental* field at ω_1 (ω_2) by "m" at the second (third) position further on in the text. For example, the "mme" term is, actually, the "m(m+q)e" one, where "q" stands for the electric quadrupolar contribution of the fundamental field at ω_1 into the nonlinear magnetic dipolar response at ω_3. We continue to keep separate, however, the "m" and "q" terms for the *first* position in the subscript, which denotes the type of the response at the newly produced frequency. In doing so, we thus distinguish between the nonlinearly generated magnetic dipolar and electric quadrupolar modes. This choice simplifies the notation, while allows us to account for contributions from both magnetic dipolar and electric quadrupolar (linearly produced) modes into various effective polarizability terms.

One fact to emphasize in Figure 7.12 is that among the polar hyperpolarizabilities, only those involving the pathway through the *magnetic* dipolar mode at ω_1 (such as α_{emm} and α_{mme}, but not α_{eee}) exhibit a resonance. This fact can clearly be tracked to the presence of the resonant peak in the magnetic, but not the electric, *fundamental* dipolar mode within the range of frequencies spanned by ω_1 (Figure 7.11). We remind that the remaining two terms involving the magnetic path through ω_1, namely α_{mmm} and α_{eme}, are of the *axial* type and vanish identically due to the polar nature of the geometry, together with the α_{mee} and α_{eem} terms. Such a

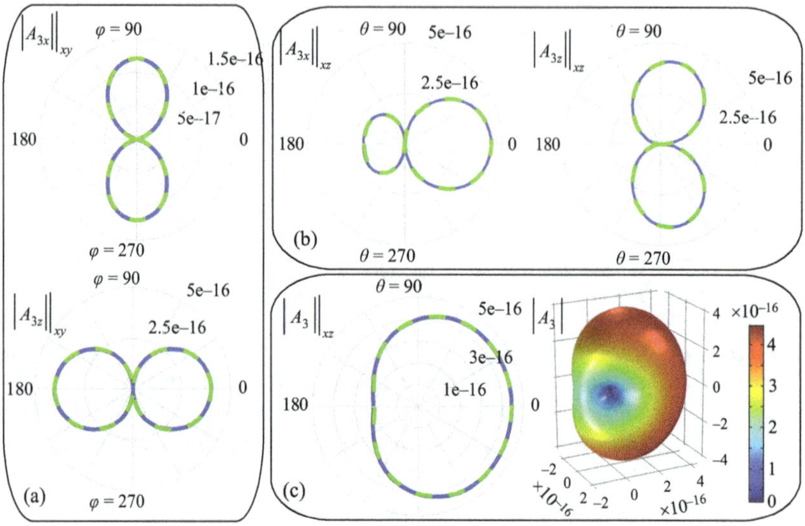

Figure 7.13 Comparison of analytical and numerical angular distributions of the nonlinearly generated far field, validating the accuracy of the nonlinear retrieval. The comparison is shown for the spectral position with $f_1 = 300$ THz *and* $f_2 = 100$ THz. *In all figures, solid green (dashed blue) lines indicate numerical (analytical) shapes.*

behavior can be seen as an extension of the Miller's rule [69] to the case of effective nonlinear magneto-electric polarizabilities. The latter rule states that intrinsic electric nonlinear susceptibilities are proportional to the product of the fundamental linear susceptibilities at the participating frequencies [69]. Such "multiplicative" character of the effective nonlinear polarizabilities comes in addition to the fact that each term in (7.5) and (7.6) involves a product of the participating incident fundamental fields. Analytical presentation of the effective hyperpolarizabilities in terms of products of the linear ones is, however, beyond the scope of the chapter.

7.4.2.1 Suppression of a nonlinearly produced dipolar mode

In Section 7.2.3.3, we discussed a situation where the nonlinear polarizability terms of the same (electric or magnetic) type within a given symmetry group are equal. We predicted that such situation would result in suppression of that type of the nonlinearly produced multipolar mode on reversal of one of the fundamental beams. As a reminder, this result is expected due to the fact that the two terms of a given type (electric or magnetic) within each symmetry group necessarily depend on a *different* type of the fundamental vector of a given frequency. For example, in the case of polar group symmetry, the two magnetic terms, "mem" and "mme," depend on the magnetic and the electric vectors, respectively, of the fundamental field at ω_2. Reversal of one of the excitation beams will thus unavoidably switch sign of one of the terms, but not the other one. Having the same parity, these two terms would interferometrically cancel when excited with the opposite phases—hence, the cancelation is expected on beam reversal. Furthermore, if both the two electric terms and the two magnetic terms within the same symmetry group have their amplitudes matched (e.g., if, simultaneously, $\alpha_{eee} = \alpha_{emm}$ and $\alpha_{mme} = \alpha_{mem}$ for a polar nanoelement), an overall suppression of the nonlinear process, for which such a relation holds, is expected when reversing one of the beams.

While we do not have the scenario where all hyperpolarizabilties would be equal in the present example (since a single, α_{mme} term is strongly dominant in the magnetic nonlinear response, with the other polar magnetic term, α_{mem}, being nearly zero), there is a spectral region, near $f_1 = 287$ THz, where the two electric nonlinear polarizability terms have similar amplitudes (Figure 7.12). One can expect, therefore, a suppression of the electric dipolar portion the DFG response when exciting the nanoelement from the opposite directions, owing to the destructive interference between the α_{eee} and α_{emm} terms. (Note that the suppression would be expected for *co-propagating* excitation if the two electric terms were matched with the *opposite* signs.) Additionally, at this spectral position, each of the electric terms is roughly half the sum of the magnetic dipolar and the electric quadrupolar modes, hence the relation $\alpha_{eee} + \alpha_{emm} \approx \alpha_{mme} + \alpha_{qme}$ roughly holds near $f_1 = 287$ THz (electric dipolar contribution is, strictly speaking, somewhat less than 50% here, as will be mentioned later). With such a balance of strengths, unidirectional generation is expected with co-propagating excitation along the z-axis: the DFG signal is expected to be suppressed in the negative z hemisphere and near-doubled in the positive z hemisphere. We can expect, therefore, the DFG far field angular distribution to switch from being unidirectional with co-propagating excitation to being symmetric and reminding that of a magnetic dipolar mode (with some contribution from the electric quadrupolar one) with contra-propagating excitation. Additionally, the DFG efficiency is expected to be reduced in the co-propagating case.

We observe precisely this response in Figure 7.14, where the DFG far field distribution is shown with co-and contra-propagating excitations for the spectral position at $f_1 = 287$ THz and $f_2 = 100$ THz. Note the toroidal shape of the DFG far field in Figure 7.14(b), characteristic of the magnetic dipolar-type response, slightly elongated due to the presence of the electric quadrupolar mode; recall that a similar shape is seen in Figure 7.7(c), which is constructed for the "magnetic" resonant position where magnetic dipolar and electric quadrupolar modes dominate. Also, integration of the DFG power flow showed about 30% reduction of the generated power with contra-propagating excitation in this example, versus the co-directional excitation case. In the present example, the total dipolar response contributes closer to 30%, rather than 50% of the total field produced by DFG; hence, the 30% of the DFG field is suppressed when cancelling the electric dipolar mode. Note also that we observe this effect even though the nanoelement is entirely symmetric with respect to the reversal along the z direction, and with the nanoelement size being less than 0.11 λ with respect to all participating wavelengths, so that there is no significant impact from phase matching.

The above example shows a possibility of controlling multipolar composition, directionality, and the overall efficiency of a certain nonlinear process by

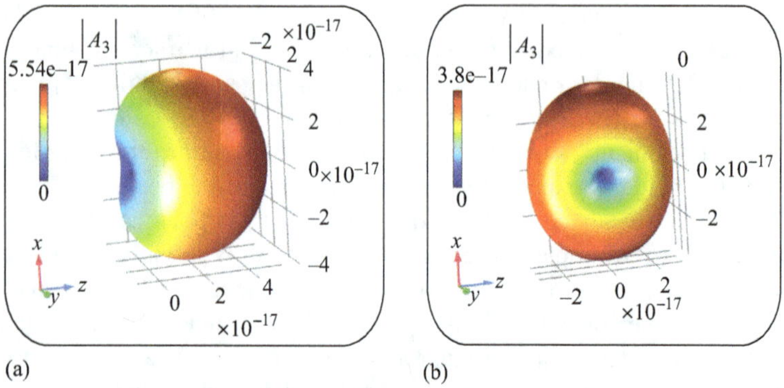

Figure 7.14 Suppression of the dipolar mode in the nonlinearly generated far field via interference of the two electric dipolar hyperpolarizability terms within the polar symmetry group, α_{eee} and α_{emm}. (a) Angular distribution of the norm of the far field generated with co-propagating excitation. Generation is seen to be directional, due to the comparable strength of nonlinearly produced even (electric dipolar) and odd (magnetic dipolar plus electric quadrupolar) multipolar modes. (b) Same as in (a), with contra-propagating excitation. Far field angular distribution is seen corresponding to that of a magnetic dipolar (plus some electric quadrupolar) mode due to suppression of the electric dipolar mode in this case.

engineering the relative strengths of various magneto-electric polarizability terms *within* nonlinearly produced multipolar modes. Note that unless a special care is taken to suppress all but two nonlinear polarizability terms, more than two magneto-electric polarizability types are usually expected to be nonzero within a given symmetry group, as soon as there is a nonzero magnetic dipolar mode in the linear response. As such, one can anticipate that, in most cases, both the efficiency and the angular distribution of nonlinear generation from a nanoelement will change when reversing the direction of *a subset* of fundamental beams, even in entirely symmetric nanoelements, and even when the nanoelement is small, such that phase matching does not make an impact. This fact makes both the efficiency and the directionality of nonlinear generation from nanoelements inherently non-reciprocal with respect to the directionality of the fundamental excitation.

In the next example, we engineer the DFG response where only two, out of the total eight, second-order effective polarizability terms are significant.

7.4.2.2 Example of nonreciprocal directionality of nonlinear generation

From Figure 7.12, one can notice that the "emm" and "mme" terms dominate when the spectral position of the fundamental beam at frequency f_1 is near "magnetic" resonance, at 300 THz. We therefore place f_1 at the magnetic resonant position,

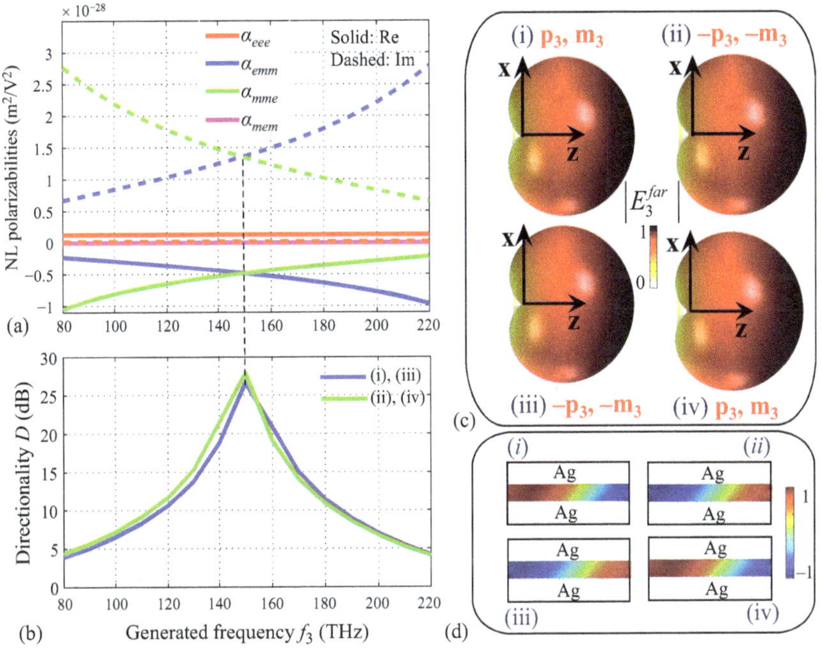

Figure 7.15 Illustration of nonreciprocal directionality of nonlinear generation achieved with dimer geometry. (a) Retrieved nonlinear hyperpolarizabilities when f_1 is fixed at 300 THz and f_2 is swept between 80 and 220 THz, as a function of f_3. (b) Directionality of nonlinear generation calculated when varying the direction of the beam at frequency ω_1 between the arrangements (i)–(iv) shown earlier in Figure 7.5, while preserving the direction of the beam at frequency ω_2. In all four arrangements, the nonlinear generation is seen suppressed (doubled) in the negative (positive) z direction, independently of the direction of the beam at frequency ω_1. Maximum directionality occurs when the two dominant hyperpolarizabilities match at $f_2 = 150$ THz. (c) Angular distribution of the norm of the generated far field at the directionality peak, for the same four orientations of the beam at frequency ω_1 as in (b). (d) x-component of the (real part of) the induced nonlinear polarization within the BBO layer, for the same four orientations of the ω_1 beam as in (b) and (c). The phase of the induced nonlinear polarization follows that of the magnetic field vector of the beam at ω_1, even though purely electric intrinsic second-order polarization is assumed.

while sweep f_2 between 80 and 220 THz. Polar magnetic and electric nonlinear polarizabilities, retrieved with such excitation, are shown in Figure 7.15(a). Axial polarizabilities again vanished identically and are not shown. when also do not show the quadrupolar terms in Figure 7.15(a): for most of the f_2 frequency range considered, nonlinearly generated quadrupolar mode contributed less than 5% to the total DFG response and, therefore, is not shown. As seen from the figure, only two polarizabilities, "emm" and "mme," are dominant among the polar ones. These polarizabilities match in strength near 150 THz ($\alpha_{emm} = \alpha_{mme}$). The latter fact is expected from to permutation symmetry [69] considerations. Indeed, due to permutation symmetry, the value of the nonlinear polarizability is preserved when interchanging the participating frequencies together with the corresponding Cartesian subscripts and, in the case case of magneto-electric nonlinear response, also with the corresponding magneto-electric subscripts [88]. Furthermore, since both the producing frequency ω_2 and the produced frequency ω_3 are far off-resonance, full permutation symmetry is in place, allowing their interchange. One can expect, therefore, the "emm" and "mme" hyperpolarizabilities to be symmetric around 150 THz where $\omega_2 = \omega_3$. This behavior is precisely observed in Figure 7.15(a). At 150 THz, the "emm" and "mme" terms cross, making $\alpha_{emm} = \alpha_{mme}$, as expected.

Besides being matched in amplitude at 150 THz, the "emm" and "mme" terms share the same "pathway" through the magnetic field at frequency ω_1. The situation where the two terms share a common "pathway" through the same type of the fundamental field, are comparable in amplitude, and where all other terms are negligible, fits precisely the scenario discussed earlier in Section 7.2.3.2, allowing a nonreciprocal directionality of nonlinear generation. Indeed, with α_{emm} and α_{mem} terms representing *the entire* electric and magnetic dipolar contributions to the DFG response, matching these hyperpolarizabilities ensures the DFG suppression in one hemisphere, due to the interference between electric and magnetic modes. Since the two polarizabilities match with the same sign, the DFG is expected to be suppressed in the negative and doubled in the positive z direction, according to (7.3). Furthermore, reversing the direction of the fundamental field at ω_1 would switch the phases of *both* α_{emm} and α_{mem} dipolar terms *simultaneously*, such that the two dipolar modes will again match with the same (now negative) sign. As a result, the direction of nonlinear generation is expected to be preserved: suppression (doubling) will again occur in the negative (positive) z hemisphere. (As discussed earlier, the situation is similar to the negative index media case [131,132], where the direction of energy propagation is preserved when phase velocities of both electric and magnetic partial waves are reversed.)

The directionality analysis of the nonlinearly generated far field, shown in Figure 7.15(b), confirms this prediction. The directionality ratio, defined as

$$D \equiv 20\log_{10}\left(\left|E_3^{far}\right|_{\theta=0^\circ} \Big/ \left|E_3^{far}\right|_{\theta=180^\circ}\right), \tag{7.23}$$

attains the 26 dB peak at 150 THz for all four combinations of the direction and phase of the beam at ω_1, shown earlier in Figure 7.4. The direction of the beam at

ω_2 is preserved along the positive z direction in all cases. A small difference between the forward ((i) and (iii)) and the reverse ((ii) and (iv)) excitations is due to the small nonzero $\alpha_{eee}^{(2)}$ term, seen in Figure 7.15(a).

Figure 7.15(c) shows the color plot of the angular distribution of the nonlinearly generated far field at the directionality peak at $f_2 = 150$ THz, for the same four orientations and phases of the fundamental field at ω_1 as in Figure 7.15(b) (the orientations shown earlier in Figure 7.4(a)). The direction of the fundamental beam at frequency ω_2 again stays fixed along positive z direction in all four cases. As seen, the generation is strongly directional, with the direction being preserved with respect to the fixed coordinate system in all four arrangements. Note that, as seen in Figure 7.15 (b), the frequency range where the DFG is directional and nonreciprocal is quite large: directionality stays above 20 dB from $f_2 \approx 140$ THz until $f_2 \approx 160$ THz. The color plot of Figure 7.15(c) looks similar within this range. Switching the direction of the beam at ω_2 would result in switching the directionality, with the DFG occurring, in that case, predominantly in the negative z direction—again independently of the orientation of the beam at ω_1 (this scenario is not presented in the figure).

Figure 7.15(d) shows the real part of the x component of the intrinsic nonlinear polarization within the spacer layer. The polarization is defined as $P_x(\omega_3) \equiv const \cdot \chi_{xij}^{(2)} : E_i(\omega_1)E_j(\omega_2)$ (with i, j standing for x, y, z) and is shown for the same spectral positions of fundamental frequencies and the same four arrangements as in Figure 7.15(c). While the analysis of the relation between the induced nonlinear polarization and the fundamental currents responsible for the excitation of certain linear multipolar modes is beyond the scope of this chapter, one can see that the phase of the induced intrinsic nonlinear polarization in sets (ii) and (iii) is opposite to that in sets (i) and (iv), following the *magnetic* vector of the excitation field at ω_1, even though purely electric intrinsic nonlinear susceptibility is assumed. In the present example, such behavior is intuitively clear: although the intrinsic nonlinear polarization responds to electric field, the majority of the local fundamental electric field at frequency ω_1 belongs to the magnetic dipolar mode, due to the resonant spectral position of that wave and, therefore, follows the phase of the magnetic excitation vector. Consequently, it is the magnetic fundamental dipolar mode that controls the phase of the induced nonlinear polarization through the above expression. The latter polarization, in turn, serves as a source for generating both electric and magnetic, nonlinearly produced, dipolar modes, which thus all happen to be controlled by the phase of the magnetic field at frequency ω_1. The resonant location may not be required if the shared "pathway" were of the electric type (e.g., if the "eee" and "mem" terms were dominant and matched in strength), due to the possibility of the nonresonant electric response. The absolute value of the induced nonlinear polarization (not shown) stays nearly identical in all four arrangement in Figure 7.15(d). While not presented here, a similar picture constructed for the combination of fundamental frequencies used in Figure 7.14, would, on the contrary, show a change in the absolute value of the nonlinear polarization between the co-propagating (set (i) or (iii)) and contra-propagating (set (ii) or (iv)) excitations, due to suppression of the portion of the nonlinear response corresponding to the electric dipolar mode.

7.5 Nonlinear scattering off a magnetoelectric metasurface

We consider now a metasurface formed by a periodic arrangement of dimer nanoelements within the *xy* plane (the coordinate notation follows that of Figure 7.9). A subwavelength period is assumed which, however, is large enough for coupling between the dimers to be negligible. We expect the nonlinear response of the metasurface to be similar, therefore, to the response of a single nanoelement, except for an interferometric cancellation of light produced in the directions other than those imposed by the momentum conservation requirements. In order to reduce the computation time and memory requirements, we consider 2D case, where each dimer is extended infinitely along the *y*-axis. The dimensions of each dimer cross section in the plane of simulations are the same as those in the *xz* plane in Figure 7.9. The metasurface is visualized later in Figure 7.19(a) in Section 7.5.2, where we consider imaging of a point source via DFG in such a metasurface. In the next section, we explain the physical mechanism allowing to produce an image of a source via DFG in a *subwavelength-thin* layer of a nonlinear material.

7.5.1 Nonlinear mirror via difference frequency generation

The imaging principle, depicted in Figure 7.16, is based on the generalized law of refraction [175], and is similar to the one reported for the process of four wave mixing (FWM) in [176]. For the DFG process at frequency $\omega_3 = \omega_1 - \omega_2$, the conservation of the tangential component of momentum at the surface of the thin nonlinear layer (depicted schematically as a yellow line in Figure 7.16) is given by $\omega_3 n_3 \sin\theta_3 = \omega_1 n_1 \sin\theta_1 - \omega_2 n_2 \sin\theta_2$. Here, n_i is the refractive index of the surrounding medium at frequency ω_i ($i = 1, 2,$ or 3) and θ_i is the angle of the beam at ω_i with respect to the surface normal (Figure 7.16), taken between 0 and $\pi/2$. For the normally incident beam at ω_1 ($\theta_1 = 0$), the above condition results in the expression resembling Snell's law, given as:

$$n_3 \sin\theta_3 = -\frac{\omega_2}{\omega_3} n_2 \sin\theta_2. \tag{7.24}$$

According to (7.24), a point source radiating at frequency ω_2 will have its image formed at frequency $\omega_3 = \omega_1 - \omega_2$ by the DFG process. Note that a subwavelength

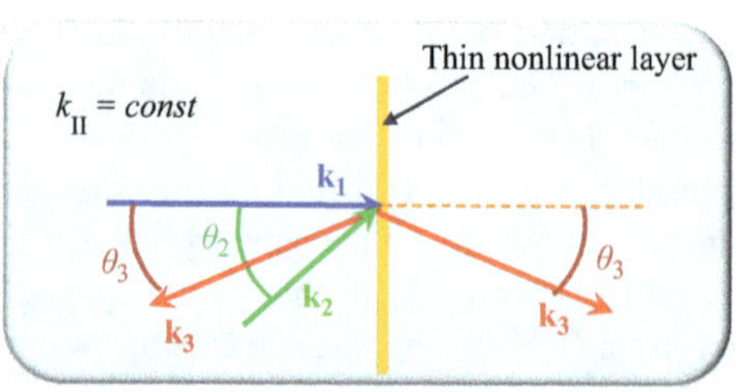

Figure 7.16 Illustration of the imaging principle based on momentum conservation in the DFG process.

thickness of the nonlinear medium is essential in producing a quality image, as it eliminates phase matching constraints, ensuring a relatively uniform DFG efficiency for all *k*-vectors emitted by the source (and captured by the nonlinear layer). In a thick crystal, momentum conservation will govern the direction of nonlinear generation at the surface [175]. Further propagation of the nonlinearly produced light inside the medium will, however, be subject to phase matching conditions. Efficient nonlinear generation will be limited, therefore, to a narrow subset of the *k*-vectors of the fundamental field satisfying these conditions. Employment of a deeply thin layer of a nonlinear medium eliminates the phase matching constraint, enabling a relatively angle-independent conversion efficiency [171].

Note additionally that, according to (7.24), some evanescent waves emitted by the source at ω_2 can be converted to propagating ones at ω_3 (as soon as the source is close enough to the nonlinear layer for the evanescent waves to be captured by the nonlinear medium), if the condition $\omega_2 < \omega_3$ is met. Although we do not explore this condition in the presented example, such evanescent to propagating waves conversion can, in general, improve the resolution of the nonlinearly produced image beyond the diffraction limit to some extent—just as in the case of up-conversion imaging with FWM [176].

Regularly, an image of a source, produced by such a deeply subwavelength layer of a nonlinear material, is expected to form on both sides of the surface with a nearly equal efficiency. The two-side generation is also a consequence of the lack of phase-matching constraints within the subwavelength-thin medium. An example of an image produced by a BBO layer of the same thickness as that of the dimer metasurface is shown in Figure 7.17. The layer is illuminated from the left by a plane wave at frequency $f_1 = 450$ THz. A point source at frequency $f_2 = 302$ THz is placed on the left side of the layer. One can indeed see an image produced at

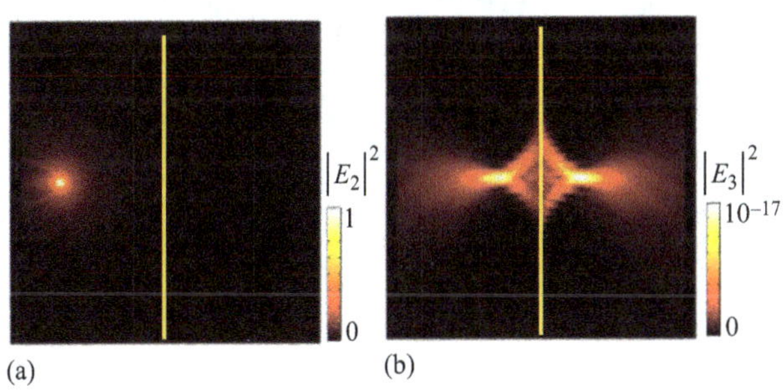

Figure 7.17 *Imaging of a point source via DFG process in a 120-nm-thin BBO layer. (a) Norm square of the electric field of the point source at* $f_2 = 302$ THz. *The yellow line indicates schematically the position of the BBO layer (not up to scale). The plane wave at frequency* $f_1 = 450$ THz *is incident from the left (not shown). (b) Norm square of the field generated at frequency* $f_3 = f_1 - f_2$. *The image of the point source at frequency* f_2 *is produced via DFG at frequency* $f_3 = 148$ *THz on both sides of the BBO layer.*

frequency $f_3 = 148$ THz on both sides with a near-identical efficiency. The choice of the excitation frequencies in this example follows the optimal spectral positions in the case of a metasurface, as discussed below.

7.5.2 One-way nonlinear mirror via multipolar interference in nonlinearly generated field

Now we replace BBO layer with the metasurface described in the beginning of the section. We aim at utilizing the nonreciprocal directionality of nonlinear generation from an individual nanoelement, described in Section 7.4.2.2, to develop a metasurface that would produce an image on *a single* side of that surface with, additionally, this side being *unchanged* when moving the source between the locations on the left and on the right of the metasurface. According to (7.24), however, if we are to produce an image of an ω_2 source on a given side independently of the source location, we need the DFG directionality to be nonreciprocal with respect to the direction of the fundamental beam at ω_2. The example presented in Section 7.4.2.2, where nonreciprocity was with respect to the ω_1 beam, would not work. For the directionality of DFG to be nonreciprocal with respect to the direction of the ω_2 beam, there should be a single dominant electric and a single dominant magnetic hyperpolarizability term in the DFG response, with these two terms sharing a "pathway" through the same type (electric or magnetic) of the ω_2 field.

We engineer such a pathway by choosing the f_2 spectral position near the magnetic resonance and sweeping the f_1 frequency over some range above that resonance. By having f_2 at the magnetic resonance position, we expect hyperpolarizability terms that share a pathway through the magnetic field of the wave at frequency f_2 (e.g., "emm", "mem", and "qem" in the case of polar symmetry group) to be dominant. We then aim at finding a spectral position for f_1 where the sum of even-parity terms sharing this pathway (e.g., "emm") would match that of the odd parity (e.g., the sum of "mem" and "qem"). Due to the 2D nature of the geometry, fundamental resonances are located at slightly higher frequencies than in Figure 7.10. We perform a retrieval similar to the one done in Section 7.3 to identify linear and nonlinear polarizabilities of a single 2D dimer structure. The retrieval procedure is adopted to the 2D case, allowing to find the electric dipolar mode by probing the scattering amplitude in the positive and negative z directions within the xz plane. The procedure is otherwise similar to the 3D case and we omit the details here. This 2D retrieval procedure has been preliminary tested by retrieving the polarizabilities of an infinite cylinder in agreement with [177]. The retrieved nonlinear polarizabilities are presented in Figure 7.18, for the case when frequency f_2 is placed at 302 THz (fundamental magnetic resonance position for the 2D geometry) and f_1 is swept between 380 and 510 THz, almost approaching fundamental electric dipolar resonance near 512 THz. As mentioned in Section 7.4.1, higher-order multipoles are more noticeable in the spectral range swept by frequency f_1 in this example, compared to the range used in Figure 7.15, especially for frequencies above 480 THz. Multipoles up to electric quadrupolar mode are, however, still dominant in the range below 450 THz which is used further in the section.

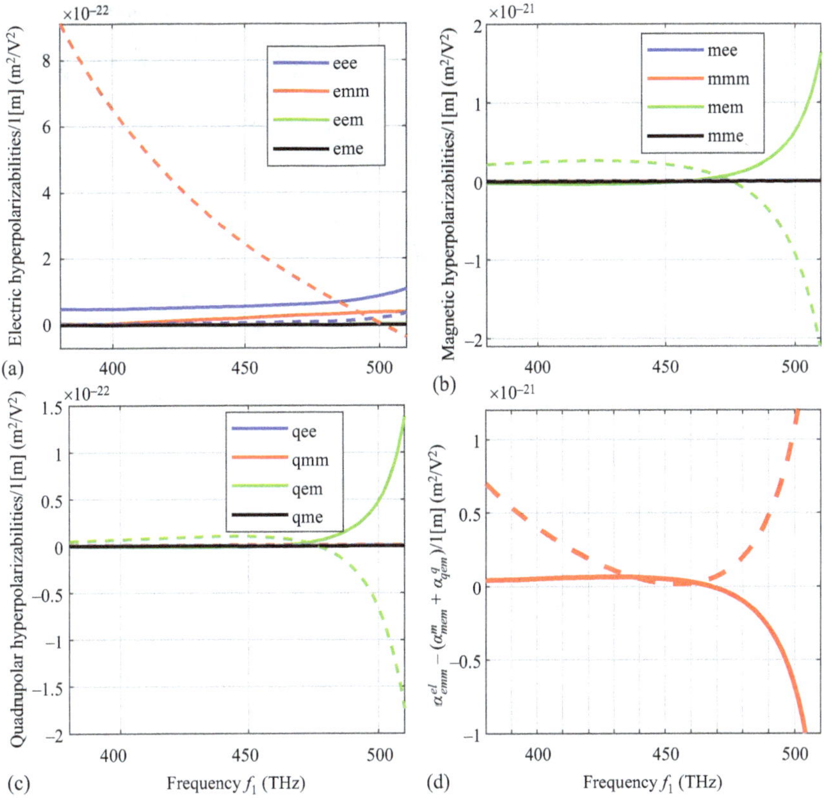

Figure 7.18 Effective nonlinear polarizabilities retrieved for the 2D dimer geometry with the xz cross section similar to that in Figure 7.9 and extended infinitely into the y direction. All graphs are shown as a function of the fundamental frequency f_1. Frequency f_2 is fixed at 302 THz. Solid (dashed) lines represent real (imaginary) parts. (a) Electric dipolar hyperpolarizabilities. (b) Magnetic dipolar hyperpolarizabilities. (c) Normalized electric quadrupolar hyperpolarizabilities. (d) The difference between the even (α_{emm}) and odd ($\alpha_{mem} + \alpha_{qem}$) parity terms sharing the same "pathway" through the magnetic field at frequency ω_2. The optimal spectral position for nonreciprocal directionality of the DFG with respect to the ω_2 beam is expected at $f_1 = 450$ THz where the difference is minimized.

As seen from the figure, the "emm," "mem," and "qem" terms dominate for most of the frequency range considered. These terms share a pathway through the magnetic field at frequency ω_2, satisfying the requirement necessary to produce a nonreciprocal image location. In order to find the spectral positions for the fundamental beams allowing to match the amplitudes of the dominant (nonlinearly produced) even-order and odd-order partial waves, we plot in Figure 7.18(d) the

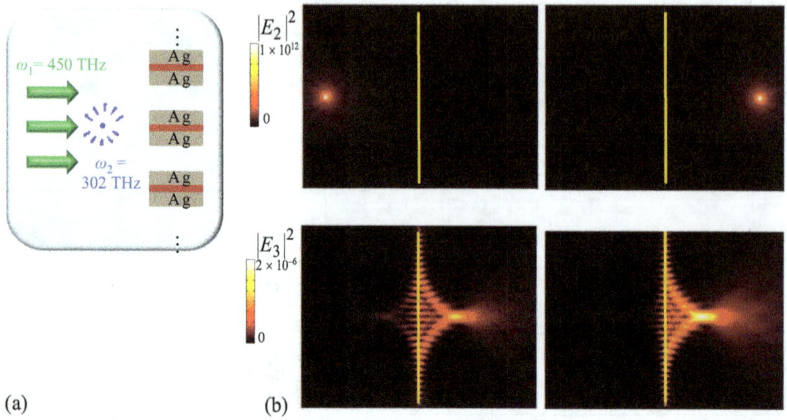

Figure 7.19 *Operation of one-way nonlinear mirror achieved via DFG in a metasurface. (a) Schematic illustration of the metasurface geometry and the illumination arrangement. The metasurface is illuminated by the plane wave at frequency* $f_1 = 450$ THz *and by a point source at frequency* $f_2 = 302$ THz. *(b) Top row: Point source at frequency* $f_2 = 302$ THz *is located on the left (left figure) and on the right (right figure) of the metasurface. Metasurface location is shown schematically as a yellow line (not up to scale). Norm square of the electric field is shown. Bottom row: Norm square of the electric field produced by the metasurface at frequency* $f_3 = 148$ THz *for the positions of the* f_2 *source on the left (left figure) and on the right (right figure). In both cases, the image is seen produced on a single (right) side which does not change when the source is moved from left to right.*

difference between the even-parity term (electric dipolar, "emm") and sum of the odd-parity terms (magnetic and electric quadrupolar, "mem"+"qem") which all share a pathway through magnetic field at frequency ω_2. As seen from the graph, the imaginary part of the difference has a clear minimum near 450 THz, while the real part stays small and relatively constant near this spectral position. We expect, therefore, the spectral position of $f_1 = 450$ THz to be optimal for achieving a nonreciprocal directionality of DFG with respect to the direction of the ω_2 beam. Note, however, that no spectral position is expected to provide a perfect nonreciprocal directionality, since there is a nonzero "eee" term, which has a larger relative contribution to the total nonlinear response than it had in Figure 7.15. This term will not follow the phase of the fundamental magnetic vector at frequency ω_2, in contrast with the rest of the dominant terms, and hence will switch from interfering constructively to interfering destructively with the "emm" term on the ω_2 beam reversal, reducing the directionality and modifying DFG efficiency.

From Figure 7.18, one can also once again notice a manifestation of Miller's rule in the effective magneto-electric terms. Only the "emm" term tends to grow toward lower frequencies in Figure 7.18(a), due to frequency f_1 moving, in this case, closer to the fundamental *magnetic* resonance near 302 THz. On the other hand, the "eee," "mem," and "qem" terms all increase toward higher frequencies in Figure 7.18(a)–(c), due to frequency f_1 now moving toward the *electric* dipolar resonance near 512 THz.

Numerical simulations confirm precisely the prediction of Figure 7.18(d). The results of the simulations, presenting an example of such one-way nonlinear imaging, are shown in Figure 7.19 for the spectral positions of the fundamental beams at $f_2 = 302$ THz and $f_1 = 450$ THz. As seen from the figure, the image of the ω_2 point source is produced via the DFG at frequency ω_3 predominantly on a single side of the metasurface, with that side being unchanged when moving the ω_2 source from left to right. We verified that the quality of the nonreciprocity (the similarity between the left and right bottom figures) was reduced if moving the f_1 spectral position away from 450 THz, predicted to be optimal in Figure 7.18(d).

7.6 Concluding remarks

The wide exploration of controlling linear directional scattering through multipolar interference has given rise to examinations of nonlinear multipolar scattering. As we have reviewed the work to date, it is clear that the flexibility offered by nonlinear excitation of multipolar components may allow for an even wider range of potential effects than in the linear case. In this chapter, we have explored the properties and engineering of multipolar nonlinear scattering from the individual particle to simple devices in an effort to show how these phenomena can be considered at each level. Key aspects of multipolar interference and multipolar component polarizability retrieval are demonstrated on simple but realistic geometries. A range of phenomena are shown including directional nonlinear generation, nonreciprocity, and effective enhancement or suppression of nonlinear processes through the engineering of multipolar interference. These exercises are meant to establish the flexibility such systems offer in controlling nonlinear pathways and processes. As was clearly shown, far subwavelength structured units can be effectively configured for a wide range of tuned nonlinear responses. The potential offered by the control of nonlinear interactions at the subwavelength scale and the extension of engineered subwavelength response to arrays and devices is a significant next step in engineered optical materials or in metasurface and metamaterials. The simple examples explored in this chapter are initial steps toward engineering large scale and integrated devices employing highly engineered nonlinear response such as compact isolators and nonlinear wavelength conversion devices with properties that advance current technical capabilities. In addition, while research in nonlinear metamaterials and metasurfaces has emphasized the enhancement of nonlinear properties, further exploring the control of nonlinear response offered through multipolar engineering will unlock more potential avenues at the fundamental level in the space of designed materials response. We look forward to further experimental work demonstrating the potential reviewed in this chapter as materials, fabrication, and characterization capabilities develop to realize the what this design approach offers.

References

[1] Landau LD, Lifshits EM, Pitaevskii LP. Electrodynamics of continuous media. 2nd ed. Teoreticheskaiya fizika. Pergamon; 1984. pp. 308–309.

[2] De Hoop AT. A reciprocity theorem for the electromagnetic field scattered by an obstacle. Appl Sci Res. 1960;8(1):135–140.

[3] Potton RJ. Reciprocity in optics. Rep Prog Phys. 2004;67(5):717–754.

[4] Carminati R, Saenz JJ, Greffet JJ, *et al.* Reciprocity, unitarity, and time-reversal symmetry of the S matrix of fields containing evanescent components. Phys Rev A. 2000;62(1):012712.

[5] Born M, Wolf E. Principles of optics: Electromagnetic theory of propagation, Interference and Diffraction of Light. 6th ed. Cambridge University Press; 1999.

[6] Mandel L, Wolf E. Optical coherence and quantum optics. Cambridge University Press; 1995. p. 897.

[7] Wax A, Backman V. Biomedical applications of light scattering. McGraw-Hill; 2010.

[8] Zhang L, Pleskow DK, Turzhitsky V, *et al.* Light scattering spectroscopy identifies the malignant potential of pancreatic cysts during endoscopy. Nat Biomed Eng. 2017;1(4).

[9] Bharadwaj P, Deutsch B, Novotny L. Optical antennas. Adv Opt Photon. 2009;1(3):438–483.

[10] Muhlschlegel P, Eisler H, Martin O, *et al.* Resonant optical antennas. Science. 2005;308(5728):1607–1609.

[11] Novotny L, van Hulst N. Antennas for light. Nat Photonics. 2011;5(2):83–90.

[12] Krasnok AE, Maksymov IS, Denisyuk AI, *et al.* Optical nanoantennas. Phys-Uspekhi. 2013;56(6):539.

[13] Li J, Salandrino A, Engheta N. Shaping light beams in the nanometer scale: A Yagi-Uda nanoantenna in the optical domain. Phys Rev B. 2007; 76(24):245403.

[14] Curto AG, Volpe G, Taminiau TH, *et al.* Unidirectional emission of a quantum dot coupled to a nanoantenna. Science. 2010;329(5994):930–933.

[15] Kosako T, Kadoya Y, Hofmann HF. Directional control of light by a nano-optical Yagi-Uda antenna. Nat Photonics. 2010;4(5):312–315.

[16] Shegai T, Chen S, Miljkovic VD, *et al.* A bimetallic nanoantenna for directional colour routing. Nat Commun. 2011;2:481.

[17] Pakizeh T, Kall M. Unidirectional ultracompact optical nanoantennas. Nano Lett. 2009;9(6):2343–2349.

[18] King NS, Li Y, Ayala-Orozco C, *et al.* Angle- and spectral-dependent light scattering from plasmonic nanocups. ACS Nano. 2011;5(9):7254–7262.

[19] Shegai T, Miljkovic VD, Bao K, *et al.* Unidirectional broadband light emission from supported plasmonic nanowires. Nano Lett. 2011;11(2):706–711.

[20] Bonod N, Devilez A, Rolly B, *et al.* Ultracompact and unidirectional metallic antennas. Phys Rev B. 2010;82(11):115429.

[21] Pakizeh T. Unidirectional radiation of a magnetic dipole coupled to an ultracompact nanoantenna at visible wavelengths. J Opt Soc Am B. 2012;29(9):2446–2452.
[22] Lavasani SHA, Pakizeh T. Color-switched directional ultracompact optical nanoantennas. J Opt Soc Am B. 2012;29(6):1361–1366.
[23] Papas CH. Theory of electromagnetic wave propagation. Mineola (NY): Dover Publications; 1988.
[24] Bohren CF, Huffman DR. Absorption and scattering of light by small particles. New York (NY): Wiley; 1998.
[25] Jackson JD. Classical electrodynamics. Wiley; 1999. Available from: http://www.wiley.com/WileyCDA/WileyTitle/productCd-047130932X.html.
[26] Kerker M, Wang DS, Giles CL. Electromagnetic scattering by magnetic spheres. J Opt Soc Am B. 1983;73(6):765–767.
[27] Alu A, Engheta N. How does zero forward-scattering in magnetodielectric nanoparticles comply with the optical theorem? J Nanophotonics. 2010;4:041590.
[28] Nieto-Vesperinas M, Saenz JJ, Gomez-Medina R, *et al.* Optical forces on small magnetodielectric particles. Opt Express. 2010;18(11):11428–11443.
[29] Garcia-Camara B, Alcaraz de la Osa R, Saiz JM, *et al.* Directionality in scattering by nanoparticles: Kerker's null-scattering conditions revisited. Opt Lett. 2011;36(5):728–730.
[30] Geffrin JM, García-Cámara B, Gómez-Medina R, *et al.* Magnetic and electric coherence in forward- and back-scattered electromagnetic waves by a single dielectric subwavelength sphere. Nat Commun. 2012;3:1171.
[31] Fu YH, Kuznetsov AI, Miroshnichenko AE, *et al.* Directional visible light scattering by silicon nanoparticles. Nat Commun. 2013;4:1527.
[32] Person S, Jain M, Lapin Z, *et al.* Demonstration of zero optical backscattering from single nanoparticles. Nano Lett. 2013;13(4):1806–1809.
[33] Hancu IM, Curto AG, Castro-Lopez M, *et al.* Multipolar interference for directed light emission. Nano Lett. 2014;14(1):166–171.
[34] Poutrina E, Urbas AM. Multipole analysis of unidirectional light scattering from plasmonic dimers. J Opt. 2014;16(11):114005.
[35] Krasnok AE, Simovski CR, Belov PA, *et al.* Superdirective dielectric nanoantennas. Nanoscale. 2014;6:7354–7361.
[36] Krasnok AE, Filonov DS, Simovski CR, *et al.* Experimental demonstration of superdirective dielectric antenna. Appl Phys Lett. 2014;104:133502.
[37] Pors A, Andersen SKH, Bozhevolnyi SI. Unidirectional scattering by nanoparticles near substrates: Generalized Kerker conditions. Opt Express. 2015;23(22):28808–28828.
[38] Poutrina E, Urbas A. Multipolar interference for non-reciprocal nonlinear generation. Sci Rep. 2016;6:25113.
[39] Lewin L. The electrical constants of a material loaded with spherical particles. J Inst Electr Eng 3. 1947;94(27):65–68.

[40] Holloway CL, Kuester EF, Gordon JA, *et al.* An overview of the theory and applications of metasurfaces: The two-dimensional equivalents of metamaterials. IEEE Antennas Propag Mag. 2012;54(2):10–35.
[41] Albooyeh M, Tretyakov S, Simovski C. Electromagnetic characterization of bianisotropic metasurfaces on refractive substrates: General theoretical framework. Ann Phys (Berl). 2016;528(9–10):721–737.
[42] Poutrina E, Rose A, Brown D, *et al.* Forward and backward unidirectional scattering from plasmonic coupled wires. Opt Express. 2013;21(25):31138–31154.
[43] Demetriadou A, Pendry JB. Taming spatial dispersion in wire metamaterial. J Phys: Condens Matter. 2008;20(29):295222.
[44] Estakhri NM, Alù A. Recent progress in gradient metasurfaces (and references therein). J Opt Soc Am B. 2016;33(2):A21–A30.
[45] Asadchy VS, Albooyeh M, Tcvetkova SN, *et al.* Perfect control of reflection and refraction using spatially dispersive metasurfaces. Phys Rev B. 2016;94: 075142.
[46] Yu N, Capasso F. Flat optics with designer metasurfaces. Nat Mater. 2014;13:139.
[47] Kildishev AV, Boltasseva A, Shalaev VM. Planar photonics with metasurfaces. Science. 2013;339(6125):1232009.
[48] Lin D, Fan P, Hasman E, *et al.* Dielectric gradient metasurface optical elements. Science. 2014;345(6194):298–302.
[49] Monticone F, Estakhri NM, Alu A. Full control of nanoscale optical transmission with a composite metascreen. Phys Rev Lett. 2013;110(20): 203903.
[50] Yang Y, Wang W, Moitra P, *et al.* Dielectric meta-reflectarray for broadband linear polarization conversion and optical vortex generation. Nano Lett. 2014;14(3):1394–1399.
[51] Pors A, Albrektsen O, Radko IP, *et al.* Gap plasmon-based metasurfaces for total control of reflected light. Sci Rep. 2013;3:2155.
[52] Hu J, Ren X, Reed AN, *et al.* Evolutionary design and prototyping of single crystalline titanium nitride lattice optics. ACS Photonics. 2017; 4(3):606–612.
[53] Urbas AM, Jacob Z, Negro LD, *et al.* Roadmap on optical metamaterials. J Opt. 2016;18(9):093005.
[54] Alu A, Engheta N. How does zero forward-scattering in magnetodielectric nanoparticles comply with the optical theorem? J Nanophotonics. 2010;4: 041590.
[55] Smirnova D, Smirnov AI, Kivshar YS. Multipolar second-harmonic generation by Mie-resonant dielectric nanoparticles. Phys Rev A. 2018;97(1):013807.
[56] Poutrina E, Huang D, Smith DR. Analysis of nonlinear electromagnetic metamaterials. New J Phys. 2010;12:093010.
[57] Huang D, Poutrina E, Smith DR. Analysis of the power dependent tuning of a varactor-loaded metamaterial at microwave frequencies. Appl Phys Lett. 2010;96(10):104104.

[58] Denk W, Strickler JH, Webb WW. Two-photon laser scanning fluorescence microscopy. Science. 1990;248:73–76.
[59] Strachan CJ, Windbergs M, Offerhaus HL. Pharmaceutical applications of non-linear imaging. Int J Pharm. 2011;417(1):163–172.
[60] Schmitt PD. Recent advances in nonlinear optical analyses of pharmaceutical materials in the solid state. Mol Pharm. 2017;14(3):555–565.
[61] Ray PC. Size and shape dependent second order nonlinear optical properties of nanomaterials and their application in biological and chemical sensing. Chem Rev. 2010;110(9):5332–5365.
[62] Agrawal GP. Applications of nonlinear fiber optics. Academic Press; 2008.
[63] Mouras R, Bagnaninchi P, Downes A, *et al.* Multimodal, label-free nonlinear optical imaging for applications in biology and biomedical science. J Raman Spectrosc. 2013;44(10, SI):1373–1378.
[64] Faccio D, Boyd RW, Bachor HA. Special issue on enhanced quantum imaging. J Opt. 2019;21(1):010201.
[65] Shih Y. Quantum imaging. IEEE J Sel Topics Quantum Electron. 2007; 13(4):1016–1030.
[66] Zeilinger A. Light for the quantum. Entangled photons and their applications: A very personal perspective. Phys Scr. 2017;92(7):072501.
[67] Saleh BEA, Teich MC. Fundamentals of photonics. New York: Wiley; 1991.
[68] Robert Alfano on the supercontinuum: History and future applications [SPIE Newsroom]. SPIE; 2014. Available from: http://spie.org/newsroom/alfano-video?ArticleID=x108037&SSO=1.
[69] Boyd RW. Nonlinear optics. 3rd ed. Academic Press; 2008.
[70] Mujumdar S, Ramachandran H. Use of a graded gain random amplifier as an optical diode. Opt Lett. 2001;26(12):929–931.
[71] Gallo K, Assanto G, Parameswaran KR, *et al.* All-optical diode in a periodically poled lithium niobate waveguide. Appl Phys Lett. 2001;79(3):314–316.
[72] Miroshnichenko AE, Brasselet E, Kivshar YS. Reversible optical nonreciprocity in periodic structures with liquid crystals. Appl Phys Lett. 2010;96:063302.
[73] Lepri S, Casati G. Asymmetric wave propagation in nonlinear systems. Phys Rev Lett. 2011;106(16):164101.
[74] Fan Y, Han J, Wei Z, *et al.* Subwavelength electromagnetic diode: One-way response of cascading nonlinear meta-atoms. Appl Phys Lett. 2011; 98(15):151903.
[75] Mahmoud AM, Davoyan AR, Engheta N. All-passive nonreciprocal metastructure. Nat Commun. 2015;6:8359.
[76] Trevino Palacios CG, Stegeman GI, Baldi P. Spatial nonreciprocity in waveguide second-order processes. Opt Lett. 1996;21(18):1442–1444.
[77] Lin XS, Yan JH, Wu LJ, *et al.* High transmission contrast for single resonator based all-optical diodes with pump-assisting. Opt Express. 2008; 16(25):20949–20954.
[78] Fan L, Wang J, Varghese LT, *et al.* An all-silicon passive optical diode. Science. 2012;335(6067):447–450.

[79] Sahoo PK, Joseph J. Optical diode using nonlinear polystyrene ring resonators in two-dimensional photonic crystal structure. Appl Opt. 2013; 52(34):8252–8257.

[80] Li N, Ren J. Non-reciprocal geometric wave diode by engineering asymmetric shapes of nonlinear naterials. Sci Rep. 2014;4:6228.

[81] Zhang Y, Li D, Zeng C, *et al.* Silicon optical diode based on cascaded photonic crystal cavities. Opt Lett. 2014;39(6):1370–1373.

[82] Shadrivov IV, Fedotov VA, Powell DA, *et al.* Electromagnetic wave analogue of an electronic diode. New J Phys. 2011;13:033025.

[83] Valev VK, Baumberg JJ, De Clercq B, *et al.* Nonlinear superchiral metasurfaces: Tuning chirality and disentangling non-reciprocity at the nanoscale. Adv Mat. 2014;26(24):4074–4081.

[84] Fan L, Wang J, Varghese LT, *et al.* An all-silicon passive optical diode. Science. 2012;335(6067):447–450.

[85] Fan S, Baets R, Petrov A, *et al.* Comment on "nonreciprocal light propagation in a silicon photonic circuit". Science. 2012;335:38.

[86] Wang J, Fan L, Varghese LT, *et al.* A theoretical model for an optical diode built with nonlinear silicon microrings. J Lightwave Technol. 2013;31(2):313–321.

[87] Sounas DL, Alu A. Time-reversal symmetry bounds on the electromagnetic response of asymmetric structures. Phys Rev Lett. 2017;118(15):154302.

[88] Rose A, Larouche S, Poutrina E, *et al.* Nonlinear magnetoelectric metamaterials: Analysis and homogenization via a microscopic coupled-mode theory. Phys Rev A. 2012;86(3):033816.

[89] Kauranen M. Freeing nonlinear optics from phase matching. Science. 2013;342:1182–1183.

[90] Kujala S, Canfield BK, Kauranen M, *et al.* Multipole interference in the second-harmonic optical radiation from gold nanoparticles. Phys Rev Lett. 2007;98(16):167403.

[91] Kujala S, Canfield BK, Kauranen M, *et al.* Multipolar analysis of second-harmonic radiation from gold nanoparticles. Opt Express. 2008; 16(22):17196–17208.

[92] Zdanowicz M, Kujala S, Husu H, *et al.* Effective medium multipolar tensor analysis of second-harmonic generation from metal nanoparticles. New J Phys. 2011;13:023025.

[93] Rose A, Huang D, Smith DR. Nonlinear interference and unidirectional wave mixing in metamaterials. Phys Rev Lett. 2013;110:063901.

[94] Canfield BK, Kujala S, Jefimovs K, *et al.* Amacroscopic formalism to describe the second-order nonlinear optical response of nanostructures. J Opt A: Pure Appl Opt. 2006;8(4, SI):S278–S284.

[95] Smirnova D, Kivshar YS. Multipolar nonlinear nanophotonics. Optica. 2016;3(11):1241–1255.

[96] Decker M, Staude I. Resonant dielectric nanostructures: A low-loss platform for functional nanophotonics. J Opt. 2016;18(10).

[97] Krasnok A, Tymchenko M, Alu A. Nonlinear metasurfaces: A paradigm shift in nonlinear optics. Mater Today. 2018;21(1):8–21.

[98] Canfield BK, Kujala S, Husu H, *et al.* Local-field and multipolar effects in the second-harmonic response of arrays of metal nanoparticles. J Nonlinear Opt Phys Mater. 2007;16(3):317–328.
[99] Czaplicki R, Zdanowicz M, Koskinen K, *et al.* Dipole limit in second-harmonic generation from arrays of gold nanoparticles. Opt Express. 2011;19(27):26866–26871.
[100] Klein MW, Enkrich C, Wegener M, *et al.* Second-harmonic generation from magnetic metamaterials. Science. 2006;313(5786):502–504.
[101] Zdanowicz M, Kujala S, Husu H, *et al.* Effective medium multipolar tensor analysis of second-harmonic generation from metal nanoparticles. New J Phys. 2011;13:023025.
[102] Makarov SV, Petrov MI, Zywietz U, *et al.* Efficient second-harmonic generation in nanocrystalline silicon nanoparticles. Nano Lett. 2017;17(5):3047–3053.
[103] Kruk SS, Camacho-Morales R, Xu L, *et al.* Nonlinear optical magnetism revealed by second-harmonic generation in nanoantennas. Nano Lett. 2017;17(6):3914–3918.
[104] Shcherbakov MR, Neshev DN, Hopkins B, *et al.* Enhanced third-harmonic generation in silicon nanoparticles driven by magnetic response. Nano Lett. 2014;14(11):6488–6492.
[105] Shcherbakov MR, Shorokhov AS, Neshev DN, *et al.* Nonlinear interference and tailorable third-harmonic generation from dielectric oligomers. ACS Photonics. 2015;2(5):578–582.
[106] Shorokhov AS, Melik-Gaykazyan EV, Smirnova DA, *et al.* Multifold enhancement of third-harmonic generation in dielectric nanoparticles driven by magnetic fano resonances. Nano Lett. 2016;16(8):4857–4861.
[107] Smirnova DA, Khanikaev AB, Smirnov LA, *et al.* Multipolar third-harmonic generation driven by optically induced magnetic resonances. ACS Photonics. 2016;3(8):1468–1476.
[108] Wang L, Shorokhov AS, Melentiev PN, *et al.* Multipolar third-harmonic generation in fishnet metamaterials. ACS Photonics. 2016;3(8):1494–1499.
[109] Rose A, Powell DA, Shadrivov IV, *et al.* Circular dichroism of four-wave mixing in nonlinear metamaterials. Phys Rev B. 2013;88(19).
[110] Dmitriev PA, Baranov DG, Milichko VA, *et al.* Resonant Raman scattering from silicon nanoparticles enhanced by magnetic response. Nanoscale. 2016;8(18):9721–9726.
[111] Shcherbakov MR, Vabishchevich PP, Shorokhov AS, *et al.* Ultrafast all-optical switching with magnetic resonances in nonlinear dielectric nanostructures. Nano Lett. 2015;15(10):6985–6990.
[112] Rahmani M, Shorokhov AS, Hopkins B, *et al.* Nonlinear symmetry breaking in symmetric oligomers. ACS Photonics. 2017;4(3):454–461.
[113] Camacho-Morales R, Rahmani M, Kruk S, *et al.* Nonlinear generation of vector beams from AlGaAs nanoantennas. Nano Lett. 2016;16(11):7191–7197.

[114] Carletti L, Locatelli A, Neshev D, *et al.* Shaping the radiation pattern of second-harmonic generation from AlGaAs dielectric nanoantennas. ACS Photonics. 2016;3(8):1500–1507.

[115] Wang L, Kruk S, Xu L, *et al.* Shaping the third-harmonic radiation from silicon nanodimers. Nanoscale. 2017;9(6):2201–2206.

[116] Wang L, Kruk S, Koshelev K, *et al.* Nonlinear wavefront control with all-dielectric metasurfaces. Nano Lett. 2018;18(6):3978–3984.

[117] Kong JA. Theory of electromagnetic waves. Wiley; 1975.

[118] Serdyukov A, Semchenko I, Tretyakov S, *et al.* Electromagnetics of bi-anisotropic materials: Theory and applications. Gordon and Breach Science Publishers; 2001.

[119] Sihvola AH, Viitanen AJ, Lindell IV, *et al.* Electromagnetic waves in chiral and bi-isotropic media. Antennas and propagation library. Artech House; 1994.

[120] Albooyeh M, Asadchy VS, Alaee R, *et al.* Purely bianisotropic scatterers. Phys Rev B. 2016;94:245428.

[121] Zhao Y, Belkin MA, Alu A. Twisted optical metamaterials for planarized ultrathin broadband circular polarizers. Nat Commun. 2012;3:870.

[122] Li J, Shen C, Diaz-Rubio A, *et al.* Systematic design and experimental demonstration of bianisotropic metasurfaces for scattering-free manipulation of acoustic wavefronts. Nat Commun. 2018;9:1342.

[123] Mirmoosa MS, Ra'di Y, Asadchy VS, *et al.* Polarizabilities of non-reciprocal bianisotropic particles. Phys Rev Appl. 2014;1(3):034005.

[124] Fan Z, Shcherbakov MR, Allen M, *et al.* Perfect diffraction with multi-resonant bianisotropic metagratings. ACS Photonics. 2018;5(11):4303–4311.

[125] Niu C, Zhao J, Du L, *et al.* Spatially dispersive dichroism in bianisotropic metamirrors. Appl Phys Lett. 2018;113(26):261102.

[126] Fiebig M, Frohlich D, Krichevtsov BB, *et al.* 2nd-harmonic generation and magnetic-dipole-electric-dipole interference in antiferromagnetic Cr_2O_3. Phys Rev Lett. 1994;73(15):2127–2130.

[127] Tanabe Y, Fiebig M, Hanamura E. Nonlinear optics of antiferromagnetic compounds. In: Sugano S, Kojima N, editors. Magneto-Optics. Springer Series in Solid-State Sciences. vol. 128. Berlin: Springer; 2000. pp. 107–136.

[128] Czaplicki R, Zdanowicz M, Koskinen K, *et al.* Dipole limit in second-harmonic generation from arrays of gold nanoparticles. Opt Express. 2011;19(27):26866–26871.

[129] Rose A, Huang D, Smith DR. Demonstration of nonlinear magnetoelectric coupling in metamaterials. Appl Phys Lett. 2012;101(5):051103.

[130] Poutrina E, Rose A, Brown D, *et al.* Forward and backward unidirectional scattering from plasmonic coupled wires. Opt Express. 2013; 21(25):31138–31154.

[131] Veselago VG. The electrodynamics of substances with simultaneously negative values of e and mu. Phys-Uspekhi. 1968;10(4):509–514.

[132] Smith DR, Padilla WJ, Vier DC, *et al.* Composite medium with simultaneously negative permeability and permittivity. Phys Rev Lett. 2000; 84(18):4184–4187.

[133] Pershan P. Nonlinear optical property of solids – energy considerations. Phys Rev. 1963;130(3):919.
[134] Jonsson F. The nonlinear optics of magneto-optic media [dissertation]. The Royal Institute of Technology, Department of Physics. Stockholm; 2000.
[135] Pan RP, Wei HD, Shen YR. Optical second-harmonic generation from magnetized surfaces. Phys Rev B. 1989;39:1229–1234.
[136] Budker D, Yashchuk V, Zolotorev R. Nonlinear magneto-optic effects with ultranarrow widths. Phys Rev Lett. 1998;81(26):5788–5791.
[137] Budker D, Kimball D, Rochester S, *et al.* Nonlinear magneto-optics and reduced group velocity of light in atomic vapor with slow ground state relaxation. Phys Rev Lett. 1999;83(9):1767–1770.
[138] Pruneri V, Longhi S. Optical parametric processes in isotropic third-order nonlinear media with magneto-optic properties. Opt Lett. 2000; 25(23):1720–1722.
[139] Hu B, Wang W. Magneto-optic effects in a nonlinear photonic crystal. Phys Rev B. 2002;65(16):165207.
[140] Boardman A, Budker D, Pisarev R. Nonlinear and integrated magneto-optics - Introduction. J Opt Soc Am B. 2005;22(1):2.
[141] Wang G, Yan X. Magneto-optic effects in subwavelength nonlinear plasmonic waveguides. Plasmonics. 2017;12(4):1131–1135.
[142] Yu W, Ma P, Sun H, *et al.* Optical tristability and ultrafast Fano switching in nonlinear magnetoplasmonic nanoparticles. Phys Rev B. 2018; 97(7):075436.
[143] Cheng JL, Guo C. Nonlinear magneto-optic effects in doped graphene and in gapped graphene: A perturbative treatment. Phys Rev B. 2018; 97(12):125417.
[144] Sipe JE, Stegeman GI. Chapter 15 - Nonlinear optical response of metal surfaces. In: Agranovich VM, Mills DL, editors. Surface Polaritons. vol. 1 of Modern Problems in Condensed Matter Sciences. Elsevier; 1982. pp. 661–701.
[145] Dadap JI, Shan J, Eisenthal KB, *et al.* Second-harmonic rayleigh scattering from a sphere of centrosymmetric material. Phys Rev Lett. 1999;83:4045–4048.
[146] Ciraci C, Poutrina E, Scalora M, *et al.* Second-harmonic generation in metallic nanoparticles: Clarification of the role of the surface. Phys Rev B. 2012;86(11):115451.
[147] Ciraci C, Poutrina E, Scalora M, *et al.* Origin of second-harmonic generation enhancement in optical split-ring resonators. Phys Rev B. 2012; 85(20):201403.
[148] de Beer AGF, Roke S, Dadap JI. Theory of optical second-harmonic and sum-frequency scattering from arbitrarily shaped particles. J Opt Soc Am B. 2011;28(6):1374–1384.
[149] Roke S, Bonn M, Petukhov A. Nonlinear optical scattering: The concept of effective susceptibility. Phys Rev B. 2004;70(11):115106.
[150] Birss RR. Symmetry and magnetism. Selected topics in solid state physics. North-Holland Pub. Co.; 1964.

[151] Kruk SS, Camacho-Morales R, Xu L, *et al.* Nonlinear optical magnetism revealed by second-harmonic generation in nanoantennas. Nano Lett. 2017;17(6):3914–3918.
[152] Poutrina E, Huang D, Urzhumov Y, *et al.* Nonlinear oscillator metamaterial model: Numerical and experimental verification. Opt Express. 2011; 19(9):8312–8319.
[153] Larouche S, Smith DR. A retrieval method for nonlinear metamaterials. Opt Commun. 2010;283:1621–1627.
[154] Rose A, Larouche S, Huang D, *et al.* Nonlinear parameter retrieval from three- and four-wave mixing in metamaterials. Phys Rev E. 2010;82:036608.
[155] Liu X, Rose A, Poutrina E, *et al.* Surfaces, films, and multilayers for compact nonlinear plasmonics. J Opt Soc Am B. 2013;30(11): 2999–3010.
[156] Kogelnik H. Coupled wave theory for thick hologram gratings. Bell Syst Tech J. 1969;48(9):2909–2947.
[157] Huang D, Poutrina E, Smith DR. Analysis of the power dependent tuning of a varactor-loaded metamaterial at microwave frequencies. Appl Phys Lett. 2010;96(10):104104.
[158] Larouche S, Rose A, Poutrina E, *et al.* Experimental determination of the quadratic nonlinear magnetic susceptibility of a varactor-loaded split ring resonator metamaterial. Appl Phys Lett. 2010;97(1):011109.
[159] Huang D, Rose A, Poutrina E, *et al.* Wave mixing in nonlinear magnetic metacrystal. Appl Phys Lett. 2011;98(20):204102.
[160] Huang D, Poutrina E, Zheng H, *et al.* Design and experimental characterization of nonlinear metamaterials. J Opt Soc Am B. 2011;28(12): 2925–2930.
[161] Liu X, Larouche S, Smith DR. Homogenized description and retrieval method of nonlinear metasurfaces. Opt Commun. 2018;410:53–69.
[162] Larouche S, Radisic V. Retrieval of all effective susceptibilities in nonlinear metamaterials. Phys Rev A. 2018;97(4):043863.
[163] Mühlig S, Menzel C, Rockstuhl C, *et al.* Multipole analysis of meta-atoms. Metamaterials. 2011;5(2):64–73.
[164] Grahn P, Shevchenko A, Kaivola M. Electromagnetic multipole theory for optical nanomaterials. New J Phys. 2012;14:093033.
[165] Arango FB, Koenderink AF. Polarizability tensor retrieval for magnetic and plasmonic antenna design. New J Phys. 2013;15:073023.
[166] Stratton JA. Electromagnetic theory. 1st ed. New York: McGraw-Hill; 1941.
[167] Mohlenkamp MJ. A fast transform for spherical harmonics. J Fourier Anal Appl. 1999;5(2):159–184.
[168] Yaghjian AD, Silveirinha M, Askarpour A, *et al.* Electric quadrupolarizability of a source-driven dielectric sphere. Prog Electromagn Res B. 2015;63:95–106.
[169] Alu A, Engheta N. Guided propagation along quadrupolar chains of plasmonic nanoparticles. Phys Rev B. 2009;79(23):235412.

[170] COMSOL Modeling Software [homepage on the Internet]. Burlington (MA): COMSOL, Inc.; c1986 -. Available from: https://www.comsol.com/.

[171] Poutrina E, Ciraci C, Gauthier DJ, *et al.* Enhancing four-wave-mixing processes by nanowire arrays coupled to a gold film. Opt Express. 2012; 20(10):11005–11013.

[172] Scalora M, Vincenti MA, de Ceglia D, *et al.* Second- and third-harmonic generation in metal-based structures. Phys Rev A. 2010;82(4):043828.

[173] Ciraci C, Scalora M, Smith DR. Third-harmonic generation in the presence of classical nonlocal effects in gap-plasmon nanostructures. Phys Rev B. 2015;91(20):205403.

[174] Dmitriev VG, Gurzadyan GG, Nikogosyan DN. Handbook of nonlinear optical crystals. Berlin: Springer; 1991.

[175] Bloembergen N, Pershan PS. Light waves at the boundary of nonlinear media. Phys Rev. 1962;128(2):606–622.

[176] Palomba S, Zhang S, Park Y, *et al.* Optical negative refraction by four-wave mixing in thin metallic nanostructures. Nat Mater. 2012;11(1):34–38.

[177] Strickland D, Ayon A, Alu A. Dynamic polarizability tensor for circular cylinders. Phys Rev B. 2015;91(8):085104.

Chapter 8
Plasmonic metasurfaces for controlling harmonic generations

Shumei Chen[1], *Guixin Li*[2], *Thomas Zentgraf*[3] *and Shuang Zhang*[4]

8.1 Introduction

In optics, harmonic generation refers to the nonlinear optical process in which two or more photons combine to form a photon of higher frequency. Harmonic generations of light play a very important role in modern optics. In the past, harmonic generations have been mainly investigated in crystals, where light can travel along a path many times of the wavelength, with accumulation of the generated nonlinear signals along the path given the phase-matching condition is satisfied. In recent years, harmonic generations in plasmonic nanoparticles have attracted growing attention, partly because of the strong field enhancement arising from the plasmonic resonances. Due to the ultrathin thickness of the plasmonic structure, the phase-matching condition along the perpendicular direction of the surface does not set a constraint. Besides, the correlation between the structural symmetry of plasmonic particles and the intensity of harmonic generation has attracted particular interest, due to the fact that even orders of harmonic generation are only allowed in systems without centrosymmetry. Plasmonic structures are interesting for the study of harmonic generations also because of the interference between contributions from different multipoles hosted by the nanoparticles. Many interesting studies have been carried along this direction, with one famous example being the role that the magnetic dipole plays in a split ring resonator. Very recently, there has been surging interest in controlling the phase, amplitude, or polarization states of the harmonic signals in an ultrathin surface consisting of an array of carefully tailored plasmonic structures. Various approaches have been proposed, such as varying the shape to control the resonance, or simply varying the orientation of identical

[1]School of Science, Harbin Institute of Technology, Shenzhen, China
[2]Department of Materials Science and Engineering, Shenzhen Institute for Quantum Science and Engineering, Southern University of Science and Technology, Shenzhen, China
[3]Department of Physics, Paderborn University, Paderborn, Germany
[4]School of Physics & Astronomy, University of Birmingham, Birmingham, UK

plasmonic structures, taking advantage of the extension of the concept of Berry phase into the nonlinear optical regime. The achievement of continuous control over the nonlinearity phase has stimulated the design of various nonlinear optical devices ranging from nonlinear lenses to nonlinear holograms. Besides plasmonic metasurfaces for harmonic generation, considerable efforts have been placed on dielectric metasurfaces that contain nanostructures of high dielectric constants. Dielectric elements can also be designed to support electric and magnetic dipole moments. The interplay between the two has led to interesting nonlinear optical effects. In this chapter, we limit our discussion to nonlinear plasmonic metasurfaces and focus on recent progress in nonlinear plasmonic metasurfaces for manipulating second harmonic generation signals. We start with the discussion of selection rules of nonlinear harmonic generation for a circularly polarized incident fundamental beam. We then proceed with the different approaches for achieving phase control in a nonlinear metasurface and the various devices based on this effect. We further discuss the nonlinear chiral effect in a metasurface with planar chirality. Finally, we conclude and give an outlook on future directions.

8.2 Selection rule in harmonic generations for circular polarizations

Symmetry plays an important role in harmonic generations. The nonlinear susceptibility tensors can be significantly simplified when considering the symmetry of the molecular structures and atomic crystal configurations [1,2]. The symmetry of the atomic structure of the nonlinear medium naturally leads to selection rules, which state that certain harmonic processes are forbidden under illumination of light of certain direction and polarization states. The investigation of the effect of symmetry on harmonic generations has been extended to structures at more macroscopic levels, such as plasmonic nanostructures, photonic crystals, etc., and certain selection rules have been routinely investigated, such as the mirror symmetry for second harmonic generation (SHG). Plasmonic structures have been one of the most widely studied systems in nonlinear optics, because of their strong light enhancement that leads to dramatically enhanced nonlinear responses. More importantly, the localized plasmonic modes hosted by the metallic structures can be very sensitive to shape, size, and surrounding medium [3]. In the past, plasmonic structures have been utilized to enhance various nonlinear processes such as SHG [4–14], third harmonic generation (THG) [15–18], and four-wave mixing [19–22]. In terms of the investigation of the effects of symmetry, many groups have looked into SHG in plasmonic structures with broken centrosymmetry, such as SRRs [4], L- and V-shaped nanorods [6,7], and G-type chiral metamolecules [8]. All these studies have focused on fundamental waves of linear polarizations.

It has been known that harmonic generations with circularly polarized fundamental waves obey certain selection rules that are determined by the rotational symmetry of the molecules [23–25]. While the selection rules for harmonic generations have been theoretically investigated based on the group theory [26], they can also be understood through formulations based on simple coordinate transformations. Here, we consider a 2D array of plasmonic structures possessing certain rotational symmetries. Due to the subwavelength size, each of the constituent plasmonic structure can be considered as a nonlinear dipole moment. For a fundamental wave of circular polarization σ along the $+z$ direction at the fundamental frequency, the electric field is expressed as, $\vec{E}^{\sigma} = \widetilde{E}_0\widehat{e}_\sigma = \widetilde{E}_0(\widehat{e}_x + i\sigma\widehat{e}_y)/\sqrt{2}$, where $\sigma = \pm 1$ represents a left- or right-handed circular polarization state. Upon excitation of the fundamental wave, a nonlinear dipole moment is induced. Here, we consider a general mth harmonic polarization process, with the nonlinear dipole moment given as,

$$\vec{p}^{\,m\omega} \equiv \vec{p}^{\,m\omega,\sigma} + \vec{p}^{\,m\omega,-\sigma} = \overset{\leftrightarrow}{\alpha}_m \left(\vec{E}^{\sigma}\right)^m \tag{8.1}$$

where $\overset{\leftrightarrow}{\alpha}_m$ is the mth harmonic nonlinear polarizability tensor of the nonlinear antenna. Since the nanostructures are very thin (usually, in the scale of a few tens of nanometers), one can assume that there is only the in-plane component in the nonlinear dipole moment. Hence, the nonlinear dipole moment can be decomposed into two in-plane rotating dipoles $\vec{p}^{\,m\omega,\sigma}$ and $\vec{p}^{\,m\omega,-\sigma}$, as shown by (8.1). Under a coordinate rotation of $2\pi/n$, where n is the order of rotational symmetry of the structure, both the incident light and the nonlinear dipole moments acquire a spin-dependent geometric phase due to the rotation spin coupling,

$$\vec{E}_T^{\sigma} = \vec{E}^{\sigma} e^{i2\pi\sigma/n}, \quad \vec{p}_T^{\,m\omega,\sigma} = \vec{p}^{\,m\omega,\sigma} e^{i2\pi\sigma/n}, \quad \vec{p}_T^{\,m\omega,-\sigma} = \vec{p}^{\,m\omega,-\sigma} e^{-i2\pi\sigma/n} \tag{8.2}$$

where $\vec{E}_T^{\sigma}$, $\vec{p}_T^{\,m\omega,\sigma}$, and $\vec{p}_T^{\,m\omega,-\sigma}$ are the electric field and the two nonlinear circular dipole moments after the coordinate transformation, respectively. After the coordinate rotation, the nonlinear polarizability α_m remains the same due to the n-fold rotational symmetry of the structure. Therefore, the nonlinear dipole moments are given by

$$\vec{p}_T^{\,m\omega,\sigma} + \vec{p}_T^{\,m\omega,-\sigma} = \overset{\leftrightarrow}{\alpha}_m \left(\vec{E}_T^{\sigma}\right)^m \tag{8.3}$$

which, in combination with (8.2), leads to

$$\vec{p}^{\,m\omega,\sigma} e^{i2\pi\sigma/n} + \vec{p}^{\,m\omega,-\sigma} e^{-i2\pi\sigma/n} = \overset{\leftrightarrow}{\alpha}_m \left(\vec{E}^{\sigma}\right)^m e^{i2m\pi\sigma/n} \tag{8.4}$$

The selection rule for the nonlinear polarization can be obtained by comparing (8.4) with (8.1),

$$\begin{aligned} \vec{p}^{\,m\omega,\sigma} &= 0 \quad \text{if } (m-1)/n \text{ is non-integer} \\ \vec{p}^{\,m\omega,-\sigma} &= 0 \quad \text{if } (m+1)/n \text{ is non-integer} \end{aligned} \tag{8.5}$$

Equation (8.5) is equivalent to the selection rule in bulk crystals, i.e., the allowed orders of nonlinear harmonic generation are given by $m = (nl + 1)$ for the same circular polarization as the incident one, and $m = (nl - 1)$ for the opposite circular polarization, where l is an arbitrary integer number so long as m is positive. Taking the THG as an example, a circularly polarized fundamental wave only produces the THG with opposite handedness in the forward direction when it propagates along the axis of a nonlinear medium with four-fold rotational symmetry [26,27]. This rule holds not only for the microscopic molecular symmetry but also at the more macroscopic symmetry level of artificial nanostructure, as verified by two independent works in 2014 [28,29].

In the work by Chen *et al.* [28], the THG in three different plasmonic metasurfaces consisting of nanoantennas with C2, C4, and C6 rotational symmetries was studied [Figure 8.1(a)]. The scanning electron micrographs of the metasurface with the three different symmetries are shown in Figure 8.1(b). In all the metasurfaces, the symmetries of the structures and the lattice were consistent with each other. For a fair comparison of the nonlinear signals, the geometries of all the samples were designed to show similar plasmonic resonance wavelengths in the near-infrared, as verified by the transmission spectra presented in Figure 8.1(c).

In order to further enhance the nonlinear conversion efficiency to facilitate the study of the selection rule, a poly(9,9-dioctylfluorence) (PFO) thin film was coated onto the metasurfaces. The THG measurements were performed at a wavelength of $\lambda = 1.25$ μm, which is close to the localized surface plasmon-polariton (LSPP) wavelengths of all three plasmonic metacrystals. The charge-coupled device (CCD) images of the THG signal at a wavelength of 417 nm from the C2, C3, and C4 symmetry samples are shown in Figure 8.2. For the bare PFO film, there is no detectable THG signal, agreeing with the selection rules for the THG for an isotropic medium [22–26]. However, once the plasmonic nanorods are embedded into the PFO film, the symmetry of the system is determined by the that of the nanorods. This is dictated by the selection rule $m = 3l \pm 1$ for C3 symmetry, according to which the allowed orders of harmonic generation with the same circular polarization as the fundamental wave are $m = [1, 4, 7, \ldots]$ and that of the opposite circular polarization are $m = [2, 5, 8, \ldots]$, with $m = 3$ not present in either circular polarization state.

More intriguing results come from the comparison of the THG signals between the C2 and C4 samples. Strong optical spin-dependent THG signals are observed when the circularly polarized fundamental wave is incident onto the C2 and C4 symmetry samples. For the C2 symmetry sample, right circularly polarized light illumination generates THG signals with both right circular polarization (RCP) and left circular polarization (LCP), as both are not forbidden by the selection rules $m = 2l \pm 1$, which allows any odd orders of harmonic generations for both circular polarizations. However, the RCP component of the THG signal is much stronger than the LCP component, with a polarization ratio between the LCP and RCP intensities of $\eta_{C2}(\mathrm{THG_{LCP}/THG_{RCP}}) = 0.121$. This ratio is not governed by the selection rule, but rather is due to the specific design of the antenna geometry. Since both LCP and RCP are allowed by the selection rule for the C2 symmetry, this ratio is in general not zero or infinity. However, their respective intensities can be controlled by the specific geometry of the antenna, such as the length and width of the gold rod, which determine the plasmonic resonances in both directions and the field distribution at both the fundamental and THG wavelengths. Therefore, it is

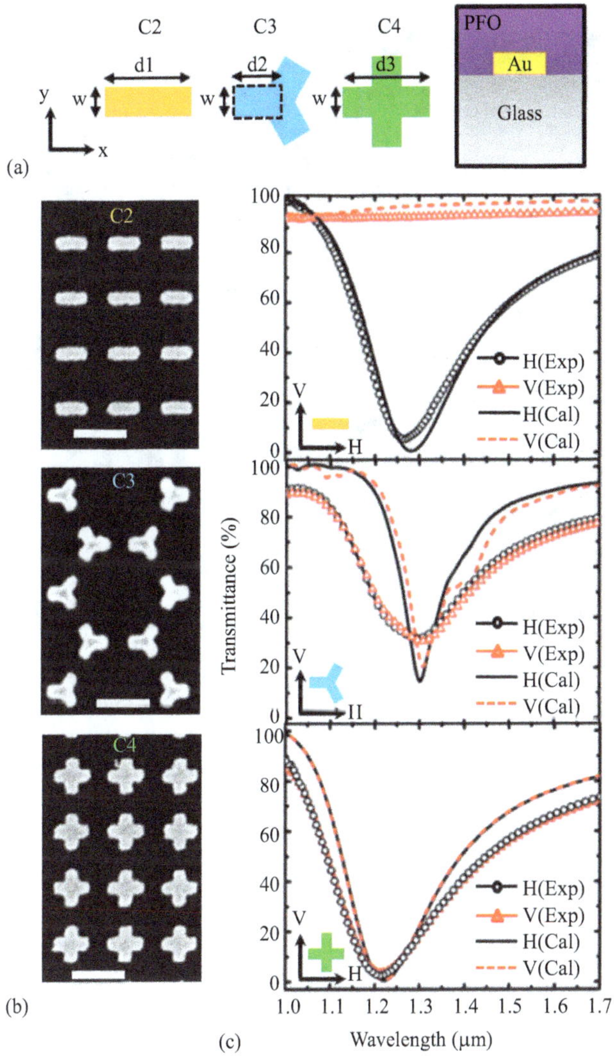

Figure 8.1 Schematic of THG with nanostructures of various rotation symmetries, adapted from [29]. (a) Geometry of plasmonic crystals with C2, C3, and C4 rotational symmetries. The unit cell consists of nanorods embedded in PFO film. The width of the nanorods is w = 50 nm, and the lengths of nanorods are d_1 = 230 nm, d_2 = 110 nm, and d_3 = 210 nm, respectively. (b) Scanning electron microscope images of the C2, C3, and C4 metasurfaces, with 400 nm scale bar. (c) Measured and simulated linear transmission spectra for the C2, C3, and C4 metasurfaces. H: horizontal polarized incident light; V: vertical polarized incident light.

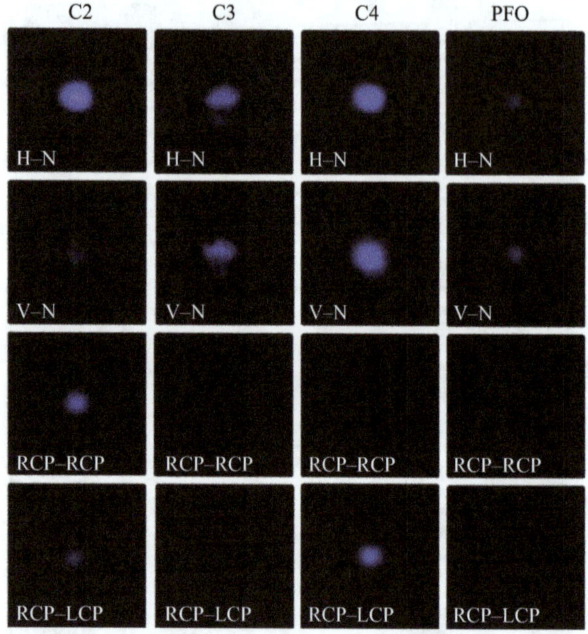

Figure 8.2 CCD images of the THG from C2, C3, C4, and bared PFO thin film under different polarization states of pumping laser, adapted from [29]. The first two rows are the results for pumping laser with linear polarization: horizontal (H) and vertical (V), respectively. The THG signals the third and fourth rows were recorded using a right-hand circularly polarized (RCP) pumping laser. The third row corresponds to the THG with RCP component, and the fourth row shows the LCP component of THG.

expected that by adjusting the geometry of the antenna with C2 symmetry, the ratio between the THG signals of the two circular polarizations can be finely tailored.

Interestingly, for the sample with C4 rotational symmetry, the LCP component dominates in the RCP-induced THG signals, with an extremely large polarization ratio η_{C4} (THG_{LCP}/THG_{RCP}) of ~29. According to the selection rule for C4 rotational symmetry $m = 4l \pm 1$, under illumination of RCP fundamental wave, the allowed harmonic generations are [1, 5, 9, ...] for RCP and [3, 7, 11, ...] for LCP. Thus, the THG signal should be solely in the LCP state and THG of the right circularly polarized light should be forbidden; however, surface roughness and the deviation from the ideal symmetry shape of the nanostructures arising from the imperfectness of nanofabrication slightly break the rotational symmetry and lead therefore to a small THG signal in the RCP state. Hence, all the experimental observations are consistent with the selection rules of THG with circularly polarized light excitation in crystal optics when generalized to 2D metacrystals [26].

8.3 Binary phase nonlinear metasurfaces

Controlling the local nonlinear polarizability is very important in nonlinear optics since the local nonlinear coefficient determines the efficiency, propagation, and pattern of the generated nonlinear light in the material. The most widely employed technique is poling, which introduces binary signs of the nonlinearity phase for SHGs. Poling technique allows quasi-phase matching scheme [30–36], which can dramatically enhance the conversion efficiency in SHGs. In poling techniques, a strong electric field is applied locally to a nonlinear crystal in the vertical direction through applied electrodes, leading to reversing of the crystal orientations. The nonlinear crystal therefore shows a spatially varying nonlinear polarization. As the symmetry breaking is along the vertical direction, the poled nonlinear crystals work most efficiently for fundamental waves propagating in the in-plane directions. There have been a few exceptions where the fundamental wave was illuminated in the vertical direction to generate diffracted nonlinear signal patterns in the transmission direction. For example, a poled crystal with a fork grating pattern has been utilized to generate vortex beams of various orders. Recently, the concept of poling was extended to plasmonic metasurfaces for controlling the nonlinear optical generation in ultrathin metasurfaces. In contrast to the poled nonlinear crystals where the nonlinear polarization is along the vertical direction, metasurfaces can be designed to have in-plane nonlinear polarizations, working perfectly for light incident at normal incidence.

A pioneering work along this direction was demonstrated by Ellenbogen *et al.* in 2015 [37]. As shown in Figure 8.3(a), the basic building block of the metasurface was a U-shaped plasmonic antenna—a so-called split ring resonator (SRR). Each SRR functions as a nonlinear meta-atom that can generate SHG which is polarized along the direction of symmetry breaking (y direction in Figure 8.3(a)) under illumination of x-polarized fundamental wave. The effective nonlinear susceptibility switches signs when the orientation of the U-shaped antennas are flipped. Figure 8.3(b) presents a nonlinear grating made from the SRRs, with each unit cell consisting of six SRRs with three facing up and three facing down. Effectively, this can be considered as a nonlinear grating with alternating signs of nonlinear susceptibilities. Experimentally, when a normally incident fundamental wave was illuminated onto the nonlinear grating, in the transmission direction, two diffraction orders were observed. The zeroth order disappears due to the destructive interference between the two subunits of each supercell [Figure 8.3(c)]. The same principle of binary nonlinearity was further applied to construct a nonlinear zone plate. As shown in Figure 8.3(d), the metasurface was divided into a series of concentric rings, with each ring consisting of many SRRs of the same orientation, while the orientations of the SRRs between neighboring rings are opposite. Under the direct SHG imaging of the surface of the zone plate, there appeared a number of dark rings at the interface between the up and down facing SRR zones [Figure 8.3 (e)]. This is due to destructive interference. At 1 mm away from the zone plate surface, a clear focusing spot was observed at the SHG frequency. This shows that the nonlinear metasurface can work effectively as a lens for the second harmonic signal.

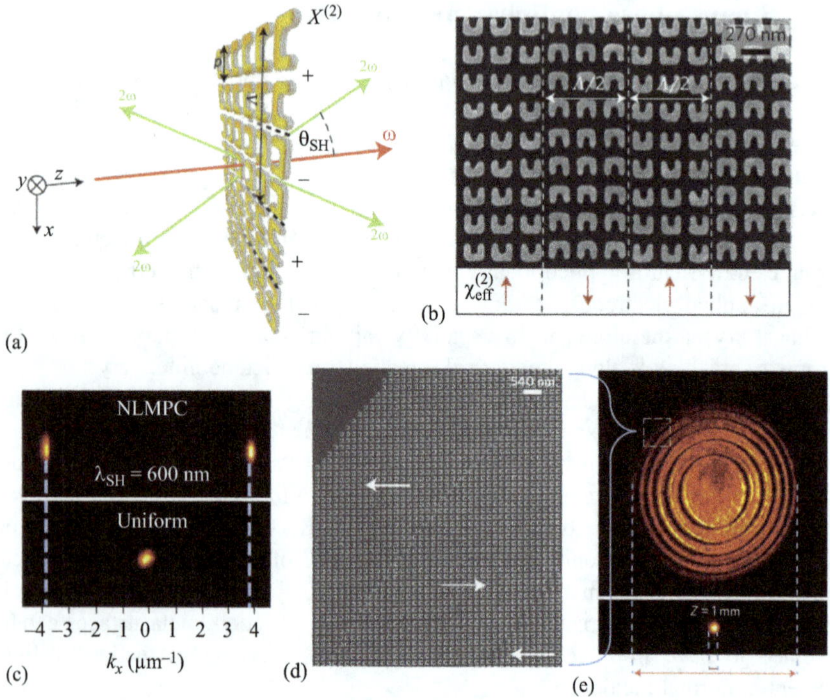

Figure 8.3 Controlling light with metamaterial-based nonlinear photonic crystals, adapted from [37]. (a) Illustration of the metamaterial-based nonlinear photonic crystals (NLMPC) consisting of SRRs. (b) The scanning electron microscope (SEM) image of the NLMPC, with a grating period Λ = 1,620 nm. (c) The SHG diffraction pattern from the NLMPC and a uniform SRR metasurface without mirror inversion of the SRR elements. (d-e) Nonlinear Fresnel zone plate from NLMPC, with (d) the zoomed SEM image of a part of the Fresnel zone plate sample and (e) the recorded SHG image with bright and dark birefringence patterns.

8.3.1 Continuous control of nonlinearity phase

Very recently, the concept of Pancharatnam-Berry phase, or geometric phase, has been utilized to design metasurfaces for controlling the wavefront of circularly polarized light. In the linear optical regime, geometric phase metasurfaces have been utilized to construct various functional ultrathin optical devices, such as flat lens imaging, generation of vortex beams, 3D holography, and optical spin-orbital interaction [38–43]. A geometric phase metasurface consists of an array of plasmonic antennas of the same geometry but with spatially varying orientations. Due to the anisotropy of each antenna, incident circularly polarized light will be partially converted into the opposite circular polarization upon transmission or

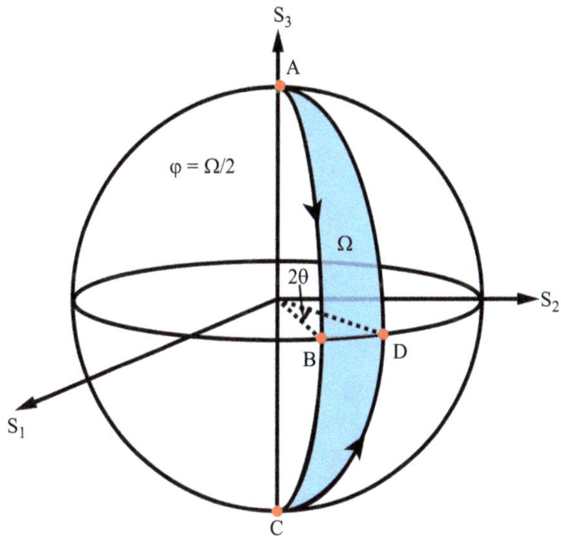

Figure 8.4 *Geometric phase indicated in Poincare sphere. When the nanostructure is rotated with an angle θ from point B to point D, the total geometric phase φ of this process is $\Omega/2$.*

reflection. Interestingly, the component opposite to that of the incident light acquires a phase that is twice the orientation angle of the antenna, regardless of the wavelength and resonant properties of the antennas. This geometric phase is usually illustrated by a Poincare sphere, as shown in Figure 8.4.

In comparison to conventional phase plate or zone plate, the geometric phase metasurface has the great advantage of precise control over the local phase, therefore exhibiting superior performance. In 2015, the concept of Berry phase was first introduced to the nonlinear optical regime to precisely control the nonlinearity phase at a subwavelength scale [44]. This novel concept significantly goes beyond the conventional nonlinear techniques due to their continuous phase control covering the whole 2π range.

The principle of the nonlinear Berry phase can be conveniently explained through a coordinate transformation in a similar way as how we illustrate the selection rule in Section 8.2. Consider a single plasmonic or dielectric nanostructure with nonlinear responses [Figure 8.5(a)]. For a fundamental beam of circular polarization state σ at normal incidence, the electric field is expressed by: $\vec{E}^{\sigma} = \widetilde{E}_0\left(\vec{e}_x + i\sigma\vec{e}_y\right)/\sqrt{2}$, where $\sigma = \pm 1$ denotes left- and right-handed circular polarization, respectively. The excitation of the nanostructures (e.g., plasmonic nanorods) represents an effective nonlinear dipole moment, which is given by:

$$\vec{p}_{\theta}^{\,n\omega} = \overleftrightarrow{\alpha}_{\theta}\left(\vec{E}^{\sigma}\right)^{n} \tag{8.6}$$

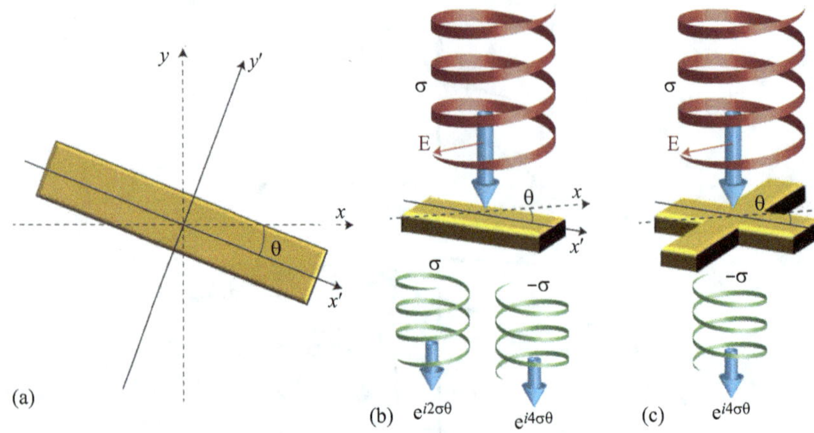

Figure 8.5 Illustration of geometric-phase-controlled nonlinear metamaterials, adapted from [44]. (a) The geometric phase is introduced by rotating a nanostructure with an angle θ with respect to the laboratory frame. (b) The nonlinear geometric phase of THG process from a C2 symmetry nanostructure. (c) The nonlinear geometric phase of THG process from a C4 symmetry nanostructure.

where α_θ is the nth harmonic nonlinear polarizability tensor of the nanostructure with an orientation angle θ. In order to investigate the dependence of the nonlinear dipole moment on θ, we define a local coordinate which is attached to the nanostructure (referred to as local frame), as shown in Figure 8.5(a). Since the local coordinate (x′, y′) axes are rotated by an angle of θ relative to the laboratory frame (x, y), the fundamental wave in the local coordinate acquires a rotational phase given by,

$$\vec{E}_L^{\sigma} = \vec{E}^{\sigma} e^{i\sigma\theta} \tag{8.7}$$

where the sub-index 'L' denotes the local coordinate. The nth harmonic nonlinear polarizability of the nanostructure in the local frame is given by $\alpha_0 = \alpha_\theta | \theta = 0$. Hence, the nth harmonic nonlinear dipole moment in the local frame is given by

$$\vec{p}_{\theta,L}^{n\omega} = \overleftrightarrow{\alpha}_0 \left(\vec{E}_L^{\sigma}\right)^n = \overleftrightarrow{\alpha}_0 \left(\vec{E}^{\sigma}\right)^n e^{in\sigma\theta} \tag{8.8}$$

Assume that the nanostructure has negligible thickness compared to the wavelength, the nonlinear polarization lies primarily in the plane, which can be decomposed into two in-plane rotating dipoles as,

$$\vec{p}_{\theta,L}^{n\omega} = \vec{p}_{\theta,L,\sigma}^{n\omega} + \vec{p}_{\theta,L,-\sigma}^{n\omega}, \text{ with } \vec{p}_{\theta,L,\sigma}^{n\omega}, \vec{p}_{\theta,L,-\sigma}^{n\omega} \propto e^{in\sigma\theta} \tag{8.9}$$

Next, we carry out an inverse coordinate transformation to return to the laboratory frame with the two rotating dipole moments given by,

$$\begin{aligned} \vec{p}^{\,n\omega}_{\theta,\sigma} &= \vec{p}^{\,n\omega}_{\theta,L,\sigma} e^{-i\sigma\theta} \propto e^{(n-1)i\sigma\theta} \\ \vec{p}^{\,n\omega}_{\theta,-\sigma} &= \vec{p}^{\,n\omega}_{\theta,L,-\sigma} e^{i\sigma\theta} \propto e^{(n+1)i\sigma\theta} \end{aligned} \tag{8.10}$$

The nonlinear polarizabilities of the nanostructure can therefore be expressed as,

$$\begin{aligned} \alpha^{n\omega}_{\theta,\sigma,\sigma} &\propto e^{(n-1)i\sigma\theta} \\ \alpha^{n\omega}_{\theta,-\sigma,\sigma} &\propto e^{(n+1)i\sigma\theta} \end{aligned} \tag{8.11}$$

The above equation shows that the nonlinear polarizabilities of the nth harmonic generation with the same or opposite circular polarization to that of the fundamental wave acquire geometric phases of $(n-1)\sigma\theta$ or $(n+1)\sigma\theta$, respectively. It was shown in Section 8.2 that the harmonic generation of circularly polarized fundamental wave is subject to selection rules based on the rotational symmetry of the structure. According to the selection rules for circular polarized pumping wave, the allowed harmonic orders for a nanostructure with m-fold rotational symmetry are given by $n = lm \pm 1$, where l is an arbitrary integer and '+' and '–' signs correspond to nonlinear signals of the same and opposite circular polarization, respectively [26,27,29].

Hence, for a nanorod structure with two-fold rotational symmetry (C2), THG signals with both the same and opposite circular polarizations as that of the fundamental wave can be generated. According to (8.11), they have a spin-dependent phase of $2\sigma\theta$ and $4\sigma\theta$, respectively [Figure 8.5(b)]. On the other hand, a nanostructure with four-fold rotational symmetry (C4) does not allow a THG process for the same polarization state as the incident polarization. Hence, only a single THG signal, that of the opposite circular polarization, is generated with a geometric phase of $4\sigma\theta$ [Figure 8.5(c)]. Importantly, due to the local isotropic response of the C4 structure, both the polarization state and the propagation of the fundamental wave will not be affected when transmitting through a metamaterial consisting of such C4 nanostructures of arbitrary orientations. Thus, by assembling the C4 nanostructures with spatially varying orientations in a 3D or 2D lattice, a nonlinear metamaterial or metasurface can be formed which shows homogeneous linear properties, but locally a well-defined nonlinear polarizability distribution for a circularly polarized fundamental wave. Moreover, the phases of the nonlinear polarizabilities of structures with different rotational symmetries for circularly polarized pumping waves are summarized in Table 8.1.

This concept of nonlinear geometric phase was verified by the characterization of a number of nonlinear binary phase gratings made from 2D arrays of plasmonic nanocrosses with a local C4 rotational symmetry [Figure 8.6(a)]. These binary nonlinear phase gratings allow one to measure the relative phase of the nonlinear polarization between C4 nanostructures of different orientations. Sample A consists of nanocrosses of identical orientation along the entire lattice and a periodicity of

Table 8.1 Nonlinear geometric Berry phase.

Harmonic orders	**Sign of σ**	C1	C2	C3 	C4 
$n = 1$	$+$				
	$-$	$2\theta\sigma$	$2\theta\sigma$		
$n = 2$	$+$	$\theta\sigma$			
	$-$	$3\theta\sigma$		$3\theta\sigma$	
$n = 3$	$+$	$2\theta\sigma$	$2\theta\sigma$		
	$-$	$4\theta\sigma$	$4\theta\sigma$		$4\theta\sigma$

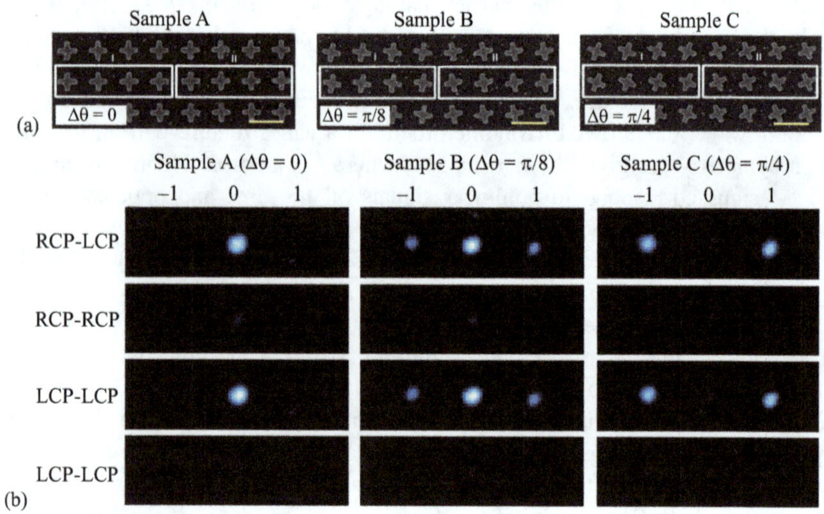

Figure 8.6 Experimental verification of the nonlinear phase with nonlinear phase gratings, adapted from [44]. (a) The SEM images of the three metasurfaces A, B, and C with different orientation angles 0, π/8, π/4 between the two subsets, respectively. Scale bar: 400 nm. (b) Measured diffraction pattern of THG signals from metasurfaces A, B, and C. The THG signals of the first and second rows correspond to a right circularly polarized (RCP) pumping laser; the third and fourth rows correspond to a left circularly polarized (LCP) pumping laser.

$a = 400$ nm in both x- and y-axis directions, while Sample B and Sample C consist of supercells of nanocrosses with two different orientations and a superperiod of $P = 3.2$ μm (eight crosses per unit cell). The relative orientation angle between the two subsets of nanocrosses within a unit cell is $\pi/8$ and $\pi/4$ for Sample B and C, respectively. According to the selection rule, only THG of the opposite circular polarization as that of the incident fundamental beam is generated with a geometric phase of $4\sigma\theta$. Thus, the introduced nonlinear phase for Sample B and Sample C corresponds to nonlinear phase gratings with phase differences of $\pi/2$ and π between the two subunits, respectively. Because the period of the binary nonlinear gratings is greater than the THG wavelength, the nonlinear signal exhibits diffraction orders in the far field, whereas the ratio between different diffraction orders can be determined by the nonlinear phase difference between the two subsets within the unit cell.

The samples were illuminated by short laser pulses at wavelength of 1,240 nm. For all the three samples, the THG signals with identical polarization states to the fundamental wave (RCP-RCP and LCP-LCP) were very weak, which was consistent with the selection rule for the THG. Because all the nanocrosses in Sample A have the same orientation, only zeroth-order THG was observed, as shown in Figure 8.6(b). For Sample B where the phase difference between the third-order nonlinearity of the two subsets of nanocrosses is $\pi/2$, the ratio between the zeroth order and the first-order THG signals is predicted to be ~ 2.4:1. The measured ratio is around 2.8:1, which agrees very well with the theoretical prediction. The π phase difference between the nonlinear polarizabilities of the two subsets of the nanocrosses in Sample C is expected to result in a complete destructive interference for the zeroth-order THG signal, which is confirmed by the measurement shown in the third column of Figure 8.6(b).

The concept of controlling the nonlinear phase is not confined to the THG, but rather can be applied to any harmonic orders. For example, a nonlinear metasurface consisting of nanostructures of C3 symmetry can operate for SHGs. Figure 8.7(a) shows a metasurface consisting of C3 rotational symmetric structures for a linear phase gradient in the second-order nonlinear polarization [Figure 8.7(b)]. The nanoantennas have a period of 400 nm in both x- and y-directions. Along the x-direction, the nanoantennas were rotated in steps 15° to generate a phase gradient in the second-order polarization. In the nonlinear measurement, a fundamental wave of wavelength 1,100 nm was illuminated onto the metasurface, which was close to the resonance wavelength of the fabricated C3 structures around 1,070 nm. Due to the C3 rotational symmetry of the structure, generation of a SHG signal with the same polarization is prohibited according to the nonlinear selection rule; whereas for the cross-polarized SHG signal (RCP-LCP and LCP-RCP), the diffraction angle is determined by the phase-gradient resulting from the orientation of the C3 nanostructures. As shown in Figure 8.7(c), for the RCP/LCP combination of the pumping wave and harmonic wave polarizations, only the -1 diffraction order was observed, verifying that the phase gradient is along the $-x$ direction. On the other hand, when the combination was changed to LCP/RCP, the nonlinear phase gradient was reversed, leading to observation of the second harmonic signal only in the $+1$ order.

The above theoretical analysis and experimental demonstration show that through the nonlinear geometric phase introduced by the spin-rotation coupling of light with subwavelength sized nanostructures, a local continuous phase can be generated. By exploiting the rotational symmetry of the nanostructure, only a single

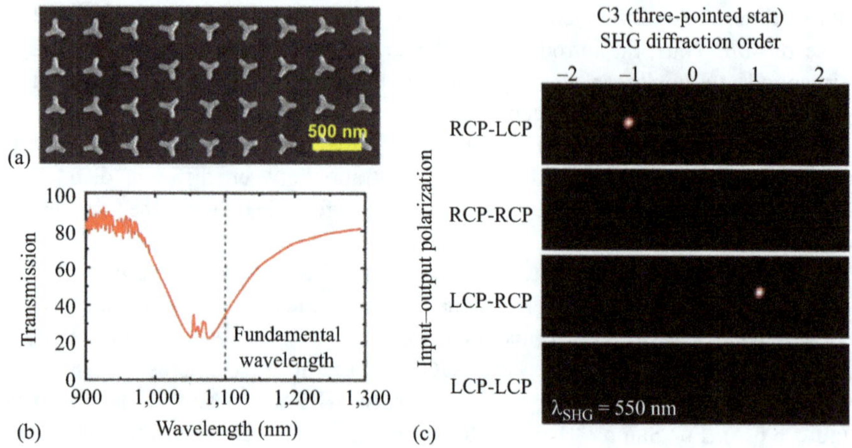

Figure 8.7 SHG measurement from a metasurfaces with nanoantennas of C3 rotational symmetry, adapted from [44]. (a) Scanning electron microscopy image of the metasurfaces. The structures are rotated along the x direction by an angle of 15°, generating a linear phase gradient along the surface. (b) Transmission spectrum of the pattern showing the resonance at 1,070 nm. The small peaks around the resonance wavelength are artifacts of our supercontinuum light source. (c) Measured SHG signal for a fundamental beam at 1,100 nm in dependence on the polarization state. The diffraction of the SHG signal appears under a diffraction angle of ±9.75° (±1 order) in correspondence with the designed phase gradient.

nonlinear signal with well-defined phase and circular polarization can be produced. More importantly, since the structures can have higher orders of rotational symmetry than 2, they show isotropic linear optical responses, meaning that their orientation angles do not affect the linear propagation of light. This decoupling between the linear and nonlinear properties is important in many nonlinear optics applications. This new approach for continuously controlling the nonlinearity can be potentially extended to 3D metamaterials made from low-loss nonlinear dielectric nanostructures, for achieving custom-defined local effective nonlinear susceptibilities, and introducing new freedoms in designing nonlinear materials for satisfying perfect phase-matching conditions, nonlinear photonic crystals with arbitrary nonlinear phase profiles, and manipulation of nonlinear signals in integrated photonic circuits or in the free space.

8.4 Nonlinear metasurface holography

Holography is one of the most important applications for metasurfaces, benefitting from their superior control over the wavefront of light. Holography is widely used to record and reconstruct 2D and 3D images by encoding phase information. In conventional holography, the phase-recording procedure is very complicated and challenging. More recently, computer generated holograms (CGH) have gained growing attention, as CGH can be designed numerically to generate arbitrary high-quality holographic images [45–48]. Metasurfaces, with their precise local phase control at the subwavelength scale, represent a new paradigm for applications in CGH. Since the first demonstration of metasurface holography in 2012 [49], there have been growing efforts along this direction, with tremendous progress made in improving both the efficiency and quality of the holographic images. With the introduction of nonlinearity phase, there has been some interests in generating holographic images in the harmonic signals. One particular application of nonlinear holography is encryption as the encoded images can only show up in the harmonic generation, which cannot be easily detected [50]. Recently, a few groups independently demonstrated nonlinear metasurface holography based on different mechanisms.

In 2016, Ye *et al.* demonstrated the first nonlinear hologram based on nonlinear geometrical phase. Interestingly, the metasurface hologram was designed to exhibit three independent sets of holographic images; one in the linear optical signal and two in second harmonic nonlinear signals, which were multiplexed by the wavelength and circular polarization state of the transmitted signals [51]. As shown in Figure 8.8, the basic building blocks of the nonlinear geometrical metasurface were SRRs, which were chosen due to their strong optical responses in both the linear and nonlinear (second-harmonic) regimes. For a circularly polarized incident beam with spin state σ, a SRR with an orientation angle of ϕ, due to its anisotropic response, can convert part of the beam into the opposite circular polarization $-\sigma$ in the transmitted beam, with a local geometric phase of $2\sigma\phi$. The SRR provides an interesting platform for investigating SHG due to its lack of centrosymmetry, and the presence of both electric and magnetic dipoles. For a pumping beam of circular polarization, the SRR without any rotational symmetry can produce SHG signals of both circular polarizations. From Section 8.3, the SHG with spin σ and $-\sigma$ acquire a nonlinear Berry phase of $\sigma\phi$ and $3\sigma\phi$, respectively (Table 8.1). Thus, the distribution of the orientation angles $\phi(r)$ of the constituent antennas in a metasurface can lead to three different phase profiles: $2\sigma\phi(r)$ for transmitted linear signal of spin $-\sigma$, $\sigma\phi(r)$ for SHG of spin σ and $3\sigma\phi(r)$ for SHG of spin $-\sigma$. However, the linear and nonlinear geometric phases, corresponding to optical signals of different frequencies and different spins, are obviously related since the three phase profiles were generated from the same single-layer metasurface. Despite this correlation, three independent holographic images can still be encoded in the metasurface by exploiting the fact that only the intensity profile of the holographic image matters and there is no constraint on the phase profile of the holographic image. Therefore, the phase profiles of the three holographic images can be arbitrarily designed to optimize their intensity patterns. Thus, this extra degree of freedom reduces the correlation between the intensity patterns of the holographic images formed by the linear and SHG signals.

A modified Fidoc algorithm can be used to design the orientation angles of the individual SRRs of the metasurface. The Fidoc algorithm is an optimized iteration

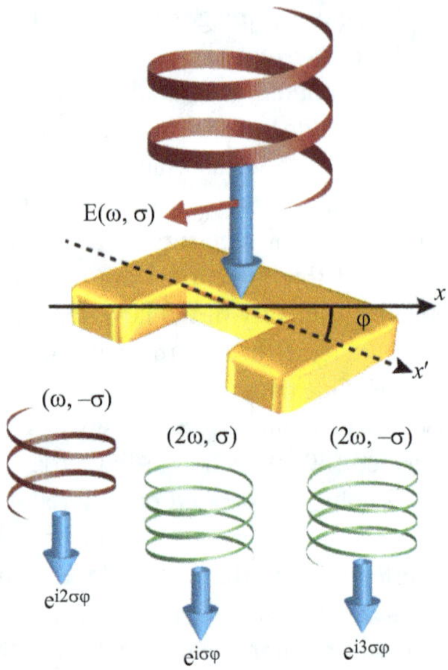

Figure 8.8 The linear and nonlinear geometric phase of a split ring resonator (SRR) unit, adapted from [51].

phase retrieval algorithm, which is widely used to obtain the phase-only CGH. Starting with a random SRR orientation angle distribution $\varphi_0(x,y)$, three linear and nonlinear channels with spatial distributions $e^{i\varphi_0(x,y)}$, $e^{i2\varphi_0(x,y)}$, and $e^{i3\varphi_0(x,y)}$ are Fourier transformed into k-space, corresponding to the field distribution of the three holographic images in the far field. These three images bear no resemblance to the target images at all. In the next step, the three amplitude distributions, corresponding to three desired holographic images, replace the amplitudes of the three k-space Fourier distributions, while their phase profiles are maintained. In the next step, they are transformed back into the space-domain (corresponding to the phase distribution at the metasurface) by an inverse Fourier transformation. The new orientation angle distribution $\varphi_1(x,y)$ of the metasurface is determined by the phase profile of the combined inverse Fourier transformed complex fields. The new phase distribution $\varphi_1(x,y)$ is used for the next iteration. After several iterations, the phase profiles converge, and one can finally acquire the desired phase-only distribution for the hologram. This phase distribution is then translated to the orientation angle of a periodic pattern of SRRs on the surface.

The design process of the nonlinear metasurface is outlined in Figure 8.9. The design starts with an initialization step by adding random phases φ_{ra}, φ_{rb}, and φ_{rc} to the amplitude distribution A_0, B_0, and C_0 of the three target images, which correspond to the frequency-helicity combination of (2ω, σ), (ω, $-\sigma$), and (2ω, $-\sigma$), respectively. After an inverse Fourier transform, the complex amplitude distributions for φ, 2φ, and 3φ on the holographic image plane are combined to form an overall complex field $H_n\exp(i\varphi_n)$. To achieve phase-only holograms, an amplitude

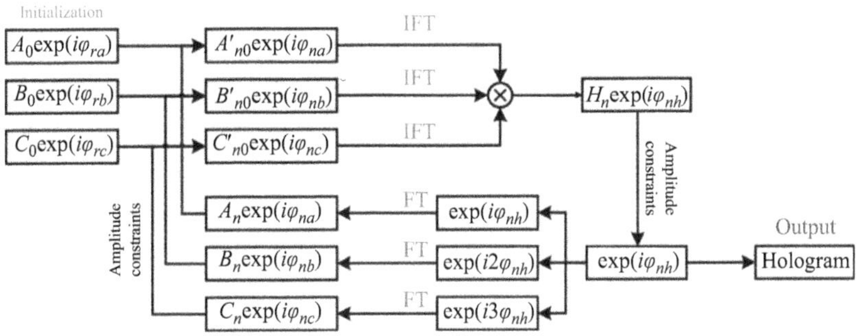

Figure 8.9 Flowchart for hologram generation with a circularly polarized incident beam of frequency ω and helicity σ, adapted from [51].

constraint is introduced, which sets the amplitude of the combined distribution to unity. φ_n now represents the orientation angle of the SRRs across the metasurface, and φ_n, $2\varphi_n$, and $3\varphi_n$ correspond to the phase distributions for the frequency-helicity combination of $(2\omega, \sigma)$, $(\omega, -\sigma)$, and $(2\omega, -\sigma)$, respectively. After several iterations, the final target phase-only distribution is acquired. A_n, B_n, C_n, A'_{n0}, B'_{n0}, and C'_{n0} refer to the amplitude distribution before and after applying the amplitude constraints in the nth iteration. FT and IFT stand for Fourier transformation and inverse Fourier transformation, respectively, whereas H_n and φ_{nh} describe the amplitude and phase after the combination of three holograms, respectively.

Considering the correlation between the three Pancharatnam-Berry phases in the linear and nonlinear signals, one certainly cannot independently optimize the holographic image carried by one channel without affecting the images carried by the other two channels. On the other hand, the iteration algorithm maximizes the quality of the images by exploiting the following fact: the quality of the images is only determined by their intensity profile, whereas the phase profiles of the holographic images can be arbitrary, and therefore can serve as free parameters for optimization of the image quality, and consequently minimizing the cross-talk between the three holographic images.

Figure 8.10 shows the experimental demonstration of the multiplexed holographic images based on the spin and wavelengths of the transmitted signals. It was observed that three independent holographic images of the letters 'X', 'R', and 'L' were constructed for the three transmitted optical beams with frequency/spin combinations of $(\omega, -\sigma)$, $(2\omega, -\sigma)$, and $(2\omega, \sigma)$, respectively. Importantly, the images formed with the SHG signal of different spins are of very good quality. Here, the design of the nonlinear hologram benefits from the unique linear and nonlinear optical properties of geometric phase metasurfaces. The different information channels can be used to encode different images within a single hologram, whereas the observation of each image depends on the read-out conditions only. Thus, with the extra degrees of freedom provided by nonlinear geometrical phases, more information channels can be utilized to provide more independent information than in linear optical holography.

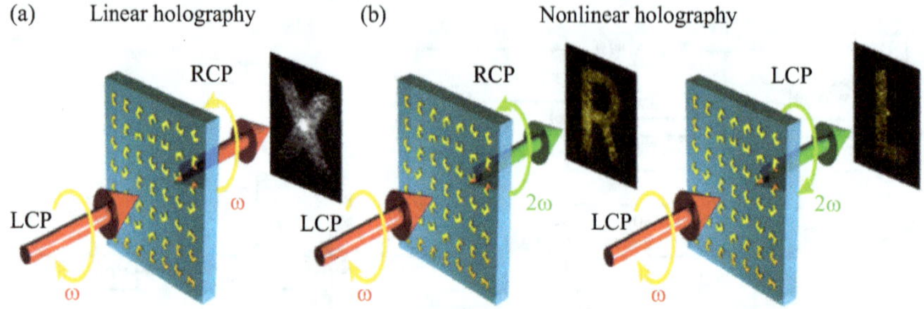

Figure 8.10 Spin and wavelength dependent holographic images, adapted from [51]. (a) Linear optical metasurface hologram for reconstruction of the letter 'X' at infrared wavelengths. (b) Second harmonic generation holographic images for the reconstruction of the letters 'R' and 'L' encoded in the different circular polarization states. While the incident wave conditions are kept fixed, the second harmonic LCP wave carries the phase –φ, while the phase of the RCP wave is –3φ.

Besides geometric nonlinear metasurfaces, holography for harmonic generations can also be realized by varying the geometry of individual unit cells. One interesting approach for generating polarization multiplexed nonlinear holography is to use a double layer configuration, with each layer only responding to a particular polarization state. Figure 8.11 shows a specific design for realizing nonlinear holographic images at two different linear polarizations. Each layer consists of an array of V-shaped antennas with different geometries [52]. By carefully designing the shape of each antenna, the phase of the effective local nonlinear polarization for SHG can be controlled. The orientations of the antennas between the two layers are orthogonal to each other such that each layer can only generate the THG for a particular linear polarization state. As such, two different holographic images can be encoded in the double-layer metasurface. Experimentally, a fundamental wave of 1,266 nm wavelength was illuminated at a normal incidence onto the metasurface, and two different holographic images were observed in the transmission direction at the third harmonic wavelength of 422 nm at different incident polarization states. In comparison to the nonlinear geometric phase–based metasurface holograms, the double layer design is more straightforward in the designing process, but requires a much more challenging fabrication process.

8.5 Nonlinear metasurface for intensity control and image encoding

In Section 8.3, it was shown that the phase of nonlinear polarization can be continuously controlled over the full range of 2π. This continuous phase engineering

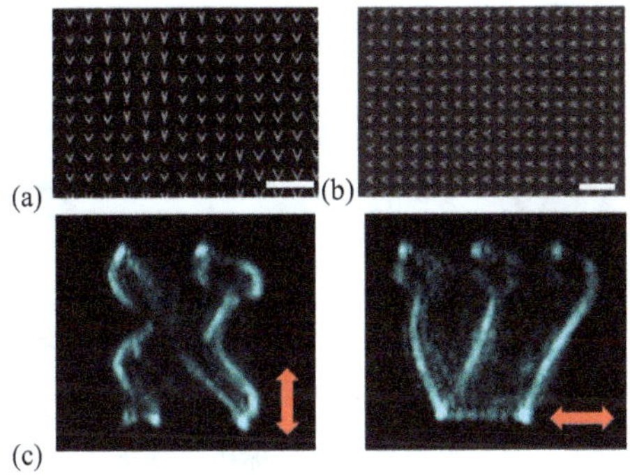

Figure 8.11 3D nonlinear metasurface holography, adapted from [52]. (a) and (b) are the SEM images the first and second layer of the 3D metasurface, respectively. Scale bar: 1 μm. (c) SHG hologram images of the Hebrew letters Alef and Shin for vertical and horizontal polarized fundamental waves, respectively.

can be readily extended to continuously control the local intensities for harmonic generations, by designing a supercell that consists of two or more nanoantennas [50]. Taking SHG as an example, an antenna of C3 rotational symmetry can be used to control the nonlinearity phase, which is three times of its orientation angle. As shown in Figure 8.12(a), for a supercell consisting of two pairs of C3 antennas, with each pair having a different orientation of θ_1 or θ_2, the nonlinear signals generated by the four antennas would interfere, with the overall local SHG intensity proportional to $\cos^2(3\theta)$. The local SHG intensity reaches maximum when the two angles are the same, while it reaches zero when their relative angle is 60° (corresponding to a nonlinear phase difference of 180°). The measured intensity of the harmonic signal confirms this relationship between the orientation angle difference and the intensity of the harmonic signal, as shown in Figure 8.12(b). Thus, by designing the orientations of the two antennas in each supercell, one can continuously engineer the intensity profile of SHG across the metasurface. Figure 8.12(c) shows the imaging at SHG wavelength of a nonlinear metasurface showing four letters, 'M', 'E', 'T', and 'A'. In the regions exhibiting these letters, the unit cells consist of antennas with the same orientation. While in the background region, the angle formed by the antennas within each unit cell is 60°. Interestingly, if one images the metasurface at the fundamental wavelength, no image is observed. Thus, the nonlinear metasurface can be applied to encode information for encryption applications.

8.6 Vortex beam generation in harmonic generation

Nonlinear metasurfaces based on Berry phase not only provide a facile and robust way for continuous manipulation of the nonlinearity phase but also introduce

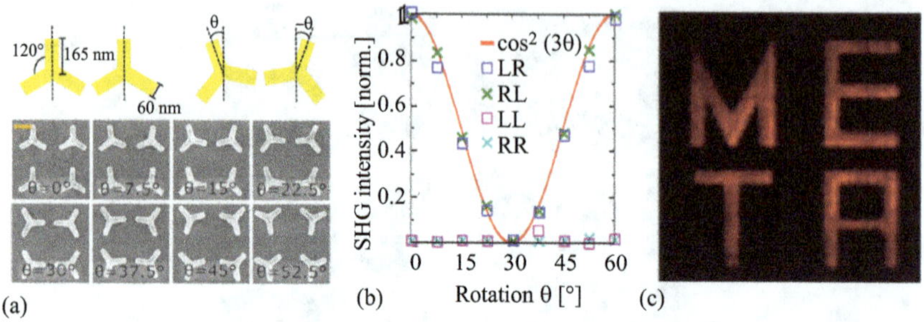

Figure 8.12 Nonlinear metasurface for intensity control, adapted from [50]. (a) Metasurfaces design and SEM images of the fabricated meta-atoms for various rotation angles θ. Scale bar: 200 nm. (b) Angle dependent SHG intensity for excitation with a fundamental wave at 1250 nm wavelength for various polarization schemes. The first letter in the legend describes the circular polarization state of the fundamental wave, the second letter the polarization state for SHG detection (L, left circularly polarized; R, right circularly polarized). (c) SHG image of the character META.

spin-orbital coupling in the nonlinear optical regime, due to the dependence of the nonlinearity phase over the spin of the incident fundamental light. Nonlinear metasurface can therefore be engineered to generate vortex beams with spin dependent orbital angular momentum [53]. Figure 8.13 shows the design of three metasurfaces for nonlinear vortex generation of different orbital angular momenta. Each metasurface consists of a series of concentric rings, with each ring made up of plasmonic antennas of C3 rotational symmetry. In the first design [Figure 8.13(d)], the orientations of the antennas slowly vary along each ring, resulting in a rotation angle of $2\pi/3$ after completing a full turn. This rotation configuration corresponds to a topological charge $q = 1/3$. Similarly, the other two samples have similar configuration, except for the different topological charges of 2/3 and 1, respectively, as shown in Figure 8.13(e) and (f).

For a circularly polarized incident fundamental Gaussian beam of spin σ, the SHG signal in the transmission direction is expected to have a spin $-\sigma$ and an angular momentum of $3\sigma q$. Thus, for the three metasurfaces with $q = 1/3$, 2/3, and 1, the corresponding angular momentum of the SHG is σ, 2σ, and 3σ, respectively. Figure 8.14 shows the experimental results. For an incident LCP fundamental beam, only SHG in RCP is generated, which exhibits a ring shape characteristic of the intensity of a vortex beam. It is observed that the size of the ring increases with q, which agrees with the theory of vortex beam. The nonlinear metasurfaces are further characterized by an incident fundamental beam of linear polarization in the horizontal direction. As the linear polarization can be decomposed into LCP and RCP waves of equal amplitude, it is expected that SHG of both circular polarizations are also generated, but with opposite orbital angular momenta l and $-l$. When the SHG signal transmits through a linear polarizer, the interference between the

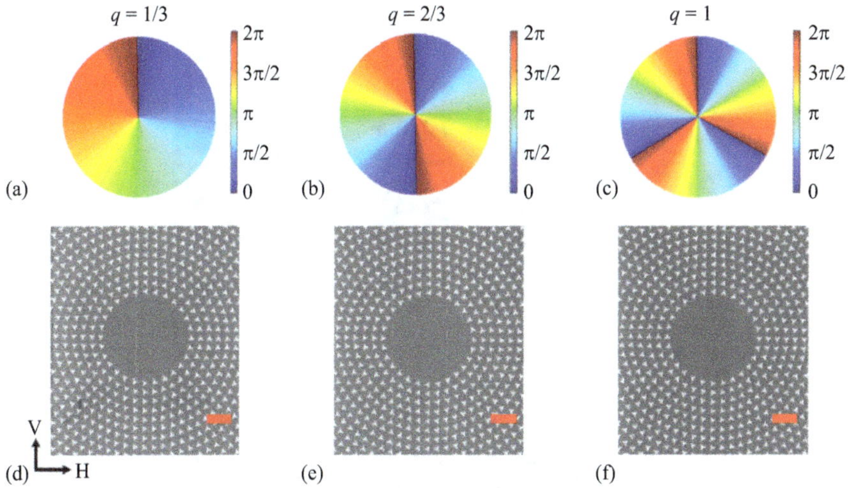

Figure 8.13 Design and optical properties of a nonlinear photonic metasurface, adapted from [53]. (a)−(c), real space phase distribution of SHG from the nonlinear photonic metasurface with various phase singularities q = 1, 2, and 3. (d)−(f), scanning electron microscope images of gold plasmonic metasurface with q = 1/3, 2/3, and 1. Scale bar: 600 nm.

two opposite orbital angular momentum leads to a Hermite–Gaussian beam with a petal number of $2l$. Thus, for the three metasurfaces, it is expected that the generated HG modes have petal numbers of 2, 4, and 6, respectively. This is confirmed by the experimental results shown in Figure 8.14. Thus, metasurfaces based on nonlinear PB phase provide an ideal platform for generating various vortex beams and vector beams in the nonlinear harmonic processes.

8.7 Nonlinear imaging

Imaging plays an important role in almost every field. In conventional systems, imaging relies on combination of lenses with curved surfaces. The emergence of metasurfaces provide a new compact solution to reduce the size of conventional imaging systems due to their flat and ultrathin features. In the past, metasurfaces have been primarily used for imaging in the linear optical regime. With the introduction of nonlinear PB phase, a metasurface can be engineered to perform imaging operations during the harmonic generation processes [54]. To achieve the nonlinear imaging operation, the nonlinear phase distribution should be given as,

$$\varphi(r) = k_0\left(\sqrt{f^2 + r^2} - |f|\right) \tag{8.12}$$

where r is the distance from the center of the metalens, f is the focal length of the lens, and $k_0 = \frac{2\pi}{\lambda}$ is the free-space wave vector, with λ being the nonlinear harmonic

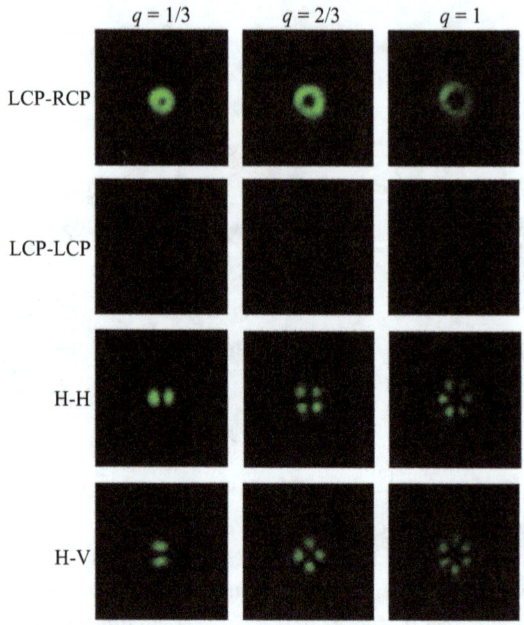

Figure 8.14 Measurement intensity distributions of SHG from metasurfaces with q = 1/3, 2/3, and 1, adapted from [53]. Under pumping of left circularly polarized FW, the SHG radiation with RCP and LCP components (first and second rows) are measured in the far field. The third and fourth rows show the results for FW with horizontal polarization (H), the SHG signals with horizontal (H) and vertical (V) polarizations were recorded by using CCD camera.

wavelength. Considering a second harmonic process, the phase distribution can be achieved by utilizing the C3 meta-atoms, with their orientation angles given as $\theta(r) = \varphi(r)/3\sigma$. Figure 8.15 shows a nonlinear metalens for SHG imaging with a diameter of 300 μm, whereas the C3 meta-atoms are aligned in polar coordinates on concentric rings with a spacing of 500 nm. The metalens is designed to have a SHG focal length $f = 500$ μm at the fundamental wavelength of 1,085 nm.

The experimental setup to characterize the focusing performance of the nonlinear lens is shown in Figure 8.16(a). A circularly polarized fundamental Gaussian beam is incident onto the metasurface. The SHG signal from the metasurface is collected in the forward direction with a microscope objective. By changing the position of the microscope objective along the optical axis, different planes perpendicular to the propagation direction are imaged onto the camera. The metasurface lens operates as a convex lens for RCP incident state. The images at different planes are shown in Figure 8.16(b). It is clearly observed that a sharp focal point is formed at a distance of 500 μm away from the metasurface. While away from this plane, the focal spot becomes more diverged.

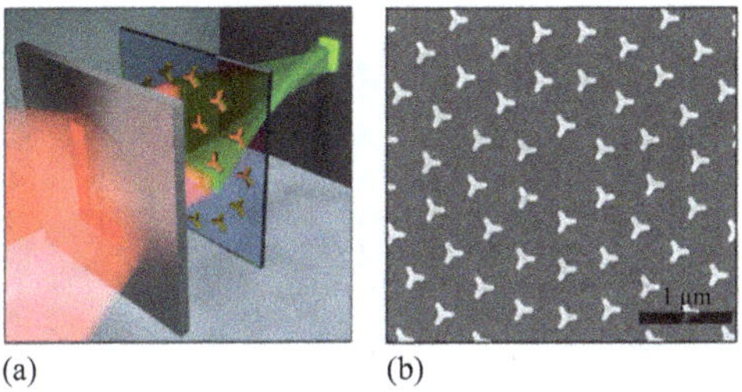

Figure 8.15 Nonlinear imaging using a metalens, adapt from [54]. (a) Illustration of the SHG imaging concept using a nonlinear metalens. (b) SEM image of the metalens sample.

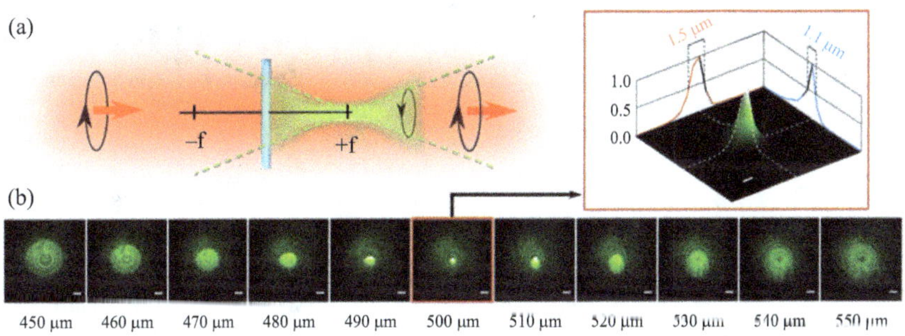

Figure 8.16 SHG Measurement of the beam propagation in focal region, adapted from [54]. (a) Schematic of the experiment setup. (b) Measured SHG intensity distribution for the nonlinear metalens with $f = 500$ µm at different transverse planes in the focal region. Scale bar: 10 µm.

The imaging performance of the lens is characterized by placing an L-shaped aperture as the object in front of the metalens at a distance of 500 µm, which is the same as the focal length. A right circularly polarized incident Gaussian beam at 1,085 nm wavelength is incident onto the L-shaped aperture, and the transmitted wave interacts with the metasurface and generates SHG signal. In the transmission direction, by adjusting the imaging plane of the microscope objective, a sharp inverted SHG image of the L-shaped aperture is formed with a magnification $\Gamma = -1$ at a location about 1,000 µm away from the metalens. Interestingly, the relationship between the object distance and the image distance violates that of the linear optical case, where the object and image distance are related by $-g = b = 2f$. In linear optics, this configuration would result in an image formation at infinity. Thus, the traditional lens equation from linear optics cannot be applied to the nonlinear imaging case. The observed image location is indeed consistent with the simulation based on a beam propagation method adapted for the nonlinear regime [Figure 8.17 (a)]. When the spin state of the fundamental wave is flipped from RCP

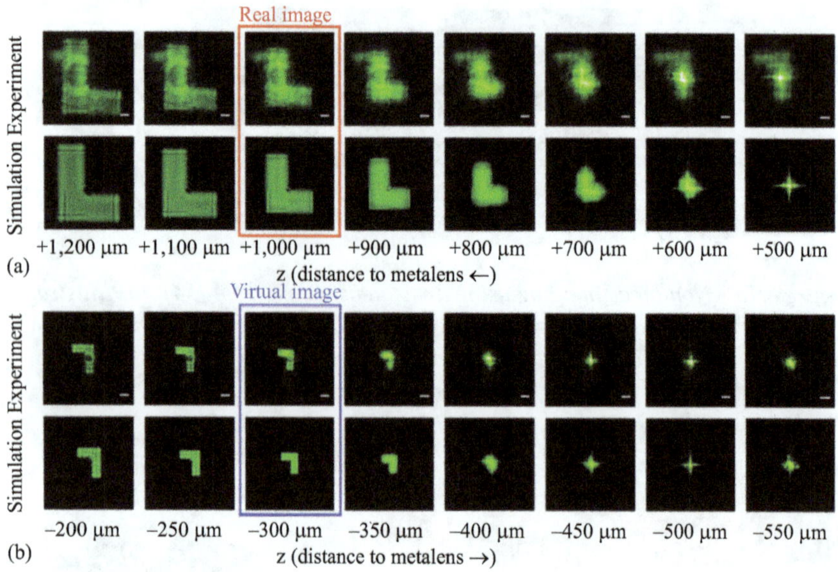

Figure 8.17 1:1 spin-dependent object imaging with the nonlinear metalens, adapted from [54]. (a) Measured (first row) and simulated (second row) SHG distribution for planes along the z-propagation direction behind the metalens showing the evolution of the image formation. The spin state of the fundamental wave is right circularly polarized (RCP), and the metalens works as a convex lens for a spin conversion from RCP to LCP. (b) By reversing the spin state of the fundamental wave from RCP to LCP, the metalens works as a concave lens for a spin conversion from LCP to RCP.

to LCP and the sign of the focal length is altered, leading to an inverted virtual image of the L located between the L-shaped aperture and the metalens [Figure 8.17(b)]. The virtual image for LCP incident light appears at a distance of $b \approx -300$ μm with a magnification of $\Gamma = 0.35$. These values are close to the simulation results that show an image distance of $\Gamma \approx -333$ μm and a magnification of $\Gamma = 0.33$.

The nonlinear imaging metasurface combines the generation and focusing of SHG into a single element. In addition, the spin-dependent focusing effect of SHG waves at both real and virtual focal planes may provide new switchable functionalities for nonlinear optical devices.

8.8 Nonlinear planar chiral metasurfaces

Optical activity has been one of the most well-adopted methods for probing the symmetry of surfaces and materials [55,56]. Optical activity comprises of two effects: optical rotatory dispersion (ORD) that rotates the linear polarization state of a beam and circular dichroism (CD) that leads to different transmission or absorption between the two circularly polarized states. ORD and CD arise from phase and absorption differences between left- and right-circularly polarized light in a chiral medium, respectively. In comparison to the usually weak chiral phenomena in linear optical regime, optical activity in nonlinear harmonic generation processes may exhibit much stronger effects [57–59]. In particular, second harmonic generation circular dichroism (SHG-CD) has become a powerful technique to probe the symmetry of organic and inorganic materials due to its high sensitivity to the asymmetry of molecules and physical interfaces [60–62]. Nonlinear optical CD effect has been demonstrated in metamaterial and plasmonic nanostructures with either chirality in 3D nanostructures, or 2D structures with the fundamental beam incident at oblique angles. Note that in some 2D chiral materials, the presence of a substrate leads to an actual 3D chiral effect. Indeed, giant circular dichroism for harmonic generations may exist on ultrathin plasmonic metasurfaces with broken in-plane mirror symmetry.

For the nonlinear metasurfaces described in preceding sections, the unit cell contains both rotational symmetry and in-plane mirror symmetry [29]. As such, for a fundamental beam at normal incidence, the mirror symmetry dictates that when the spins of incident beam and the detected harmonic signal are simultaneously reversed, the intensity of the harmonic generation remains the same. Therefore, there is no nonlinear chiral effect in such structures. Here, we look into metasurfaces consisting of structures that maintain the rotational symmetry but lack in-plane mirror symmetry [63]. In such metasurfaces, the nonlinear selection rule for circular polarized beam still holds; however, the nonlinear CD effect can become very pronounced.

Figure 8.18 shows the schematics of the metasurface for SHG study. Since the planar structure has C3 rotational symmetry, for an incident circular polarized beam, only the SHG with opposite circular polarization to that of the fundamental beam is generated. However, due to the broken in-plane mirror symmetry, it is expected that the intensities of the SHG signals for the opposite circular polarization states of the fundamental beam are generally different.

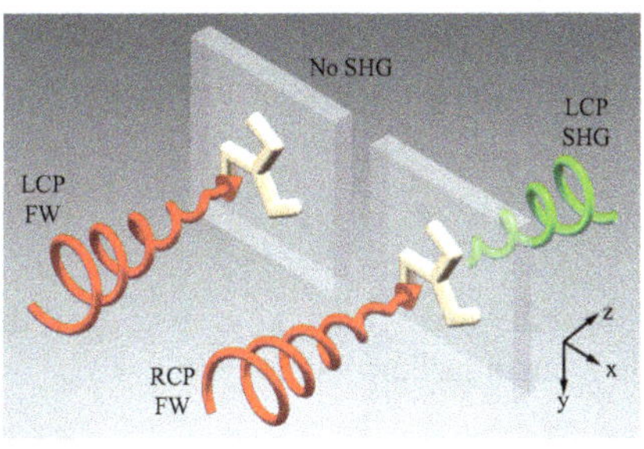

Figure 8.18 Nonlinear planar chiral metasurface, adapted from [63].

For the ultrathin C3 structure in the *x-y* plane as shown in Figure 8.18, the relevant effective nonlinear susceptibility tensor elements are those with sub-indices consisting only of *x* and *y*. The rotational symmetry of both the individual structure and the lattice reduces the effective nonlinear susceptibility tensor to just two independent elements: $\chi_{yyy} = -yxx = -xxy = -xyx = \chi_1$, $xxx = -xyy = -yyx = -yxy = \chi_2$ [23]. Consequently, the effective nonlinear polarization can be expressed as,

$$P_x^{2\omega} = \varepsilon_0\chi_2\left(E_x^2 - E_y^2\right) - 2\varepsilon_0\chi_1 E_x E_y \tag{8.13}$$

$$P_y^{2\omega} = -2\varepsilon_0\chi_2 E_x E_y - \varepsilon_0\chi_1\left(E_x^2 - E_y^2\right) \tag{8.14}$$

Based on the above equations, for a circularly polarized incident beam with the electric field vector given by $\vec{E} = E_0\left(\widehat{e}_x \pm i\widehat{e}_y\right)/\sqrt{2}$ ('+' and '–' sign correspond to LCP and RCP, respectively), the effective nonlinear polarization can be expressed as,

$$\vec{P}^{2\omega} = (\chi_2 \mp i\chi_1)\left(\widehat{e}_x \mp i\widehat{e}_y\right)\varepsilon_0 E_0^2 \tag{8.15}$$

which agrees with the selection rule—a circularly polarized fundamental light can only produce the SHG with an opposite polarization for a structure with C3v symmetry. Equation (8.15) also shows that the intensities of the SHG signals with oppositely circularly polarized incident light are different if χ_1 and χ_2 have a phase difference. On the other hand, if χ_1 and χ_2 are in phase, one can show that an appropriate coordinate transformation reduces the in-plane nonlinear coefficients to a single one $\chi = \sqrt{\chi_1^2 + \chi_2^2}$, which is the same as for a structure with C3 symmetry. The maximum contrast between the two circular polarization states can be achieved when χ_1 and χ_2 are equal in amplitude and $\pi/2$ out of phase. This would lead to a unity SHG-CD, which is defined by $CD = \frac{I_{LCP}^{2\omega} - I_{RCP}^{2\omega}}{I_{LCP}^{2\omega} + I_{RCP}^{2\omega}}$. The unity SHG-CD indicates the complete cancellation of the nonlinear contributions from χ_1 and χ_2 for one circular polarization state, but a constructive interference between their contributions for the other circular polarization.

For a linearly polarized incident wave $(E_x, E_y) = (E_0, 0)$, one gets $P_x^{2\omega} = \varepsilon_0\chi_2 E_0$ and $P_y^{2\omega} = -\varepsilon_0\chi_1 E_0$. To identify the polarization state of the SHG signal, one can place another linear polarizer (analyzer) in the transmission direction. For a given orientation angle θ of the analyzer, the measured SHG intensity would be

$$I_\theta^{2\omega} \propto \left|P_\theta^{2\omega}\right|^2 \propto |\chi_2 \cos\theta - \chi_1 \sin\theta|^2 \tag{8.16}$$

By normalizing χ_1 to unity and setting $\chi_2 = a\exp(i\varphi)$, the above equation can be rewritten in a simple sinusoidal form as,

$$I_\theta^{2\omega} \propto 1 - \cos\beta \sin(2\theta + \delta) \tag{8.17}$$

where

$$\beta = \sin^{-1}\left(\frac{2a \sin \varphi}{1+a^2}\right) \text{ and } \delta = \tan^{-1}\left(\frac{a^2-1}{2a \cos \varphi}\right) \tag{8.18}$$

The SHG-CD can be further simplified as CD = sin(β). Thus, the circular dichroism for SHG can be independently obtained by two different measurements—the ellipticity of the SHG for a linearly polarized fundamental beam and a direct measurement of the relative difference between the SHG signals of the two opposite circularly polarized incident fundamental beams.

Figure 8.19(a) shows the fabricated gold nanostructures with thickness of 30 nm arranged in a triangular lattice with periods of 400 nm. By linear transmission measurement, one measures the resonance wavelength of the sample to be 1,165 nm. The measurement results of the SHG with circular polarized incident beam are shown in Figure 8.19(b). We found that SHG signals from the plasmonic nanostructures have opposite polarization states as compared to that of the illuminating laser. However, we also measured extremely low SHG intensity with the same polarization state as that of the illuminating laser although they are theoretically forbidden in the dipole approximation but can be observed due to the imperfections of nanofabrication [64,65]. For the SHG measurements, both the L-R and R-L spectra increase with wavelength up to 1,160 nm. However, the L-R intensity drops abruptly with further increasing wavelength above 1,160 nm, while the R-L intensity further increases and peaks around 1,220 nm. As a result, there exists a large difference between the SHG intensities for the two circularly polarized incident beams for wavelength above 1,160 nm. Particularly, at around 1,280 nm wavelength, the SHG for the illumination with LCP light (L-R) drops

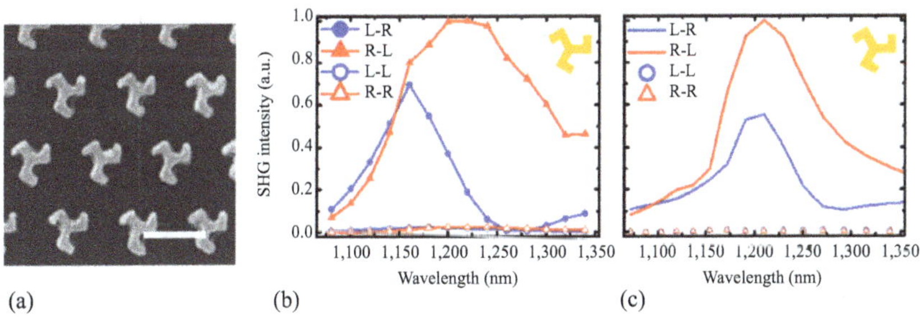

Figure 8.19 Geometry and polarization dependent SHG spectra from the planar chiral metasurface with C3 symmetry, adapted from [63]. (a) SEM image of the C3 metasurface with broken mirror symmetry. (b) Measured and (c) simulated polarization dependent SHG spectra. The first letter in the legend describes the circular polarization state of the fundamental wave, the second letter the polarization state for SHG detection (L, left circularly polarized; R, right circularly polarized).

almost to zero while that for the illumination with RCP light (R-L) is very strong, leading to near unity value of 0.98 for SHG-CD at this wavelength.

Our measurements nicely confirm numerical simulations of the SHG signals from such structures, which also predict large differences between the two polarized signals [Figure 8.19(c]. The measured nonlinear spectra of the SHG intensity show similar features as those of calculated SHG spectra, with similar peak locations of the SHG for both R-L and L-R, and similar dip locations for L-R. The slight discrepancy between the experiment and the simulation might be due to both the deviation of the fabricated sample from the ideal design and the limited precision of the simulation of the nonlinear signals due to their extreme sensitivity to the local field distribution.

8.9 Summary and outlook

In this chapter, we have discussed the principle underlying the control of the phase, intensity, and polarization states in harmonic generation processes with plasmonic metasurfaces. We have shown that the introduction of PB phase in the nonlinear optical regime provides an entirely new and powerful route toward achieving this goal. The control over nonlinear processes enables the design and implementation of many interesting nonlinear photonic devices, ranging from nonlinear holography to imaging lenses. However, the further progress toward practical applications of these nonlinear devices is limited by the conversion efficiency. The combination of the metasurface with materials with high nonlinear coefficients such as polymers and quantum wells may provide a solution to overcome this. Thus far, the nonlinear PB phase has only been used for plasmonic metasurfaces, whereas it is possible to extend this important concept to dielectric metasurfaces. In addition, the nonlinear PB phase may also be extended into 3D systems consisting of multiple layers of metasurfaces for further enhancing the nonlinear conversion.

References

[1] Shen, Y. R. 'The principles of nonlinear optics'. John Willey & Sons, New York, USA (1991).

[2] Boyd, R. W. 'Nonlinear optics'. Academic Press, San Diego, USA (2008).

[3] Kauranen M., Zayats A. V. 'Nonlinear plasmonics'. *Nature Photonics.* 2012; **6**:737–748.

[4] Klein M. W., Enkrich C., Wegener M., Linden S. 'Second-harmonic generation from magnetic metamaterials'. *Science.* 2006; **313**:502–504.

[5] Fan W., Zhang S., Panoiu N. C., *et al.* 'Second harmonic generation from a nanopatterned isotropic nonlinear material', *Nano Letters.* 2006; **6**:1027–1030.

[6] Kujala S., Canfield B. K., Kauranen M., Svirko Y., Turunen J. 'Multipole interference in the second-harmonic optical radiation from gold nanoparticles', *Physical Review Letters.* 2007; **98**:167403.

[7] Czaplicki R., Husu H., Siikanen R., *et al.* 'Enhancement of second-harmonic generation from metal nanoparticles by passive elements', *Physical Review Letters*. 2013; **110**:093902.
[8] Valev V. K., Silhanek A. V., Verellen N., Gillijns W., P. Van Dorpe, Aktsipetrov O. A., Vandenbosch G. A. E., Moshchalkov V. V., Verbiest T. 'Asymmetric optical second-harmonic generation from chiral G-shaped gold nanostructures', *Physical Review Letters*. 2010; **104**:127401.
[9] Valev V. K., Silhanek A. V., Gillijns W., *et al.* 'Plasmons reveal the direction of magnetization in nickel nanostructures', *ACS Nano*, 2011; **5**:91–96
[10] Zhang Y., Grady N. K., Ayala-Orozco C., Halas N. J. 'Three-dimensional nanostructures as highly efficient generators of second harmonic light'. *Nano Letters*. 2011; **11**:5519–5523.
[11] Ko K. D., Kumar A., Fung K. H., *et al.* 'Nonlinear optical response from arrays of Au bowtie nanoantennas'. *Nano Letters*. 2011; **11**:61–65.
[12] Cai W., Vasudev A. P., Brongersma M. L. Electrically controlled nonlinear generation of light with plasmonics, *Science*. 2011; **333**:1720–1723.
[13] Aouani H., Navarro-Cia M., Rahmani M., *et al.* 'Multiresonant broadband optical antennas as efficient tunable nanosources of second harmonic light'. *Nano Letters*. 2012; **12**:4997–5002.
[14] Linden S., Niesler F. B. P., Förstner J., Grynko Y., Meier T., Wegener M. 'Collective effects in second-harmonic generation from split-ring-resonator arrays'. *Physical Review Letters*. 2012; **109**:015502
[15] Hentschel M., Utikal T., Giessen H., Lippitz M. 'Quantitative modeling of the third harmonic emission spectrum of plasmonic nanoantennas'. *Nano Letters*. 2012; **12**:3778–3782.
[16] Utikal T., Zentgraf T., Paul T., *et al.* 'Towards the origin of the nonlinear response in hybrid plasmonic systems'. *Physical Review Letters*. 2011; **106**:133901.
[17] Hanke T., Krauss G., Träutlein D., Wild B., Bratschitsch R., Leitenstorfer A. 'Efficient nonlinear light emission of single gold optical antennas driven by few-cycle near-infrared pulses'. *Physical Review Letters*. 2009; **103**:257404.
[18] Liu H., Li G. X., Li K. F., *et al.* 'Linear and nonlinear Fano resonance on two-dimensional magnetic metamaterials'. *Physical Review B*. 2011; **84**:235437.
[19] Chen S., Wong W., Pun Y., Cheah K., Li G. 'Surface plasmon-enhanced third harmonic generation from gold-polymer hybrid plasmonic crystal'. *Advanced Optical Materials*. 2013; **1**:522–526.
[20] Renger J., Quidant R., N. van Hulst, Novotny L. 'Surface-enhanced nonlinear four-wave mixing'. *Physical Review Letters*. 2010; **104**:046803.
[21] Genevet P., Tetienne J.-P., Gatzogiannis E., *et al.* 'Large enhancement of nonlinear optical phenomena by plasmonic nanocavity gratings'. *Nano Letters*. 2010; **10**:4880–4883.
[22] Rose A., Powell D. A., Shadrivov I. V., Smith D. R., Kivshar Y. S. 'Circular dichroism of four-wave mixing in nonlinear metamaterials'. *Physical Review B*. 2013; **88**:195148.

[23] Alon O. E., Averbukh V., Moiseyev N. 'Selection rules for the high harmonic generation spectra'. *Physical Review Letters*. 1998; **80**:3743–3746.
[24] Alon O. E., Moiseyev N. 'Balslev-Combes theorem within the framework of the finite-matrix approximation'. *Physical Review A*. 1992; **46**:3807–3811.
[25] Zdanska P., Averbukh V., Moiseyev N. 'High harmonic generation spectra of aligned benzene in circular polarized laser field'. *The Journal of Chemical Physics*. 2003; **118**:8726–8738.
[26] Bhagavantam S., Chandrasekhar P. 'Harmonic generation and selection rules in nonlinear optics'. *Proceedings of the Indian Academy of Sciences – Section A*. 1972; **76**:13–20.
[27] W. K., Bloembergen N. 'Third-harmonic generation in absorbing media of cubic or isotropic symmetry, *Physical Review B*. 1971; **4**:3437–3450.
[28] Konishi K., Higuchi T., Li J., Larsson J., Ishii S., Kuwata-Gonokami M. 'Polarization-controlled circular second-harmonic generation from metal hole arrays with threefold rotational symmetry'. *Physical Review Letters*. 2014; **112**:135502.
[29] Chen S., Li G., Zeuner F., *et al.* 'Symmetry selective third harmonic generation from plasmonic metacrystals'. *Physical Review Letters*. 2014; **113**:033901.
[30] Armstrong J. A., Bloembergen N., Ducuing J., Pershan P. S. 'Interactions between light waves in a nonlinear dielectric'. *Physical Review*. 1962; **127**:1918–1939.
[31] Fejer M. M., Magel G. A., Jundt D. H., Byer R. L. 'Quasi-phase-matched second harmonic generation: Tuning and tolerances'. *IEEE Journal of Quantum Electronics*. 1992; **28**:2631–2654.
[32] Imeshev G., Galvanauskas A., Harter D., Arbore M. A., Proctor M., Fejer M. M. 'Engineerable femtosecond pulse shaping by second-harmonic generation with Fourier synthetic quasi-phase-matching gratings'. *Optics Letters*. 1998; **23**:864–866.
[33] Zhang X., Lytle A. L., Popmintchev T., *et al.* 'Quasi-phase-matching and quantum-path control of high-harmonic generation using counterpropagating light'. *Nature Physics*. 2007; **3**:270–275.
[34] Ellenbogen T., Voloch-Bloch N., Ganany-Padowicz A., Arie A. 'Nonlinear generation and manipulation of airy beams'. *Nature Photonics*. 2009; **3**:395–398.
[35] Xu P., Zhu S. N. 'Quasi-phase-matching engineering of entangled photons'. *AIP Advances*. 2012; **2**:041401.
[36] Zhu S. N., Zhu Y. Y., Qin Y. Q., Wang H. F., Ge C. Z., Ming N. B. 'Harmonic generation in a Fibonacci optical superlattice of $LiTaO_3$'. *Physical Review Letters*. 1997; **78**:2752–2755.
[37] Segal N., Keren-Zur S., Hendler N., Ellenbogen T. 'Controlling light with metamaterial-based nonlinear photonic crystals'. *Nature Photonics*. 2015; **9**:180–184.

[38] Yu N., Genevet P., Kats M. A., *et al.* 'Light propagation with phase discontinuities: Generalized laws of reflection and refraction'. *Science.* 2011; **334**:333–337.

[39] Ni X., Emani N. K., Kildishev A., Boltasseva V. A., Shalaev V. M. 'Broadband light bending with plasmonic nanoantennas'. *Science.* 2012; **335**;427.

[40] Chen X., Huang L. L., Mühlenbernd H., *et al.* 'Dual-polarity plasmonic metalens for visible light'. *Nature Communications.* 2012; **3**:1198.

[41] Huang L. L., Chen X., Mühlenbernd H., *et al.* 'Three-dimensional optical holography using a plasmonic metasurface'. *Nature Communications.* 2013; **4**:2808.

[42] Li G. X., Kang M., Chen S. M., *et al.* 'Spin-enabled plasmonic metasurfaces for manipulating orbital angular momentum of light'. *Nano Letters.* 2013; **13**:4148–4151.

[43] Yin X. B., Ye Z. L., Rho J., Wang Y., Zhang X. 'Photonic spin Hall effect at metasurfaces'. *Science.* 2013; **339**:1405–1407.

[44] Li G., Chen S., Pholchai N., *et al.* 'Continuous control of the nonlinearity phase for harmonic generations', *Nature Materials.* 2015; **14**:607–612.

[45] Akahori H. 'Spectrum leveling by an iterative algorithm with a dummy area for synthesizing the kinoform'. *Applied Optics.* 1986; **25**:802–811.

[46] Georgiou A., Christmas J., Collings N., Moore J., Crossland W. A. 'Aspects of hologram calculation for video frames'. *Journal of Optics A: Pure and Applied Optics.* 2008; **10**:035302.

[47] Jia J., Wang Y., Liu J., *et al.* 'Reducing the memory usage for effective computer-generated hologram calculation using compressed look-up table in full-color holographic display', *Applied Optics.* 2013; **52**:1404–1412.

[48] Pan Y., Wang Y., Liu J., Li X., Jia J. 'Fast polygon-based method for calculating computer-generated holograms in three-dimensional display', *Applied Optics.*, 2013; **52**:A290-A299.

[49] Larouche S., Tsai Y. J., Tyler T., Jokerst N. M., Smith D. R. 'Infrared metamaterial phase holograms'. *Nature Materials.* 2012; **11**:450–4.

[50] Walter F., Li G., Meier C., Zhang S., Zentgraf T. 'Ultrathin nonlinear metasurface for optical image encoding'. *Nano Letters.* 2017, **17**:3171–3175.

[51] Ye W., Zeuner F., Li X., Reineke B., *et al.* 'Spin and wavelength multiplexed nonlinear metasurface holography'. *Nature Communications.* 2016; **7**:11930.

[52] Almeida E., Bitton O., Prior Y. 'Nonlinear metamaterials for holography'. *Nature Communications.* 2016; **7**:12533.

[53] Li G., Wu L., Li K. F., *et al.* 'Nonlinear metasurface for simultaneous control of spin and orbital angular momentum in second harmonic generation'. *Nano Letters.* 2017; **17**:7974–7979.

[54] Schlickriede C., Waterman N., Reineke B., Georgi P., Li G., Zhang S., Zentgraf T. 'Imaging through nonlinear metalens using second harmonic generation'. *Advanced Materials.* 2018; **30**:1703843.

[55] Hazen R. M., Sholl D. S. 'Chiral selection on inorganic crystalline surfaces'. *Nature Materials*. 2003; **2**:367.
[56] Ernst K.-H. 'Molecular chirality at surfaces'. *Physica Status Solidi B: Basic Solid State Physics*. 2012; **249**:2057–2088.
[57] Verbiest T., Kauranen M., Persoons A. 'Light-polarization-induced optical activity'. *Physical Review Letters*. 1999; **82**:3601–3604.
[58] Petralli-Mallow T., Wong T. M., Byers J. D., Yee H. I., Hicks J. M. 'Circular dichroism spectroscopy at interfaces: A surface second harmonic generation study'. *The Journal of Physical Chemistry*, 1993; **97**:1383–1388.
[59] Byers J. D., Yee H. I., Hicks J. M. 'A second harmonic generation analog of optical rotatory dispersion for the study of chiral monolayers'. *The Journal of Chemical Physics*, 1994; **101**:6233–6241.
[60] Maki J. J., Kauranen M., Persoons A. 'Surface second-harmonic generation from chiral materials'. *Physical Review B*. 1995; **51**:1425–1434.
[61] Verbiest T., Kauranen M., Van Rompaey Y., Persoons A. 'Optical activity of anisotropic achiral surfaces'. *Physical Review Letters*. 1996; **77**:1456–1459.
[62] Fischer P. 'Nonlinear optical spectroscopy of chiral molecules, in comprehensive chiroptical spectroscopy: Instrumentation, methodologies, and theoretical simulations'. Volume 1 (eds Berova N., Polavarapu P. L., Nakanishi K., Woody R. W.) John Wiley & Sons, Inc., Hoboken, New Jersey, USA (2012).
[63] Chen S., Zeuner F., Weismann M., *et al.* 'Giant nonlinear optical activity of achiral origin in planar metasurfaces with quadratic and cubic nonlinearities'. *Advanced Materials*. 2016; **28**:2992–2999.
[64] Kolkowski R., Petti L., Rippa M., Lafargue C., Zyss J. 'Octupolar plasmonic meta-molecules for nonlinear chiral watermarking at subwavelength scale'. *ACS Photonics*. 2015; **2**:899–906.
[65] Kuwata-Gonokami M., Saito N., Ino Y., *et al.* 'Giant optical activity in quasi-two-dimensional planar nanostructures'. *Physical Review Letters*. 2005; **95**:227401.

Chapter 9

Optical nanoantennas for enhanced THz emission

Andrei Gorodetsky[1,2], Sergey Lepeshov[1], Alex Krasnok[3] and Pavel Belov[1]

Terahertz (THz) technology is now standing at the lab doorstep to the real world applications such as spectroscopy, biological sensing, security imaging, detection of dangerous or illicit substances, and ultrafast data transfer. Photoconductive antennas (PCAs) are one of the most frequently used devices for THz-wave generation and detection in established THz measurement and spectroscopy setups owing to their compactness and operation at room temperatures. However, the performance of PCAs is limited by low absorption of pump laser radiation in the antenna gap, long lifetime of photoexcited carriers, and antenna's low thermal tolerance and instability. The recently proposed approach to overcome these obstacles relies on optical nanoantennas, i.e., resonant nanostructures that are capable to transform incident light into a strongly localized near-field within THz antenna gap. This chapter is intended to systematize the main results obtained in this promising area of hybrid optical-to-terahertz PCAs and photomixers. We summarize the main results on hybrid THz antennas, compare the approaches to their implementation, and offer further perspectives of their development.

9.1 Introduction

Many elementary excitations in semiconductors and dielectrics [1–5] such as rotational and vibrational resonances of complex, including biological molecules [6–8], lie in the terahertz (THz) range (0.1–10 THz) of the spectrum. As a result, THz spectroscopy finds a vast number of applications in areas ranging from the detection of dangerous or illicit substances [7,9,10], and biological sensing [11–13] to diagnosis and treatment of diseases in medicine [14–17]. Additionally, the THz band is of special importance for spectroscopy in astrophysics [18–20] because

[1]Department of Physics and Engineering, ITMO University, St. Petersburg, Russia
[2]School of Physics and Astronomy, University of Birmingham, Birmingham, UK
[3]Advanced Science Research Center, City University of New York, New York, USA

many astrophysical objects often emit in this spectral range. Moreover, the THz technology is promising for wireless communication systems [21,22] as a complementary technology to transmitters and receivers operating in RF and microwave ranges, because it can potentially increase the data transfer rate by hundreds of times [23–25]. Although, there are some obstacles on the way to the real-life application of THz technologies such as strong THz absorption by metals and water [26,27], a significant progress has already been made [28–30].

For the past few decades, the generation of coherent THz waves remained a very challenging task. However, research in light-matter interaction of short (compared with all relaxation times) optical pulses with semiconductors [31–34] and nonlinear materials stimulated the development of compact low-power THz radiation transmitters. Nowadays, the commonly used method for a low-power THz generation relies on optical down conversion effect which involves semiconductor structures with a large surface conductivity [5,35–37]. Schematically, this kind of THz antennas is depicted in Figure 9.1 (left). Here, two or more conductive electrodes spaced by a certain gap are deposited onto a semiconductor surface. The electrodes are biased by an external voltage. The resulting structure is referred to as *THz photoconductive antenna (PCA)* if pumped by femtosecond (fs) optical pulses, or *THz photomixer* in the case of continuous (CW) pump by two lasers with spectral detuning lying in the THz range. The gap being excited by an fs-laser gets the free carrier increasing for a short period of time, resulting in the THz pulse generation. The duration of emitted THz pulse and its spectrum are mainly determined by the carrier concentration decay rate. In the case of THz photomixer, the carrier concentration changes with the frequency difference between the two pump lasers causing the occurrence of CW THz signal.

The efficiency of such THz sources is strongly limited by absorption of the laser radiation in the PCA gap, which is usually rather small because of the high reflection coefficient caused by a high refractive index of the semiconductors at optical frequencies. Moreover, the conventional PCA's efficiency is limited by low drift velocities of the photoinduced carriers in semiconductor substrates [38–40] and material breakdown threshold [41].

It has been shown recently that so-called optical nanoantennas (NAs) [42–44] placed into the photoconductive gap (Figure 9.1, right) can significantly increase

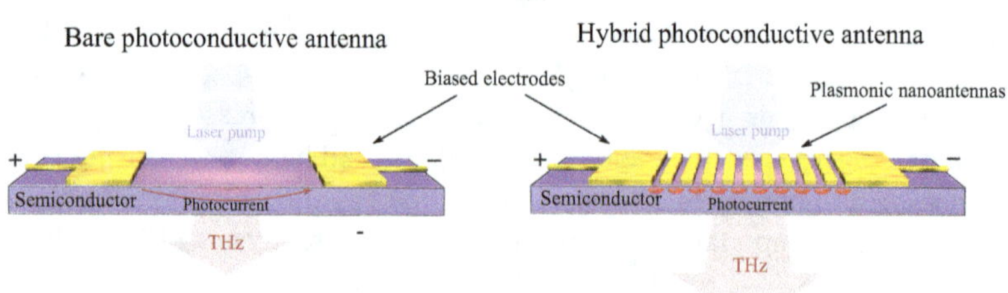

Figure 9.1 A standard and plasmonic-enhanced THz photoconductive antenna. Reproduced with permission. [37] Copyright 2017, Wiley-VCH.

the efficiency of the THz pulse generation. The resulting hybrid antenna comprises a photoconductive THz antenna and an array of optical NAs [37,45–72] and is referred to as a *hybrid THz-optical PCA*. The efficiency of the entire hybrid antenna that is determined by the efficiencies of both THz PCA and an optical NA array is usually at least tenfold higher than that of a conventional THz antenna. Besides the pump field enhancement, the plasmonic NAs also improve the antenna thermal stability due to the large thermal conductivity of metals.

This chapter summarizes the main results in the field of simulation and implementation of hybrid optical-to-terahertz PCAs and photomixers. We start with a brief introduction to the main operating principles of THz generation in PCAs and photomixers pumped with fs-optical pulses and two CW lasers. This description provides an overview of how plasmonics can help enhance the conversion efficiency of the antennas. We discuss the general operating principles of optical NAs and the prospects they offer for THz emission, and the types of antennas that have been used to date. Finally, we summarize and compare the main approaches with the hybrid antennas and give a short outlook of possible further studies.

9.2 Principles of THz photoconductive antennas and photomixers

9.2.1 Methods of coherent THz generation

There are several well-established approaches to THz generation. As THz frequency range is situated between microwaves and the infrared radiation, and there are no materials that provide direct coherent THz generation, the most developed methods include charged particles deceleration [73–75], quantum cascade lasers [76], electronic sub-THz sources [77], nonlinear difference frequency generation in crystals [78], waveguides [79], and by accelerated motion of short-living carriers in laser filaments [80] and semiconductors [31]. All methods possess their own advantages and downsides, but this discussion is beyond the scope of this chapter, and we advise readers to refer to well-presented reviews [1,4,5] and books [81,82] on the subject.

Apart from these well-established methods, novel approaches of optical-to-THz downconversion that employ an artificial electromagnetic media, such as metamaterials and metasurfaces [83–86], spintronic emitters [87], graphene [88], carbon nanotubes [89], and 3D topological insulators [90] have been recently proposed.

In this chapter, we focus on the CW and pulsed THz downconversion in PCAs, and the generation enhancement enabled by nanophotonics, namely plasmonic NAs and metasurfaces. PCAs are biased sources of THz radiation, thus, in principle, their efficiency can overcome the Manley–Rowe limit, as every electron-hole pair generated by optical pump photon can potentially produce more than one THz photon. To introduce the reader to the subject, we start from a brief introduction to CW and pulsed THz generation in PCAs.

9.2.2 Pulsed THz generation in photoconductive antennas

The simplest layout for submillimetre and THz waves generation in semiconductor-based PCAs is shown in Figure 9.2(a). A fs-laser (usually Ti:Sapphire [5,91] or much cheaper fiber-based or semiconductor laser systems such as second or fundamental harmonic of Er-doped fiber lasers) serves as a source of ultrashort light pulses with a duration ranging from several tens to hundreds of femtoseconds.

THz generation in a semiconductor surface due to the accelerated motion of nonequilibrium charge carriers induced by fs-laser pulses and driven by semiconductor surface field has a low optical-to-terahertz conversion efficiency. To overcome these limits, metal electrodes (usually a nickel-gold-platinum alloy) are deposited onto the substrate surface and are biased with an external voltage to create a high electric field in the gap between them, where the photocarriers are generated, and thus enhancing the generation of the THz waves [31,37]. Such structures together with a photoconductive semiconductor substrate are referred to as *THz PCAs*. The operation principles of such antenna are shown in Figure 9.2(a).

Ultrashort optical or IR pump pulses with photon energy above the bandgap excite the photocarriers in the semiconductor. The carriers are driven by the applied DC (or rectangular pulsed) bias, thus leading to the rise of the photocurrent. As the carriers recombine, the current drops depending on the lifetime of the semiconductor (usually <500 fs). As the far field THz signal is proportional to the time derivative of the current, a bipolar ultrabroadband THz pulse can be observed. The output power of THz generation in a semiconductor P_{THz} at the given frequency of Ω_{THz} is given by [37]

$$\boldsymbol{P}_{\mathbf{THz}} = \frac{\mathbf{1}}{\mathbf{1} + (\boldsymbol{\Omega}_{\mathbf{THz}}\boldsymbol{\tau}_{\mathbf{eff}})^2} \tag{9.1}$$

where τ_{eff} is the effective lifetime of photocarriers. From (9.1), it is clearly seen that the shorter the lifetime τ_{eff} is, the greater power P_{THz} can be obtained. Gallium arsenide grown by molecular beam epitaxy on a low-temperature substrate (LT-GaAs) is usually chosen as a photoconductive material, as this growth method allows obtaining a material with a larger number of defects and hence with shorter lifetimes of nonequilibrium charge carriers ($\leq 200\,fs$).

Silicon lens, shown in Figure 9.2(a), serves to collect and collimate the generated THz radiation [92] and at the same time as a heatsink to transfer heat excess away from the substrate. Moreover, it works as an anti-reflective element, since the refractive index of Si in THz range is equal to $n_{\mathrm{Si}} = 3.4$, which is about the same as one of high-resistive GaAs ($n_{\mathrm{GaAs}} \approx 3.6$) [93].

Thus, under an ultrashort laser pulse, electron-hole pairs are formed in the surface layer. The conductivity of the semiconductor as a consequence increases proportionally to the electrons and holes concentration: $\sigma \sim qn_e\mu_e + qn_p\mu_p$, where q is the elementary charge, n_e (n_p) is the concentration of electrons (holes), and μ_e (μ_p) is their mobility. The increase of conductivity results in a giant current pulse of photoinduced charges, electrons, and holes, being accelerated in opposite directions by the external electric field in the PCA gap. The acceleration of charge carriers by

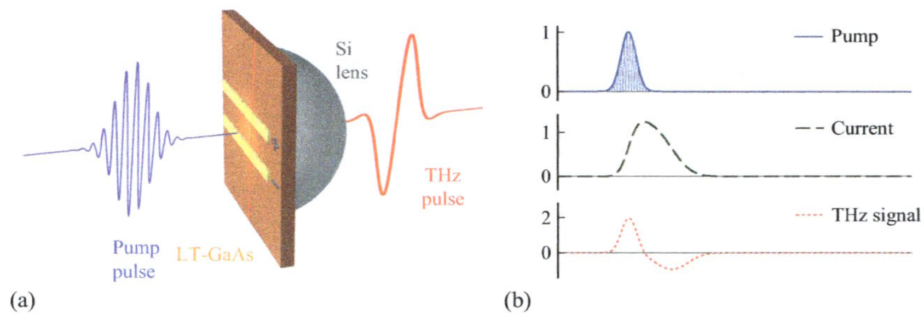

Figure 9.2 (a) Schematic view of a setup used to generate pulsed THz radiation by photoconductive LT-GaAs based antenna and (b) time evolution of pump radiation envelope, antenna photocurrent and generated THz signal. Reproduced with permission. [37] Copyright 2017, Wiley-VCH.

the external field E_{bias}, in this case, can be described by the differential equation:

$$\frac{dv_{\text{e,p}}}{dt} = -\frac{v_{\text{e,p}}}{\tau_{\text{eff}}} + \frac{qE}{m_{\text{e,p}}}, \tag{9.2}$$

where $v_{\text{e,p}}$ is the drift velocity of electrons and holes, $m_{\text{e,p}}$ are their masses, τ_{eff} is approximately 30 fs in low-temperature GaAs, and E is the local field. The latter is related to E_{bias} by $E = E_{\text{bias}} - \frac{P}{\alpha\varepsilon_0}$, where α is the static dielectric susceptibility of semiconductor (equal to 3 for the LT-GaAs), ε_0 is the vacuum permittivity, and P is the polarization caused by the separation of electrons and holes. This ratio allows the estimation of the LT-GaAs bias threshold to be 300 $kV \cdot cm^{-1}$. The temporal evolution of the polarization is presented by the following inhomogeneous differential equation:

$$\frac{dP}{dt} + \frac{P}{\tau_{\text{rec}}} = J(t), \tag{9.3}$$

where τ_{rec} is the recombination time of holes and electrons, and $J(t)$ is the current density in the surface layer of the semiconductor, which under DC bias is proportional to the surface conductivity: $J(t) \propto \sigma(t) \sim qn_{\text{e}}\mu_{\text{e}} + qn_{\text{p}}\mu_{\text{p}}$. From (9.2) and (9.3), using Maxwell's equations, one can derive a formula for the radiated THz electric field of the radiated THz at a distance z much larger than THz wavelength λ (i.e., far field) from the source [37]:

$$E(z,t) = -\frac{A}{4\pi\varepsilon_0 c^2 z} \times \frac{dJ}{dt} = -\frac{A}{4\pi\varepsilon_0 c^2 z} \times \left(qv\frac{dn}{dt} + qn\frac{dv}{dt}\right), \tag{9.4}$$

where n is the total carrier concentration, v is their velocity, A is the area of the irradiated surface (the area of the gap between the electrodes), and c is the speed of light in vacuum. The external field increases the energy of charge carriers, and hence the output power of THz radiation, not influencing, however, the spectral characteristics, as they are determined by the lifetime of electrons and holes. Pump pulse envelope, photocurrent in the surface layer, and resulting THz field examples obtained with this model are shown in Figure 9.2(b).

Some researchers have observed that various pump spot size on the semiconductor in the THz PCA gap can affect the output power [94]. Moreover, there are several options for photoconductor pump: uniform illumination of the antenna gap, nonuniform pump, and partial asymmetric illumination (Figure 9.3(a)). In the case of uniform illumination, the photocurrent is proportional to the optical pump power. In the case of asymmetric illumination, the photocurrent and hence the THz signal depend on the pump position, leading to significantly higher signal when the pump is closer to the anode [95]. The increased electric field near the anode, and, consequently, increased electrons concentration in this region (Figure 9.3(b)) leads to an increase of the photocurrent, resulting in an enhanced THz signal power. Asymmetric illumination of the PCA gap can give more than fourfold enhancement compared to the uniform illumination [96].

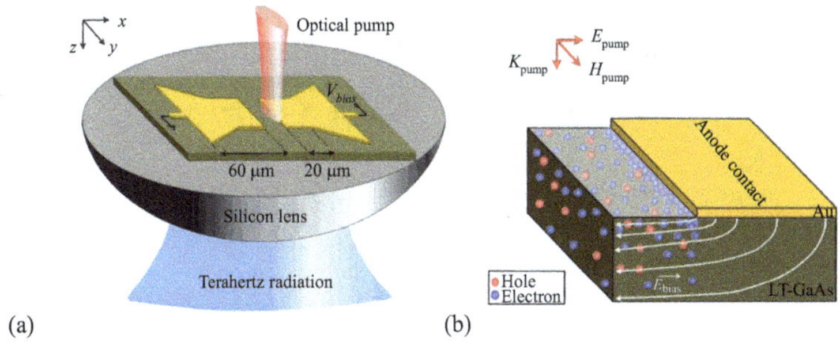

Figure 9.3 (a) Photoconductive antenna consisting of a GaAs substrate and electrodes arranged in the bowtie geometry. (b) Photoinduced electrons distribution of near the anode when the bias voltage is applied to the electrodes [56]. Reproduced with permission. [37] Copyright 2017, Wiley-VCH.

Thus, emitted THz power depends on three main parameters of the setup: (i) absorbed optical power in the surface layer, (ii) applied bias value in the excited semiconductor volume, and (iii) high dark resistance and the ratio of carrier mobility to carrier lifetimes that is responsible for the derivative of the photocurrent. Some of these values can be addressed by plasmonic optical NAs, as we will show below.

9.2.3 Pulsed THz detection in photoconductive antennas

THz detection uses the same approach as in generation. As optical pump pulse arrives to the PCA, electron-hole pairs are formed in the surface layer. The conductivity of the semiconductor as a consequence increases proportionally to the electrons and holes concentration, THz pulse, arriving simultaneously, drives the carriers and creates a photocurrent between antenna electrodes. The photocurrent can be measured directly, and it is proportional to the THz field at the pulse overlap timestamp.

By varying the time delay between THz and probe IR optical pulse, the whole pulse profile in time domain can be registered. The measured photocurrent is actually a convolution between the photoconductivity of the PCA detector substrate and the arriving THz pulse. Usually, the THz pulse is significantly longer than the photocurrent, and deconvolution is not performed. Thus, the measured photocurrent is considered to represent the arriving THz pulse waveform. For better signal-to-noise ratio, THz signal is usually modulated by the pulsed bias applied to the antenna to allow lock-in detection. As the design of the THz PCA detector is nearly identical to the emitter, optical NAs can potentially enhance the PCA detector operation in a similar way as they do for transmitters.

9.2.4 Continuous wave THz generation in photomixers

THz heterodyne generation process in semiconductors is similar to that discussed in the preceding section, with the only difference in the pump laser regime. Under

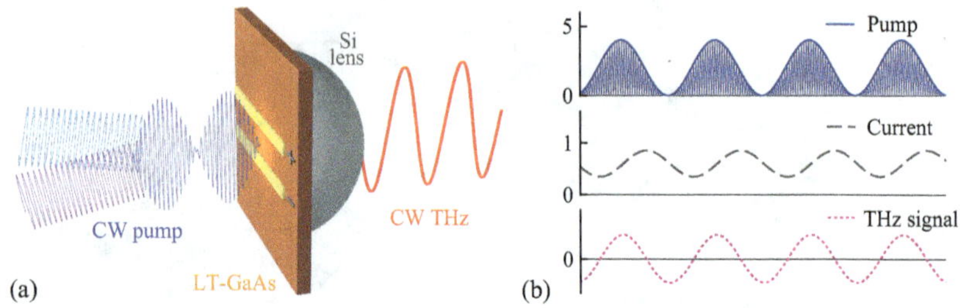

Figure 9.4 *(a) Schematic illustration of CW THz generation by a LT-GaAs-based photomixer and (b) time evolution of pump radiation envelope, antenna photocurrent, and generated THz signal. Reproduced with permission. [37] Copyright 2017, Wiley-VCH.*

continuous double-wavelength pump with a frequency difference lying in the THz region, electron-hole pairs are formed in the surface layer and the conductivity of the semiconductor is modulated proportionally to the electrons and holes concentration. Such devices for photoconductive CW THz generation are usually called *photomixers* [97]. The schematic layout of CW THz generation in a photomixer and its principle of operation are shown in Figure 9.4(a).

The surface conductivity and charge carrier acceleration in a photomixer are similar to those in a PCA, and the number of carrier pairs similarly follows the pump evolution:

$$\frac{dn(t)}{dt} = \frac{\eta P_{\mathrm{pump}}(t)}{h\nu} - \frac{n}{\tau_{\mathrm{rec}}}, \tag{9.5}$$

where η is the external quantum efficiency, $P_{\mathrm{pump}}(t)$ is the pump intensity, τ_{rec} is the carrier recombination time, and $h\nu$ is the pump photon. The pump radiation in the case of photomixing comprises two laser beams bringing to the surface average intensities P_1 and P_2 at the frequencies ω_1 and ω_2, respectively. Thus, the intensity at the surface can be described as:

$$P_{\mathrm{pump}}(t) = P_1 + P_2 + 2\sqrt{P_1P_2}(\cos(\omega_1 - \omega_2)t + \cos(\omega_1 + \omega_2)t). \tag{9.6}$$

The second oscillating term in the sum changes within a timescale that is significantly shorter than the carrier effective recombination time τ_{rec}, hence not affecting the photocurrent. Neglecting this term, after substitution of (9.6) into (9.5), one can obtain the evolution of carrier density [97]:

$$n(t) = \frac{\eta\tau_{\mathrm{rec}}}{h\nu}\left(P_1 + P_2 + \frac{2\sqrt{P_1P_2}\sin(\omega t + \phi)}{\sqrt{1 + \omega^2\tau_{rec}{}^2}}\right), \tag{9.7}$$

where $\omega = \omega_1 - \omega_2$ is the THz frequency, and $\phi = \tan^{-1}\left(\frac{1}{\omega\tau_{\mathrm{rec}}}\right)$ is the phase delay due to the nonzero carrier lifetime. Since the applied field is constant, for the photocurrent we obtain [82]:

$$J(t) = I_{DC} + \frac{I_{AC}}{\sqrt{1+\omega^2 {\tau_{rec}}^2}} \sin(\omega t + \phi), \tag{9.8}$$

where $I_{DC} = U_{bias}\mu_e \tau_{rec}(P_1 + P_2)$ is the DC photocurrent and $I_{AC} = 2U_{bias}\mu_e\tau_{rec}\sqrt{P_1 P_2}$ is the AC component of the current, which is responsible for the generation of the CW THz signal:

$$E_{THz}(t) \propto \frac{dJ}{dt} = \frac{\omega I_{AC}}{\sqrt{1+\omega^2 {\tau_{rec}}^2}} \cos(\omega t + \phi). \tag{9.9}$$

It can be seen that for the maximum conversion efficiency, when the photomixer is pumped near to its optical breakdown limit, the powers of both wavelength modes should be as close to each other as possible. The pump intensity profile, the corresponding photocurrent, and far-field THz signal of the photomixer calculated with the described model are shown in Figure 9.4(b).

Just like PCA, three main parameters of the setup define the efficiency of the photomixer: (i) absorbed optical power in the surface layer, (ii) applied bias value in the excited semiconductor volume, and (iii) the ratio of carrier mobility to carrier lifetimes that is responsible for the derivative of the photocurrent. Again, most of these values can be addressed by plasmonic optical NAs, as we will show below.

9.2.5 Effect of the contact electrodes shape on the radiative characteristics of THz photoconductive antennas and photomixers

One of the most important characteristics of all antennas, including radio, microwave, and THz, is the amount of emitted power. However, there is a significant difference between the radio or microwave antenna, and THz PCA—the nature of the input signal. In the case of THz PCA, it is the laser radiation that generates the nonequilibrium electrons and holes in the surface layer of the photoconductor. Therefore, the THz PCA is usually considered as a system consisting of two elements: the photoconductor and the antenna. These elements, in turn, are described by the traditional physical quantities, such as active and reactive conductivity, capacitance, and input impedance [96,98].

The output power $P_{THz}(\omega)$ of THz PCA can be expressed in terms of antenna resistance and reactance [96]:

$$P_{THz}(\omega) = \frac{1}{2}\left|S_p(\omega)P_{opt}(\omega)\right|^2 \xi \frac{G_A}{(G_A + G_P)^2 + (\omega C + B_A)^2}, \tag{9.10}$$

where $S_p(\omega)$ is the current susceptibility of the photoconductor, $P_{opt}(\omega)$ is the optical envelope pump power, i.e., single frequency in the case of CW pump, or a continuum broadband spectrum of femtosecond pulse pump, ξ is the antenna efficiency (which describes the dissipative losses in the antenna material) G_A and B_A are, respectively, the active and reactive conductances of the antenna, G_P is the conductivity of the photoconductor, and C is its capacitance. Regardless of the THz photoconductive antenna geometry, in order to maximize its radiated power, its stray conductance must be much less than its active conductance $\omega C + B_{\mathrm{A}} \ll G_{\mathrm{A}} + G_{\mathrm{P}}$. Full or partial compensation of reactive conductance can be achieved by maintaining the inductive load of the antenna at the same level as the capacitive load of the photoconductor, or by adding a negative reactive conductance shunt into the antenna [99].

The geometry of PCA electrodes affects both the spectrum width and the frequency of the maximum THz radiation intensity. Usually, PCA are fabricated in the shape of dipole, logarithmic, spiral, and bowtie antennas [37]. The consideration of the real antenna geometry can be done numerically or experimentally. In [66], a comparative analysis of two common THz antenna geometries, spiral, and bowtie, is performed.

The dependence of radiation resistance and reactance of these antennas on the frequency in Figure 9.5(a) demonstrates the broadband radiative nature of spiral PCA electrodes, and highly resonant behavior of bowtie PCA. The dependence of the radiated THz power on the bias applied to electrodes in Figure 9.5(b) shows the superiority of spiral antenna over bowtie, having a larger radiation resistance and a smaller reactance within a broadband frequency range of THz pulsed signal. Thus, the maximum active resistance and the minimum absolute value of the reactive resistance are prerequisite for efficient THz radiation generation and can be adjusted by choosing of the antenna geometry.

Since active and reactive resistances affect the power emitted by a THz antenna, the spectral characteristics of the antenna impedance and reactance determine the spectral characteristics of the THz signal. It has been observed that the antenna with radiation resistance slightly depending on frequency in the THz range has a wider frequency range of an effective operation [66,98]. The spiral and bowtie geometries discussed above are examples of such antennas. Figure 9.6

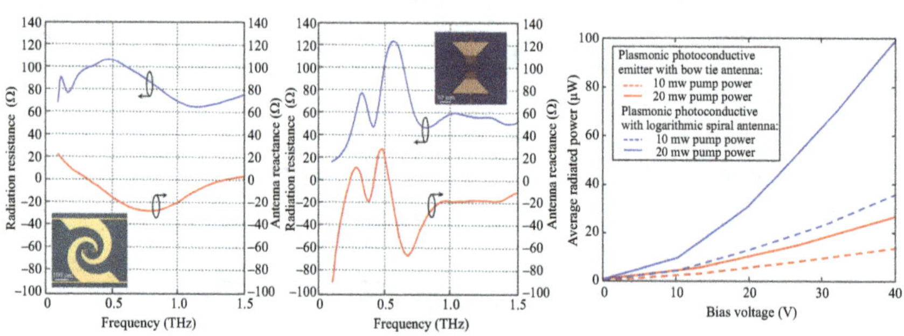

(a) (b)

Figure 9.5 (a) Frequency dependence of impedance (blue) and reactance (red) of two different THz antennas, spiral, and bowtie, respectively. (b) Dependencies of the radiated power of the spiral (blue) and bowtie (red) antennas on the bias voltage between electrodes for different powers of fs-pump [66]. Reproduced with permission. [37] Copyright 2017, Wiley-VCH.

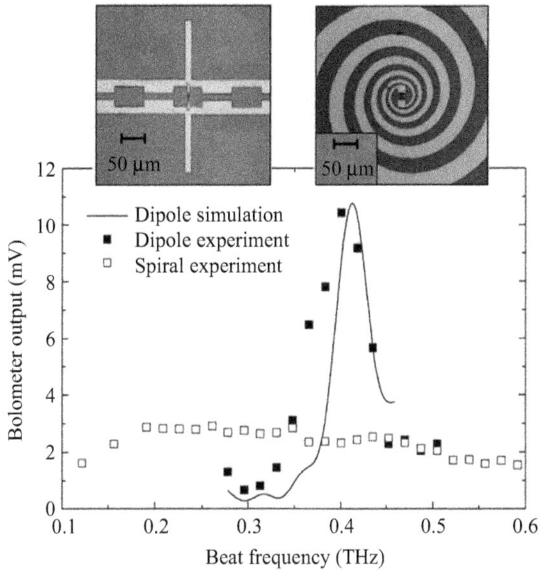

Figure 9.6 Frequency dependence of the output bolometer signal for the spiral and full-wavelength dipole THz photomixers. Dots represent experimental data, solid lines = numerical calculations [98]. Reproduced with permission. [37] Copyright 2017, Wiley-VCH.

shows the spectrum of the THz signal from the spiral photomixer antenna (white squares) measured by a bolometer. At frequencies between 0.1 and 0.6 THz the signal power is almost constant. On the other hand, antennas with a pronounced resonant impedance, such as full-wavelength dipole antennas (black squares), have narrow-band emission spectrum [99,100].

As plasmonic metal elements possess very high conductance, which can potentially affect both resistance and reactance, their effect must be accounted when designing a plasmon-enhanced PCA [57].

9.2.6 How plasmonic optical nanoantennas can enhance THz generation and detection

Optical NAs possess all the inherent characteristics of their radio and THz counterparts, except of course that they operate with fields in the optical spectral range. The dimensions of NAs are typically of the order of several hundreds of nanometers. NAs have been utilized for boostings photodiodes, LEDs, and photovoltaics [44,101,102]. Recently, a number of papers have demonstrated the enhancement of the optical-to-terahertz conversion efficiency in THz PCAs and photomixers with plasmonic NAs embedded within the gap [37,45–61,65–71]. Here, we consider the benefits of the optical NAs from the perspective of application for hybrid THz antennas.

To understand the advantage of optical NAs for THz PCAs, let us recall that the radiated THz field amplitude is proportional to photocurrent density [see (9.4) and (9.9)]. The photocurrent depends on the density of carriers that is a function of time, carrier mobility, and the bias field. Hence, it is needed to maximize the photocurrent density to enhance the THz field generated by the PCA. However, the bias field and charge density are limited by the material breakdown, charge screening, and thermal damage factors. Recently, it has been demonstrated that these factors can be considerably improved with the application of optical NAs. The benefits of the optical NAs application include a strong electric field enhancement within semiconductor substrate, reducing the carrier screening effect, and thermal stability enhancement, as we discuss below.

9.2.6.1 Strong electric field enhancement

An optical NA, as a part of a hybrid THz-optical PCA or photomixer, converts the pump optical radiation into a strong near-field distribution and redistributes the energy within the semiconductor layer. The enhancement of an electric field is described by the local field enhancement factor $\delta = \frac{|E|}{|E_0|}$, the ratio of the moduli of the electric field $|E|$ increased locally (i.e., at a certain point) in the presence of a NA compared to its value $|E_0|$ in the absence of the NA.

The local electric field enhancement occurs due to the plasmon resonance of the NA [101,102]. At the same time, the plasmon resonance excited in the metal NA asymmetrically scatters the incident wave, and the field is redistributed in the semiconductor substrate so that the local electric field enhancement factor near the electrodes becomes much greater than the electric field of the incident wave. Therefore, the optical power absorbed by the semiconductor P_{opt} in the volume V:

$$P_{opt} = \frac{1}{2}\int \sigma\delta^2 |E_0|^2 dV \tag{9.11}$$

increases proportionally to δ^2. Here, the value σ is a medium conductivity. As a result, the concentration of nonequilibrium charge carriers near the electrodes $n_{\mathrm{e,p}}$ increases, and, hence the efficiency of optical-to-terahertz conversion increases too, according to (9.4) and (9.9). The physical dimensions and material of the NAs play an important role in the excitation of the plasmon resonance and the electric field enhancement in the substrate.

Moreover, as the applied to the PCA DC bias creates the electric field in the nonmetal part of the PCA gap, the electric field of the plasmonic NA enhanced hybrid PCA is dependent on the filling ratio of the NAs.

9.2.6.2 Reducing of the carrier screening effect

The performance of conventional photoconductive THz antennas is strongly limited by low drift velocities of the photoinduced charge carriers in semiconductor substrates [103,104]. For an efficient optical-to-terahertz conversion in the frequency range above ~0.5 THz, the carriers relaxation time τ_{rec}, as discussed earlier in Sections 9.2.2 and 9.2.4 must be subpicosecond. The mean free path at these times, $l = v_{\mathrm{e,p}}\tau_{\mathrm{rec}}$, is about 100 nm. Under higher pumps, the concentration of nonequilibrium electrons and holes becomes large enough to have an impact on the photocurrent. The mean free path is reduced mainly due to the screening of the applied external field by the intrinsic field. Accordingly, the overall signal drops, while signal at higher frequencies is slightly enhanced because of the mean free path shortening. This effect is called *carrier screening effect*. The diffraction limit for electromagnetic waves does not allow focusing the laser pump into the spot of 100 nm in diameter. Therefore, in order to reduce the effect of screening and to improve the properties of PCAs, optical NAs can be employed [50].

9.2.6.3 Thermal stability enhancement

The excess of heat in conventional THz antennas leads to the thermal generation of carriers after optical pulse excitation [37]. Recombination of thermal and photo-induced carriers in subsequent cycles of charge carrier generation prevents the stable operation of the PCA. Recently, it has been shown that the optical NAs embedded in a THz PCA can improve the thermal stability of the latter due to the large thermal conductivity of metals [52]. For example, gold has the thermal conductivity of 3.14 $W \cdot cm^{-1}K^{-1}$ that is six times larger than the thermal conductivity of GaAs (0.55 $W \cdot cm^{-1}K^{-1}$). Such higher thermal efficiency also makes the optical NAs very attractive for use in THz sources.

Thus, the optical NAs being embedded in a THz PCA can maximize the transfer of light energy into the absorptive semiconductor substrate and minimize the path length of charge carriers traveling between metallic electrodes, simultaneously driving them with higher acceleration. As a result, the optical NAs enhance the performance of the photoconductive THz antennas. Below, we show the approach to the design of optical NAs for THz enhancement, review the most recent works on hybrid THz PCAs, and organize them according to the approach to optical NA realization.

9.3 Design of optical plasmonic nanoantennas

Now, we focus on the analytical methods of plasmonic NA modeling. Let's consider a spherical metallic nanoparticle in the nonmagnetic homogeneous environment irradiated by a plane wave. The scattered electric field of the particle can be found by solving Maxwell's equations in the spherical coordinates in the form of Mie expansion [105]:

$$\boldsymbol{E}(\boldsymbol{r},\theta,\varphi) = \sum_{l=1}^{\infty} a_l \boldsymbol{N}_{l1}(\boldsymbol{r},\theta,\varphi) + b_l \boldsymbol{M}_{l1}(\boldsymbol{r},\theta,\varphi), \tag{9.12}$$

where $\boldsymbol{N}_{l1}$ and $\boldsymbol{M}_{l1}$ are the vector spherical functions, a_l and b_l are the expansion coefficients corresponding to the electric (a) and magnetic (b) spherical multipoles given by:

$$a_l = \frac{R_l^{(a)}}{R_l^{(a)} + iT_l^{(a)}}, \qquad b_l = \frac{R_l^{(b)}}{R_l^{(b)} + iT_l^{(b)}}. \tag{9.13}$$

Functions R_l and T_l can be rewritten in the following form:

$$\begin{aligned} R_l^{(a)} &= n\psi_l'(kR)\psi_l(nkR) - \psi_l(kR)\psi_l'(nkR), \\ T_l^{(a)} &= n\chi_l'(kR)\psi_l(nkR) - \chi_l(kR)\psi_l'(nkR), \end{aligned} \tag{9.14}$$

$$\begin{aligned} R_l^{(b)} &= n\psi_l'(nkR)\psi_l(kR) - \psi_l(nkR)\psi_l'(kR), \\ T_l^{(b)} &= n\chi_l(kR)\psi_l'(nkR) - \chi_l'(kR)\psi_l(nkR). \end{aligned} \tag{9.15}$$

Here, $\psi_l(x) = \sqrt{\frac{\pi x}{2}} J_{l+1/2}(x)$, $\chi_l(x) = \sqrt{\frac{\pi x}{2}} N_{l+1/2}(x)$, $J_{l+1/2}(x)$ and $N_{l+1/2}(x)$ are the spherical Bessel functions of the first and third kind, respectively. By calculating the expansion coefficients one can obtain scattering (Q_{scs}), extinction (Q_{ext}), and absorption (Q_{abs}) cross sections:

$$\begin{aligned} Q_{scs} &= \frac{2}{(kR)^2}\sum_{l=1}^{\infty}(2l+1)(|a_l|^2 + |b_l|^2), \\ Q_{ext} &= \frac{2}{(kR)^2}\sum_{l=1}^{\infty}(2l+1)\mathrm{Re}(a_l + b_l), \\ Q_{abs} &= Q_{ext} - Q_{scs}, \end{aligned} \tag{9.16}$$

where, R is a nanoparticle radius and k is a wavenumber. Figure 9.7(a) shows the scattering cross section spectrum calculated for a silver spherical particle with the radius of 145 nm and partial contributions of the different multipoles in the total scattering. It is clearly seen that the dominant multipoles are electric dipole (a_1) and electric quadrupole (a_2). This follows from the plasmonic nature of the modes exciting in the metal nanoparticle. The electric field distributions demonstrate the strong enhancement of electric field at the surface of plasmonic nanoparticle with the enhancement factor of 4 (Figure 9.7(b)). This increases the absorption in the hot spots of the localized surface plasmon (LSP) by 16 times.

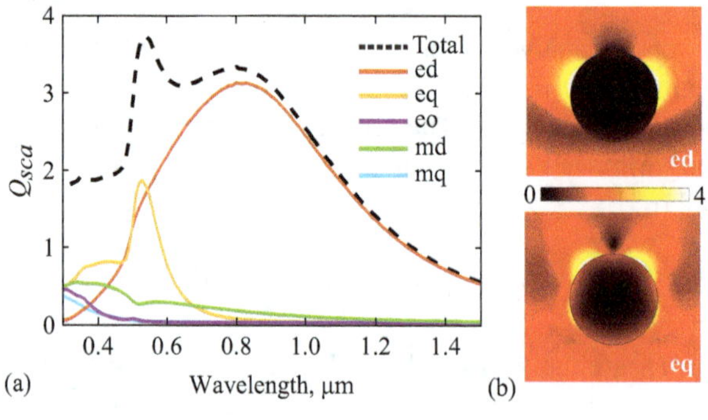

Figure 9.7 (a) Normalized scattering cross section spectrum of a spherical gold nanoparticle and different multipole contributions to the scattering. (b) Distributions of the electric field amplitude normalized by the plane wave amplitude at electric dipole (ed, 0.8 μm) and electric quadrupole (eq, 0.57 μm).

In the case of small nanoparticles (radius of the particle much smaller than the wavelength), one can apply the quasi-static approximation (QSA) to model their electromagnetic properties [106]. The scattering and absorption cross sections in QSA are derived as:

$$\begin{aligned} Q_{sca} &= \frac{8}{3}(kR)^4 \left| \frac{\varepsilon - \varepsilon_h}{\varepsilon + 2\varepsilon_h} \right|^2 \\ Q_{abs} &= 4kR\text{Im}\left(\frac{\varepsilon - \varepsilon_h}{\varepsilon + 2\varepsilon_h} \right) \end{aligned} \tag{9.17}$$

where ε is nanoparticle permittivity and ε_h is permittivity of the environment. For the air environment, we can rewrite these expressions as:

$$\begin{aligned} Q_{sca} &= \frac{8}{3}(kR)^4 \left| \frac{\varepsilon - 1}{\varepsilon + 2} \right|^2 \\ Q_{abs} &= 4kR\text{Im}\left(\frac{\varepsilon - 1}{\varepsilon + 2} \right) \end{aligned} \tag{9.18}$$

The amplitude of field close to the sphere surface in the air is defined by field enhancement factor [106]:

$$\delta = \left| \frac{3}{\varepsilon + 2} \right|. \tag{9.19}$$

This immediately gives us the condition for the resonant field enhancement:

$$Re(\varepsilon) + 2 = 0. \tag{9.20}$$

The condition can be satisfied only in the spectral region near the plasma frequency of a metal [107]. For the noble metals such as gold and silver, the wavelengths of the resonant field enhancement are 510 and 355 nm, respectively, that is far from the fs-laser pumping of PCAs (785 nm). Moreover, the enhancement is strongly restricted by the dissipation losses of the metal [108]. In resonant regime, (9.19) is given by:

$$\delta = \frac{3}{Im(\varepsilon)}. \tag{9.21}$$

These limitations can be partially overcome by using of ellipsoidal nanoparticles. The scattering and absorption cross sections for ellipsoidal nanoparticles with semi-axes R_x, R_y, and R_z and electric field polarized along x direction in QSA are defined by:

$$\begin{aligned} Q_{sca} &= \frac{2k^4 R_y^2 R_z^2}{3\pi\varepsilon_h} \left| \frac{\varepsilon - \varepsilon_h}{3\varepsilon_h + 3L_x(\varepsilon - \varepsilon_h)} \right|^2 \\ Q_{abs} &= \frac{4kR_y R_z}{R_x \varepsilon_h} \text{Im}\left(\frac{\varepsilon - \varepsilon_h}{3\varepsilon_h + 3L_x(\varepsilon - \varepsilon_h)} \right) \end{aligned} \tag{9.22}$$

where L_x is a geometrical factor equal to 1/3 for spherical particles and less than 1/3 for ellipsoidal particles. By changing L_x, one can tune the resonance wavelength even to the near-infrared spectral range. For the wavelength of 785 nm, this L_x reaches 0.01 that corresponds to a prolate ellipsoid.

The discussed analytical methods allow the precise and quick modeling of simple plasmonic NAs. However, these methods are not applicable for calculations involving complex geometries of NAs such as triangles, rectangles, and hexagons [109]. Also, in most analytical methods taking a substrate into account is problematic, while it is crucial for PCA modeling, especially when high index semiconductor substrates are used. For these reasons in the overwhelming majority of cases, the numerical methods and packages (CST Microwave studio, COMSOL, and Lumerical FDTD) are used to calculate and optimize the plasmonic NA characteristics, while analytical methods can be used for preliminary estimations.

As an example, we bring the results of numerical calculations of the electric field distribution at the air-GaAs interface with an Ag oblate semi-spheroid NA and without the NA from [57], which are presented in Figure 9.8(a). The results are obtained with the software package CST Microwave Studio. The structure is excited by normally propagated plane electromagnetic wave with the wavelength of 800 nm. The geometrical parameters of the NA correspond to the optimal ones for the excitation of plasmon resonance in such structure at the wavelength of 800 nm. The large semi-axis of the spheroid is $a = 170$ nm, minor semi-axis of the spheroid is $b = 100$ nm. The magnitude of the electric field under the NA is about three times larger versus the case of the NA absence (Figure 9.8(a)). The map of absorption enhancement at the wavelength of 800 nm as a function of the NA radius and distance between NA centers is presented in Figure 9.8(b). Here, the minor radius of the oblate spheroidal nanoparticle has been fixed to 55 nm. The

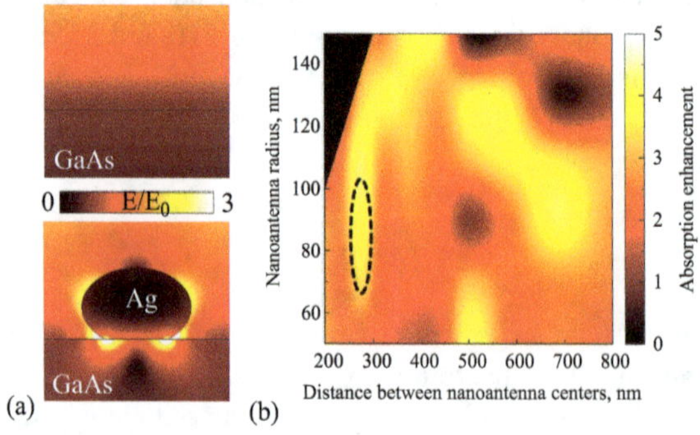

Figure 9.8 (a) Numerically calculated electric field distribution profiles at the interface between air and the GaAs substrate without and with a silver nanoantenna. (b) Dependence of the absorption enhancement of the electromagnetic wave in GaAs with nanoantenna array on the nanoantenna radius and distance between nanoantenna centers (a typo in the label of colorbar).

absorption enhancement has been calculated as the power absorbed in 100-nm GaAs surface film with silver NA array normalized to the power absorbed in the GaAs film without NAs. Therefore, the growth of the absorbed power and, thus, the carrier density in the 100-nm surface layer has the highest impact on average photocurrent which depends on the external electric field. It is clearly seen that there are two strong local maxima of the absorption enhancement in the simulation map accompanied by few "hot spots" in the map. The first maximum is in the narrow range of the distances near 280 nm and in the wide range of radii from 70 to 115 nm. The second absorption maximum is at the distances of 650–730 nm for the particles with radii ranging between 85 and 105 nm. Thus, from these numerical calculations one can expect at least fourfold enhancement of the THz PCA performance with such optimized plasmonic nanostructures.

9.4 Results of plasmonic nanoantennas implementation for THz generation enhancement

9.4.1 Interdigitated electrodes

The first approaches to THz generation enhancement included the use of plasmonic-like interdigitated electrodes in the PCAs. In the following works [110–115], even the first photomixers [97,116–118] had the interdigitated electrodes merely for DC electric field enhancement.

Such interdigitated dipole electrode NAs consist of two interdigitating combs. Such structures provide a stronger field localization in the gaps of the so-called "nano-fingers" and larger values of the absorbed power. The metal dipole NAs are mostly used in THz photomixers, enhancing an optical field of pump laser and redistributing it into the surface layer of the semiconductor substrate near the electrodes.

Figure 9.9(a) presents an interdigitated nanostructure embedded in the gap between the electrodes of the logarithmic photomixer. THz photomixers with such nanostructures feature lower parasitic capacitance and, consequently, lower reactance if compared with conventional THz photomixers, as well as higher optical field localization inside the working area of the antenna (see Figure 9.8(b)), similar structures of a larger size were used in pioneering interdigitated photomixers. Photomixers with nanostructured dipole antennas in the photoconductive gap are even more effective sources of CW THz radiation [98] if compared to microstructured ones. In some works [115,120], it has been shown that with the increasing number of dipoles (fingers), the optical-to-THz conversion efficiency increases (Figure 9.9(c)). This dependence can be explained by the smaller distance between the dipoles in THz photomixer with a greater number of fingers, hence increased the degree of the optical field localization in the gap and decreased the effective lifetime of photoinduced charge carriers.

9.4.2 Plasmon monopole nanoantennas

The simplest type of NAs applied for hybrid THz antennas is monopole one. The monopole NAs are able to provide a strong electric field enhancement of laser

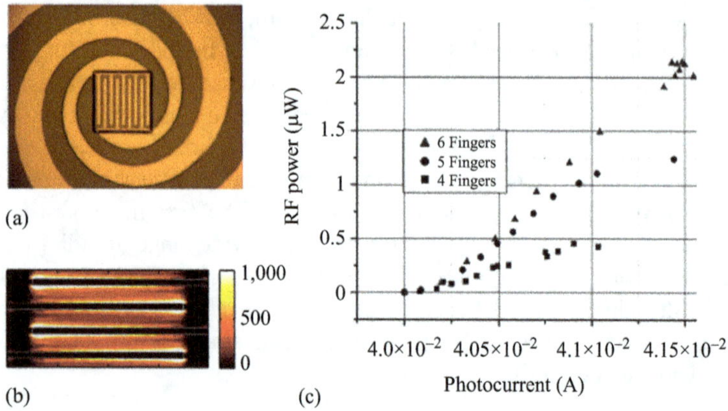

Figure 9.9 (a) Spiral THz photomixer with built-in finger NA. (b) Electric field distribution in finger nanostructure of THz photomixer. (c) Comparison of the radiated power in nanostructured THz photomixers with a different number of fingers [53,98,119,120]. Reproduced with permission. [37] Copyright 2017, Wiley-VCH.

radiation in their vicinity [44]. The optical properties of such NAs depend on their size, shape, and material. Thus, they can be easily tuned to any optical wavelength of operation. The arrays of the monopole NAs can be easily fabricated by means of chemical technics or thermal dewetting [121] even inside the gap of THz PCAs [57].

The authors of [49] propose silver plasmonic monopole NAs in the form of the so-called *nanoislands* (Figure 9.10(a)). The fabrication of such monopole NAs is done by thermal evaporation of silver (Ag), with subsequent deposition onto LT-GaAs antenna substrate surface. This method can be used to pattern nanostructures inside a large area, because of lower costs compared to, for example, electron beam lithography (EBL) used for the precise fabrication of metal nanostructures. Silver in the hybrid THz-optical antenna is used due to the high quality factor of the silver nanostructures in the visible range, which allows for achieving maximum localization of the field in the semiconductor surface layer with minimal dissipated losses of NAs. In addition, Ag has a relatively low melting temperature and therefore its thermal evaporation is possible without thermal damage to the semiconductor GaAs substrate [49]. The geometrical parameters of the monopoles are not ideally optimized so that the plasmon resonances are excited in the structure under an optical pump, but still provide a reasonable optical pump absorption enhancement. For a pump radiation of 800 nm wavelength, which is typical for Ti:Sapphire fs-lasers, nanoislands should have a thickness of 20 nm and a diameter of 173 nm. Later, this approach was adopted to obtain even higher generation enhancement [57].

Figure 9.10(b) compares the emission characteristics of a conventional THz PCA and of a hybrid optical-THz antenna containing nanoislands. It is clearly seen

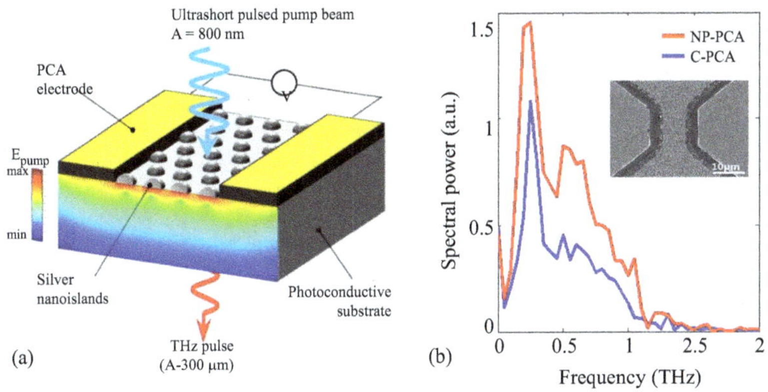

Figure 9.10 *(a) Schematic representation of the photoconductive THz antenna with a large area of the silver nanoislands (monopole NAs). (b) The power spectra of the THz radiation for the conventional photoconductive antenna (C-PCA) and hybrid photoconductive antenna (NP-PCA) THz photoconductive antennas. Inset: the SEM image of the silver nanoislands [49]. Reproduced with permission. [37] Copyright 2017, Wiley-VCH.*

that the hybrid structure has a 1.5–2 times higher conversion efficiency, particularly enhancing the medium-frequency part of the generated spectrum. This enhancement can be explained by (9.4), as the monopole NAs concentrate the light within the gaps of adjacent nanoparticles, so that the absorption of optical radiation in these areas increases, growing the concentration of nonequilibrium carriers $n_{e,p}$, which increases the amplitude of the output THz signal E_{THz}. However, this field redistribution leads to a decrease in its value in the vicinity of nanoislands and significantly reduces the density of electrons and holes in these areas. Nevertheless, this does not affect the total current of photoinduced charges, as silver monopoles are almost perfect conductors for an external electric field.

Similar fabrication method has been applied for engineering the hybrid PCA enhanced by monopole NAs shown in Figure 9.11(a). In this work [57], authors carry out a comprehensive analysis of the plasmonic modes excited in the array of silver oblate spheroid nanoparticles and optimize its parameters. The numerical calculations from this work are discussed in detail in Section 9.3 and presented in Figure 9.11(c). For the PCA geometry, the authors choose a log-periodic antenna characterized by a broadband radiation spectrum. Figure 9.8(b) demonstrates the spectral radiation amplitude of such a PCA with silver NAs (red) and without NAs (black). It is interesting that the amplitude amplification is uneven with respect to the frequency change. In the frequency range from 0.05 to 0.4 THz, one can observe a twofold increase of the radiation amplitude; while in the range from 0.7 to 1.4 THz, the enhancement reaches the maximum value of 6 (Figure 9.11(c)). The authors explain this dependence by the modification of the PCA impedance by the

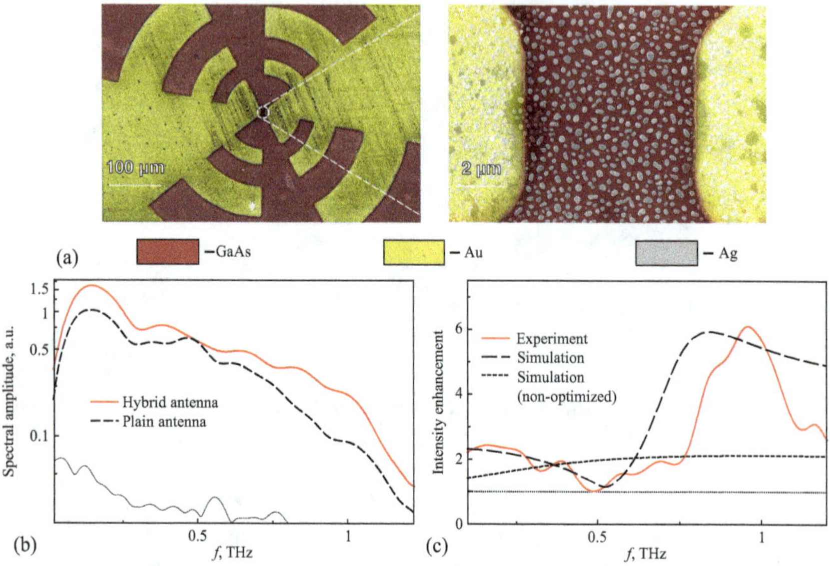

Figure 9.11 Typical SEM images of the fabricated log-periodic THz PCA and its central part (a); optimized design silver nanoantenna enhanced hybrid PCA performance and the plain PCA spectral amplitudes (b); and spectral intensity enhancement gained due to optimized silver nanoantennas in the gap of the PCA (red curve) (c), compared with simulated signal intensity enhancement from hybrid PCA (black dashed curve) and simulated enhancement, corresponding to the geometry from [49] (black dotted curve) [57].

NA array. According to (9.10), such a modification is caused by the change of the surface conductivity of the GaAs substrate and by the PCA capacitance increase due to the NA array at the surface of the substrate. Figure 9.11(c) contains spectral intensity enhancement obtained from the spectral amplitude ratio and calculated by using (9.10) for optimized NA array and nonoptimized array from [49].

Another possible realization of hybrid THz antennas is an array of rectangular monopole NAs arranged inside the photoconductive gap of PCAs and photomixers [60]. Figure 9.12(a) schematically shows a hybrid THz antenna with an array of rectangular nanostructures as well as a SEM image of these nanostructures. The nano-pattern consists of AuGe squares having a diagonal of 500 nm and a lattice period of 600 nm. This nano-pattern is incorporated in the THz antenna on the SI-GaAs region between the two electrodes. The mechanism of optical local field enhancement in the vicinity of these NAs is similar to the one in previously described, namely the plasmon resonance excitation. This structure has been realized utilizing the EBL. This nanostructured hybrid THz-optical PCA is able to enhance the emitted THz power of conventional PCAs up to the factor of 2 (Figure 9.12(b)).

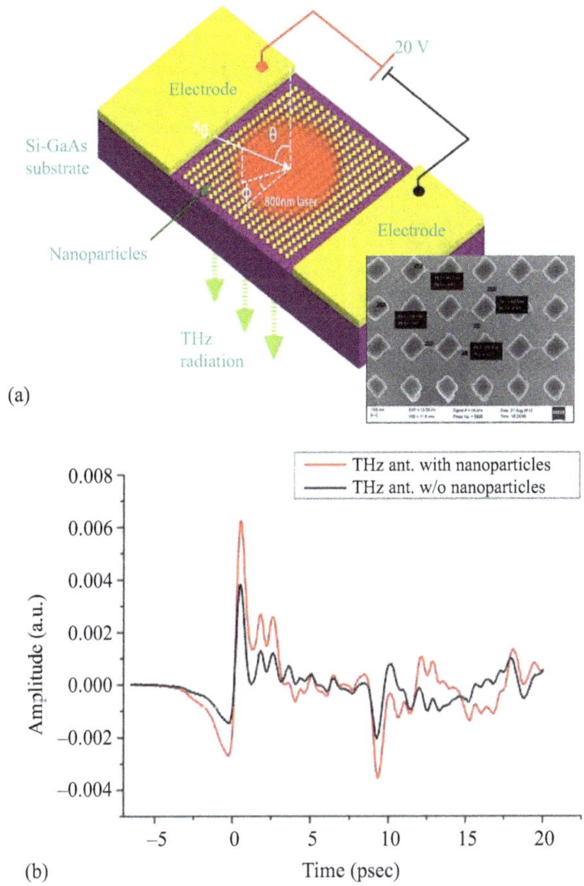

Figure 9.12 (a) Hybrid THz-optical photoconductive antenna with an array of monopole NAs. The array is fabricated on a substrate of semi-insulating GaAs (SI-GaAs). (b) Output signal of THz photoconductive antennas with (red curve) and without (black curve) monopole [60]. Reproduced with permission. [37] Copyright 2017, Wiley-VCH.

In [52], the hybrid THz antenna with hexagonal metal nanostructures has been proposed and realized. It has been shown that these antennas are superior to plasmonic nanorods-based ones (described below) in THz generation enhancement. Figure 9.13(a) shows an array of such monopole NAs fabricated inside the photoconductive gap of a dipole THz antenna using focused ion beam technology. It is known that the increase of pumping intensity increases the photoinduced current, which leads to heating of the semiconductor substrate and eventually to its thermal breakdown. However, this type of the PCA was shown to possess high resistance to thermal breakdown.

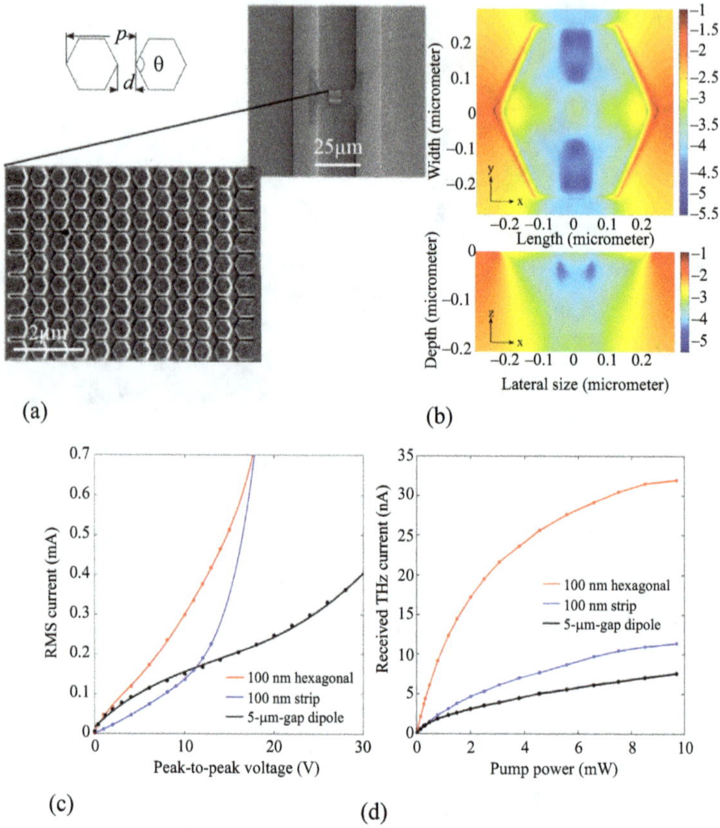

Figure 9.13 *(a) SEM image of periodic array (with period p) of hexagonal NAs arranged on a SI-GaAs substrate. (b) Electric field distribution profile near the hexagonal NA. (c) Peak-to-peak voltage dependence of photo-excited current in the hybrid THz antenna with hexagonal NAs (red curve), strip plasmonic gratings (blue curve), and the conventional dipole antenna (black curve). (d) Comparison of received THz currents from the hybrid THz antennas with hexagonal monopoles, strip plasmonic gratings, and conventional dipole antenna in depends on the pump optical power [52]. Reproduced with permission. [37] Copyright 2017, Wiley-VCH.*

Figure 9.13(c) demonstrates the advantage of the hexagonal NAs over grating NAs and conventional THz antennas. To achieve similar photoinduced currents, different antennas require the application of different biases between the electrodes. Moreover, maintaining a predetermined value of the current in the structure with hexagonal monopoles requires less external stress than in other antenna realizations. This suggests that conventional antennas and plasmonic gratings generate more heat than hexagonal NAs, and, accordingly, have a greater risk of experiencing thermal breakdown. Figure 9.13(d) shows the dependence of radiated THz power on the average power of the fs-laser pump. When comparing the grating (this type of the hybrid PCA will be considered later) and hexagonal NAs, the superior performance of the latter is clearly seen. This is due to a combination of the external field generated by the electrodes, and localization of photoinduced charges in the vicinity to the hexagons' tops (Figure 9.13(b)). Apart from the mentioned geometries, in a theoretical work, embedded plasmonic nanocylinders for THz generation enhancement have been proposed [58].

9.4.3 Plasmon dipole nanoantennas

Embedded dipole tip-to-tip NAs [62] represents another well-developed approach to the enhancement of PCAs and photomixers. Two most commonly used realizations of such nanostructures embedded in a THz photomixer are shown in Figure 9.14(a).

On one side, conventional rectangular dipole nanoparticles are shown and trapezoidal bowtie NAs on the other. Being a continuation of the metal electrodes, these tip-to-tip NAs are designed to reduce the path length of photoinduced electrons and holes. Moreover, due to the nanoscale gap between the dipoles, tip-to-tip NAs enhance the localization of the external optical field [53,61]. High localization of the field in the gap (see Figure 9.14(c), left) occurs because of the surface plasmon excitation due to plasmon resonance at the antenna excitation wavelength [122]. This increases a laser radiation absorption in the semiconductor substrate within the gap and, consequently, the concentration of charge carriers in the gap. The rapid increase in the charge carrier concentration proportional to the pump

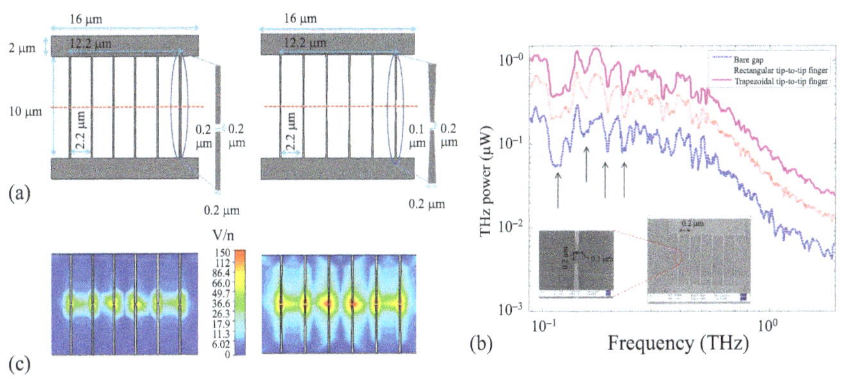

Figure 9.14 (a) Geometry and (c) electric field distribution profiles of rectangular (left) and trapezoidal (right) dipole NAs embedded in a THz photoconductive antenna. (b) The spectra of the THz radiation power for hybrid-optical photoconductive dipole antennas with built-in trapezoidal and rectangular NAs and conventional THz photoconductive antenna [62]. Reproduced with permission. [37] Copyright 2017, Wiley-VCH.

leads to a huge impulse of the photocurrent that is much greater than in conventional THz antennas. The amplitude of THz radiation is directly proportional to the rate of change of the photocurrent density, according to (9.9). So, THz dipole antenna with integrated optical NAs demonstrates higher output power of THz signal than conventional one at the same pump rate, only because of such nanoscale extensions of the electrodes (Figure 9.14(b)).

In [62], it has been noticed that due to lower capacitive reactive resistance, receiving ability of the bowtie antenna is much better than one of the rectangular-shaped NAs. As a result, the degree of the optical field localization in the trapezoidal NA gap is significantly higher (Figure 9.14(c), right). In addition, a smaller capacitance increases the efficiency of generation of THz radiation by such antenna, just as interdigitated nanofingers [98].

One order of magnitude higher (ca. 100-fold) enhancement has been demonstrated in similar structures, if compared to interdigitated fingers [53], due to the extremely high field enhancement in the nanosized NA gaps.

9.4.4 2D plasmonic gratings

Numerous researchers [50,51,54,56,59,63,66,68,69,71] have shown that one of the most effective implementations of dipole NAs in the THz PCAs and photomixers is the array of metal dipoles whose length is of the order of the gap size between the electrodes. Arrays of dipole plasmonic NAs have received the name of plasmonic gratings or plasmonic nanorod gratings [50].

Usually, the penetration of light through the apertures is limited by the diffraction. However, the geometrical parameters of the gratings can be chosen so that the radiation of an fs-laser excites the surface plasmon wave in this periodic structure. This process is rather resonant, and the geometry of electrodes must be matched to the excitation wavelength. The excitation of plasmon waves allows the transfer of the most part of the optical pump energy into a semiconductor [123], and the pump energy is confined in the semiconductor, as a result of asymmetric scattering [50], thus substantially increasing pump absorption locally in the substrate surface layer, inherently generating a larger number of photocarriers.

The optical properties of this structure are very sensitive to the polarization of the incident light. When the electric field of the optical pulse is parallel to the plasmonic lattice nanorods, the reflection from the grating surface is sufficiently large. On the contrary, in the case of perpendicular orientation of the electric field vector and nanorods, maximum conversion of incident light into plasmon waves is achieved.

The most direct approach to realization of 2D grating enhanced PCA is demonstrated in [50,59]. In both papers, the authors independently suggest to use unincorporated plasmonic grating with stripes perpendicular to the gap (and, respectively, the applied bias) of the THz photoconductive switch (Figure 9.15(a)). The grating enhances the pump absorption, and concentrates the carriers in the surface layer, thus leading to higher photocurrent. In both works, it is shown that when using this structure, more than twofold increase of the amplitude of THz signal can be achieved (see Figure 9.15(b)). However, conversion enhancement in

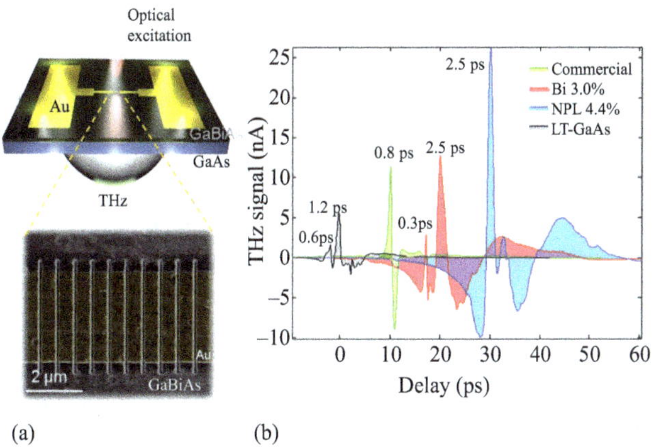

Figure 9.15 (a) Geometry and SEM image of the THz dipole antenna with the unincorporated plasmonic grating. (b) THz signals of commercial THz antenna (green curve), dipole antenna with the GaAsBi semiconductor substrate (red curve), dipole antenna with the LT-GaAs substrate (black curve), and hybrid THz-optical antenna with the nanoplasmonic grating (blue curve) [59]. Reproduced with permission. [37] Copyright 2017, Wiley-VCH.

this type of plasmonic grating is still low if compared to other methods, due to the large capacitance of the resulting structure.

Similar approach is used in [54] to demonstrate plasmon-enhanced LT-GaAs PCA operation under below bandgap 1,570 nm pump. Due to strong field concentration in the substrate surface layer under the plasmonic grating, a probability of two-photon absorption is substantially increased, and free carriers are efficiently generated. Obviously, this gives an order of magnitude increase in generated THz electric field and, consequently, two orders of magnitude higher intensity, if compared with bare LT-GaAs substrate pumped below bandgap. More importantly, these plasmon-enhanced hybrid PCAs outperform even specially designed for 1.5 μm pump commercial InGaAs emitters both in terms of power and bandgap, due to shorter carrier lifetimes of LT-GaAs. The results of THz generation in these antennas are presented in Figure 9.16. Over tenfold enhancement of THz conversion in LT-GaAs substrate, plotted in Figure 9.16(b), is quite coherent with over threefold field enhancement in the surface layer, demonstrated earlier in this chapter. Broader spectral width of LT-GaAs PCAs (Figure 9.16(c)) is associated with shorter carrier lifetimes in this material, if compared with InGaAs.

Later, this approach was enhanced even further, after 2D plasmon grating was replaced by rectangular plasmon monopoles (see Section 9.4.3 for other plasmon-monopole enhanced PCAs), and even more efficient and polarization independent hexagonal monopoles [55]. Simulations showed 30% pump reflection from grating, reduced to 21% in rectangular, and to below 10% in hexagonal structures. As a

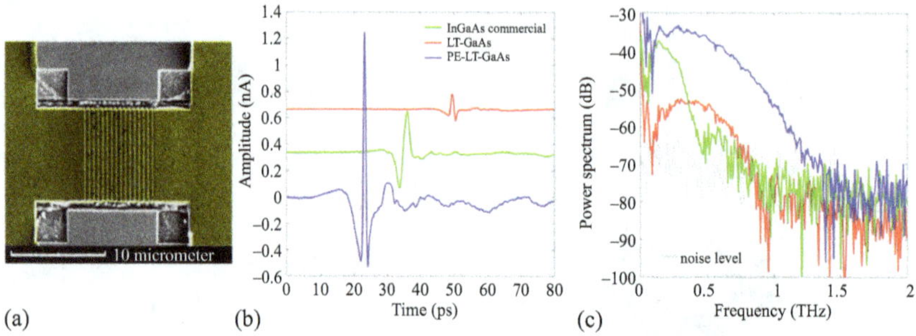

Figure 9.16 *(a) SEM image of the hybrid PCA used for below bandgap pump, (b) THz time-domain signals from plasmon enhanced (blue), InGaAs commercial (green) and bare LT-GaAs emitters under the same 1,570 nm pump, and (c) their corresponding spectra [54].*

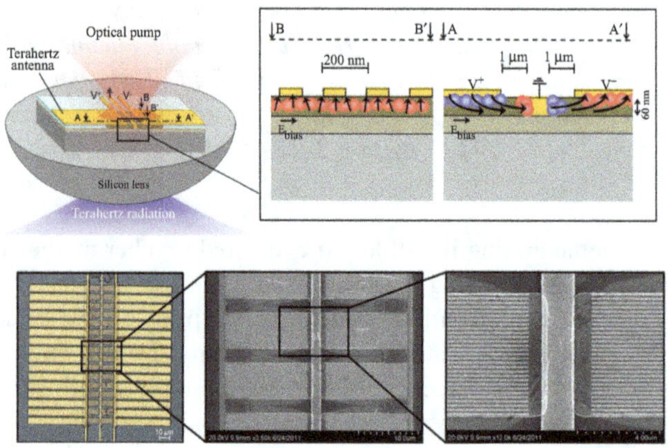

Figure 9.17 *Structural diagram of the THz PCA with embedded plasmonic gratings and ground electrode, its principle of operation (top). Top view of an array of such antennas and their SEM image (bottom) [45]. Reproduced with permission. [37] Copyright 2017, Wiley-VCH.*

result, conversion efficiency in hexagonal structures doubled, if compared with grating. However, since this approach involves a nonlinear process of two photon absorption, it is rather qualitative than quantitative enhancement.

Figure 9.17 depicts another possible implementation of plasmonic nanorod 2D grating embedded in dipole photoconductive THz antenna. Unlike the previous design, the rods are oriented along the gap, and are connected together. Such a layout, fulfilled with the use of thin (<100 nm) photoconducting substrate, significantly shortens the path of the charge carriers to the electrodes, thus allowing for use of longer carrier lifetime substrates.

Moreover, this design reduces the capacitive reactance of the THz antenna and accordingly improves its emitting ability and increases the antenna gain coefficient. Slow electrons and holes present in a long-lifetime substrate do not contribute to THz generation, and are collected by the ground electrode in the middle. In the absence of a grounding electrode, these unwanted carriers recombine with charges of opposite sign of the subsequent cycles of generation, thereby reducing the total number of particles that reach electrodes. A disadvantage of this plasmonic grating realization is that only half of nonequilibrium electrons and holes reach the electrodes within subpicosecond time. This limits the effectiveness of such structures to 50%.

To overcome this limit, an alternative hybrid PCA design with similar plasmonic grating placement was proposed [56]. This design is shown in Figure 9.18(a). In this hybrid PCA, the embedded plasmonic gratings are placed along the gap, as in previous design, and are attached to both anode and cathode contact. The surface plasmon excited by the laser pump (Figure 9.18(b)) greatly enhances the intensity of the optical field near the metallic electrodes (Figure 9.18 (c)). Upon pumping in the vicinity of the anode, the path length of photoinduced electrons, i.e., carriers that predominantly contribute to the THz generation, is significantly reduced in comparison with conventional THz antennas, and even with the initially proposed design, and the optical absorption enhancement is further enhanced by the 150-nm-thick dielectric SiO_2 passivation layer.

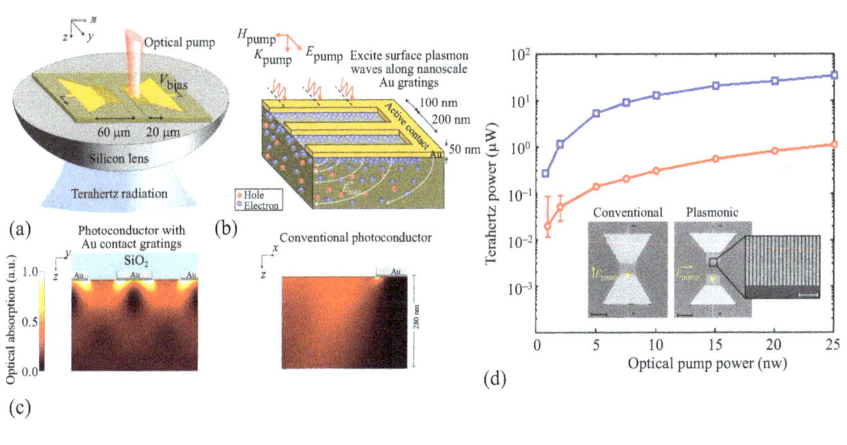

Figure 9.18 *(a) Geometry of the hybrid THz-optical photoconductive antenna with 2D plasmonic grating and (b) distribution of electrons and holes near plasmonic nanorods. (c) Image of the distribution of absorbed optical energy in GaAs substrates with plasmonic gratings on the surface and standard THz photoconductive antenna, (d) THz output power vs. pump power for THz hybrid optical photoconductive antenna (blue) and a standard THz photoconductive antenna of the same geometry (red) [56]. Reproduced with permission. [37] Copyright 2017, Wiley-VCH.*

The combined use of the plasmonic electrodes, suggested earlier [45] with sub-picosecond LT-GaAs substrate, and extra optical absorption enhancement by SiO_2 passivation layer allowed the authors of [56] to achieve almost two orders of magnitude of THz power enhancement (Figure 9.18(d)). The enhancement reached 50× at lower pumps, and dropped down to 30× at higher. This approach demonstrates significant enhancement, and yet allows for higher pump powers (up to 100 mW). Similar NA design coupled with spiral photomixer was demonstrated to enhance THz generation in CW regime [69], polarization insensitive option with cross-shaped plasmonic metasurface was proposed later [47]. The only drawback of such hybrid PCA design is the complexity of manufacturing.

Later, this layout of hybrid PCA was advanced further and applied for high-power efficient THz emitters [46]. As it may have been understood by a discriminating reader, plasmonic electrodes at the cathode are not really necessary and can be removed. Anode plasmonic electrodes, covering the larger portion of the emitter, and dielectric anti-reflection layer are enough to ensure conversion enhancement. To employ higher-power pumps, a large area interdigitated approach with half of the gaps covered to prevent photocurrent in the unwanted direction [124] was used.

Large area allowed for the use of optical pumps of up to above 700 mW, and conversion efficiency reached ~3E-4, with broadband (up to 4 THz) operation. The layout of such antenna is shown in Figure 9.19(a). Further optimization involved the addition of the resonant subcavity by incorporating a distributed Bragg reflector (DBR) under the photoconducting layer. The DBR assured the reflection of the unabsorbed radiation coming into the substrate, thus forming a standing wave in the subcavity formed in the surface layer. This ensures the absence of long living carriers deep in the slower and unbiased semiconductor and the recycling of residual pump power, thus further enhancing the pump absorption (Figure 9.19(b)). With this improvement, the photocurrent and hence the THz emission received an extra ~6-fold boost, resulting in ~30 times higher generated THz power. In

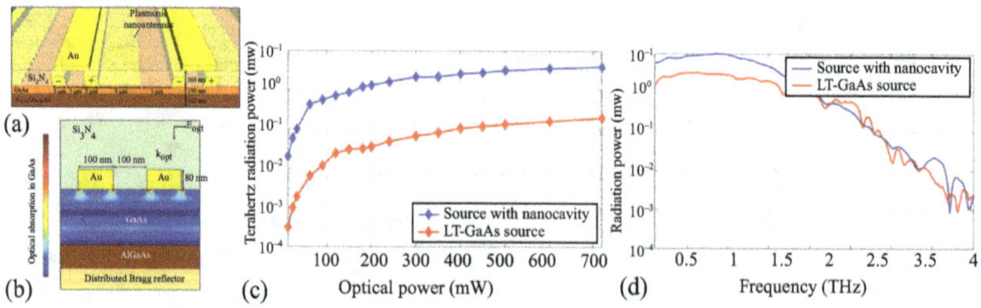

Figure 9.19 (a) Geometry of the hybrid THz-optical photoconductive array antenna with 2D plasmonic grating; (b) optical absorption in the photoconductive substrate subcavity; (c) THz radiation power of plasmonic enhanced array antenna (red) and subcavity enhanced array (blue); and (d) their corresponding spectra.

Figure 9.19(c), the dependence of THz power on the laser pump power for DBR-enhanced large area plasmonic PCA and similar antenna on bare LT-GaAs plasmonic-only are shown. A ~30-fold enhancement can be seen across all pump powers. The resulting spectra are shown in Figure 9.19(d). The absence of enhancement above 2 THz is observed because only the initially generated in the close vicinity (tens of nm) to anode photocarriers contribute to higher frequencies, while the reflected pump is reabsorbed in the photoconducting layer [46].

To manufacture such structures, the methods of electron-beam lithography, accompanied by the deposition of Au on the semiconductor substrate are used. A dielectric layer covering NA reduces Fresnel reflection from its surface, thereby increasing the absorption coefficient of the optical pump into the substrate. Usually, few hundred nm thick SiO_2 or Si_3N_4 is used for these purposes. This coating provides up to 70% absorption of pump light into the semiconductor material.

Comparison of conventional THz PCAs and hybrid optical-to-THz antennas with integrated plasmonic gratings reveals at least one order of magnitude superiority of the latter. Such a significant increase in the output power of THz radiation is predominantly achieved by reduction of the charge carriers' path length to the electrodes, thus greatly increasing the photocurrent and decreasing the number of recombined electron-hole pairs.

An important role in THz hybrid optical PCAs is played by the thickness of the semiconductor substrate, since at a sufficiently large depths (over 100 nm) photo-induced charge carriers that do not participate in the effective generation of THz radiation start to accumulate. These electrons and holes rise from the depths of the substrate to the surface, where they recombine with other charges, thus significantly reducing the ratio of optical-to-terahertz conversion. To prevent negative effects caused by the nonequilibrium carriers accumulated in the lower layers of the substrate, its thickness should not exceed several hundreds of nm, as it is proposed in the latter design. However, there is another workaround for this problem without reducing the thickness of semiconductor by using 3D plasmonic gratings [67].

9.4.5 3D plasmonic gratings

Plasmonic gratings described above are 2D nanoscale structures on the semiconductor surface. They provide a high absorption of light quanta by a semiconductor material and significantly increase the concentration of photoinduced charge carriers near metal electrodes. However, the effectiveness of optical-to-terahertz conversion is still limited by the depth at which charge carriers are generated (~100 nm). To overcome this limitation, 3D plasmonic gratings embedded in the semiconductor substrate are proposed. It has been shown that such structures allow achieving 7.5% optical-to-THz conversion efficiency [67]. Figure 9.20(a)–(d) shows a schematic of the spiral PCA containing 3D plasmonic grating in the gap, the schematic, and SEM images of 3D plasmonic grating, and distribution pattern of the absorbed optical radiation, correspondingly. Because of its complexity, the manufacturing of such antennas should be discussed in more detail. The process of 3D plasmonic gratings fabrication starts with chemical vapor deposition

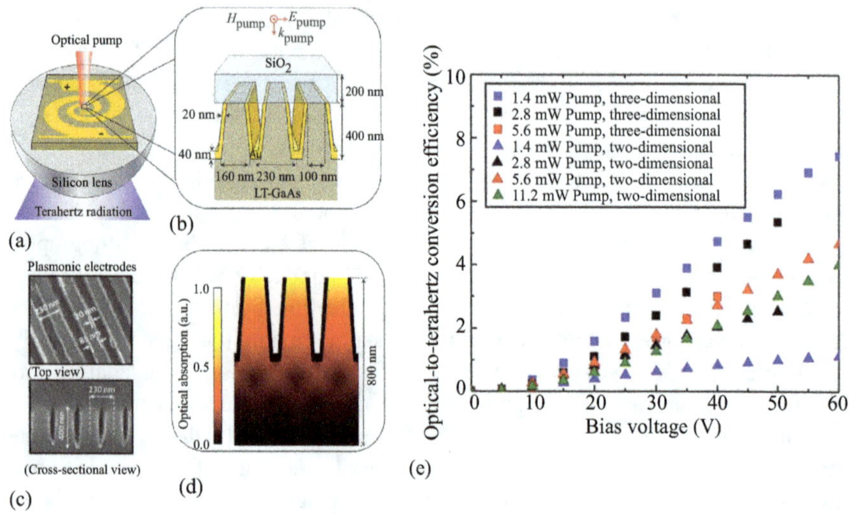

Figure 9.20 (a) Schematic representation of the spiral hybrid PCA, (b) schematic of the integrated 3D plasmonic structure, and (c) its SEM picture. (d) Simulated field distribution in the substrate and (e) dependence of the optical-to-terahertz conversion on bias applied to antenna electrodes at different pump powers for THz photoconductive antennas with built-in 2D and 3D plasmonic gratings [67]. Reproduced with permission. [37] Copyright 2017, Wiley-VCH.

of a 200 nm film of SiO_2 onto the surface of the semiconductor. Next, on the surface of SiO_2 film, a nanoscale Nickel grid is applied by the EBL. The grid serves as a template layer for etching of SiO_2. When using reactive ion etching with a source of inductively coupled plasma, a SiO_2 grating is formed, which serves as a mask to create nano-walls made of semiconductor material. The plasmonic grating is obtained by deposition of gold on the sidewalls of the resulting semiconductor nanostructures. The height of the walls of the nanostructure determines the order of the modes excited in the resulting subwavelength waveguides. In such a structure, waveguide TEM modes of fourth order can be excited by p-component of the fs-laser [125]. A SiO_2 mask is then removed by wet etching. Additionally, to reduce the losses associated with reflection, the surface of the antenna is covered with a 200 nm layer of SiO_2. The resulting structure absorbs up to 70% of the incident laser radiation.

Figure 9.20(d) demonstrates the advantages of 3D gratings over 2D. One can see the large coefficient of optical-to-THz conversion in 3D plasmonic grating in comparison with the 2D analogue, at all pump powers. At the lowest optical pump power of 1.4 mW, optical-to-THz conversion efficiency reaches 7.5%. Such an increase in efficiency is due to the fact that 3D plasmonic grating is embedded into the surface layer of the semiconductor substrate; hence, the contact surface of the grating with the semiconductor is much greater than that of the 2D plasmonic gratings. This increases the concentration of the photoinduced charges near the metal electrodes, which are involved in the effective generation of THz radiation [67]. However, with increasing optical pump power, optical-to-THz conversion efficiency in 3D structures decreases. This happens because a large number of separated electron-hole pairs generate a significant electric field with the direction opposite to the external field. The resulting field turns out to decrease; hence, the efficiency of THz generation is reduced [103].

Thus, THz PCAs with integrated 3D plasmonic gratings provide greater efficiency of optical-to-THz conversion in comparison with 2D plasmonic gratings. In the future, despite the complexity of fabrication technology, these 3D nanoscale structures can replace 2D in hybrid THz-optical PCAs.

9.4.6 Comparison of the reviewed approaches

In this section, we have distinguished five main approaches to plasmonic NAs for THz generation enhancement design. All approaches possess their advantages and drawbacks, demonstrate significant performance enhancement, and are suitable for certain THz applications. Though the approaches described above appear chronologically, the fabrication complexity and, in most cases, efficiency, developed in the similar way. Plasmonic gratings offer highest enhancement, but are most hard in production. The main mechanisms that are engaged by the involvement of plasmonic NAs are the reduction of reflection, pump scattering into the surface layer, electric field concentration, and carrier-free mean path shortening. One outstanding effect is the triggering of the nonlinear two-photon absorption of the pumps with phonon energy below the bandgap, due to extremely high field concentration, thus allowing the use of telecom wavelength pump lasers for LT-GaAs substrate PCA pumping [54,55]. Once, the researchers noticed uneven conversion enhancement across the THz spectrum, and explained the effect by the impedance change as a result of metallic NAs introduction [57], in other works, the resonant change in impedance, was situated outside the operational band.

To summarize all results presented, we have gathered together studies on various approaches to the creation of hybrid PCAs for THz generation, in order to facilitate a direct comparison of the proposed methods (see Table 9.1). We conclude that the most high-performance hybrid THz PCAs are the ones based on 3D plasmonic gratings with the maximum conversion efficiency of 7.5%, at the pump power of 1.4 mW. For comparison, hybrid THz PCAs based on 2D plasmonic gratings have the maximum conversion efficiency of about 0.2% only, but at a higher pump power of 25 mW.

Gray-shaded rows depict nonlinear two-photon absorption enhancement, i.e., unlike other cases, a nonlinear process, hence giving larger values for similar structures.

9.5 Enhancement of THz detection with nanoantennas

Hybrid PCAs were predominantly studied as THz generators, because the photocurrent and electrical field concentration they provide are the key enabling

Table 9.1. Comparison of the proposed fabrication methods of hybrid photoconductive antennas for efficient THz generation.

Method of plasmon resonance obtaining	Regime	Fabrication complexity	Enhancement factor	Spectral range, THz	Maximum efficiency	At pump power (mW)	# Ref.
Silver nanoparticles	Pulse	Relatively low	1.6×	0.1–1.2	–	50	[49]
Optimized silver nanoparticles	Pulse	Relatively low	2–6×	0.1–1.2	–	50	[57]
AuGe nanoparticles	Pulse	Moderate	4×	0.1–1.5	–	–	[60]
Hexagonal nanoantennas	Pulse	High	20×	0.1–2.4	–	2	[52]
Hexagonal nanoantennas	Pulse	High	300×	0.1–2.4	–	50	[55]
Rectangular nanoantennas	Pulse	High	220×	0.1–2.4	–	50	[55]
Interdigitated metallic dipole nanoantenna (six-nanofinger)	CW	Moderate	–	0.1–1.1	0.0002%	26	[115]
Interdigitated metallic dipole nanoantenna (five-nanofinger)	CW	Moderate	–	0.1–1.1	0.0006%	50	[126]
Interdigitated metallic dipole nanoantenna (eight-nanofinger)	CW	Moderate	–	0.1–1.1	0.007%	90	[99]
Rectangular metal dipole nanoantennas	CW	High	4×	0.1–1.1	0.0005%	55	[62]
Trapezoid metal dipole nanoantennas	CW	High	7×	0.1–1.1	0.0008%	55	[62]
Rectangular metal dipole nanoantennas	CW	High	100×	0.3–1.6	–	90	[53]
Unincorporated plasmonic grating	Pulse	High	2.5×	0.1–1	–	80	[50]
Unincorporated plasmonic grating	Pulse	High	4×	0.1–2.1	–	17	[59]
Unincorporated plasmonic grating	Pulse	High	160×	0.1–2.4	–	50	[54]
Plasmonic grating with grounding electrode	Pulse	High	15×	0.1–1.5	0.005%	85	[45]
Plasmonic grating	Pulse	High	35–50×	0.1–1.5	0.2–3%	25	[56]
Plasmonic grating	CW	High	8×	0.3–2.4	0.08%	100	[69]
Plasmonic grating	CW	High	–	0.3–2.4	0.5%	150	[66]
Plasmonic grating	CW	High	–	0.1–1.5	0.0003 %	10	[63]
3D plasmonic grating	Pulse	Extremely high	400×	0.1–2.1	7.5%	1.4	[67]
Large area plasmonic gratings	Pulse	High	–	0.2–5	0.1%	200	[65]
Polarization insensitive plasmonic metasurface	Pulse	High	–	0.1–1.6	0.3%	8	[47]
Nanocavity enhanced plasmonic grating	Pulse	Very high	20–60×	0.1–4	0.6%	720	[46]

mechanisms for efficient THz generation. In THz photoconductive detectors, the use of optically resonant nanoparticles in the photoconductive gap, just as in emitters, provides higher carrier concentration, and, consequently, higher photocurrent response to the THz field. The use of nanoparticle enhanced hybrid detectors is not as crucial as hybrid transmitters, because detector's laser pump powers are generally lower than transmitters'. Several works have demonstrated, obviously, more efficient operation of hybrid PCA detectors, but similar enhancement could have been obtained by bringing the pump power up to achieve the same substrate conductivity.

The most significant progress of hybrid THz detectors is the demonstration of the near-field THz imaging resolution enhancement, up to $\lambda/150$ [127]. This unprecedented resolution has become available due to the extremely high scattering provided by NAs into the semiconductor, i.e., low (<200°C) temperature grown gallium arsenide (LT-GaAs) detector surface layer and hence high enough concentration of carriers in the ultrathin semiconductor layer next to subwavelength detector aperture for evanescent THz field registration (Figure 9.21).

Plasmonic NAs in this case drastically reduce the pump reflection from the photogap (Figure 9.21(b)), because of the scattering they introduce. All incoming radiation is scattered into the LT-GaAs layer, and is redistributed inside it by the DBR, instead of normal reflection, as in the case of the bare substrate [128]. This is the only application of plasmonic grating that makes a crucial change in detector operation. Thanks to absorption and photocurrent formation in the thin layer behind the subwavelength aperture, it has become possible to shift the detector closer to the aperture, thus achieve drastic enhancement of detector sensitivity to evanescent THz wave, and, inherently, make the aperture even smaller, thus achieving even better resolution.

Another intriguing phenomenon was demonstrated, similarly to generation, at pumping the PCA with photons having the energy below the bandgap [54]. In this case, even at 25 mW pump, optical field concentration by the plasmonic grating in the PCA gap triggers nonlinear two-photon absorption, thus allowing for the use of

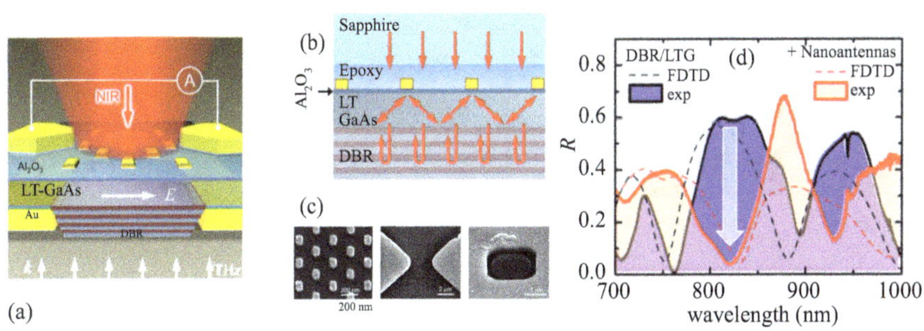

Figure 9.21 Plasmon monopole nanoantenna enhanced THz near-field detector. (a) The layout of detector, (b) illustration of the pump beam behavior in the antenna, (c) SEM image of the fabricated detector, and (d) reflection reduction due to nanoantenna presence [127].

telecom laser pump in conjunction with LT-GaAs PCA substrate. Just like the analogous generator, such a detector provided 160-fold enhancement of the photocurrent response in comparison with bare LT-GaAs antenna. And, if compared with commercial InGaAs detector, designed for this wavelength, much broader operational spectrum is achieved, due to faster recombination times in LT-GaAs substrate. Relative operational enhancement is very similar to that for transmitter shown in Figure 9.16(c).

Large area plasmonic THz detector was demonstrated very recently. A key novelty of this detector that makes it superior to the state-of-the-art is its specific large-area device geometry. Strong interaction between the incident THz and optical beams at the nanoscale enables strong response, while maintaining a broadband operation. The large active area of the device allows robust operation against optical and THz beam misalignments [129].

The proposed device and its operational characteristics are shown in Figure 9.22. In order to achieve highly sensitive detection in the PCA, the NA geometry should be selected such that the THz field and optical pump intensity are maximized in close proximity to the NA tips. By maximizing the concentration of the photocarriers and the electric field of the THz pulse, driving them, higher photocurrent levels are induced and higher detection sensitivity levels are achieved, consequently. Therefore, designs with longer NAs are preferred as far as the required detection bandwidths are supported by the device [129].

However, competition between factors sets the optimum NA tip-to-tip gap size. The narrower tip-to-tip gap the stronger is the THz field inside the device active area, but at the same time, the wider the tip-to-tip gap, the higher is the photocarrier concentration. As both parameters are favorable for efficient PCA detector performance, a careful study of the parameters effect on the sensitivity was carried out. As a result, a detector with over 5 THz operational bandgap and over 107 dB of signal-to-noise ratio was developed. Such sensitivity exceeds typical THz time domain systems by two orders of magnitude, and this unprecedented result was enabled by the plasmonic NAs in the PCA gaps.

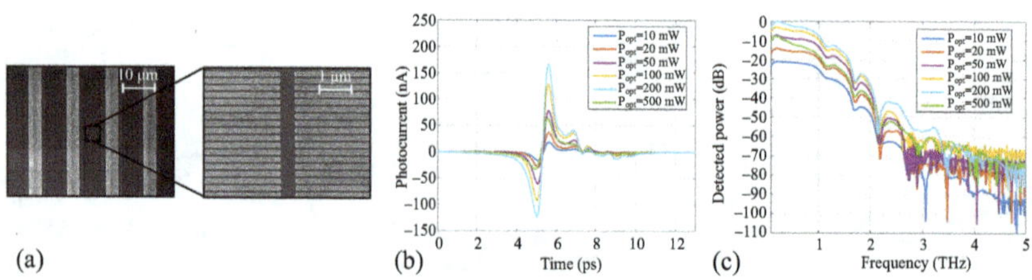

Figure 9.22 (a) Scanning electron microscope image of a plasmonic nanoantenna detector array fabricated on a LT-GaAs substrate with a nanoantenna length of 2 μm and tip-to-tip gap size of 500 nm; (b) and (c) are the measured photocurrent of a detector prototype with a nanoantenna length of 4 μm and tip-to-tip gap size of 500 nm in the time-domain and the corresponding power spectra, respectively [129].

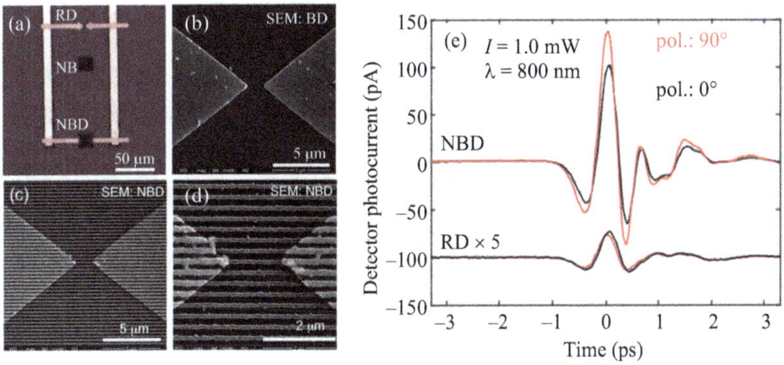

Figure 9.23 Photoconducting THz detector incorporating nanobeams (NBD) and reference THz detector (RD) fabricated on the same chip. (a) Optical image showing both detectors connected to the same transmission line, and a 20 × 20 μm² patch with nanobeams (NB) used for correlation of the detector performance with the reflectivity spectrum. SEM images of the two detectors [(b) and (c)] and (d) a close-up image of the antenna gap (the active region) of NBD. (e) THz pulse detected by the metasurface detector (NBD) and the reference detector (RD) signal [128].

Before conclusion, we wanted to mention one breakthrough result that may pave the way for future hybrid PCA research—dielectric enhanced hybrid PCA detector. As the plasmonic approach has provided a significant boost to the PCA technology, and resulted in numerous efficient photoconductive sources and detectors, there is always a question about further steps. A very recent paper by Thompson et al. [128] demonstrated the enhancement operation of the detector PCA by dielectric NAs, the so-called nanobeams (NB). The design and the results of such antenna operation are shown in Figure 9.23.

The stripes were patterned using the EBL. Then, the surface was dry-etched to the depth of ~70 nm to form the nanobeams with vertical walls [128]. The resulting PCA contained only the LT-GaAs nanobeams on top of the 90 nm thick LT-GaAs layer and the AlGaAs/GaAs DBR attached to the surface of the sapphire substrate. Electrodes were fabricated using the EBL by depositing metallic contacts with 1.5μm gap directly over the nanobeams Figure 9.23(d). Similar waveforms were recorded by bare and metasurface enhanced PCAs; however, the metasurface detector exhibited ~15 times higher photocurrent (Figure 9.23(e)).

9.6 Outlook

This chapter has demonstrated obvious advantages of hybrid THz PCAs and photomixers over conventional ones, especially in their conversion efficiency at low pump powers. With the development of compact ultrafast fiber [130] and semiconductor [131,132] lasers, followed by the use of new photoconductive materials suitable for the longer-wavelength, e.g., Yb-based lasers (1,030 nm), Er-doped fiber lasers (1,550 nm) or semiconductor quantum dot lasers (950–1,300 nm). These materials include GaBiAs [34], InGaAs [133], and also ones containing ErAs [134], and InAs [33,35] quantum dots in GaAs substrates. Joint use of compact lasers with efficient plasmonic enhanced converters can lead to the creation of highly effective, miniature THz transmitters, spectroscopic, imaging, and data transfer devices, working at room temperatures both in pulsed and CW regimes. Despite the complexity and high cost of hybrid electrodes fabrication, the total setup cost, that will exclude the extremely expensive titanium-sapphire laser, should significantly go down.

On the other hand, despite a number of plasmonic NAs' advantages associated with their small size and strong electric field localization, such NAs have large dissipative losses resulting in their low efficiency. To overcome such limitations, a new type of NAs based on dielectric nanoparticles (e.g., silicon) with a high dielectric constant has been recently proposed [135]. Such NAs are very promising for hybrid THz-optical PCAs. Namely, they also have an ability to enable enhancement of local electric (and magnetic) field [136]. Moreover, they have a much higher damage threshold due to low dissipative losses, thus allowing for high power pumping which is promising for effective THz generation. Dielectric NAs have been demonstrated as THz detection amplifiers, but have not yet been applied for emitters.

The future studies in this promising area may also be focused on further electric field enhancement through the Fano resonance in periodic structures supporting Wood's anomaly by a similar way it has been done in photovoltaics and electron photoemission [137]. It turns out that when the localized surface plasmon resonance or Mie resonance of an individual particle in the periodic array is close to the frequency of the Wood's anomaly, the collective effects lead to narrow-band Fano resonance with a quality factor much larger compared to that of the single particle. This is accompanied by a strong electric field enhancement nearby the particles and may lead to further enhancement in generated free carrier concentrations [138].

Bound states in the continuum open one more possibility to further increase the absorptivity and free currents generation with extremal spectral sensitivity. These states are localized modes in the continuous spectrum of an electromagnetically open resonant nanostructure. An existence of these states allows designing photonic resonators with diverging and potentially infinite lifetimes and quality factors [139–141].

Finally, modern approaches of coherent photonics and nanooptics such as coherent perfect absorption and virtual perfect absorption [142,143] open up new opportunities for efficient THz radiation because of their ability to absorb light perfectly even in thin films. Another benefit of these approaches is that due to their potential to achieve all-optical modulation, they make the system tunable or reconfigurable.

Thus, despite a large number of research works in this area and very encouraging results, consensus about the advantage of a particular structure and/or material as the absorption enhancer for the pump radiation, and the magnifier for effective generation area in a semiconductor is not achieved yet. This is a very young research area where more studies and experiments are needed. Current research is aimed primarily at the optimization of the proposed methods and quest for new types of NAs. In addition, the search for more simple solutions to the problem of transferring optical radiation into the largest possible nonequilibrium carrier pairs number in a semiconductor, increasing the speed of their movement through the creation of more intense local electric and magnetic fields and reduction of their lifetime by reducing their free path lengths is also needed.

References

[1] B. Ferguson and X.-C. Zhang, Nat. Mater. 1, 26 (2002).
[2] P. H. Siegel, IEEE Trans. Microw. Theory Tech. 50, 910 (2002).
[3] C. A. Schmuttenmaer, Chem. Rev. 104, 1759 (2004).
[4] M. Tonouchi, Nat. Photonics 1, 97 (2007).
[5] P. U. Jepsen, D. G. Cooke, and M. Koch, Laser Photon. Rev. 5, 124 (2011).
[6] Y. Jin, G. Kim, and S. Jeon, J. Korean Phys. Soc. 49, 513 (2006).
[7] J. A. Zeitler, P. F. Taday, D. A. Newnham, M. Pepper, K. C. Gordon, and T. Rades, J. Pharm. Pharmacol. 59, 209 (2007).
[8] T. R. Globus, D. L. Woolard, T. Khromova, et al., J. Biol. Phys. 29, 89 (2003).
[9] M. Massaouti, C. Daskalaki, A. Gorodetsky, A. D. Koulouklidis, and S. Tzortzakis, Appl. Spectrosc. 67, 1264 (2013).
[10] J. F. Federici, B. Schulkin, F. Huang, et al., Semicond. Sci. Technol. 20, S266 (2005).
[11] H.-R. Park, K. J. Ahn, S. Han, Y.-M. Bahk, N. Park, and D. Kim, Nano Lett. 13, 1782 (2013).
[12] S. J. Park, J. T. Hong, S. J. Choi, et al., Sci. Rep. 4, 4988 (2014).
[13] L. Xie, W. Gao, J. Shu, Y. Ying, and J. Kono, Sci. Rep. 5, 8671 (2015).
[14] A. G. Davies, E. H. Linfield, and M. B. Johnston, Phys. Med. Biol. 47, 3679 (2002).
[15] Q. Sun, Y. He, K. Liu, S. Fan, E. P. J. Parrott, and E. Pickwell-MacPherson, Quant. Imaging Med. Surg. 7, 345 (2017).
[16] K. Shiraga, Y. Ogawa, T. Suzuki, N. Kondo, A. Irisawa, and M. Imamura, J. Infrared, Millimeter, Terahertz Waves 35, 493 (2014).
[17] L. Rong, T. Latychevskaia, C. Chen, et al., Sci. Rep. 5, 8445 (2015).
[18] K. E. K. Coppin, J. E. Geach, I. Smail, et al., Mon. Not. R. Astron. Soc. 416, no (2011).
[19] F. Wyrowski, R. Güsten, K. M. Menten, H. Wiesemeyer, and B. Klein, Astron. Astrophys. 542, L15 (2012).
[20] U. U. Graf, C. E. Honingh, K. Jacobs, and J. Stutzki, J. Infrared, Millimeter, Terahertz Waves 36, 896 (2015).
[21] I. F. Akyildiz, J. M. Jornet, and C. Han, Phys. Commun. 12, 16 (2014).
[22] T. Kürner and S. Priebe, J. Infrared, Millimeter, Terahertz Waves 35, 53 (2014).

[23] J. Federici and L. Moeller, J. Appl. Phys. 107, 111101 (2010).
[24] S. Koenig, D. Lopez-Diaz, J. Antes, et al., Nat. Photonics 7, 977 (2013).
[25] C. Wang, B. Lu, C. Lin, et al., IEEE Trans. Terahertz Sci. Technol. 4, 75 (2014).
[26] J. Xu, K. W. Plaxco, and S. J. Allen, J. Chem. Phys. 124, 036101 (2006).
[27] G. A. Siles, J. M. Riera, and P. Garcia-del-Pino, IEEE Antennas Propag. Mag. 57, 48 (2015).
[28] G. Ducournau, P. Szriftgiser, A. Beck, et al., IEEE Trans. Terahertz Sci. Technol. 4, 328 (2014).
[29] A. J. Seeds, H. Shams, M. J. Fice, and C. C. Renaud, J. Light. Technol. 33, 579 (2015).
[30] T. Nagatsuma, G. Ducournau, and C. C. Renaud, Nat. Photonics 10, 371 (2016).
[31] D. H. Auston, K. P. Cheung, and P. R. Smith, Appl. Phys. Lett. 45, 284 (1984).
[32] K. L. Vodopyanov, Laser Photon. Rev. 2, 11 (2008).
[33] R. R. Leyman, A. Gorodetsky, N. Bazieva, et al., Laser Photon. Rev. 10, 772 (2016).
[34] K. Bertulis, A. Krotkus, G. Aleksejenko, et al., Appl. Phys. Lett. 88, 201112 (2006).
[35] A. Gorodetsky, A. Yadav, E. Avrutin, K. A. Fedorova, and E. U. Rafailov, IEEE J. Sel. Top. Quantum Electron. 24, 1 (2018).
[36] V. G. Bespalov, V. N. Krylov, S. E. Putilin, and D. I. Stasel'ko, Opt. Spectrosc. 93, 148 (2002).
[37] S. Lepeshov, A. Gorodetsky, A. Krasnok, E. Rafailov, and P. Belov, Laser Photon. Rev. 11, 1600199 (2017).
[38] K. J. Siebert, A. Lisauskas, T. Löffler, and H. G. Roskos, Jpn. J. Appl. Phys. 43, 1038 (2004).
[39] T. Hattori, K. Egawa, S. Ookuma, and T. Itatani, Jpn. J. Appl. Phys. 45, L422 (2006).
[40] W. Shi, X. F. Sun, J. Zeng, and W. L. Jia, in Jpn. J. Appl. Phys., edited by L. Zhou (2007), p. 662228.
[41] M. Beck, H. Schäfer, G. Klatt, et al., Opt. Express 18, 9251 (2010).
[42] P. Bharadwaj, B. Deutsch, and L. Novotny, Adv. Opt. Photonics 1, 438 (2009).
[43] L. Novotny and N. Van Hulst, Nat. Photonics 5, 83 (2011).
[44] A. E. Krasnok, I. S. Maksymov, A. I. Denisyuk, et al., Physics-Uspekhi 56, 539 (2013).
[45] C. W. Berry and M. Jarrahi, New J. Phys. 14, 105029 (2012).
[46] N. T. Yardimci, S. Cakmakyapan, S. Hemmati, and M. Jarrahi, Sci. Rep. 7, 4166 (2017).
[47] X. Li, N. T. Yardimci, and M. Jarrahi, AIP Adv. 7, (2017).
[48] N. T. Yardimci and M. Jarrahi, Small 1802437, 1802437 (2018).
[49] S.-G. Park, Y. Choi, Y.-J. Oh, and K.-H. Jeong, Opt. Express 20, 25530 (2012).

[50] S.-G. Park, K. H. Jin, M. Yi, J. C. Ye, J. Ahn, and K.-H. Jeong, ACS Nano 6, 2026 (2012).

[51] B. Heshmat, H. Pahlevaninezhad, Y. Pang, et al., Nano Lett. 12, 6255 (2012).

[52] A. Jooshesh, L. Smith, M. Masnadi-Shirazi, et al., Opt. Express 22, 27992 (2014).

[53] H. Tanoto, J. H. Teng, Q. Y. Wu, et al., Sci. Rep. 3, 2824 (2013).

[54] A. Jooshesh, V. Bahrami-Yekta, J. Zhang, T. Tiedje, T. E. Darcie, and R. Gordon, Nano Lett. 15, 8306 (2015).

[55] F. Fesharaki, A. Jooshesh, V. Bahrami-Yekta, et al., ACS Photonics 4, 1350 (2017).

[56] C. W. Berry, N. Wang, M. R. Hashemi, M. Unlu, and M. Jarrahi, Nat. Commun. 4, 1622 (2013).

[57] S. Lepeshov, A. Gorodetsky, A. Krasnok, et al., Sci. Rep. 8, 6624 (2018).

[58] T. Gric, A. Gorodetsky, A. Trofimov, and E. Rafailov, J. Infrared, Millimeter, Terahertz Waves 1 (2018).

[59] B. Heshmat, M. Masnadi-Shirazi, R. B. Lewis, et al., Adv. Opt. Mater. 1, 714 (2013).

[60] H. Surdi, A. Singh, and S. S. Prabhu, Proceeding 2013 COMSOL 1 (2013).

[61] G. Seniutinas, G. Gervinskas, E. Constable, et al., in *Proc. SPIE - Int. Soc. Opt. Eng.*, edited by J. Friend and H. H. Tan (2013), p. 892322.

[62] N. Khiabani, Y. Huang, L. E. Garcia-Munoz, Y.-C. Shen, and A. Rivera-Lavado, IEEE Trans. Terahertz Sci. Technol. 4, 501 (2014).

[63] R. A. Mohandas, J. R. Freeman, M. Natrella, et al., Opt. Express 25, 10177 (2017).

[64] N. T. Yardimci and M. Jarrahi, in *2016 IEEE MTT-S Int. Microw. Symp.* (IEEE, 2016), pp. 1–4.

[65] N. T. Yardimci, H. Lu, and M. Jarrahi, Appl. Phys. Lett. 109, 191103 (2016).

[66] C. W. Berry, M. R. Hashemi, and M. Jarrahi, Appl. Phys. Lett. 104, (2014).

[67] S. Yang, M. R. Hashemi, C. W. Berry, and M. Jarrahi, IEEE Trans. Terahertz Sci. Technol. 4, 575 (2014).

[68] C. W. Berry, M. R. Hashemi, S. Preu, H. Lu, A. C. Gossard, and M. Jarrahi, Appl. Phys. Lett. 105, 011121 (2014).

[69] C. W. Berry, M. R. Hashemi, S. Preu, H. Lu, A. C. Gossard, and M. Jarrahi, Opt. Lett. 39, 4522 (2014).

[70] N. T. Yardimci, S. Yang, C. W. Berry, and M. Jarrahi, IEEE Trans. Terahertz Sci. Technol. 5, 223 (2015).

[71] M. Jarrahi, IEEE Trans. Terahertz Sci. Technol. 5, 391 (2015).

[72] N. T. Yardimci and M. Jarrahi, in *2016 IEEE Int. Symp. Antennas Propag.* (IEEE, 2016), pp. 1187–1188.

[73] G. L. Carr, M. C. Martin, W. R. McKinney, K. Jordan, G. R. Neil, and G. P. Williams, Nature 420, 153 (2002).

[74] M. Mineo and C. Paoloni, IEEE Trans. Electron Devices 57, 1481 (2010).

[75] W. He, C. R. Donaldson, L. Zhang, K. Ronald, P. McElhinney, and A. W. Cross, Phys. Rev. Lett. 110, 165101 (2013).

[76] B. S. Williams, Nat. Photonics 1, 517 (2007).
[77] A. Khalid, N. J. Pilgrim, G. M. Dunn, et al., IEEE Electron Device Lett. 28, 849 (2007).
[78] K. L. Yeh, M. C. Hoffmann, J. Hebling, and K. A. Nelson, Appl. Phys. Lett. 90, 1 (2007).
[79] K. Vijayraghavan, Y. Jiang, M. Jang, et al., Nat. Commun. 4, 2021 (2013).
[80] X. Xie, J. Dai, and X.-C. Zhang, Phys. Rev. Lett. 96, 075005 (2006).
[81] X.-C. Zhang and J. Xu, Introduction to THz Wave Photonics (Springer US, Boston, MA, 2010).
[82] Y. S. Lee, Principles of Terahertz Science and Technology (Springer US, Boston, MA, 2009).
[83] M. W. McCutcheon, D. E. Chang, Y. Zhang, M. D. Lukin, and M. Loncar, Opt. Express 17, 9 (2009).
[84] L. Luo, I. Chatzakis, J. Wang, et al., Nat. Commun. 5, 3055 (2014).
[85] D. K. Polyushkin, E. Hendry, E. K. Stone, and W. L. Barnes, Nano Lett. 11, 4718 (2011).
[86] A. Atrashchenko, A. Arlauskas, R. Adomavičius, et al., Appl. Phys. Lett. 105, 191905 (2014).
[87] T. Seifert, S. Jaiswal, U. Martens, et al., Nat. Photonics 10, 483 (2016).
[88] Y. Bahk, G. Ramakrishnan, J. Choi, et al., ACS Nano 8, 9089 (2014).
[89] O. V. Kibis, M. Rosenau da Costa, and M. E. Portnoi, Nano Lett. 7, 3414 (2007).
[90] L.-G. Zhu, B. Kubera, K. Fai Mak, and J. Shan, Sci. Rep. 5, 10308 (2015).
[91] A. A. Andreev, V. G. Bespalov, A. Gorodetsky, et al., Opt. Spectrosc. 107, 538 (2009).
[92] J. Van Rudd and D. M. Mittleman, J. Opt. Soc. Am. B 19, 319 (2002).
[93] D. Grischkowsky, S. Keiding, M. van Exter, and C. Fattinger, J. Opt. Soc. Am. B 7, 2006 (1990).
[94] P. C. Upadhya, W. Fan, A. Burnett, et al., Opt. Lett. 32, 2297 (2007).
[95] S. E. Ralph and D. Grischkowsky, Appl. Phys. Lett. 59, 1972 (1991).
[96] C. W. Berry and M. Jarrahi, J. Infrared, Millimeter, Terahertz Waves 33, 1182 (2012).
[97] E. R. Brown, F. W. Smith, and K. A. McIntosh, J. Appl. Phys. 73, 1480 (1993).
[98] I. S. Gregory, C. Baker, W. R. Tribe, et al., IEEE J. Quantum Electron. 41, 717 (2005).
[99] S. M. Duffy, S. Verghese, A. McIntosh, A. Jackson, A. C. Gossard, and S. Matsuura, IEEE Trans. Microw. Theory Tech. 49, 1032 (2001).
[100] I. S. Gregory, W. R. Tribe, B. E. Cole, M. J. Evans, E. H. Linfield, a. G. Davies, and M. Missous, Appl. Phys. Lett. 85, 1622 (2004).
[101] V. Giannini, A. I. Fernández-Domínguez, S. C. Heck, and S. A. Maier, Chem. Rev. 111, 3888 (2011).
[102] S. Lal, S. Link, and N. J. Halas, Nat. Photonics 1, 641 (2007).
[103] G. C. Loata, M. D. Thomson, T. Löffler, and H. G. Roskos, Appl. Phys. Lett. 91, 232506 (2007).

[104] J. E. Pedersen, V. G. Lyssenko, J. M. Hvam, et al., Appl. Phys. Lett. 62, 1265 (1993).
[105] G. Mie, Ann. Phys. 330, 377 (1908).
[106] P. K. Jain, X. Huang, I. H. El-Sayed, and M. A. El-Sayed, Plasmonics 2, 107 (2007).
[107] L. Tonks and I. Langmuir, Phys. Rev. 33, 195 (1929).
[108] A. Boltasseva and H. A. Atwater, Science (80). 331, 290 (2011).
[109] E. Hao and G. C. Schatz, J. Chem. Phys. 120, 357 (2004).
[110] P. J. Hale, J. Madeo, C. Chin, S. S. Dhillon, J. Mangeney, J. Tignon, and K. M. Dani, Opt. Express 22, 26358 (2014).
[111] J. Madéo, N.Jukam, D.Oustinov, et al., Electron. Lett. 46, 611 (2010).
[112] S. Winnerl, J. Infrared, Millimeter, Terahertz Waves 33, 431 (2012).
[113] Y. Zeng, T. Kreouzis, X. Chen, and R. S. Donnan, IET Semin. Dig. 2015, 4 (2015).
[114] E. A. Michael, B. Vowinkel, R. Schieder, M. Mikulics, M. Marso, and P. Kordoš, Appl. Phys. Lett. 86, 111120 (2005).
[115] R. Mendis, C. Sydlo, J. Sigmund, M. Feiginov, P. Meissner, and H. L. Hartnagel, Solid. State. Electron. 48, 2041 (2004).
[116] K. A. McIntosh, E. R. Brown, K. B. Nichols, O. B. McMahon, W. F. Dinatale, and T. M. Lyszczarz, Appl. Phys. Lett. 67, 3844 (1995).
[117] E. R. Brown, K. A. McIntosh, K. B. Nichols, and C. L. Dennis, Appl. Phys. Lett. 66, 285 (1995).
[118] K. A. McIntosh, E. R. Brown, K. B. Nichols, O. B. McMahon, W. F. DiNatale, and T. M. Lyszczarz, Appl. Phys. Lett. 69, 3632 (1996).
[119] Y. Hou, J. R. Liu, M. Buchanan, et al., Nano-Micro Lett. 4, 10 (2012).
[120] I. Camara Mayorga, M. Mikulics, A. Schmitz, et al., in *Proc. SPIE - Int. Soc. Opt. Eng.*, edited by J. Zmuidzinas, W. S. Holland, and S. Withington (2004), p. 537.
[121] S. V. Makarov, V. A. Milichko, I. S. Mukhin, et al., Laser Photon. Rev. 10, 91 (2016).
[122] M. D. Wissert, A. W. Schell, K. S. Ilin, M. Siegel, and H. J. Eisler, Nanotechnology 20, (2009).
[123] C. W. Berry and M. Jarrahi, Lasers Electro-Optics Quantum Electron. Laser Sci. Conf. (QELS), 2010 Conf. 1 (2010).
[124] A. Dreyhaupt, S. Winnerl, T. Dekorsy, and M. Helm, Appl. Phys. Lett. 86, 121114 (2005).
[125] S.-H. Yang and M. Jarrahi, Opt. Lett. 38, 3677 (2013).
[126] I. S. Gregory, C. Baker, W. R. Tribe, et al., Appl. Phys. Lett. 83, 4199 (2003).
[127] O. Mitrofanov, I. Brener, T. S. Luk, and J. L. Reno, ACS Photonics 2, 1763 (2015).
[128] R. J. Thompson, T. Siday, S. Glass, et al., Appl. Phys. Lett. 110, 041105 (2017).
[129] N. T. Yardimci and M. Jarrahi, Sci. Rep. 7, 42667 (2017).
[130] M. E. Fermann and I. Hartl, Nat. Photonics 7, 868 (2013).

[131] E. U. Rafailov, M. A. Cataluna, and W. Sibbett, Nat. Photonics 1, 395 (2007).

[132] M. A. Gaafar, A. Rahimi-Iman, K. A. Fedorova, W. Stolz, E. U. Rafailov, and M. Koch, Adv. Opt. Photonics 8, 370 (2016).

[133] M. Suzuki and M. Tonouchi, Appl. Phys. Lett. 86, 051104 (2005).

[134] A. Krotkus and J.-L. Coutaz, Semicond. Sci. Technol. 20, S142 (2005).

[135] A. E. Krasnok, A. E. Miroshnichenko, P. A. Belov, and Y. S. Kivshar, Opt. Express 20, 20599 (2012).

[136] R. M. Bakker, D. Permyakov, Y. F. Yu, et al., Nano Lett. 15, 2137 (2015).

[137] S. V. Zhukovsky, V. E. Babicheva, A. V. Uskov, I. E. Protsenko, and A. V. Lavrinenko, Plasmonics 9, 283 (2014).

[138] V. G. Kravets, F. Schedin, and A. N. Grigorenko, Phys. Rev. Lett. 101, 087403 (2008).

[139] L. Li and H. Yin, Sci. Rep. 6, 26988 (2016).

[140] C. W. Hsu, B. Zhen, A. D. Stone, J. D. Joannopoulos, and M. Soljačić, Nat. Rev. Mater. 1, 16048 (2016).

[141] A. Krasnok and A. Alú, J. Opt. 20, 064002 (2018).

[142] D. G. Baranov, A. Krasnok, T. Shegai, A. Alù, and Y. Chong, Nat. Rev. Mater. 2, 17064 (2017).

[143] D. G. Baranov, A. Krasnok, and A. Alù, Optica 4, 1457 (2017).

Chapter 10

Active photonics based on phase-change materials and reconfigurable nanowire systems

Lei Kang[1], Liu Liu[1], Sarah J. Boehm[2], Lan Lin[1], Theresa S. Mayer[1], Christine D. Keating[2] and Douglas H. Werner[1]

Metamaterials that derive their properties from the subwavelength structure of meta-atoms offer unprecedented flexibility for manipulating light-matter interaction. To promote the development of practical metadevices, considerable efforts have been made to achieve photonic metamaterials that exhibit an active response under external stimuli. The primary strategy is to incorporate active materials that possess variable refractive index into the design of metamaterials. Among the various active materials, phase-change materials have recently attracted particular attention due to their variation of material properties in a broad frequency band. On the other hand, as structures and geometric arrangement of building blocks (i.e., artificial atoms or meta-atoms) determine the properties of metamaterials, reconfigurable metamaterials based on mechanical tuning have also been intensively studied. Unlike most reported mechanically tunable metamaterials based on elastic and electromagnetic forces, directed self-assembly (DSA) of nanoparticles opens a new path for low-cost, large-area reconfigurable photonic systems. In this chapter, we present our findings on active photonics metadevices based on phase transition of vanadium dioxide (VO_2) and reconfigurable nanowire assemblies. Our work shows the potential of the proposed metamaterial systems for applications ranging from electro-optical information process, storage, and display to energy-efficient smart windows.

10.1 Introduction

The emerging field of artificial electromagnetic media has undergone an explosive growth since the pioneering work in metamaterials at microwave frequencies [1,2]. Indeed, metamaterials have shown fascinating capability in manipulating

[1]Department of Electrical Engineering and Center for Nanoscale Science, The Pennsylvania State University, University Park, PA, USA
[2] Department of Chemistry, The Pennsylvania State University, University Park, PA, USA

electromagnetic waves based on the engineering of meta-atoms, i.e., the subwavelength building blocks that are used to construct the media in a periodic manner. In particular, a series of unusual physical phenomena including artificial magnetism [3], negative refraction [2], and invisibility cloaking [4] have been observed in electromagnetic metamaterials. Given the geometric scalability of meta-atoms, the concept of metamaterials has also enabled photonic nanostructures that are capable of realizing exotic material properties such as negative refractive index [5], optical magnetism [6], giant optical chirality [7], and hyperlensing effects [8] in the optical regime, although the dispersion of metals causes significant Ohmic loss at optical frequencies. More recently, researchers have shown that optical metasurfaces based on single-layer or few-layer planar nanostructures can be used to achieve efficient manipulation of optical waves [9,10], overcoming the high optical loss and fabrication complexity that are generally associated with 3D metamaterials. Nevertheless, arising from the structural properties of meta-atoms, the generally fixed response of metamaterials presents a severe caveat that will have to be addressed to enable practical applications.

Metamaterials with dynamically tunable responses have been intensively studied since the emergence of the field [11–16]. In brief, two primary strategies have been employed to achieve tunability, i.e., incorporating active materials that possess variable refractive index into the design of metamaterials and mechanically tuning the structure of the metamaterials. However, compared with the counterparts at lower (microwave and THz) frequencies, optical metamaterials that can be dynamically tuned over a large scale remain rare. This can be attributed to three factors. First, there are rather limited active materials that can provide a significant refractive index variation under application of external fields. Second, given the fact that optical meta-atoms are generally on a nanometer scale, integration between the nanostructures and active media becomes particularly difficult. Third, changing the geometric structure, including the arrangement of the nano-resonators in a controllable manner, is extremely challenging.

Based on these understandings, we recently investigated two distinct photonic metamaterial systems for the realization of metadevices operating in the mid- and near-infrared (IR) regimes. In particular, our study shows that the hybridization of vanadium dioxide (VO_2) and plasmonic nanostructures can be delicately balanced to give rise to new devices with a series of mid-IR optical functionalities. Moreover, based on the gold nanowires fabricated with the bottom-up method, we show that electric field directed assembly could facilitate reconfigurable photonic devices in a broad wavelength range. Our findings further demonstrate the potential for the development of practical metadevices capable of converting electrical stimuli to multifunctional control of light.

10.2 Phase transition enabled tunable metadevices

The phase-change phenomenon is ubiquitous in nature—all substances experience phase transition at certain temperature and pressure conditions. As phase change is in general accompanied by heat absorption and release, phase-change materials (PCMs) with a high heat of fusion have been widely used for a variety of thermal energy related applications. Besides inducing a change in thermal properties, phase transition generally also has a significant influence on many other physical properties of materials, such as density, volume, electrical conductivity, and refractive index, to name a few. In a stark contrast to those (such as water or bee wax) that experience a transition in their state of matter at a certain temperature, PCMs that solely change their crystalline structure when undergoing a phase transition have a great potential for applications beyond thermal energy manipulation. An important example is GeSbTe (germanium–antimony–tellurium or GST) which has been successfully employed to enable information storage devices such as rewritable optical discs and memory devices, in which the phase transition of GST between an amorphous state and a crystalline state provides the basis for rewritable information storage.

Besides GST, vanadium dioxide (VO_2), a classical transition metal oxide, has also attracted great attention due to its insulator-to-metal transition (IMT). In contrast to the crystallization process of GST that requires a temperature variation between 150 and 600°C [17,18], the IMT of VO_2 occurs around room temperature ($T_{IMT} \sim 67$°C in bulk crystals) [19]. In particular, the atomic structure of VO_2 that can be described by a monoclinic lattice at low temperature (insulator phase) is rearranged to a tetragonal structure at the transition temperature (metallic phase), resulting in strong lattice vibrations and corresponding mobile electrons. Using scanning near-field IR microscopy, Qazilbash *et al.* have directly imaged the emergence and evolution of the nanoscale metallic puddles during the IMT of VO_2 [19]. These observations unambiguously revealed the development of the transformation of VO_2 from an insulator to a metal and can also explain the charge dynamics and other inhomogeneities of VO_2 at the nanoscale. The phase transition of VO_2 gives rise to a series of dramatic variations in its material properties, which has led to an increasing interest in the use of VO_2 for a variety of applications in a number of widely differing technologies. For instance, due to the abrupt change in the electronic properties (e.g., conductivity) at the IMT, VO_2 has been considered as a promising candidate for the key material in next-generation transistors [20,21]. Owing to its near-room-temperature IMT, VO_2 in the form of thin films has also been used to facilitate IR radiation management [22,23]. Moreover, VO_2 has also been proposed to enable devices for computing and information storage [24].

Besides the interesting electronic properties, the IMT may also lead to drastic variations in the refractive index of VO_2 over a broad spectral range. This characteristic, which is distinct from the index change of most other active materials, makes VO_2 particularly attractive for use in devices for active electromagnetics and photonics. Another important fact is that the IMT of VO_2 can be triggered by a variety of mechanisms, including thermal excitation [19], optical pumping [25,26], charge injection [27], electrical signals [28], etc. These intriguing properties show that VO_2 is ideal for low-power consuming and ultrafast photonic devices. Indeed, by introducing VO_2 into metallic structure-based metamaterial/metasurface systems, researchers have demonstrated a number of active metadevices for a variety of applications from microwave to optical frequencies. For instance, Driscoll *et al.* have reported memory metamaterials at THz [29], and Liu *et al.* have demonstrated that the IMT of VO_2 can be induced by the largely enhanced THz field in the gap of split ring resonators (SRRs) [30]. In the optical regime, Dicken *et al.* have shown that near-IR reflection spectra of an array of SRRs fabricated on top of a VO_2 film is strongly dependent on temperature [31]. More recent studies have demonstrated that the IMT of VO_2 can be used to manipulate a series of properties of metasurfaces, including the plasmonic color [32,33], extinction ratio [34], and polarization sensitivities [35].

The electrically triggered IMT of VO_2 actually allows electrically tunable metadevices that are essential in practical applications. Despite controversy over their microscopic origins, electrical current and field-effect induced phase transitions in VO_2 have been reported previously [20,21]. It has been shown that the observed IMT in VO_2 for these devices only occurs at localized regions, where electrical current filaments are generated under an applied electric field. Consequently, this may not lead to sufficient macroscopic change in the material properties to produce detectable modulations in metadevices. On the other hand, based on a comprehensive use of nano-engineered metals, researchers have demonstrated electro-optic plasmonic and photonic systems in which the metallic structures simultaneously support the optical resonances and other functionalities such as electric field application [19–21]. The corresponding systems are referred to as "self-sufficient" or "self-contained." Actually, other conventional properties such as high thermal conductivity can be further exploited to expand regulating approaches for tunable metadevices based on different active materials [36–38].

10.2.1 An electrically actuated VO_2-hybrid metadevice

We recently demonstrated a VO_2-hybrid metamaterial platform for electrically triggered multifunctional control in the mid-IR region [39]. Figure 10.1 illustrates the schematic of the unit cell and the corresponding hybrid metamaterial that consist of a mesh-patterned gold film, which is separated from an optically thick gold back plane by a thin VO_2 film. A thin (50 nm) alumina spacer is introduced at both the VO_2 and Au interfaces to reduce the influence of the VO_2's loss on the resonance behavior of the hybrid metamaterials at room temperature. It can be seen that, in contrast to most of the reported VO_2-based metamaterials in which the VO_2 serves as a bulk substrate, in our design a subwavelength-thick VO_2 film is incorporated into the metamaterial as part of the resonating structure. This design strategy allows the metadevice to leverage the response regulation provided by the index change of VO_2 during its IMT to the greatest extent. Furthermore, besides supporting the so-called perfect absorber optical resonance mode [40], the top mesh-patterned Au layer, which is topologically continuous and connected to an external circuit, can simultaneously facilitate Joule heating as a result of the in-plane current flow. Therefore, the optical responses (reflection/absorption) of the proposed metamaterial can be tuned by applying electrical current that thermally triggers the IMT in the VO_2.

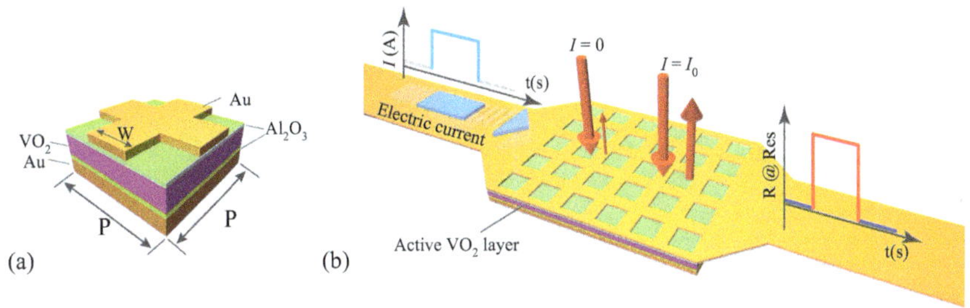

Figure 10.1 *(a) Schematic of one lateral unit cell of the VO_2-hybrid metamaterial composed of a sandwich system with a patterned-mesh top gold (Au) layer (100 nm), a VO_2 layer (260 nm), and an optically thick Au ground plane (P = 1,550 nm, w = 600 nm). To optimize device performance, a 50 nm-thick Al_2O_3 layer was applied in between both gold/VO_2 interfaces. (b) Configuration of the electrically active metadevice. Besides supporting optical resonances, the top patterned gold layer was connected to an external circuit to enable computer-controlled Joule heating that triggers the IMT in the VO_2 layer. Consequently, the optical responses (e.g. reflection characteristics) of the metadevice may vary as a function of the electrical current flowing through the layer. Reproduced with permission [39]. Copyright 2016, Nature Publishing Group.*

The dielectric property ($\varepsilon = \varepsilon_1 + i\varepsilon_2$) of a VO_2 film was firstly measured using ellipsometry. As shown in Figure 10.2(b), in the wavelength range of interest, VO_2 can be treated as a low-loss dielectric with ε_1 around 8 at room temperature; while at a temperature higher than T_{IMT}, VO_2 may behave like a "poor" metal possessing negative ε_1 with high loss. Using the measured permittivity of VO_2, full-wave electromagnetic simulations were performed to elucidate the optical properties and the corresponding response modulations of the proposed hybrid metamaterial. Figure 10.2(a) shows that, at room temperature, two sharp dips can be identified in the reflectance spectrum of the metamaterial for normal incidence at wavelengths of 2.95 and 3.58 μm. As the reflectance, transmittance, and absorption properties of the metamaterials satisfy the relation $R(\omega) = 1 - A(\omega)$ and $T(\omega) = 0$, the observed reflectance dips represent the near-unity absorption at the two wavelengths. In addition, the electric field distribution in the spacer (Figure 10.2(c) and (d)) clearly illustrates the profile of the absorber modes at the two resonances. Furthermore, from the reflectance spectrum at 85°C (Figure 10.2(a)), one can see that at temperatures higher than T_{IMT}, the metamaterial becomes highly reflective in the entire wavelength range of interest. This indicates that the IMT-enabled large variation in the reflectance spectra of the system. Considering the subwavelength thickness of the VO_2 film (~260 nm, i.e., less than $\lambda_{res}/12$) in the hybrid metamaterial, the proposed design strategy provides an opportunity for creating highly sensitive and ultra-compact electro-optic metadevices.

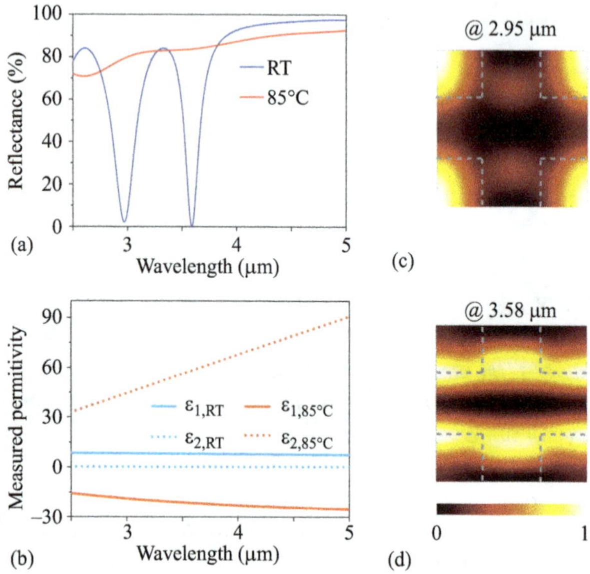

Figure 10.2 (a) Simulated reflectance spectra of the VO_2-hybrid metamaterial at room temperature and 85°C. (b) Measured permittivity of a VO_2 film on top of Au at room temperature and 85°C. Simulated electric field distribution ($|E|^2$) within the VO_2 layer for a unit cell at the wavelengths of (c) 2.95 and (d) 3.58 μm. Reproduced with permission [39]. Copyright 2016, Nature Publishing Group.

To demonstrate the proposed tunable response, the electrically active VO_2-hybrid metadevice was fabricated on a Si substrate. The photo image of the device and the scanning electron microscope (SEM) image of the metamaterial area are shown in Figure 10.3(a) and (b), respectively. It can be seen that the top patterned Au layer was extended to two self-supporting electrodes that can be connected to an external circuit for the application of electrical current. It should be noted that high-quality VO_2 films are the key factors in realizing the desired tunability of the proposed metadevices. Although the growth of VO_2 films has been intensively studied in the past decades and a number of preparation procedures (such as epitaxial growth, pulsed laser deposition (PLD) and sputtering) have been developed, the fabrication of high-quality VO_2 films on an amorphous substrate remains a challenging task. Using a two-step sputter/anneal process, we successfully grew a single phase polycrystalline VO_2 thin film on top of a 50-nm-thick Al_2O_3 barrier layer that was obtained by atomic layer deposition (ALD) on an unpatterned gold film. Figure 10.3 (c) depicts the cross-sectional SEM image of the film stack after the VO_2 film was grown and Figure 10.3(d) shows the atomic force microscopy (AFM) surface height profile of the fabricated VO_2 film. We note that the microscopic structural properties, including the grain size of the VO_2 film, can be further improved through fabrication process optimization. Nevertheless, the characterization results, including

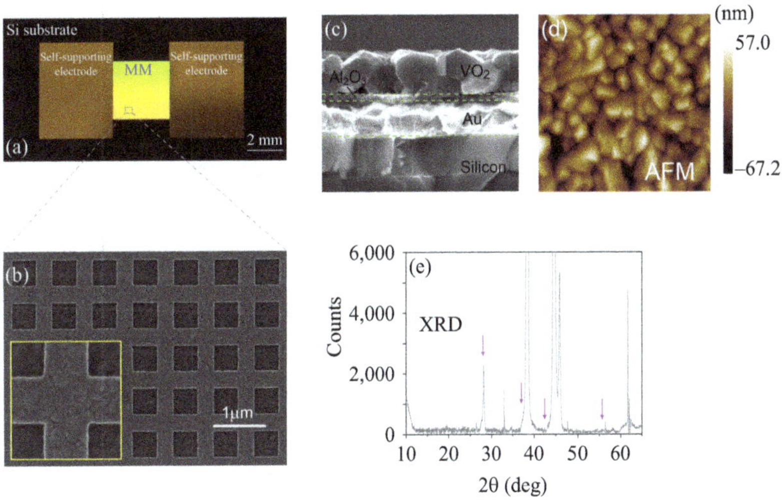

Figure 10.3 *(a) Photo of the device. (b) SEM image of the metamaterial area. Inset shows an enlarged SEM image of a unit cell. (c) Cross-sectional SEM image of the film stack after growth of the VO_2 film. (d) AFM surface height profile and (e) XRD spectrum of the fabricated VO_2 film. Reproduced with permission [39]. Copyright 2016, Nature Publishing Group.*

the dielectric properties (Figure 10.2(b)) and the X-ray powder diffraction (XRD) spectrum (Figure 10.3(e)), indicate that the obtained VO_2 films are of good quality and can be used for studying the IMT-induced tunable response from the hybrid metamaterials.

In order to clearly identify the tuning mechanism, before examining the electrically tunable behavior of the proposed metamaterials, the reflection behavior of the device as a function of temperature was firstly characterized. As the schematic in Figure 10.4(a) shows, the fabricated metadevice was mounted on a thermoelectric Peltier heater but not connected to the external electrical circuitry. Figure 10.4(b) shows the reflectance spectra that were collected by using a Fourier transform IR (FTIR) spectrometer at a series of heater temperature values after the sample reached thermal equilibrium. It is seen that at room temperature (24°C) two sharp dips can be identified in the reflectance spectra around 3.05 and 3.85 μm; similarly at 74°C, the spectral features disappear when the metamaterial becomes highly reflective in the entire wavelength range of interest. These observations agree well with the simulations. More importantly, the reflectance at the two resonances continues to increase when the device temperature increases, before being saturated at around 70°C. Based on the data collection for a cyclic change of the device temperature, the on-resonance reflectance as a function of increasing and then decreasing temperature was obtained (Figure 10.4(c) and (e)), illustrating a complete picture of the temperature-dependent tuning effect. It can be seen that the

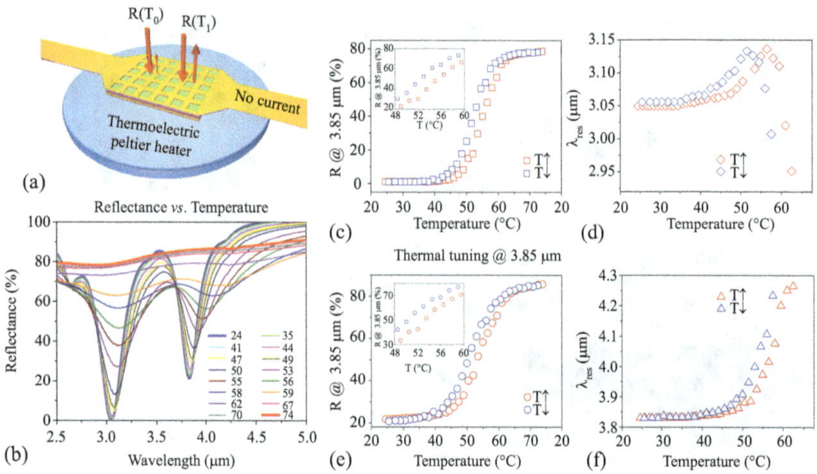

Figure 10.4 (a) Schematic of the testing system for thermal tuning of the hybrid metamaterial. (b) Measured reflection spectra at a series of device temperatures. (c) and (e) The on-resonance reflectance for a cyclic change of the device temperature. (d) and (f) The corresponding resonance wavelength (λ_{res}) as a function of the temperature. The temperature sweeping rate was 2°C min^{-1}. Reproduced with permission [39]. Copyright 2016, Nature Publishing Group.

reflectance at the two resonances starts to increase only when the temperature exceeds 40°C. This can be understood as a result of the gradual development of the metallic phase in VO_2 [19]: the nanoscale metallic puddles grow as the temperature increases and only start to significantly change the macroscopic properties (such as permittivity) when their size (volumetric factor) reached a certain value. Another observation is that in the temperature range from ~40 to ~60°C, the reflectance values at the same temperature are different for the temperature increasing and decreasing processes. This effect, i.e., the dependence of device response on the change history of the device temperature, arises from the intrinsic hysteresis property of VO_2. Besides the change in the on-resonance reflectance value, the temperature variation also leads to changes in the resonance wavelength λ_{res}. Figure 10.4(d) and (f) shows the dependence of λ_{res} of the two resonances as a function of temperature, in which the continuous tuning and hysteresis characteristics are also identified.

Based on these observations, the electrical tuning of the metamaterials' optical response in *thermal equilibrium* was further determined by performing FTIR reflection measurements for a series of electrical currents (I) applied to the top mesh-patterned Au layer. Figure 10.5(a) depicts the corresponding current-dependent reflectance spectra. It can be seen that the reflectance around the two resonances increases with the increasing value of applied current and saturates when the intensity of I exceeds 1.0 A. The similarity found between Figures 10.4(b)

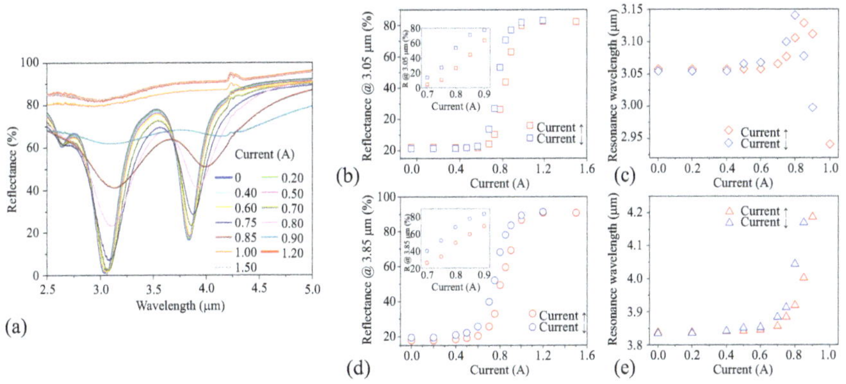

Figure 10.5 *(a) Measured reflection spectra for a series of intensities of electrical current applied. (b) and (d) The on-resonance reflectance for a cyclic change of the electrical current applied. (c) and (e) The corresponding resonance wavelength (λ_{res}) as a function of current. Reproduced with permission [39]. Copyright 2016, Nature Publishing Group.*

and 10.5(a) reveal the origin of the electrical current-based tuning effect, i.e., the nano-engineered metals enable the electrical signal controlled phase transition through the Joule heating effect. Moreover, collecting data for a cyclic change of the electrical current, in Figure 10.5(b)–(e) we illustrate the on-resonance reflectance as a function of increasing and then decreasing I. Similar to the case in direct thermal modulation (Figure 10.4(c)–(f)), a current threshold of ~0.5 A is identified and, for intensity of I in the range from 0.5 to 1.0 A, the reflection behavior (reflectance and resonance wavelength) at the two resonances was continuously tuned. Hysteresis characteristics in the reflection modulation are also seen. These observations provide the basis for constructing a self-sufficient platform that supports electrically tunable properties enabled by the IMT of VO_2. In the following Sections from 10.2.1.1 to 10.2.1.3, a series of proof-of-concept demonstrations are presented to highlight the versatility of the proposed metadevice for multi-functional control.

10.2.1.1 Electrically switchable reflection

Temporal characteristics of active devices under external stimulus are of great importance for practical applications. Therefore, based on the response tuning behavior in *thermal equilibrium* shown in Figure 10.6, on-resonance reflectance modulations are further examined by applying a series of electrical current pulses while recording the magnitude of the reflectance (Figure 6(a)–(d)). A 4.0-A-high and 0.25-s-wide electrical current pulse train was used to "trigger" the phase transition of VO_2 in the metadevice via a Joule heating effect, leading to the on-resonance reflectance switching as a function of time. It can be seen that the reflectance at 3.05 μm (3.85 μm) "jumps" from ~0% (~25%) to ~80% IMT when

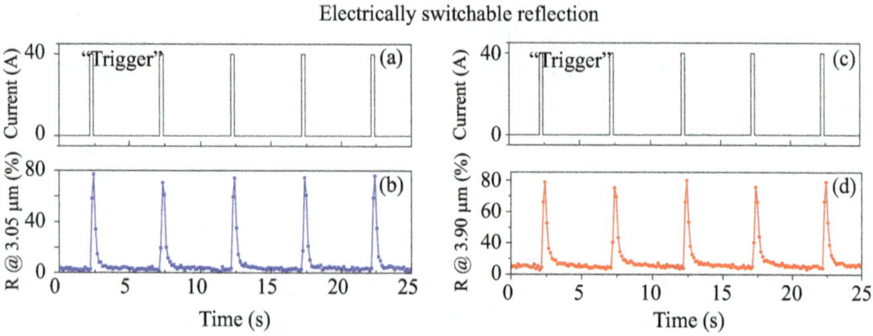

Figure 10.6 Electrically switchable reflection. (a) and (c) Electrical current pulses (0.25 s width) used to "trigger" localized MIT in the VO_2 film. The observed electrical switching of on-resonance reflectance at wavelengths of (b) 3.05 μm and (d) 3.90 μm. Reproduced with permission [39]. Copyright 2016, Nature Publishing Group.

the trigger current pulse was implemented. It should be noted that the time resolution (~0.12 s) available in the employed FTIR system limits the determination of the electro-optic response time of the metadevice. To overcome this, the response of 5 pulses was measured and superimposed into one 5 s window, which allowed us to better resolve the electrical modulation with a finer time step. Rising and falling edges of ~0.2 and ~0.5 s, respectively, are identified from the measured time trace of the on-resonance reflectance. However, the limited time-sampling rate of the spectroscopy system still causes a slight fluctuation of the maximum tuning magnitude observed in Figure 10.6(b) and (d). Nevertheless, these results clearly indicate the capability of the proposed VO_2-hybrid metadevices to enable electrically switchable reflection. It should be noted that although the most significant modulations are found at the two resonances, the IMT of VO_2 actually results in reflection switching in the entire frequency band of interest. The temperature fluctuations of the device arising from environmental temperature changes were kept within 0.3°C during the measurements.

Thermal analysis shows that the current intensity, which is associated with the Joule heating power, primarily determines the temperature rise of the VO_2 film, while the pulse duration controls the total power applied to the device and the actual temperature that the film can reach. It should be noted that as the layered structure of the VO_2-hybrid metamaterial was fabricated on top of a silicon substrate and measured in air, the thermal energy may simultaneously dissipate to the adjacent environment. Therefore, the application of current pulses of larger magnitude and shorter duration will be helpful to achieve more localized temperature change in the device. On the contrary, electrical pulses of lower magnitude and longer duration may lead to global Joule heating that causes a temperature increase not only for the metamaterials but also for the environment, causing unnecessary energy consumption and slower modulation of the device. Therefore, by applying

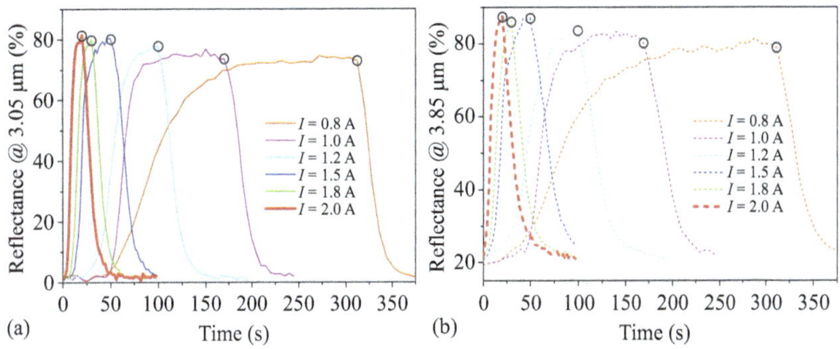

Figure 10.7 An additional examination of the electrically switchable reflection of the VO_2-hybrid metamaterial. (a) and (b) On-resonance reflectance as functions of time for different applied current pulses. Reproduced with permission [39]. Copyright 2016, Nature Publishing Group.

an electrical current pulse train of different intensity and duration, we further conducted a more detailed study of the temporal performance of metadevices.

Figure 10.7 shows the corresponding time trace of the on-resonance reflectance obtained from a sequential measurement of 50 reflectance spectra during the heating-cooling cycle under each experimental condition. In our experiments, the resistance of the Au mesh of the sample used was $R = 1\ \Omega$, and accordingly, the Joule heating power was 0.64, 1, 1.44, 2.25, 3.24, and 4 W, respectively. Clearly, the "on" and "off" reflectance switching time at both resonances decreases with an increase in the Joule heating power. Moreover, no drop of the modulation depth was observed with a 2.0-A-high and 20-s-wide current pulse. We emphasize that VO_2-hybrid metamaterials have the potential for ultrafast modulation and switching. In fact, the IMT process of VO_2 at a subpicosecond time scale was previously reported in different systems involving thermal processes [26,30]. Nevertheless, it should also be noted that, besides the Joule heating process, the temporal performance of the proposed metadevice is also constrained by the thermal properties at the device level. Therefore, the dynamic response of the device can be further improved through thermal engineering, e.g., reducing the thermal capacity of the device. This can be achieved through, e.g., better crystallization control of VO_2 films and thickness re-optimization of the low thermal conductivity buffer layer (Al_2O_3).

10.2.1.2 Electrically programmable memory effect

Compared with other active media that enable tunability of metamaterials by providing a variable dielectric environment, VO_2, exhibiting the unique characteristic of hysteresis, offers the possibility of constructing metadevices with more complex functionality beyond only magnitude tunability. Indeed, using a hysteretic dielectric response at THz frequencies, Driscoll *et al.* have demonstrated a THz metamaterial that exhibits a memory effect [29]. A close inspection of the results shown in Figure 10.5(b) and (d) indicates that, besides the reversible reflection switching that can be obtained by applying I at low (or zero) and saturated levels, the proposed VO_2-hybrid metadevices show the potential for realizing an electro-optic memory device. Figure 10.5(b) and (d) also reveal that the maximum hysteresis point occurs at $I = 0.8$ A, which can be used as the ideal initial state for an electrical current triggered memory effect. Accordingly, a group of electrical current pulses was implemented to study the hysteresis of the VO_2-enabled memory effect in the proposed metadevice. In particular, as shown in Figure 10.8(a), positive (0.8 A, 0.25 s) and negative (–0.8 A, 0.75 s) current pulses referred to as the "write" and "erase" inputs of this memory process, respectively, were applied in a circular sequence on top of a 0.8 A bias current.

Figure 10.8(b) and (d) shows the corresponding time-trace of on-resonance reflectance. It can be seen that, in each sequence of current pulse application, the device shows steady on-resonance reflectance at two distinct levels before and after the "write" pulses, whereas "erase" pulses clean the influence of the "write" pulse and bring the reflectance back to the initial level. As indicated by the dashed lines, the application of electrical current pulses actually created two reflection states, i.e., the "0" and "1" states, at the two resonances with a ~10 and ~5% reflectance contrast, respectively. This flip-flop operation in reflectance indicates that the metadevice can be used as an electrically erasable and programmable memory device, in which the programmed reflectance can be read out as long as the information is stored.

Three distinct advantages distinguish the proposed metadevice from the previously reported metamaterials with memory effect: electrical biasing without an external temperature control, operating in the mid-IR wavelength region and exhibiting dual-band binary reflectance contrast for storing and reading information. In addition, the subwavelength thickness of the VO_2 film and the planar structure allow the proposed metamaterials to achieve good energy efficiency in the

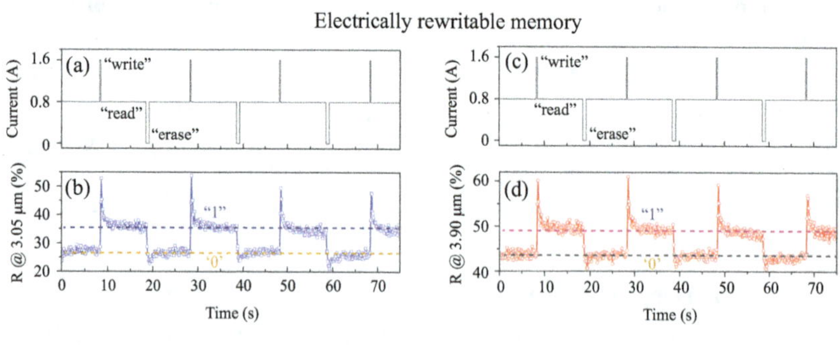

Figure 10.8 Electrically rewritable memory effect. (a) and (c) Electrical current pulses used for the study of memory effect. (b) and (d) The measured time trace of the on-resonance reflectance indicating the memory effect in the metadevice. Reproduced with permission [39]. Copyright 2016, Nature Publishing Group.

response modulation. In particular, experimental data shows that the input power level in the switching measurements was ~1 mJ per unit cell per pulse, which may be further reduced by reducing the device footprint and through thermal engineering. Moreover, recent studies have shown that the phase transition temperature of VO_2 can be tailored by various materials engineering approaches such as chemical doping. For instance, Paik *et al.* reported near-room-temperature IMT of VO_2 thin films obtained from tungsten-doped colloidal VO_x nanocrystals [41]. These studies show the potential of the proposed VO_2-hybrid metadevices for nonvolatile memory devices that require no bias.

10.2.1.3 Electrically tunable IR images

As mentioned above, the optical responses of the proposed hybrid metamaterial primarily arise from the excitation of the perfect absorber modes in the structure. Owing to their flexibility in engineering optical properties, static perfect absorbers have been demonstrated as spatial light modulators based on spatially selected absorptivity [42]. Beyond that, based on the electrically active responses, the proposed metadevice can enable even more sophisticated spatial modulation schemes of IR signals. As the optical microscope and SEM images in Figure 9(a)–(c) shows, a metadevice with spatial information was fabricated to investigate this capability. In particular, a metamaterial with the same structure as studied above was firstly fabricated and then, some area of the top gold mesh layer was removed to form three letters, "PSU." Moreover, to enable the electrical control, the top gold layer was again extended and connected to an external circuit. It should be noted that, compared to the metamaterial with complete structure, the VO_2/ground-plane layered structure (in the absence of the patterned gold layer) may exhibit a distinct reflection behavior. In particular, the reflection from the letters' regions is expected to decrease with increasing device temperature due to dielectric property variations in VO_2 [43]. Consequently, the corresponding metadevice can potentially be used for a complicated spatial information display.

Figure 9(d)–(f) shows the measured 2D IR images at a series of electrical currents. In particular, it can be seen that the letters "PSU" were highly reflective in a low-reflective background when no current was applied to the system (Figure 10.9(d)). When $I = 2.03$ A (Figure 10.9(e)), the three letters faded out in the IR image and became invisible. Moreover, when the applied current was further increased (Figure 10.9(f)), the areas corresponding to the three letters turned out to be an *intaglio* on a highly reflective background. To quantitatively assess these IR images, the spatial distribution of reflectance across each of the letters (shown in Figure 10.9(d)) are plotted and displayed in Figure 9(g)–(i). It should be noted that the IR images were acquired at a wavelength of 2.67 μm, which is off from the resonance wavelengths identified in Figure 10.5(a) discussed above. This deviation of spectroscopic response is due to the off-normal incidence in the imaging process (with effective numerical aperture (NA) = 0.5) and can be clearly seen from the corresponding reflectance spectrum (Figure 10.10). Furthermore, we note that at wavelengths (such as 10 μm) far away from the resonance mode, no current-

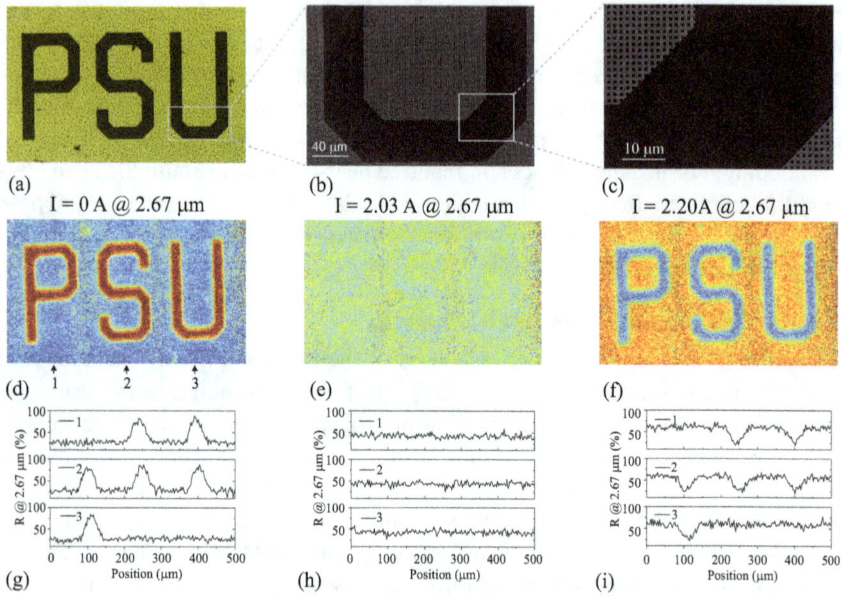

Figure 10.9 Electrically tunable IR images. (a) Microscopic image of the fabricated metamaterials with an area of the top Au layer removed to form the letters "PSU." (b and c) SEM images of a corner of the patterned letter "U." (d)–(f) IR images at a wavelength of 2.67 mm, when a series of electrical currents were applied. (g)–(i) The reflectance distribution across the images at locations labeled by "1," "2," and "3." Reproduced with permission [39]. Copyright 2016, Nature Publishing Group.

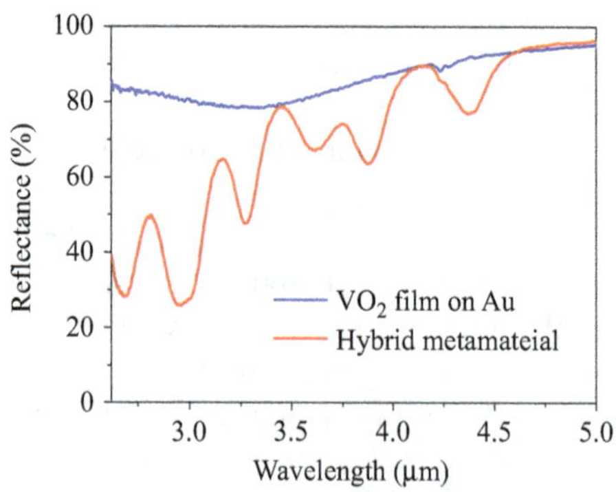

Figure 10.10 Room temperature reflectance spectra of a VO_2 film on Au and the hybrid metamaterial measured with an FTIR imaging microscope. Reproduced with permission [39]. Copyright 2016, Nature Publishing Group.

dependent variations were observed in the IR images, further revealing the resonance-based spatial light modulation capability of the proposed metadevices.

10.2.2 Conclusions and future studies

A series of proof-of-concept studies presented here show that electrified metamaterial absorbers integrated with VO_2 could be utilized as self-sufficient multifunctional metadevices. In particular, the intrinsic characteristics of the proposed metadevices, such as temporal, frequency, and spatial optical responses, can be tuned through the application of electrical current. This study reveals that the demonstrated VO_2-hybrid metadevices have the potential to be used as an active photonic platform for electrically triggered optical information process, storage, and display. After the publication of our findings, researchers have also demonstrated the electrical signal-controlled response from VO_2-hybrid metamaterials in another wavelength range [44], further indicating the design flexibility of the proposed metadevices.

Given the fact that emissivity of an object at equilibrium should be equal to its absorptivity (Kirchhoff's law), metamaterial absorbers have been used as thermal emitters with structure-determined emission properties [45]. In addition, based on the temperature-sensitive optical properties of VO_2, unusual emission behavior of the material in the form of thin films was recently reported [43]. Therefore, we envision that the proposed VO_2-hybrid metamaterials can enable IR emission management, including electrically controlled spatial and temporal emissivity. In particular, as the emissivity (absorption) of the proposed VO_2-hybrid metadevice decreases with the increase of device temperature; from the point of view of thermal detectors the device may look cooler when it is electrically heated up. Taking into account the metamaterial enabled flexibility in the design of resonance behaviors, this property provides the potential for manageable IR camouflage. Primarily determined by its phase transition temperature, VO_2-loaded metadevices are suitable for the management of thermal emission at far-IR wavelengths, for example, 5–15 μm. To enable control over the thermal emission in the near- and mid-IR regions, PCMs with higher transition temperature, such as NbO_2 ($\sim$800°C) [46], are expected to be introduced into the metamaterials.

Although the response tuning of the hybrid metamaterials observed in our study was triggered by electrical signals, its origin is the thermally induced phase transition in VO_2. As we discussed in Section 10.2.1, there has been evidence that the IMT of VO_2 can also be triggered by electric fields [20,21]. Given the fact that the proposed metamaterials are formed by two topologically continuous metallic layers, they can naturally support the electro-optic functionalities based on field effects, implying the potential of the metadevices to support ultrafast response modulations. It should be noted that constructing such electro-optic VO_2-hybrid metadevices may be a challenging task. This is because it may involve both a deeper theoretical understanding of the electronic dynamics during VO_2's phase-change process, which may help to avoid the localized transition, and difficulties in fabrication due to the properties (such as high-quality film growth, electrical insulation, optimized circuit response, etc.) required by electro-optic applications. Nevertheless, given the versatility and practicality of electro-optic metadevices, the corresponding investigations may be worth the effort because they can help to establish a new paradigm in metamaterial-based systems for electrically active photonics.

10.3 Nanoparticle assembly-based metadevices

Nano- and microparticles, referred to as natural or artificially synthesized particles with nanometer- or micrometer-size features, provide a means to manipulate light through controls such as their geometrical structures and arrangement. For instance, gold nanoparticles have a long history of optical use as the color tuner of glass, which has recently been shown to be a result of surface plasmon resonance (SPR), a phenomenon that is increasingly understood as one of the most important phenomena of modern optics [47]. Supporting SPR, the metallic particles are accordingly termed as plasmonic particles. On the other hand, it has been further found that creating subwavelength dimensions of plasmonic structures is critical in enabling strong SPR in the optical regime. To this end, top-down methods such as electron-beam lithography (EBL) and focused-ion beam (FIB) have been widely used in fabrication of nanophotonics systems consisting of nano- and microscale structures. Although top-down approaches enable fabrication of nanostructures with precise control over size and spatial arrangement, they generally suffer from certain disadvantages, including high-cost, low-yield, low-areal density, etc. In a stark contrast, nano- and microparticles with controlled size and material composition can be chemically synthesized using high-yield and inexpensive bottom-up approaches, which provides the basis for large-area plasmonic devices based on collective properties of particle assemblies for practical applications [48,49]. More importantly, it has been learned that adjustment of characteristics of the assemblies such as density, arrangement, and orientation can be achieved by using external stimulus, which is expected to be the key driver behind the efforts in nano- and microparticle research owing to the potential for realizing material systems with reconfigurable responses [50]. However, the fields of particle synthesis, assembly, and device-level design are largely disconnected, posing considerable challenges.

Directed self-assembly (DSA) refers to the process by which nano- or microparticles organize into certain ordered structures exploiting an external field to modulate the thermodynamic forces. In spite of the employment of the basic principles of conventional self-assembly, DSA can give rise to highly ordered nonequilibrium structures because the corresponding systems can "reside" in a state of local equilibrium [50]. Indeed, leveraging the collective response of DSA-enabled structural materials researchers have demonstrated devices for a variety of applications ranging from electronics [51,52] and chemical sensors [53] to photonics systems [54–56]. Advantages in component diversity offered by DSA-facilitated structures due to the diverse materials are also of great interest. Nevertheless, achieving specific assembly structures for desired switchable materials that could serve as components in on-chip devices remains challenging, because it requires DSA that achieves reconfigurability goals and simultaneously provides high location/orientation accuracy.

Among the different DSA strategies, approaches that combine top-down lithography-based surface patterning and field applications are considered to be the most promising for production of assemblies that are well-organized and actively controllable. Dielectrophoresis (DEP) force and the electrohydrodynamic (EHD) flow effect arising from electric field application across on-chip micropatterned electrodes has been used to provide the requisite control over particle orientation and placement [57–59]. Since the corresponding forces can be tuned in real time by changing the applied field conditions (such as the magnitude and frequency), this method can actually be implemented to facilitate assembly, reorganization, and optimization. In light of this, electric field-based DSA expands the opportunities for the development of active electro-optic devices such as ultra-compact polarization components, electro-optic switches, and on-chip modulators. It should be noted that compared with the reconfigurable metamaterials based on elastic and electromagnetic forces [16], bottom-up nanoparticle assembly provides an alternative route for low-cost, large-area photonic metadevices with reconfigurable functionalities [60,61].

Here, we present a reconfigurable nanowire lattice polarizer as a proof-of-concept demonstration of a DSA-enabled metadevice. Information about synthesis of the employed nanowire, full-wave simulations, device-level design and fabrication, and the optical measurements and analyses is presented to provide the complete picture of the proposed metadevice system. Finally, conclusions and a scope for future research are provided.

10.3.1 Reconfigurable IR-polarizer based on nanowire assemblies

10.3.1.1 The synthesis of nanowires

Templated electrodeposition is used to synthesize the gold nanowires. This method has been previously reported by different groups [62–66]. Briefly, particles were prepared by sequential electrochemical reduction of metal ions into the pores of membrane templates [65,66], followed by particle release in solution (Figure 10.11) [67]. As shown in Figure 10.11, first, using E-beam evaporation, a silver film was deposited on the backside of an anodic aluminum oxide (AAO) membrane. In the following electrodeposition process, this silver film serves as the working electrode for reduction of metal ions from solution. Second, metal salts were electrically plated into the porous AAO membrane with a nominal pore diameter of 200 nm [64]. Consequently, the dimensions of the synthesized nanowires are determined by two factors, i.e., membrane pore diameter and deposition duration, which govern the particle width and length, respectively. Next, nanowires are released by dissolution of the template membrane. Last, through a modified sol-gel method, nanowires are coated with an amorphous silica shell [68]. It should be noted that, different from the previously reported

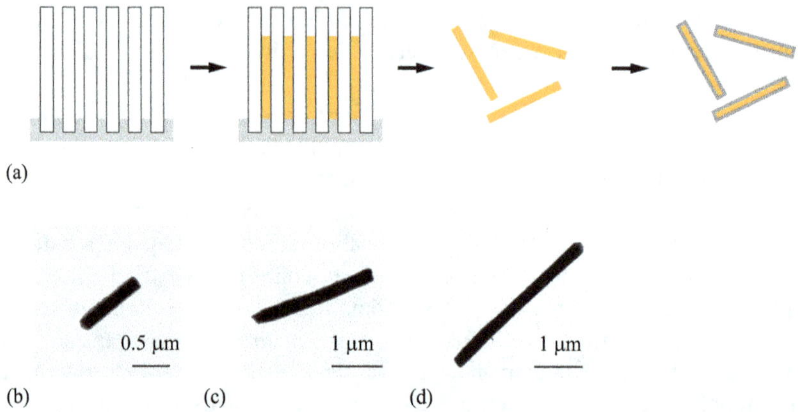

Figure 10.11 Synthesis of gold nanowires. (a) Schematic of the synthesis process using templated electrodeposition. (b)–(d) Transmission electron microscopy (TEM) images of 1.1, 2.2, and 3.7 μm Au nanowires. Reproduced with permission [75]. Copyright 2017, Wiley-VCH.

"surface sol-gel" approach [69], in our case the SiO_2 coating is synthesized in a single step and covers the entire wire surface with a thickness controlled by the reaction time. We note that the dielectric shell is important, because metallic particles assembled across bare electrodes can cause electric shorting that may lead to particle damages, field dropping, and poor assembly quality. Consequently, in our studies, the individual metallic particles were coated by a thin (30 nm) silica layer. Experiments demonstrated that the silica coating can well prevent electric shorting and particle aggregation, enabling high-quality and reversible assemblies. It was previously reported that this electrical shorting can also be prevented by covering the electrodes with a thin dielectric layer [70,71]. It should be noted that, besides the alumina film used in our experiments, there are other well-defined alternative templates, such as the commercially available polycarbonate, nano-scale hole arrays obtained from interference photolithography [72], self-organized copolymers [73], and nanoindented bulk Al from electrochemical oxidation [74].

10.3.1.2 The design and fabrication of microelectrodes

In order to provide the local alternating current (AC) electric fields for the alignment of NWs, on-chip electrodes are fabricated and then integrated with the assemblies. In our experiments, gold electrodes of two types of geometries (stem/bulb and interdigitated) of micrometer scale were fabricated by photolithography on glass coverslips. Specific fabrication parameters can be found in [75]. Beyond the single-layer planar electrode, to achieve reconfigurability for the NW assemblies, a dual electrode design including two sets of electrodes that are orthogonal to one another was developed to realize on-demand rotation of the lattice. As we will see in Section 10.3.1.4, this alignment and orientation control of the lattices offered by the geometry and configuration optimization of electrode systems is critical in achieving field-switchable behavior of the NW-based metadevices.

10.3.1.3 The assembly mechanisms

It has been well known that control of orientation and placement of particles can be achieved by applying a local electric field, a phenomenon referred to as dielectrophoresis (DEP). In particular, as illustrated in Figure 10.12(a), the DEP force that is exerted on particles originates from the non-uniformity of the electric field, i.e., the field gradient. Accordingly, by using on-chip electrodes with predetermined features, researchers have successfully realized control over density, orientation, and position of particle assemblies. It is worth noting that since the DEP force has no requirement on the charge characteristics of the particles, DEP has been implemented to control the assembly of a wide variety of polarizable nanomaterials such as carbon nanotubes and nanowires, both metal and semiconductor. For instance, carbon nanotubes can be assembled between parallel electrodes and the corresponding assembly density is closely dependent on the electrode spacing and the concentration of the carbon nanotube solution [58].

Recent studies have shown that the strength of the DEP force is mainly determined by several factors like the shape and size of the particles and the electrical properties (including the dispersion characteristics) of both the particle and the suspending medium. In particular, assuming that the initial electric field is not changed significantly by the presence of the particle, the time-averaged DEP force on a particle can be expressed as $F_{DEP} = \frac{\pi r^2 L}{6} Re(\varepsilon_m) Re\left[\frac{\varepsilon_p - \varepsilon_m}{\varepsilon_p + 2\varepsilon_m}\right] \nabla |E|^2$, where r, L, and ε_p are the radius, length, and dielectric constant of the particle, respectively, while ε_m denotes the dielectric constant of the suspending medium. It can be seen that, besides the geometrical factor, F_{DEP} is dominated by three factors: the

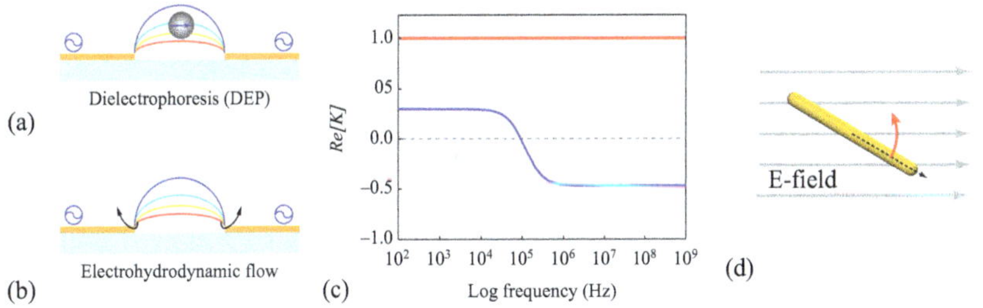

Figure 10.12 Particle location and organization control based on electric field enabled by micropatterned electrodes. Schematic of (a) dielectrophoretic (DEP) force and (b) EHD flow facilitated by electric field application across the electrodes. (c) Dispersion of Re[K] of gold (red) and silica (blue) particles suspended in water. (d) Schematic of the torque applied on a gold nanowire in an electric field.

permittivity of the suspending solution ($Re(\varepsilon_m)$), the polarization factor ($K = \frac{\varepsilon_p - \varepsilon_m}{\varepsilon_p + 2\varepsilon_m}$) and the field gradient ($\nabla|E|^2$). The field gradient-based electric-field force on the particle has been widely used to achieve particle assemblies with controllable alignment. Furthermore, given that dielectric constants are in general dispersive, the polarization factor can be strongly dependent on the frequency (*f*) of the AC applied field, i.e., $K(f) = \frac{\varepsilon_p(f) - \varepsilon_m(f)}{\varepsilon_p(f) + 2\varepsilon_m(f)}$. Accordingly, the magnitude, and more importantly, the sign of $Re(K)$, can be a function of *f*, which offers a great potential for frequency selective assembly. Figure 10.12(c) shows the dispersion of $Re(K)$ of gold (red) and silica (blue) particles suspended in water, in which $Re(K)$ of gold particles can be regarded as a positive constant, while that of silica particles dramatically decreases and flips sign at frequencies around 10^5 Hz. This is because, with the increasing frequency, the permittivity of silica dramatically decreases to a value smaller than that of the solution (water) at a frequency around 10^5 Hz, leading to the sign flip of $Re(K)$. In contrast, in the same frequency range, gold's permittivity is dominated by its extremely large negative real part, resulting in the much less dispersive $Re(K)$. Consequently, when an AC electric field is applied, as illustrated in Figure 10.12(c), gold nanowires in water have a tendency to align along the electric field, while the orientation of silica nanowires is expected to vary as a function of the electric field frequency. Besides enabling orientation manipulation, the frequency-dependent DEP forces may also allow the separation of nanoparticles of different types based on their distinct electrophysiological characteristics.

As illustrated in Figure 10.12(b), EHD, which is referred to as the dynamics of charged fluids under an electric field, is the other effect that has significant influence on the particle assembly. Though as mentioned above the particles in our study are uncharged, whereas most solutions suitable for nanoparticle assembly are electrically charged. As a result, electric fields shaped by the electrodes may produce flows in the solution due to the EHD created hydrostatic pressure. Therefore, besides the DEP force, assembly of particles under an electric field will also be regulated by the EHD flow effect.

10.3.1.4 Gold nanowire assemblies based on electric-field DSA

Exploiting control over assembly formation, researches have achieved organized nanomaterials for a variety of functionalities. In general, two strategies have been employed to design devices based on nanomaterials with electric field controlled location and organization. One is to achieve dry nanoparticle assemblies of a desired formation [51,52]. For instance, after spatially positioning the nanowires using electric-field assembly, post-assembly drying was performed to realize the unusual electrical characteristic of the resulting assemblies [71]. The other strategy is to use nanoparticle assemblies that remain in their assembly liquid to realize electric field controlled reconfigurable devices. For example, turning on and off the electric field in a cyclical manner can anneal nanoparticle assembly and achieve well-organized lattice structures with low-defect density [76–78]. Moreover, the electric field selectivity mentioned above has been utilized to switch nanoparticle assemblies between different organizations. Structure switching of polystyrene nanoparticle assemblies between three distinct organizations has been reported [79].

Figure 10.13(a) and (b) illustrates a coplanar electrode setup of a typical assembly experiment, in which an aqueous suspension of particles is placed within a silicone spacer on top of the patterned microelectrodes of an alternating stem and bulb geometry [80]. Figure 10.13(c) shows the simulated electric field gradient in a plane on top of the electrodes at constant field conditions (800 V/cm, 400 kHz, for the stem region). Clearly, the field gradient is highest in the stem region. Accordingly, as shown in Figure 13(d)–(h), the DEP force causes the gold NWs to concentrate in the stem region as the voltage/frequency of the applied bias increases, resulting in NW assemblies that span the gap. In the experiments, a function generator was used to create the AC electric field in the stem region that ranged from 200 to 2,000 V/cm for peak field strength and 100 kHz–10 MHz for frequency. The electrode gap size (~50 μm) is more than 10 times of the NW length. Given the fact that immediate application of a field of high magnitude/ frequency may create significantly more defects, voltage and frequency were gradually increased in the assembly process to achieve lattices with higher uniformity. The formation evolution of the lattices as a function of the electric field condition is

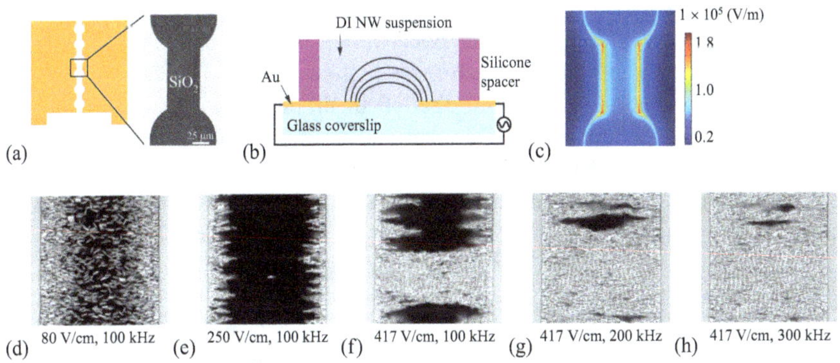

Figure 10.13 Assembly approach using stem/bulb (S/B) geometry. (a) Top view schematic and optical microscopic image of the coplanar stem/bulb electrode setup. (b) The corresponding side view schematic. (c) Simulated electric field gradient at the top surface of the patterned stem/bulb electrodes at constant field conditions (800 V/cm, 400 kHz, for the stem region). (d)–(h) Sequential images taken at the same assembly location of solid 2.5 μm Au nanowires under a series of different field conditions. Applied field strength was increased over ca. 13-min time period. The 2D lattice formation was observed in the gap region between electrodes as peak field strength and frequency increased. Reproduced with permission [80]. Copyright 2015, American Chemical Society.

shown in Figure 13(d)–(h). It can be seen that, in the initial stages of assembly (Figure 10.13(d) and (e)), NWs prefer to gather along the electrode edges, i.e., the region of highest field gradient, which can be attributed to the DEP effect when a field of low magnitude and frequency is applied. Then, as the magnitude and frequency of the field was gradually increased, additional NWs were attracted to the ends of the assembled particles due to the so-called mutual DEP [81,82] and thus generated lattices assembled across both the stem and bulb areas (Figure 10.3(f)–(h)). An additional annealing process (cycling the field on and off) allows particles within the lattice to relocate to more favorable locations. Furthermore, a close inspection of the lattice structure shows that under a field of high magnitude/frequency the particles were staggered and slightly overlapping, leading to a "running-bond" brickwork pattern. Simulations of the electric-field force revealed that this lattice structure corresponds to the most energetically stable configuration. The staggered formation results in relatively "open" lattices, which, as shown in the following discussion, has significant influence on the optical response of the assemblies.

10.3.1.5 Optical simulations

To explore the optical response of nanowire lattices, full-wave simulations were performed. To better correlate with experimental conditions, the NWs were assumed to be embedded in D_2O ($n = 1.33$) and experimental gold material properties were adopted. Given the fact that the resonances of the individual nanoparticles generally determine the collective responses of nanoparticle clusters, the scattering of an individual nanowire was first calculated. The configuration of the scattering simulation is shown on the left of the first row of Figure 10.14, in which an isolated nanowire, with dimensions used in experimental assemblies (Figure 10.11), is illuminated by a linearly polarized wave. Polarization of the light was either along (TE) or perpendicular (TM) to the axis of the NW. The normalized scattering cross-section (Q_{Scat}) spectra of an individual nanoparticle of length 1.1, 2.2, and 3.7 μm are shown in Figure 14 (a)–(c), respectively. It can be seen that for TE polarization, Q_{Scat} of the 1.1 μm long NW peaks at around 3.8 μm, while that for TM polarization is rather low in the entire wavelength range of concern, indicating a strong polarization selectivity in the light scattering behavior of the NW. This optical anisotropy can also been clearly seen in the corresponding scattered electric field distributions ($|E_{scat}|$) at 3.74 μm (Figure 10.15(a) and (b)). Moreover, for NWs of fixed diameter, the Q_{Scat} peak for TE polarization redshifts with the increase in the NW's length. Specifically, the Q_{Scat} peak of the 3.7-μm-length NW moves out of the longer wavelength edge (6.0 μm) of the simulation band, implying that strong scattering anisotropy can be achieved in a broad wavelength range by simply varying the dimensions of the NWs.

Consequently, it can be expected that assemblies of NWs with well-aligned orientation exhibit anisotropic collective optical responses (e.g. transmission and reflection) originating from the resonances of the individual NWs. The polarization-dependent transmittance of the aligned gold NW lattices was further simulated to determine the collective optical response of the NW lattices achieved through electric

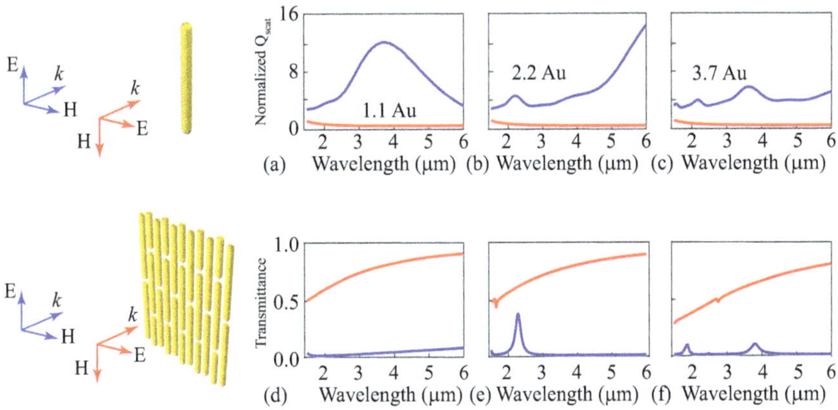

Figure 10.14 Simulated scattering and collective optical anisotropy of gold nanowire systems. Normalized scattering cross section (Q_{Scat}) of an isolated gold nanowire of length (a) 1.1 μm, (b) 2.2 μm, and (c) 3.7 μm. (a)–(c) Q_{Scat} values were normalized to the nanowire's cross-sectional area. Transmittance of linearly polarized light through the lattice formed by gold nanowires of length (d) 1.1-μm, (e) 2.2-μm, and (f) 3.7-μm. The configurations of the scattering and transmittance simulation are shown on the left of the figure. Reproduced with permission [75]. Copyright 2017, Wiley-VCH.

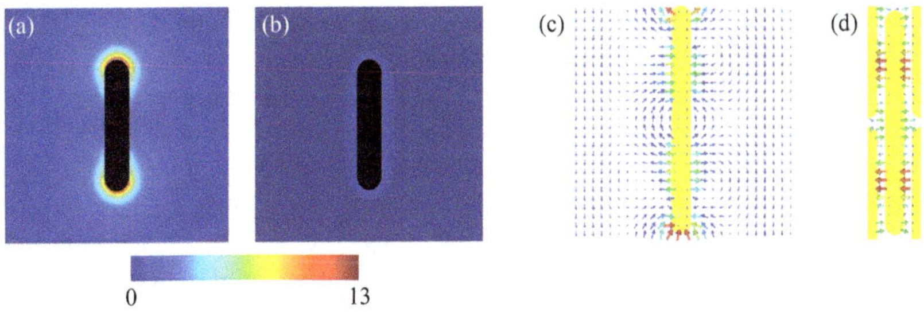

Figure 10.15 Simulated field distributions of an individual nanowire and the nanowire lattices. The scattering electric field ($|E_{scat}|$) distributions of a 1.1-μm-long nanowire under (a) TE and (b) TM excitation at a wavelength of 3.74 μm. The electric field vector distributions of a lattice formed by 2.2-μm-long nanowires under (d) TE (side view) and (c) TM (front view) excitations. Reproduced with permission [75]. Copyright 2017, Wiley-VCH.

field assembly. Transmittance of a single layer (along the propagation direction of light) of the gold NW lattice was considered, as shown in the schematic on the left of the second row of Figure 10.14. Staggered alignment along the direction of the NW's axes was used and, gaps between neighboring NWs were set as 60 nm which is two times the thickness of the silica coating on the NWs used in the experiments. As can be seen from Figure 10.14(d)–(f), in the near- and mid-IR wavelength region of interest, the transmittance spectra of TE and TM polarized light show a pronounced contrast. In particular, the lattice of 1.1-μm-long NWs allows more than 60% transmittance of TM-polarized light at a wavelength of 2.0 μm, while it almost completely blocks the TE-polarized light at the same wavelength, indicating an extremely large extinction ratio between the two orthogonal polarizations. In addition, the high transmission band of the TM mode redshifts with the increasing length of the NWs, while lattices of all three NW dimensions exhibit extremely low transmittance for the TE mode. The peaks observed at shorter wavelengths in the transmission spectra of the TE mode arise from the dark mode resonance (Figure 10.15(c) and (d)) supported by the NWs [83], while they have limited influence on the broadband polarizer behavior of the NW lattices. Furthermore, the optical response of the NW lattices are consistent with the scattering behavior of individual NWs, which in turn verifies that the shape anisotropy of the nanoparticles determines both the scattering behavior of individual NWs and the collective optical anisotropy of the NW systems. Simulation results here show the ability of gold NW lattices to engineer the strong and broadband optical responses for most polarization sensitive applications in the near- and mid-IR, which provides a pathway to reconfigurable photonic metadevices exploiting the intrinsic structural flexibility of nanoparticle assemblies.

10.3.1.6 Optical characterization

Although assemblies formed using stem/bulb electrodes show slightly fewer defects, a parallel electrode geometry appears to be more suitable than a more complicated assembly setup owing to its simple structure. Consequently, as illustrated in Figure 10.16(a), interdigitated electrodes (IDEs) were employed in the optical characterization of NW assemblies and an external polarizer was used as a means to measure the corresponding optical anisotropy of the samples. This setup allows spectra collection at a location of the assembly area of interest; whereas, when a certain voltage is applied across the coplanar electrodes, IDEs with varied gap size readily offer an AC electric field of different magnitude. As shown in the inset of Figure 10.16(a), the NW solutions were contained within a double-sided adhesive spacer, atop a glass cover slip patterned with the gold IDEs. In the experiments, transmission spectra of linearly polarized light parallel (TE) and perpendicular (TM) to the long axes of the nanowires (i.e., direction across the gap of the IDEs) through the lattice were collected using an FTIR coupled microscope. Experimental spectra were normalized to that of an analogous sample chamber filled with deuterium oxide. Figure 10.16(e)–(g) illustrates the reflected light optical microscope images of the resultant assembly of each nanowire type. In the experiments, nanowires were allowed to sediment from suspension onto the substrate surface for 2 min before the external AC voltage was applied. Moreover, each

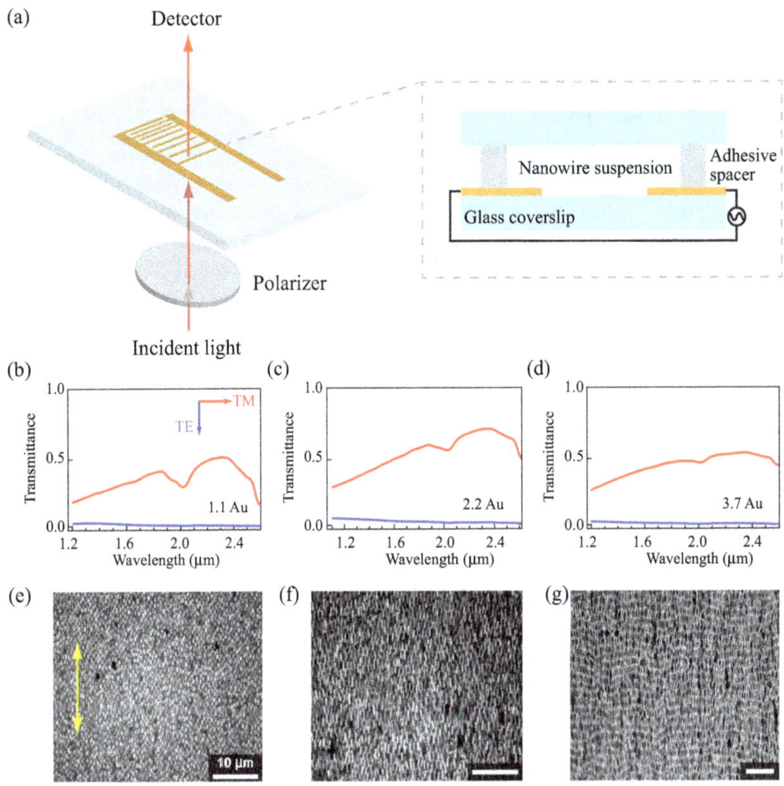

Figure 10.16 *Characterization of the optical anisotropy of the gold nanowire lattices. (a) Schematic of experimental setup for FTIR transmission measurements. A sodium-silicate glass linear polarizer is used to generate linearly polarized light that is normally incident on the lattice, and the transmitted wave was collected by an IR detector of the FTIR microscope. Inset: Schematic of a cross-sectional view of experimental setup. (b)–(d) Transmittance spectra for a lattice of 1.1, 2.2, and 3.7 μm Au nanowires. (e)–(g) The corresponding reflected light optical microscopic images. Reproduced with permission [75]. Copyright 2017, Wiley-VCH.*

microscope image shown in Figure 10.16 was collected 1 min after the application of the electric field. Figure 10.16(b)–(d) shows the corresponding anisotropic transmission spectra of assemblies of gold NWs for three typical lengths (Figure 10.11), i.e., 1.1, 2.2, and 3.7 μm. It can be seen that all three lattice types exhibited strong polarization selectivity in a wavelength range from 1.2 to 2.6 μm. In particular, the lattices effectively block TE polarized light but simultaneously allow the transmission of TM polarized light, leading to an extinction ratio up to 17 dB. The characterization wavelength range was primarily limited by the transparent range of the solvent, the substrate, and superstrate used. Deuterium oxide was substituted for deionized (DI) water to improve sample transparency in the IR region, while the high absorption of glass coverslips prevented quantitative measurements at wavelengths longer than 2.6 μm. It should be noted that optical diffraction was expected when incident wavelengths were comparable with or even smaller than the feature dimension (length) of the nanowires. As shown in the simulated results (Figure 10.14), the NW lattices exhibit large optical anisotropy in the mid-IR region. Nevertheless, the observed transmission contrast in the near-IR regime demonstrates the broadband nature of the nanowire polarizer.

A comparison between Figures 10.14 and 10.16 reveals a fairly good agreement between the simulations and experiments. In particular, besides the high-contrast transmission of the two orthogonal polarizations, both measured and simulated transmission spectra show an overarching trend to decrease the TM transmittance at shorter wavelengths as the NW length increases. Nevertheless, notable deviations between simulated and experimental data are also observed. For instance, compared with the simulation prediction, in experiments the 1.1-μm-long NW lattice showed lower TM transmittance. We note that a few possible factors, including the particle density, actual lattice formation, the number of nanowire layers, defect sites, etc., impede a quantitative comparison between the simulations and experiments in this case. For instance, a close inspection of the microscopic images shown in Figure 10.16(e)–(g) shows that the lattice of the 1.1-μm-long NW is obviously more particle-dense than the other two wire types. Moreover, lattices formed by longer (e.g. 3.7-μm-long) NWs have more open area due to the more observable stagger lattice formation and also are prone to high defect density. Furthermore, the NWs can possibly form lattices of multiple layers of particles, while, since the gold NWs are opaque, it is difficult to determine the number of the particle layers in the devices. We note that, despite the fact that these uncertainties are difficult to include in the simulation, the simulated data can still provide insight into the optical anisotropy of the gold NW lattices.

As lattice orientation determines the optical anisotropy of the NW assemblies, on-demand rotation of the lattice is desired to realize reconfigurability of the proposed lattice systems. As the schematic in Figure 10.17 illustrates, inspired by complex multielectrode designs for manipulation of nano- and microstructures such as single cells [84], liquid droplets [85], and dielectric particles [86], a dual-electrode design consisting of two interdigitated electrodes was developed for control over the positive DEP of gold NWs. In particular, separated by a double-sided adhesive spacer, the two electrodes were assembled orthogonally to one another, offering independent electric field control for the assembly of nanowire suspension in deuterium oxide. Accordingly, as depicted in the inset of Figure 10.17, in the ideal case the gold NW lattice can switch between the initial and the reconfigured orientation when the electric field switches direction by selectively applying voltage to one of the interdigitated electrodes. To examine the reconfigurable optical response of the system, transmittance spectra of linearly polarized light with its polarization parallel to arms of either the bottom (TE) or the

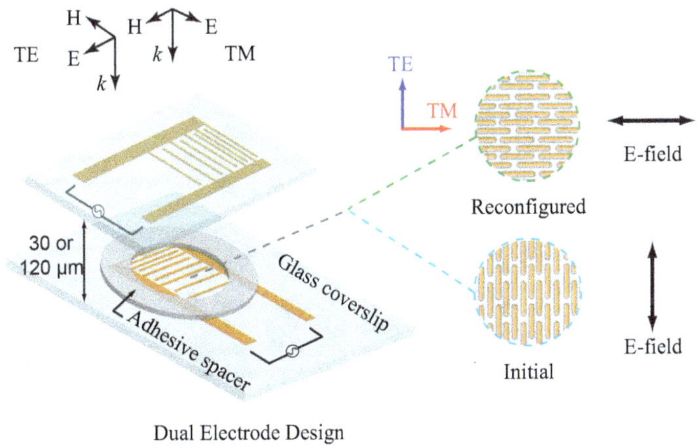

Figure 10.17 Schematic illustrating the dual electrode design. Two patterned interdigitated electrodes face the nanowire solution, contained within an adhesive spacer. The top electrode is placed orthogonal to the bottom one, enabling the rotation of the electric field direction by 90°. Inset: Idealized nanowire lattice rotates when the electric field switches direction by selective voltage application to one of the electrodes. In particular, the NWs are oriented parallel to the incident polarization (TE) in the initial state, and the NWs are aligned perpendicular to TE in the reconfigured state. Reproduced with permission [75]. Copyright 2017, Wiley-VCH.

top (TM) electrode were measured when the gold NWs were assembled into lattices by an electric field applied across one of the two electrodes.

Figure 10.18 shows the optical microscope images of initial and reconfigured nanowire lattices and the corresponding FTIR spectra. The schematics depicting the directivity of the electric field used at the two states and the corresponding expected lattice orientations are also illustrated on the left. The assemblies of 1.1-, 2.2-, and 3.7-μm-long Au NWs in experiments were recorded at the initial and then the reconfigured states using a reflected light optical microscope and shown in the central two columns of Figure 10.18. It can be seen that when voltage is selectively applied to one of the two interdigitated electrodes, the NWs may rotate and tend to align along the electric field. Accordingly, as the transmission spectra in Figure 10.18(c), (f) and (i) show, all the lattices exhibit electric field controlled reconfigurable anisotropy. In particular, in the wavelength range of interest, for initial orientation (i.e., when the voltage was applied to the bottom electrode), all three lattices specifically block TE polarized light but allow high transmission of TM polarized light. Moreover, in all cases, for reconfigurable orientation (i.e., when the voltage was applied to the top electrode), transmittance of TE polarization dramatically increases and that of TM polarization decreases simultaneously. These

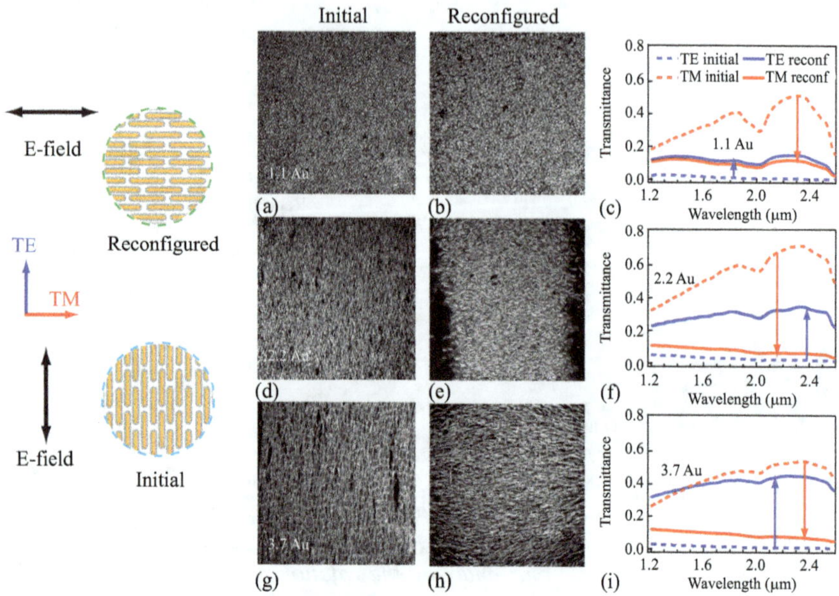

Figure 10.18 NW assembly-based reconfigurable polarizers. Reflected light optical microscope images of NW lattices in initial and reconfigured states for (a) and (b) 1.1-μm-long Au NWs, (d) and (e) 2.2-μm-long Au NWs, and (g) and (h) 3.7-μm-long Au NWs. The corresponding FTIR spectra are shown in (c), (f), and (i). Electric field conditions used for initial: (a) 300 V/cm, 10 MHz; (d) 150 V/cm, 500 kHz; and (g) 100 V/cm, 250 kHz, and reconfigured: (b) 1,000 V/cm, 1 kHz; (e) 500 V/cm, 1 kHz; and (h) 1,000 V/cm, 10 kHz. Initial spectra (Figure 10.16(b)–(d)) are also presented for comparison. The schematics of electric field used at the two states and the corresponding ideal lattice orientations are also illustrated on the left. Reproduced with permission [75]. Copyright 2017, Wiley-VCH.

observations clearly indicate the reconfigurable optical response of the proposed metadevice.

A close examination of the transmittance spectra illustrated in Figure 10.18 shows that the observed reconfigurability is closely dependent on the length of the NWs forming the lattices. Specifically, in the reconfigured state, the transmittance contrast between TE and TM polarization increases with the length of the NWs. For instance, the lattice formed by 3.7-μm-long Au NWs maintains ≈30% transmittance contrast across the entire wavelength range of interest at both initial and reconfigured states, while only limited extinction ratio was found in the reconfigured assembly of 1.1-μm-long Au NWs. This can be attributed to the different mechanisms that dominate the corresponding assembly processes when voltage is selectively applied across the bottom and top electrodes. For the initial state, the assemblies are primarily driven by DEP forces that produce well-aligned NW lattices as well as the correspondingly large optical anisotropy (dashed lines in Figure 10.18(c), (f), and (i)). However, when the top electrode is in use, the corresponding electric field may not be able to effectively align the NWs due to the sediment effect of the NWs. Instead, the assemblies can be directed by the EHD flow produced by the low-frequency electric field across the top electrodes [57,58], resulting in less ordered lattices. Consequently, optimized field conditions were used to form the reconfigured assembly (see the caption of Figure 10.18 for details). Accordingly, the 3.7-μm-long Au NWs that are most responsive in the field form more uniform lattices owing to their larger gold volume, especially when voltage was applied to the top electrodes. Furthermore, the corresponding successful 90° rotation of the NWs relative to the initial state enables the large extinction ratio (Figure 10.18(i)). For 1.1-μm-long Au NWs which are least responsive to the electric field, ordered alignment cannot be achieved in the reconfigured state, leading to an observable decrease in the extinction ratio compared with that of the initial lattice (Figure 10.18(c)). For the lattice formed by 2.2-μm-long Au NWs, high TM transmission and good extinction ratio were found in the initial state, and in the reconfigured state the well aligned NW structure facilitates selectivity switching with a 15% transmittance difference between two incident polarizations.

10.3.2 Conclusions and future studies

A field-switchable broadband polarizer based on electrically reconfigurable gold NW assemblies was presented. By introducing a dual electrode design that enables reversible rotation of the lattice by 90° on-demand, the electric field controlled reconfigurable transmission anisotropy arising from the collective photonic response of gold NWs was experimentally observed. Our work further reveals that electric field directed assembly is a useful strategy to enable active nanosystems that provide potential for electrically reconfigurable photonic devices, such as smart windows.

Consequently, the ability to form and rapidly reconfigure assemblies of colloidal nanowires at predetermined locations should be of high priority for future studies. Studies of directed assembly show that the control over lattice structures has at least three degrees of freedom, i.e., particle segmentation pattern, frequency-dependent dipolar coupling and the field switching. Frequency-dependent dipolar coupling can vary the lattice periodicity in real time, while field switching, as shown above, offers a means of controlling the organization of the nanowires. Besides the solid nanowires discussed above, in our experiments segmented Au nanowires obtained from partially etching have also been used to achieve field control assemblies, wherein the segmentation pattern provides more opportunities for lattice formation control. Our observations show that different assembly formations could be achieved by designing the segment structure of partially etched nanowires (PENs) [80].

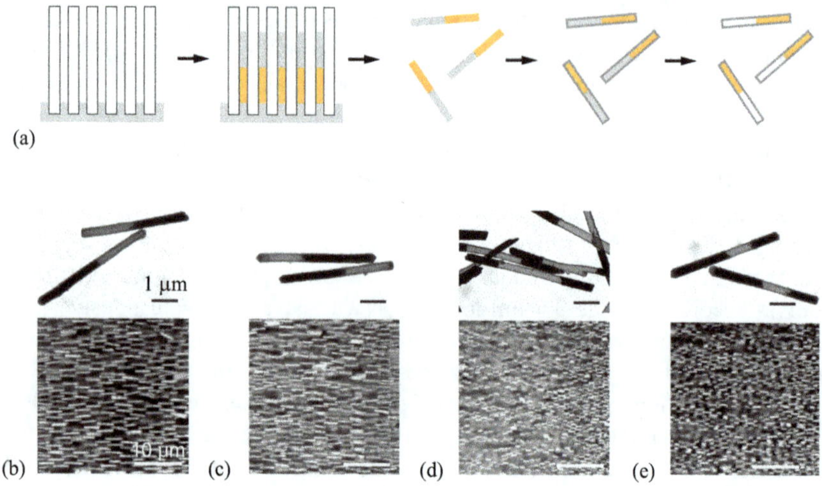

Figure 10.19 Synthesis and assembly of partially etched gold nanowires. (a) Schematic of the synthesis process using templated electrodeposition and acid etching. (b)–(e) TEM images of four types of the fabricated PENs and reflectance optical micrographs of the corresponding assemblies under an applied electric field. Field conditions are as follows: (b) 800 V/cm, 400 kHz; (c) 500 V/cm, 750 kHz; (d) 2 kV/cm, 3.5 MHz; and (e) 700 V/cm, 1 MHz. Reproduced with permission [80]. Copyright 2015, American Chemical Society.

Figure 10.19(a) illustrates a typical fabrication process we used for PENs, which was reported previously [62–66]. Different from the fabrication of solid gold nanowires (Figure 10.11), suspensions of segmented nanowires can be fabricated by a sequential electrodeposition of gold and silver in a porous alumina membrane, followed by particle release. The template pore diameter determines the nanowire width, while the length of each metal segment can be independently controlled by the amount of electric current passed in different plating solutions. A thin amorphous silica shell ($\approx$30 nm) can then be completed through a modified sol−gel process. Finally, the PENs that are comparable to those fabricated via lithography [87] can be obtained from selective etching of multiple metal segments. Besides preventing electrical breakdown when the nanowires are assembled between bare electrodes, the cylindrical silica coating of PENs also provides mechanical support holding the core segments together after the selective etching process. Figure 10.19(b)–(e) illustrates the TEM images of four types of typical PENs and the microscopic images of their assemblies under an applied electric field. It can be seen that the segmentation pattern including the thickness and the number of the gold segments in the PENs can be well-controlled. Moreover, the corresponding field-driven assemblies observed show good alignment along the field directivity, which is similar to that observed in systems formed by solid gold NWs. This indicates that the assembly process is driven by the Au segments and the influence of etched segments is negligible. Nevertheless, the assemblies formed by PENs reveals two unique features. First, in contrast to solid gold nanowire systems, the formation of assemblies of PENs is less sensitive to the frequency of the applied field. For instance, for assembly of 2.5-μm-1etch Au NWs, no clear frequency-dependent change in particle overlap was found in the experiments. This can be attributed to the asymmetric structure of the Au segment. Second, the lengths of Au segment may vary the periodicity of the assemblies in a direction parallel to the electric field. This behavior potentially offers the opportunity for customized PEN-enabled metadevices with engineered optical response based on finely tuned segment lengths.

PENs actually provide us with great flexibility in the design of reconfigurable metadevices. As illustrated in Figure 10.12(d), the Re[K] of gold and silica indicates the distinct dispersion properties of the two materials. In particular, for frequencies higher than 10^5 Hz, gold and silica have positive and negative DEP, respectively; while, at lower frequencies, they both exhibit positive DEP. Consequently, by engineering the length of the gold segments as well as the thickness of the silica coating, it is possible to achieve 90° alignment rotation of PENs with a single pair of electrodes by varying the frequency of the assembly field, which may dramatically improve the practicality of the DSA-based reconfigurable devices. Furthermore, PENs can be used to achieve plasmonic response at visible and near-IR regions. The anisotropic optical response seen in solid gold nanowires arises from their geometrical anisotropy, i.e., the dimension contrast between the length and diameter of individual NWs. It has been observed that, under the current experimental conditions, NWs of larger diameter (e.g. >100 nm) exhibit better assembly dynamics under an external electric field, which, however, indicates that long NWs are required for pronounced optical anisotropy, as shown in Figures 10.16 and 10.18. It is not difficult to intuit that to enable a similar NW-based phenomenon in the near-IR and visible regimes, one should take advantage of their radial plasmon resonance (i.e., the resonance geometrically determined by the diameter of the NWs). Therefore, the PENs are ideal for optical anisotropy arising from the radial plasmon resonance, because each PEN can be viewed as a composite consisting of multiple subunits (i.e., Au disks whose radial dimension may facilitate optical resonance in the near-IR and visible regions). Given these characteristics, we envision the PENs formed assemblies as good candidates for electrically controlled metadevices that can enable more sophisticated modulation schemes for optical signals.

References

[1] Smith, D.R., Padilla, W.J., Vier, D.C., Nemat-Nasser, S.C., Schultz, S.: 'Composite Medium with Simultaneously Negative Permeability and Permittivity' *Phys. Rev. Lett.*, 2000, **84**, (18), pp. 4184–4187.

[2] Shelby, R.A., Smith, D.R., Schultz, S.: 'Experimental Verification of a Negative Index of Refraction' *Science*, 2001, **292**, (5514), pp. 77–79.

[3] Pendry, J.B., Holden, A.J., Robbins, D.J., Stewart, W.J.: 'Magnetism from Conductors and Enhanced Nonlinear Phenomena' *IEEE Trans. Microw. Theory Tech.*, 1999, **47**, (11), pp. 2075–2084.

[4] Schurig, D., Mock, J.J., Justice, B.J., *et al.*: 'Metamaterial Electromagnetic Cloak at Microwave Frequencies' *Science*, 2006, **314**, (5801), pp. 977–980.

[5] Valentine, J., Zhang, S., Zentgraf, T., *et al.*: 'Three-Dimensional Optical Metamaterial with a Negative Refractive Index' *Nature*, 2008, **455**, (7211), pp. 376–379.

[6] Yuan, H.-K., Chettiar, U.K., Cai, W., *et al.*: 'A Negative Permeability Material at Red Light' *Opt. Express*, 2007, **15**, (3), p. 1076.

[7] Gansel, J.K., Thiel, M., Rill, M.S., *et al.*: 'Gold Helix Photonic Metamaterial as Broadband Circular Polarizer' *Science*, 2009, **325**, (5947), pp. 1513–1515.

[8] Liu, Z., Lee, H., Xiong, Y., Sun, C., Zhang, X.: 'Far-Field Optical Hyperlens Magnifying Sub-Diffraction-Limited Objects' *Science*, 2007, **315**, (5819), pp. 1686–1686.

[9] Yu, N., Genevet, P., Kats, M.A., *et al.*: 'Light Propagation with Phase Discontinuities: Generalized Laws of Reflection and Refraction' *Science*, 2011, **334**, (6054), pp. 333–337.

[10] Ni, X., Emani, N.K., Kildishev, A.V., Boltasseva, A., Shalaev, V.M.: 'Broadband Light Bending with Plasmonic Nanoantennas' *Science*, 2012, **335**, (6067), pp. 427–427.

[11] Boardman, A.D., Grimalsky, V.V., Kivshar, Y.S., *et al.*: 'Active and Tunable Metamaterials' *Laser & Photonics Rev.*, 2011, **5**, (2), pp. 287–307.

[12] Zheludev, N.I., Kivshar, Y.S.: 'From Metamaterials to Metadevices' *Nat. Mater.*, 2012, **11**, (11), pp. 917–924.

[13] Hess, O., Pendry, J.B., Maier, S.A., Oulton, R.F., Hamm, J.M., Tsakmakidis, K.L.: 'Active Nanoplasmonic Metamaterials' *Nat. Mater.*, 2012, **11**, (7), pp. 573–584.

[14] Turpin, J.P., Bossard, J.A., Morgan, K.L., Werner, D.H., Werner, P.L.: 'Reconfigurable and Tunable Metamaterials: A Review of the Theory and Applications', *Int. J. Antenn. Propag.* 2014, **2014**, p. 429837.

[15] Fan, K., Padilla, W.J.: 'Dynamic Electromagnetic Metamaterials' *Mater. Today*, 2015, **18**, (1), pp. 39–50.

[16] Zheludev, N.I., Plum, E.: 'Reconfigurable Nanomechanical Photonic Metamaterials' *Nat. Nanotechnol.*, 2016, **11**, (1), pp. 16–22.

[17] Wuttig, M., Yamada, N.: 'Phase-change Materials for Rewriteable Data Storage' *Nat. Mat.*, 2007, **6**, (11), pp. 824–832.

[18] Raoux, S.: 'Phase Change Materials' *Annu. Rev. Mater. Res.*, 2009, **39**, (1), pp. 25–48.

[19] Qazilbash, M.M., Brehm, M., Chae, B.-G., *et al.*: 'Mott Transition in VO_2 Revealed by Infrared Spectroscopy and Nano-Imaging' *Science*, 2007, **318**, (5857), pp. 1750–1753.
[20] Kim, H.-T., Chae, B.-G., Youn, D.-H., *et al.*: 'Mechanism and Observation of Mott Transition in VO_2-Based Two- and Three-Terminal Devices' *New J. Phys.*, 2004, **6**, (1), p. 52.
[21] Ruzmetov, D., Gopalakrishnan, G., Ko, C., Narayanamurti, V., Ramanathan, S.: 'Three-Terminal Field Effect Devices Utilizing Thin Film Vanadium Oxide as the Channel Layer'*J. Appl. Phy.*, 2010, **107**, (11), p. 114516.
[22] Wang, C., Zhao, L., Liang, Z., Dong, B., Wan, L., Wang, S.: 'New Intelligent Multifunctional SiO_2/VO_2 Composite Films with Enhanced Infrared Light Regulation Performance, Solar Modulation Capability, and Superhydrophobicity' *Sci. Technol. Adv. Mater.*, 2017, **18**, (1), pp. 563–573.
[23] Sun, K., Riedel, C.A., Urbani, A., *et al.*: 'VO_2 Thermochromic Metamaterial-Based Smart Optical Solar Reflector' *ACS Photon.*, 2018, **5**, (6), pp. 2280–2286.
[24] Zhou, Y., Ramanathan, S.: 'Mott Memory and Neuromorphic Devices' *Proceedings of the IEEE*, 2015, **103**, (8), pp. 1289–1310.
[25] Cavalleri, A., Dekorsy, T., Chong, H.H.W., Kieffer, J.C., Schoenlein, R. W.: 'Evidence for a Structurally-Driven Insulator-to-Metal Transition in VO_2: A View from the Ultrafast Timescale' *Phys. Rev. B*, 2004, **70**, (16), p. 161102.
[26] Cavalleri, A., Tóth, C., Siders, C.W., *et al.*: 'Femtosecond Structural Dynamics in VO_2 During an Ultrafast Solid-Solid Phase Transition' *Phys. Rev. Lett.*, 2001, **87**, (23), p. 237401.
[27] Appavoo, K., Wang, B., Brady, N.F., *et al.*: 'Ultrafast Phase Transition via Catastrophic Phonon Collapse Driven by Plasmonic Hot-Electron Injection' *Nano Lett.*, 2014, **14**, (3), pp. 1127–1133.
[28] Qazilbash, M.M., Li, Z.Q., Podzorov, V., *et al.*: 'Electrostatic Modification of Infrared Response in Gated Structures Based on VO_2' *Appl. Phys. Lett.*, 2008, **92**, (24), p. 241906.
[29] Driscoll, T., Kim, H.-T., Chae, B.-G., *et al.*: 'Memory Metamaterials' *Science*, 2009, **325**, (5947), pp. 1518–1521.
[30] Liu, M., Hwang, H.Y., Tao, H., *et al.*: 'Terahertz-Field-Induced Insulator-to-Metal Transition in Vanadium Dioxide Metamaterial' *Nature*, 2012, **487**, (7407), pp. 345–348.
[31] Dicken, M.J., Aydin, K., Pryce, I.M., *et al.*: 'Frequency Tunable Near-Infrared Metamaterials Based on VO_2 Phase Transition' *Opt. Express*, 2009, **17**, (20), pp. 18330–18339.
[32] Shu, F.-Z., Yu, F.-F., Peng, R.-W., *et al.*: 'Dynamic Plasmonic Color Generation Based on Phase Transition of Vanadium Dioxide' *Adv. Opt. Mater.*, 2018, **6**, (7), p. 1700939.
[33] Song, S., Ma, X., Pu, M., *et al.*: 'Tailoring Active Color Rendering and Multiband Photodetection in a Vanadium-Dioxide-Based Metamaterial Absorber' *Photonics Res.*, 2018, **6**, (6), pp. 492–497.

[34] Sun, M., Taha, M., Walia, S., *et al.*: 'A Photonic Switch Based on a Hybrid Combination of Metallic Nanoholes and Phase-Change Vanadium Dioxide' *Sci. Rep.*, 2018, **8**.

[35] Jia, Z.-Y., Shu, F.-Z., Gao, Y.-J., *et al.*: 'Dynamically Switching the Polarization State of Light Based on the Phase Transition of Vanadium Dioxide' *Phys. Rev. Appl.*, 2018, **9**, (3), p. 034009.

[36] Kang, L., Lan, S., Cui, Y., *et al.*: 'An Active Metamaterial Platform for Chiral Responsive Optoelectronics' *Adv. Mater.*, 2015, **27**, (29), pp. 4377–4383.

[37] Li, W., Valentine, J.: 'Metamaterial Perfect Absorber Based Hot Electron Photodetection' *Nano Lett.*, 2014, **14**, (6), pp. 3510–3514.

[38] Li, W., Coppens, Z.J., Besteiro, L.V., Wang, W., Govorov, A.O., Valentine, J.: 'Circularly Polarized Light Detection with Hot Electrons in Chiral Plasmonic Metamaterials' *Nat. Commun.*, 2015, **6**, p. 8379.

[39] Liu, L., Kang, L., Mayer, T.S., Werner, D.H.: 'Hybrid Metamaterials for Electrically Triggered Multifunctional Control' *Nat. Commun.*, 2016, **7**, p. 13236.

[40] Watts, C.M., Liu, X., Padilla, W.J.: 'Metamaterial Electromagnetic Wave Absorbers' *Adv. Mater.*, 2012, **24**, (23), pp. OP98–OP120.

[41] Paik, T., Hong, S.-H., Gaulding, E.A., *et al.*: 'Solution-Processed Phase-Change VO_2 Metamaterials from Colloidal Vanadium Oxide (VO_x) Nanocrystals' *ACS Nano*, 2014, **8**, (1), pp. 797–806.

[42] Liu, X., Starr, T., Starr, A.F., Padilla, W.J.: 'Infrared Spatial and Frequency Selective Metamaterial with Near-Unity Absorbance' *Phys. Rev. Lett.*, 2010, **104**, (20), p. 207403.

[43] Kats, M.A., Blanchard, R., Zhang, S., *et al.*: 'Vanadium Dioxide as a Natural Disordered Metamaterial: Perfect Thermal Emission and Large Broadband Negative Differential Thermal Emittance' *Phys. Rev. X*, 2013, **3**, (4), p. 041004.

[44] Zhu, Z., Evans, P.G., Haglund, R.F., Valentine, J.G.: 'Dynamically Reconfigurable Metadevice Employing Nanostructured Phase-Change Materials' *Nano Lett.*, 2017, **17**, (8), pp. 4881–4885.

[45] Liu, X., Tyler, T., Starr, T., Starr, A.F., Jokerst, N.M., Padilla, W.J.: 'Taming the Blackbody with Infrared Metamaterials as Selective Thermal Emitters' *Phys. Rev. Lett.*, 2011, **107**, (4), p. 045901.

[46] O'Hara, A., Nunley, T.N., Posadas, A.B., Zollner, S., Demkov, A.A.: 'Electronic and Optical Properties of NbO_2' *J. Appl. Phy.*, 2014, **116**, (21), p. 213705.

[47] Maier, S.A.: 'Plasmonics: Fundamentals and Applications' (Springer US, 2007).

[48] Lal, S., Link, S., Halas, N.J.: 'Nano-Optics from Sensing to Waveguiding' *Nat. Photonics*, 2007, **1**, (11), pp. 641–648.

[49] Sargent, E.H.: 'Infrared Photovoltaics Made by Solution Processing' *Nat. Photonics*, 2009, **3**, (6), pp. 325–331.

[50] Grzelczak, M., Vermant, J., Furst, E.M., Liz-Marzán, L.M.: 'Directed Self-Assembly of Nanoparticles' *ACS Nano*, 2010, **4**, (7), pp. 3591–3605.

[51] Long, Y.-Z., Yu, M., Sun, B., Gu, C.-Z., Fan, Z.: 'Recent Advances in Large-Scale Assembly of Semiconducting Inorganic Nanowires and Nanofibers for Electronics, Sensors and Photovoltaics' *Chem. Soc. Rev.*, 2012, **41**, (12), pp. 4560–4580.

[52] Liu, X., Long, Y.-Z., Liao, L., Duan, X., Fan, Z.: 'Large-Scale Integration of Semiconductor Nanowires for High-Performance Flexible Electronics' *ACS Nano*, 2012, **6**, (3), pp. 1888–1900.

[53] Wang, L., Ma, W., Xu, L., *et al.*: 'Nanoparticle-Based Environmental Sensors' *Mater. Sci. Eng. R Rep.*, 2010, **70**, (3–6), pp. 265–274.

[54] Heiss, M., Fontana, Y., Gustafsson, A., *et al.*: 'Self-Assembled Quantum Dots in a Nanowire System for Quantum Photonics' *Nat. Mater.*, 2013, **12**, (5), pp. 439–444.

[55] Ge, J., He, L., Hu, Y., Yin, Y.: 'Magnetically Induced Colloidal Assembly into Field-Responsive Photonic Structures' *Nanoscale*, 2011, **3**, (1), pp. 177–183.

[56] Yan, R., Gargas, D., Yang, P.: 'Nanowire Photonics' *Nat. Photonics,*, 2009, **3**, (10), pp. 569–576.

[57] Velev, O.D., Gangwal, S., Petsev, D.N.: 'Particle-Localized AC and DC Manipulation and Electrokinetics' *Annu. Rep. Prog. Chem., Sect. C: Phys. Chem.*, 2009, **105**, (0), pp. 213–246.

[58] Smith, B.D., Mayer, T.S., Keating, C.D.: 'Deterministic Assembly of Functional Nanostructures Using Nonuniform Electric Fields' *Annu. Rev. Phys. Chem.*, 2012, **63**, (1), pp. 241–263.

[59] Edwards, T.D., Bevan, M.A.: 'Controlling Colloidal Particles with Electric Fields' *Langmuir*, 2014, **30**, (36), pp. 10793–10803.

[60] Thorkelsson, K., Bai, P., Xu, T.: 'Self-Assembly and Applications of Anisotropic Nanomaterials: A Review' *Nano Today*, 2015, **10**, (1), pp. 48–66.

[61] Hsu, S.-W., Rodarte, A.L., Som, M., Arya, G., Tao, A.R.: 'Colloidal Plasmonic Nanocomposites: From Fabrication to Optical Function' *Chem. Rev.*, 2018, **118**, (6), pp. 3100–3120.

[62] Sioss, J.A., Keating, C.D.: 'Batch Preparation of Linear Au and Ag Nanoparticle Chains via Wet Chemistry' *Nano Lett.*, 2005, **5**, (9), pp. 1779–1783.

[63] Smith, B.D., Kirby, D.J., Keating, C.D.: 'Vertical Arrays of Anisotropic Particles by Gravity-Driven Self-Assembly' *Small*, 2011, **7**, (6), pp. 781–787.

[64] Nicewarner-Peña, S.R., Freeman, R.G., Reiss, B.D., *et al.*: 'Submicrometer Metallic Barcodes' *Science*, 2001, **294**, (5540), pp. 137–141.

[65] Al-Mawlawi, D., Liu, C.Z., Moskovits, M.: 'Nanowires Formed in Anodic Oxide Nanotemplates' *J. Mater. Res.*, 1994, **9**, (4), pp. 1014–1018.

[66] Hulteen, J.C., Martin, C.R.: 'A General Template-Based Method for the Preparation of Nanomaterials' *J. Mater. Chem.*, 1997, **7**, (7), pp. 1075–1087.

[67] Martin, B.R., Dermody, D.J., Reiss, B.D., *et al.*: 'Orthogonal Self-Assembly on Colloidal Gold-Platinum Nanorods' *Adv. Mater.*, 1999, **11**, (12), pp. 1021–1025.

[68] Yin, Y., Lu, Y., Sun, Y., Xia, Y.: 'Silver Nanowires Can Be Directly Coated with Amorphous Silica to Generate Well-Controlled Coaxial Nanocables of Silver/Silica' *Nano Lett.*, 2002, **2**, (4), pp. 427–430.

[69] Kovtyukhova, N.I., Mallouk, T.E., Mayer, T.S.: 'Templated Surface Sol–Gel Synthesis of SiO_2 Nanotubes and SiO_2-Insulated Metal Nanowires' *Adv. Mater.*, 2003, **15**, (10), pp. 780–785.

[70] Smith, P.A., Nordquist, C.D., Jackson, T.N., *et al.*: 'Electric-Field Assisted Assembly and Alignment of Metallic Nanowires' *Appl. Phys. Lett.*, 2000, **77**, (9), pp. 1399–1401.

[71] Li, M., Bhiladvala, R.B., Morrow, T.J., *et al.*: 'Bottom-Up Assembly of Large-Area Nanowire Resonator Arrays' *Nat. Nanotechnol.*, 2008, **3**, (2), pp. 88–92.

[72] Savas, T.A., Schattenburg, M.L., Carter, J.M., Smith, H.I.: 'Large-Area Achromatic Interferometric Lithography for 100 nm Period Gratings and Grids' *J. Vac. Sci. Technol. B: Microelectronics and Nanometer Structures Processing, Measurement, and Phenomena*, 1996, **14**, (6), pp. 4167–4170.

[73] Thurn-Albrecht, T., Schotter, J., Kästle, G.A., *et al.*: 'Ultrahigh-Density Nanowire Arrays Grown in Self-Assembled Diblock Copolymer Templates' *Science*, 2000, **290**, (5499), pp. 2126–2129.

[74] Masuda, H., Asoh, H., Watanabe, M., Nishio, K., Nakao, M., Tamamura, T.: 'Square and Triangular Nanohole Array Architectures in Anodic Alumina' *Adv. Mater.*, 2001, **13**, (3), pp. 189–192.

[75] Boehm, S.J., Kang, L., Werner, D.H., Keating, C.D.: 'Field-Switchable Broadband Polarizer Based on Reconfigurable Nanowire Assemblies' *Adv. Funct. Mater.*, 2017, **27**, (5), p. 1604703.

[76] Lumsdon, S.O., Kaler, E.W., Williams, J.P., Velev, O.D.: 'Dielectrophoretic Assembly of Oriented and Switchable Two-Dimensional Photonic Crystals' *Appl. Phys. Lett.*, 2003, **82**, (6), pp. 949–951.

[77] Papadakis, S.J., Gu, Z., Gracias, D.H.: 'Dielectrophoretic Assembly of Reversible and Irreversible Metal Nanowire Networks and Vertically Aligned Arrays' *Appl. Phys. Lett.*, 2006, **88**, (23), p. 233118.

[78] Chaudhary, K., Juárez, J.J., Chen, Q., Granick, S., Lewis, J.A.: 'Reconfigurable Assemblies of Janus Rods in AC Electric Fields' *Soft Matter*, 2014, **10**, (9), pp. 1320–1324.

[79] Hernández-Navarro, S., Ignés-Mullol, J., Sagués, F., Tierno, P.: 'Role of Anisotropy in Electrodynamically Induced Colloidal Aggregates' *Langmuir*, 2012, **28**, (14), pp. 5981–5986.

[80] Boehm, S.J., Lin, L., Guzmán Betancourt, K., *et al.*: 'Formation and Frequency Response of Two-Dimensional Nanowire Lattices in an Applied Electric Field' *Langmuir*, 2015, **31**, (21), pp. 5779–5786.

[81] Gierhart, B.C., Howitt, D.G., Chen, S.J., Smith, R.L., Collins, S.D.: 'Frequency Dependence of Gold Nanoparticle Superassembly by Dielectrophoresis' *Langmuir*, 2007, **23**, (24), pp. 12450–12456.

[82] Pohl, H.A.: 'Dielectrophoresis: The Behavior of Neutral Matter in Nonuniform Electric Fields' (Cambridge University Press, 1978).

[83] Liu, M., Lee, T.-W., Gray, S.K., Guyot-Sionnest, P., Pelton, M.: 'Excitation of Dark Plasmons in Metal Nanoparticles by a Localized Emitter' *Phys. Rev. Lett.*, 2009, **102**, (10), p. 107401.

[84] Schnelle, T., Müller, T., Fiedler, S., Fuhr, G.: 'The Influence of Higher Moments on Particle Behaviour in Dielectrophoretic Field Cages' *J. Electrostat.*, 1999, **46**, (1), pp. 13–28.

[85] Müller, T., Gradl, G., Howitz, S., Shirley, S., Schnelle, T., Fuhr, G.: 'A 3-D Microelectrode System for Handling and Caging Single Cells and Particles' *Biosens. Bioelectron.*, 1999, **14**, (3), pp. 247–256.

[86] Guzman, A.R., Kim, H.S., de Figueiredo, P., Han, A.: 'A Three-Dimensional Electrode for Highly Efficient Electrocoalescence-Based Droplet Merging' *Biomed. Microdevices*, 2015, **17**, (2), p. 35.

[87] Qin, L., Park, S., Huang, L., Mirkin, C.A.: 'On-Wire Lithography' *Science*, 2005, **309**, (5731), pp. 113–115.

Chapter 11

Dancing angels on the point of a needle: nanofabrication for subwavelength optics

Mikhail Y. Shalaginov[1], Fan Yang[1], Juejun Hu[1] and Tian Gu[1,2]

11.1 Opening remarks

The point of a needle is often used as a metaphor for miniscule objects and has hence inspired the classical question: How many angels can dance on the point of a needle? In the emerging world of nanotechnology, however, there is a plenty of room on the point of a needle and it can easily be a dancing stage for engineers equipped with state-of-the-art nanofabrication tools (literally—as we shall show later).

In optics, the capability to produce and manipulate nanoscale structures has culminated in subwavelength optics that provide unprecedented control over light propagation compared to their traditional bulk or wavelength-sized counterparts. Other chapters of this book have unveiled the fundamental theory as well as the fascinating prospect of leveraging subwavelength nanoantennas and nano-plasmonic structures for light manipulation. In this chapter, we will focus on the technologies that are used to create subwavelength optical devices. Since there are already a number of well-written textbooks and review articles describing nano-fabrication methods in general [1–5], we feel that repeating another exhaustive review about nanofabrication offers little value to the community. Instead, here we choose to highlight the difficulties associated with subwavelength structure processing and give a snapshot of the ingenious solutions that have been developed to tackle these challenges.

Compared to processing of semiconductor nanoelectronics, subwavelength optics fabrication is unique in two ways. First, nanofabrication of subwavelength optics often involves a wide range of materials not necessarily within the repertoire of standard Complementary Metal-Oxide-Semiconductor (CMOS) processing to achieve desired optical functions. The examples include noble metals for plasmonics, transparent

[1]Department of Materials Science & Engineering, Massachusetts Institute of Technology, Cambridge, MA, USA
[2]Materials Research Laboratory, Massachusetts Institute of Technology, Cambridge, MA, USA

conducting oxides for active photonics, and chalcogenide compounds for high-index-contrast metasurfaces. Second, even when only standard materials are utilized, the fabrication processes often have to be customized to meet the unique requirements set forth by optical applications: fabrication of metasurfaces operating in the visible spectrum based on high-aspect-ratio (HAR) TiO_2 nanostructures is a case in point [6]. In light of these unique features, the remainder of this chapter is organized into two sections. We will start by evaluating various material platforms and standard planar nanofabrication methods in the context of subwavelength optical structure definition. The second part of the chapter covers strategies to realize three classes of structures that often present significant fabrication challenges to experimentalists, yet are highly relevant for subwavelength optics: HAR structures, 3D components, and devices on unconventional, non-planar substrates.

11.2 Standard planar nanofabrication technologies applied to subwavelength optics

Subwavelength optics deals with artificial optical interfaces that provide on-demand control of phase, amplitude, and polarization of light, a capability not available in natural materials. By arranging nanostructured optical elements in flat subwavelength arrays, planar subwavelength optics (such as metasurfaces) can leverage standard nanofabrication techniques for potential low-cost, scalable manufacturing [5,7–20]. In addition to material property and process optimization, key considerations for fabricating such nanostructures include resolution, precision, reproducibility, throughput, reliability, cost effectiveness, and CMOS compatibility. With the development of the semiconductor industry, current lithographic fabrication techniques nowadays offer high resolution, complex geometry definition, high throughput, and high reproducibility for manufacturing planar nanoparticles. In order to achieve desired optical functionalities at the subwavelength scale, non-traditional materials or customized processes are usually needed. In this section, traditional planar nanofabrication approaches adapted for fabricating planar subwavelength optics are reviewed and discussed. Additionally, prospects for scaled-up manufacturing which leverages available tools from the semiconductor industry are further discussed.

11.2.1 Materials for subwavelength optics

11.2.1.1 Plasmonic subwavelength optics

Subwavelength optics has been initially based on metallic nanoparticles made of noble metals (e.g., gold and silver) to enhance light concentration at the nanoscale, which, in turn, enables versatile manipulation of electromagnetic waves [21–25]. Among metals, gold and silver exhibit lower loss in the visible and near-infrared spectral range. Both of them can be deposited using physical vapor deposition techniques, such as electron beam evaporation, thermal evaporation, or sputtering. A typical method for patterning metallic nanoparticles involves electron beam lithography (EBL) to define the structures. As shown in an exemplary fabrication

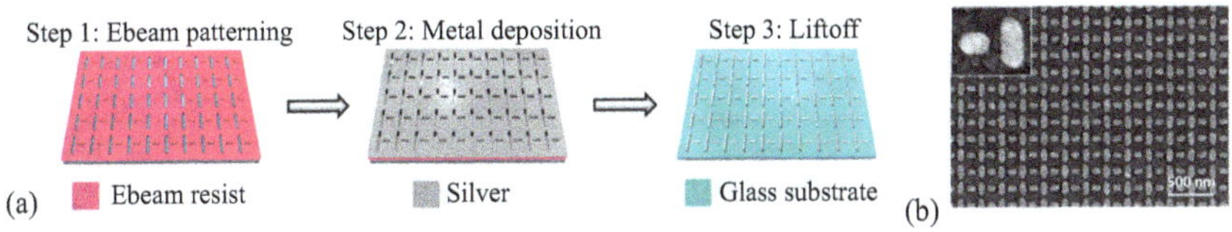

Figure 11.1 Typical fabrication flow of a plasmonic metasurface [25].

process flow in Figure 11.1 [25], a substrate is first coated with an electron beam resist layer, and subsequently exposed and patterned by a scanning focused electron beam. After development of the resist, a metal layer is deposited and patterned on the substrate via liftoff. Multiple metallic layers can be integrated in stacks to allow individual tuning of the induced radiation coming from both sides of the meta-device [16,26–28]. This enables, for example, plasmonic metasurface designs combining resonant phase tuning and the Pancharatnam–Berry (PB) phase.

The excessive losses of metals have severely limited the device performance at optical frequencies. Potential alternative plasmonic materials have been explored [29,30], such as transparent conducting oxides (TCOs) [30], intermetallics (e.g., silicides, germanides, borides, and nitrides) [31], metallic alloys [32], graphene [33], etc. In particular, TCOs have been widely used in optoelectronic devices with well-developed CMOS-compatible processes. When employed in nanophotonics, TCO nanostructures based on ITO, ICO, or ZnO doped with Al or Ga allow tailorable device functionalities [34–38]. For instance, ITO is a potential candidate for plasmonic materials operating in NIR [39,40]. Due to its non-stoichiometric nature, the material deposition process and annealing conditions can play major roles in determining the material optical performance. High-quality ITO films can be produced by sputtering or laser ablation [41]. Significant variance of dielectric functions have been observed when ITO films were annealed at different conditions, including both annealing temperature and ambient gas environment [30,42].

Al- or Ga-doped ZnO films can also be deposited by sputtering or laser ablation. The deposition conditions (temperature and oxygen partial pressure) and dopant concentration are important to obtain low-loss, high-quality films [30,43,44]. In one example, TCO metasurfaces made of Ga:ZnO were developed and used as a quarter-wave plate (QWP) for polarization control in the NIR spectral region [45]. The metasurface consisted of orthogonally patterned nanorod arrays designed to provide a phase difference of $\pi/2$ for perpendicularly polarized light. As schematically depicted in Figure 11.2, the Ga:ZnO film was deposited on a glass substrate using pulsed laser deposition. A dry-etching method with bilayer resists involving hydrogen silsesquioxane (HSQ) for masking and poly(methyl methacrylate) (PMMA) as the sacrificial layer was developed to fabricate the metasurface. Owing to Ga:ZnO's weak dispersion of dielectric function, a broadband QWP functionality was realized.

11.2.1.2 Silicon-based materials

Plasmonic subwavelength optics are inherently restricted by the large Joule losses. They typically operate in the reflection mode, which further restricts the application

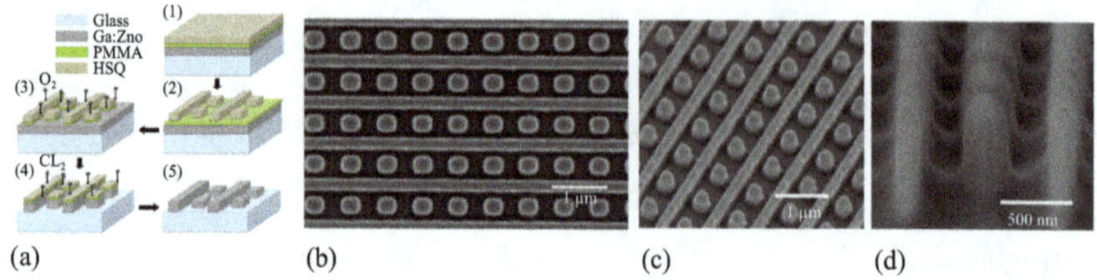

Figure 11.2 (a) Process flow for fabrication of Ga:ZnO metasurfaces using reactive-ion etching (RIE) with bilayer resist. (1) Spin-coating of PMMA and HSQ layer over Ga:ZnO film, (2) e-beam lithography of HSQ resist, (3) O_2 RIE etch of PMMA with HSQ as the etch mask, (4) CL_2 (which stands for chlorine) RIE etch of Ga:ZnO with HSQ and PMMA as the etch mask, and (5) removal of PMMA and HSQ etch mask with acetone. (b) Top view, (c) 30° tilted view, and (d) cross-sectional (75° tilted) view of the FESEM image of a fabricated Ga:ZnO metasurface [45].

space. All-dielectric nanoparticles can eliminate such unfavorable aspects and enable a large array of meta-optics with performance comparable to commercial bulk optical components [46–49]. Such high-index nanoparticles support high-quality Mie resonances and low-profile nanostructure designs [5,49]. The low optical loss, enhanced electric and magnetic resonances, tunability of scattering directionality, and CMOS compatibility make all-dielectric metasurfaces a desirable platform to realize high-performance flat optical devices with scalable manufacturing capability. Here, we review planar nanofabrication techniques adopted for fabricating dielectric metasurfaces.

The utilization of silicon in subwavelength optics has been investigated extensively owing to several beneficial properties, such as high refractive index and low absorption loss within telecommunication wavelengths [50], as well as strong electric and magnetic multipolar responses associated with Si nanoparticles. Furthermore, Si is the backbone of state-of-the-art electronic and photonic technologies, offering compatibility with standard CMOS processes.

Silicon-based subwavelength nanostructures can be fabricated using lithographic methods, which provide precise control of the optical responses with a variety of nanoparticle geometries [51–53]. Typical process steps based on EBL for fabricating silicon-based nanoparticles are schematically illustrated in Figure 11.3(a) [12]. Monocrystalline silicon-on-insulator wafers or low-index substrates (such as quartz) coated with amorphous or polycrystalline silicon films deposited via low-pressure chemical vapor deposition (LPCVD) or plasma-enhanced chemical vapor deposition (PECVD) are typically used. The film is then patterned by EBL and reactive-ion etching (RIE). Examples of fabricated silicon nanodisk structures are shown in Figures 11.3(b) and (c).

Functional meta-optic devices further require the integration of multiple metasurfaces, optical apertures, and stops, as well as alignment features to form an

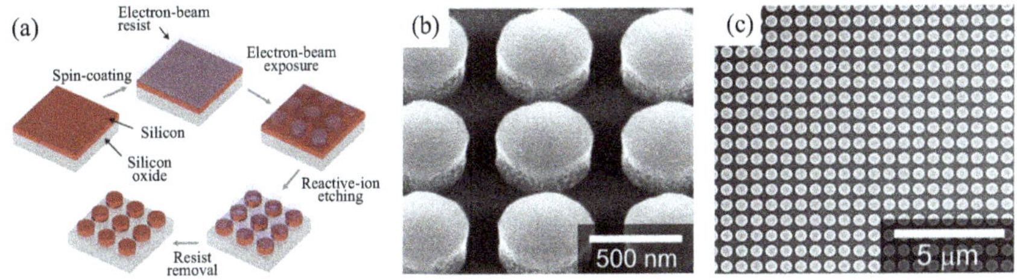

Figure 11.3 *(a) Typical process steps for fabricating high-index dielectric subwavelength nanoparticles using e-beam lithography [12]; (b) and (c) scanning electron micrographs of fabricated silicon nanodisks [51].*

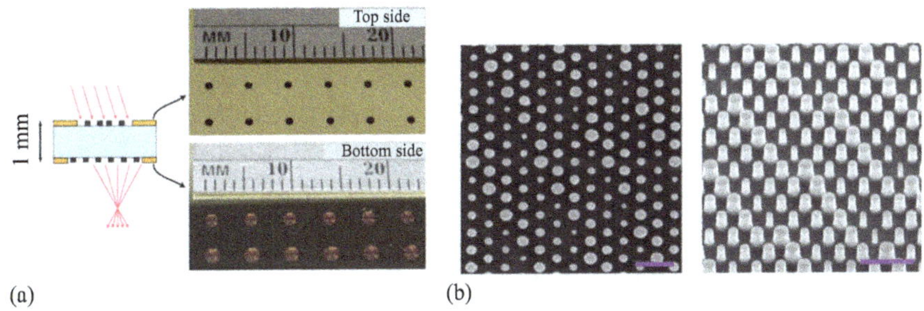

Figure 11.4 *(a) Schematic and optical images of a doublet metalens consisting of two metasurfaces formed on both sides of a fused silica substrate. (b) SEM images of fabricated metasurfaces consisting of a-Si nanoposts [54]. Scale bars, 1 μm.*

optical system. Along this direction, a doublet lens with metasurfaces fabricated on both sides of a fused silica substrate was demonstrated, as shown in Figure 11.4 [54]. Hydrogenated amorphous Si (a-Si) layers were deposited on both sides of the substrate via PECVD and defined into nanopost patterns via EBL and dry etching. After the first metasurface was patterned on one side of the substrate, the second metasurface was lithographically aligned and patterned on the other side. The patterned side of the substrate is protected by a polymer layer while the other side is being processed. Apertures were created by patterning opaque metal layers on both sides of the wafer, followed by coating of antireflection layers.

It is also possible to fabricate metasurfaces on membranes to take advantage of their nanoscale flat architecture. A MEMS-actuated dielectric metalens was demonstrated consisting of a movable metasurface on a silicon nitride membrane and a fixed metasurface on a glass substrate. The distance between the two metasurfaces was controlled by electrostatic actuation of the membrane [53]. To fabricate such a movable metasurface, a SiO_2 layer was first deposited on a SiN coated

Si wafer, followed by the deposition of an a-Si layer. After the metasurface was patterned into the a-Si layer and metal contacts were defined, the remaining part of the wafer under the membrane was removed via wet etching. Then, the metasurface was released after the membrane was patterned and dry etched. To provide alignment between the front and back sides of the a-Si layer, alignment marks were etched through the layer.

In [55], substrate conformal soft-imprint lithography was combined with RIE to fabricate silicon nanocylinders used as subwavelength Mie resonators, which provided efficient broadband omnidirectional antireflection coatings on a silicon wafer. The soft-imprint lithography process employed a PDMS stamp molded from a master Si wafer to replicate patterns onto a sol–gel layer spin-coated on the target Si wafer. RIE was subsequently utilized to define Si nanoparticles using the sol–gel patterns as a mask. By coating the silicon nanocylinders with an additional Si_3N_4 coating using LPCVD, the reflection reduction was further enhanced.

Silicon nitride (SiN) possesses a broadband transparency ranging from near UV to infrared and is compatible with CMOS processes allowing integration leveraging existing foundry infrastructures [53,56–58]. A low-contrast dielectric metasurface based on SiN was demonstrated in [58]. The authors designed and fabricated vortex beam generators using SiN operating in the visible spectral range, leveraging low visible absorption of SiN and CMOS compatibility of the material.

11.2.1.3 III–V-based materials

III–V semiconductors (such as GaAs, GaN, and GaP) are of great interest to subwavelength optics due to their high refractive indices and excellent optoelectronic properties [59–64]. In particular, GaAs nanoparticles with zero optical backscattering have been demonstrated using a hybrid integration process [63]. As shown in Figure 11.5(a)–(c), epitaxial liftoff was utilized to transfer a 1 μm-thick GaAs membrane onto a fused silica substrate. The film was then thinned by RIE and patterned into nanostructure arrays with EBL and RIE. Furthermore, multistep lithographic processes for creating GaAs metasurfaces were developed based on selective wet oxidation (Figure 11.5(d)–(f)) [60–62]. Here, HAR etching was performed along with selective wet oxidation of AlGaAs underlayers to fabricate GaAs metasurfaces. As a result, multilayer GaAs metamaterials were realized with an embedded low-index native oxide layer between the nanoresonators and the substrate designed for broadband high reflectivity. In another example [62], the fabrication process began with epitaxial growth using MOCVD or MBE. This was followed by e-beam lithography and ICP dry etch to pattern the etch masks. Subsequently, the selective wet oxidization of AlGaAs was performed to form $(Al_xGa_{1-x})_2O_3$.

11.2.1.4 Chalcogenide compounds

Chalcogenide alloys, compounds containing one or multiple chalcogen elements (e.g., sulfur (S), selenium (Se), and tellurium (Te)), are attractive candidates for building subwavelength IR optics, owing to their exceptional optical properties, such as broadband infrared transparency, large refractive indices, tailorable

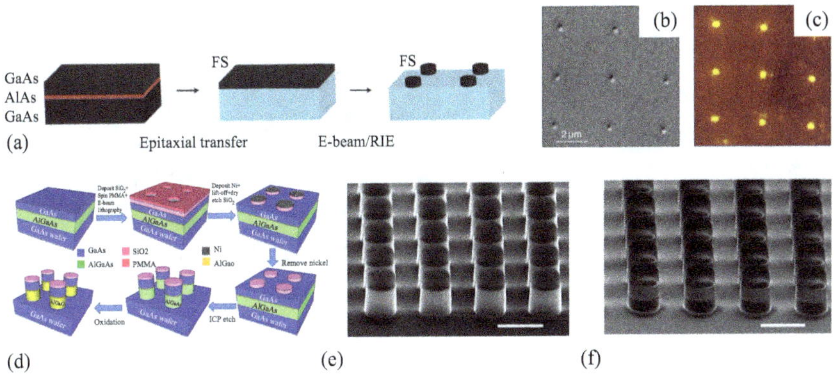

Figure 11.5 *(a) Fabrication process of GaAs nanopillars and (b) SEM image of GaAs nanopillar arrays (radius ~ 90 nm). (c) Confocal photoluminescence image of the nanopillar arrays [63]. (d) Fabrication process for constructing AlGaAs-based dielectric metasurfaces. SEM images showing GaAs dielectric nanocylinders (e) before and (f) after the selective wet oxidation process. The scale bar corresponds to 1 μm [62].*

optical properties, large optical nonlinearity, and versatile fabrication processes with epitaxy-free deposition of high-quality thin films [65–70]. Furthermore, chalcogenide-based phase-change materials (PCMs) undergo dramatic, nonvolatile electrical and optical property change between their amorphous and crystalline states under external thermal, optical, or electrical stimuli, qualifying them as promising materials enabling active reconfigurable photonics [71].

Tellurium dielectric metamaterials and metasurfaces have been developed [72–74]. Te nanocubes have been fabricated on a barium fluoride (BaF_2) substrate and designed to center the magnetic resonance reflection peak at 10 μm [72]. In this work, a 1.7-μm-thick Te film was deposited on a BaF_2 substrate using electron-beam evaporation, with the nanostructures patterned by EBL followed by RIE. A multi-cycle deposition-etch process was utilized to fabricate the Te dielectric metamaterial structures [74], as shown in Figure 11.6(a). The multiple cycles of the deposition-etching-liftoff process reduce the deposition thickness of Te each time and effectively mitigate pinch-off issues during deposition. The fabricated nanocubes are shown in Figure 11.6(b), in contrast to the structures fabricated from a single deposition/liftoff step (Figure 11.6(c)), which suffer from severe pinch-off effect resulting in tapered meta-atom geometry.

Besides Te, PbTe is another material boasting a high refractive index larger than 5. It is thus ideally suited for forming dielectric meta-atoms supporting high-quality Mie resonances [75,76]. By monolithically integrating a nanocrytalline PbTe thin film on a low-index CaF_2 substrate ($n = 1.4$) and utilizing a dielectric Huygens meta-atom design, high-efficiency mid-infrared (mid-IR) transmissive meta-optic devices have been demonstrated [76]. The combination of PbTe and

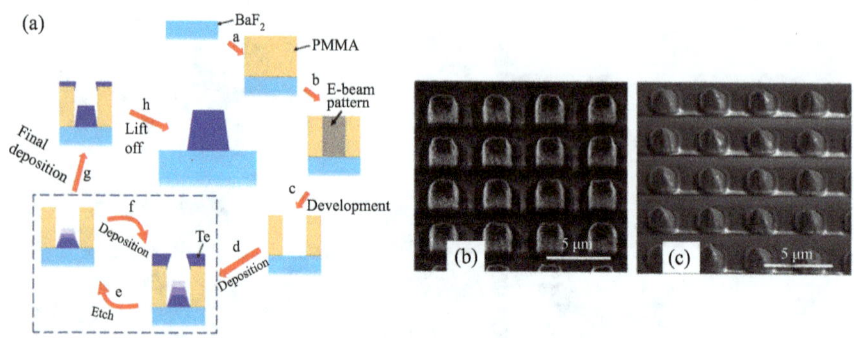

Figure 11.6 *(a) Multi-cycle deposition-etch process for patterning Te films; (b) Te dielectric metamaterial structure fabricated using multi-cycle deposition-etching-liftoff technique; (c) Te dielectric metamaterial structure fabricated using a single deposition-liftoff step [74].*

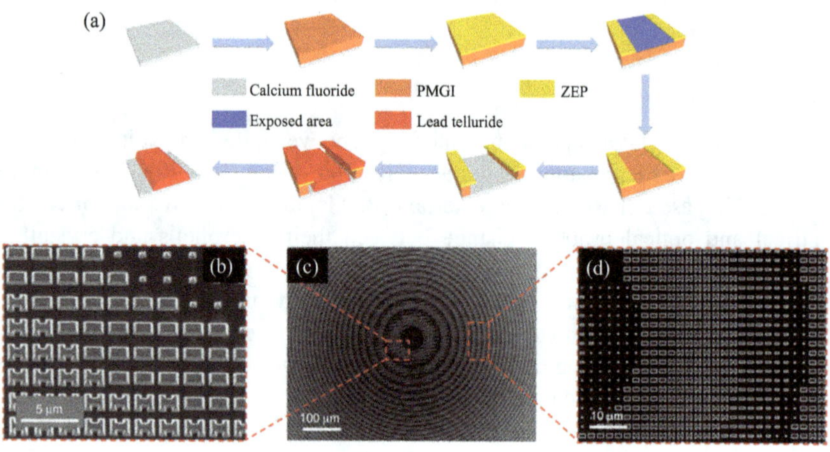

Figure 11.7 *(a) Schematic fabrication process flow of the meta-optical devices; (b)–(d) Top-view SEM micrographs of the fabricated aspheric metalens.*

CaF_2 contributes to an ultra-thin profile of the meta-atoms due to their large index contrast, with a thickness of only 1/8 of the operating wavelength in free-space (i.e., 650 nm at the 5.2 μm wavelength), as well as low optical attenuation from 4 to 9 μm wavelengths. The double-layer liftoff process and devices fabricated using this method are shown in Figure 11.7. This work marks the first experimental realization of a Huygens metasurface in the mid-IR and the first demonstration of diffraction-limited focusing in the mid-IR using transmissive metalenses. The material selection further foresees scalable manufacturing of meta-optics with large-area, high-throughput (growth rate ~100 nm/min) PbTe film deposition via simple single-source thermal evaporation and wafer-scale lithographic patterning of the film [77–80].

11.2.1.5 Phase-change materials

Optical phase-change materials (O-PCMs) exhibit drastic optical property changes upon undergoing solid-state phase transformations [71]. Such transitions can be triggered by thermal, optical, or electrical stimuli and are attractive for photonic applications to realize reconfigurable optical devices. A subset of chalcogenide alloys possesses such phase-change properties, such as the commonly used GeSbTe (GST) alloy, a popular material for nonvolatile, tunable subwavelength optics [81–87]. In [81], GST films with a thickness of 70 nm were sputtered on a glass substrate coated with a $ZnS–SiO_2$ layer. After deposition of the GST film, a protective $ZnS–SiO_2$ layer is subsequently applied. Then, a reversible index modification from phase transition is induced optically by applying trains of femtosecond pulses onto the GST film (Figure 11.8(a)). As a result, reconfigurable flat optics with binary or grayscale patterns are created by using a diffraction-limited resolution optical writing process (Figure 11.8(b)–(f)).

Despite the large index modulation Δn of GST, they also exhibit considerable optical losses due to the relatively small bandgap and excessive free carrier absorption (FCA), limiting their optical performance as well as application space. Recently, a new group of O-PCMs, Ge–Sb–Se–Te (GSST), has been developed and implemented in integrated photonic and meta-optic devices [88]. GSST features

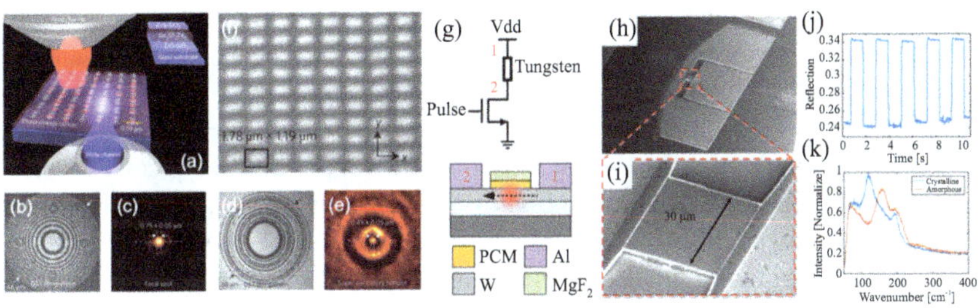

Figure 11.8 *(a) Laser writing of reconfigurable photonic devices in GST phase-change film to locally vary the film property between its amorphous and crystalline states. (b) Fresnel zone plate written in phase-change film and (c) optical hotspot focused by this lens at 730 nm wavelength. (d) Binary super-oscillatory lens patterned in the phase-change film and (e) optical hotspot focused by this lens at 730 nm wavelength. (f) Dielectric metamaterials written in the phase-change film. (g) Schematic of electrothermal switching of GSS_4T_1. (h) SEM of the full device used to switch a 30 μm × 30 μm pixel. (i) Zoom-in on the pixel with a square pattern of GSS_4T_1. (j) Time-dependent absolute reflection measurements of a 1,550-nm laser focused onto the pixel. (k) Raman measurements of the GSS_4T_1 film after being triggered by amorphization and crystallization pulses.*

unprecedented broadband optical transparency covering the telecommunication bands to the long-wave infrared (LWIR). Thin films of GSST can be deposited by thermal evaporation from a single $Ge_2Sb_2Se_4Te_1$ source and patterned using standard RIE or liftoff fabrication processes. Record low losses in nonvolatile photonic circuits and reconfigurable metasurfaces have been demonstrated. As an example, electrical pixelated switching of meta-optic devices was realized by integrating micro-heaters underneath thin-film GSST, as shown in Figure 11.8(g)–(i). Complete, reversible, and repeatable switching of the GSS_4T_1 film between its amorphous and crystalline states has been demonstrated (Figure 11.8(j) and (k)).

Another approach for constructing reconfigurable nanophotonic devices involves the integration of volatile phase-transition materials, such as vanadium dioxide (VO_2) [89–91], which also exhibit large optical property modification under external stimuli, albeit in a volatile way. The nanostructures were usually either directly patterned in VO_2 [90,92,93], or formed from plasmonic metals positioned on top of or embedded in a tunable VO_2 layer [90], [94–98]. In addition, the epitaxial growth of VO_2 on a sapphire substrate to form deep subwavelength VO_2 nanobeams was demonstrated [99]. This approach potentially allows fabrication of large-area tunable metasurfaces without resorting to nanostructuring processes.

11.2.1.6 2D van der Waals materials

2D materials (e.g., isolated single-layer graphene) are another attractive platform for subwavelength optics [7], [100–105]. Graphene exhibits highly variable dielectric permittivity useful for realizing active photonic devices. Nanostructures such as nanodisks, nanorods, and nanoholes have been demonstrated with broadband tunable resonances [101,106–108]. It also supports plasmonic resonance in the mid-IR spectral regime with high-quality factors and tight field confinement [101,102,109].

So far, the integration of 2D materials onto photonic devices typically employs layer transfer of exfoliated 2D membranes onto another substrate containing photonic devices [110]. This "hybrid" integration approach has been widely adopted but is limited by the topology of underlying structures or would require an additional planarization process to avoid rupture of the 2D membranes at structure corners [111,112]. Meanwhile, the interaction region between the optical mode and 2D layer is limited to the top of the optical structure. A monolithic integration approach is more desirable in terms of improving manufacturing yield, throughput, integration precision, and flexibility of device design. Such an approach would involve the growth of optical films on 2D materials and subsequent lithographic patterning of photonic devices [113–115]. However, it is not trivial to grow optically thick dielectric films on 2D materials due to its chemically inert surface, which makes nucleation and growth of uniform dielectric films challenging [116]. Surface modification could catalyze nucleation via ozone [117], NO_2 [118], or perylene tetracarboxylic acid [119], while sacrificing carrier mobility in graphene. While the former process tends to severely degrade graphene quality and the latter is not practical for growing optically thick layers, without the surface functionalization, continuous dielectric films could be formed via electron beam evaporation or atomic layer deposition (ALD). PECVD is another potential alternative approach, while low-density, low-power plasma is required to mitigate plasma damage to graphene surface [120]. In general, the integration of 2D materials requires special attention on protecting the material's structural integrity from high temperatures, plasma, and reactive chemicals used during the process.

Amorphous chalcogenide glasses have recently been demonstrated as an ideal material candidate for integration with 2D materials. The glasses' amorphous nature and good van der Waals adhesion to different substrates allow epitaxy-free and substrate-agnostic deposition with minimal induced thermal and structural damages. Optically thick chalcogenide glass films have been deposited on a variety of 2D materials and patterned into functional optical devices without degrading the material and device properties [100]. Figure 11.9(a) schematically depicts a generic fabrication process to integrate chalcogenide glass photonic devices with graphene. Multi-layer photonic structures with 2D materials embedded inside optical structures were also realized, which optimized the interaction between the optical mode and 2D layer (Figure 11.9(b)).

11.2.2 Large-scale manufacturing: a case study of optical metasurfaces

Optical metasurfaces have revolutionized how optical components are designed and fabricated by completely deviating from the conventional operation principles

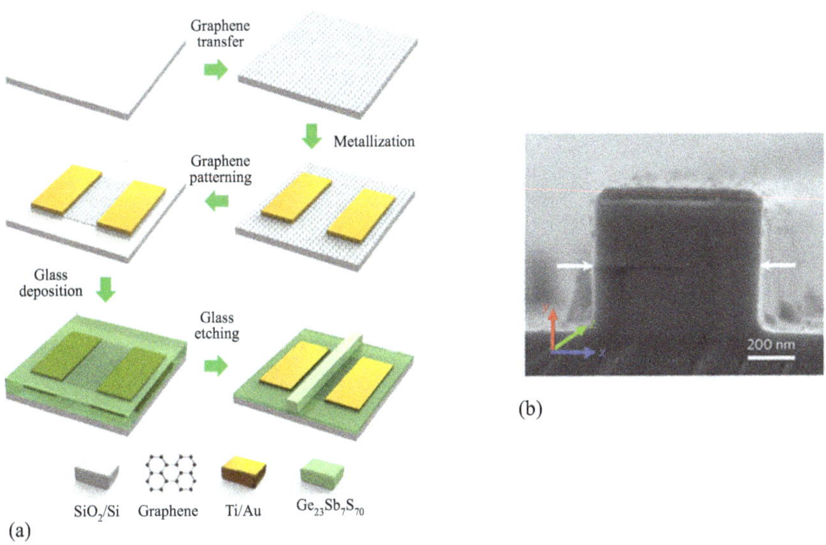

Figure 11.9 (a) Generic schematic fabrication process to integrate chalcogenide glass (ChG) photonic devices with graphene; (b) SEM image of the fabricated ChG waveguide structure embedding graphene; white arrows mark graphene location [100].

based on bulk optical materials. Instead, the wavefront of light is efficiently "molded" by a subwavelength array of nanostructures. Such an optically thin and planar architecture can potentially leverage microfabrication processes matured in the semiconductor industry to enable large-volume, low-cost, and scalable manufacturing. While EBL is a versatile fabrication tool widely used in the research community and well suited for prototyping purposes, the process is relatively costly and time consuming. For large-scale manufacturing, alternative high-throughput, low-cost fabrication approaches, such as photolithography, interference lithography, nanoimprinting, etc., are preferred. Meanwhile, wet etching processes can be utilized to potentially replace RIE processes for select applications [121].

Toward the goal of scalable fabrication, stepper lithography has been employed to fabricated large-area (a few centimeters in diameter) flat lenses based on metasurfaces with high yield [15]. The fabrication methodology and fabricated devices are shown in Figure 11.10. Here, an a-Si layer (600 nm thick) was first deposited on a 4-inch fused silica wafer using PECVD. After applying the photoresist and contrast-enhancement layers, the pattern of the metalens, which was carried out by a quartz photomask, was projected onto the wafer via an i-line stepper. After development, RIE was used to create the nanopost arrays in the a-Si layer. The use of a stepper allows the rapid replication of the metalens patterns over the entire wafer area by incrementally stepping the wafer position and repeating the exposure process, allowing the fabrication of hundreds of wafers per hour. Lastly, the wafer was diced into individual metalenses with diameters of 2 cm each. High-quality imaging performance has been achieved using the metalens fabricated via this approach.

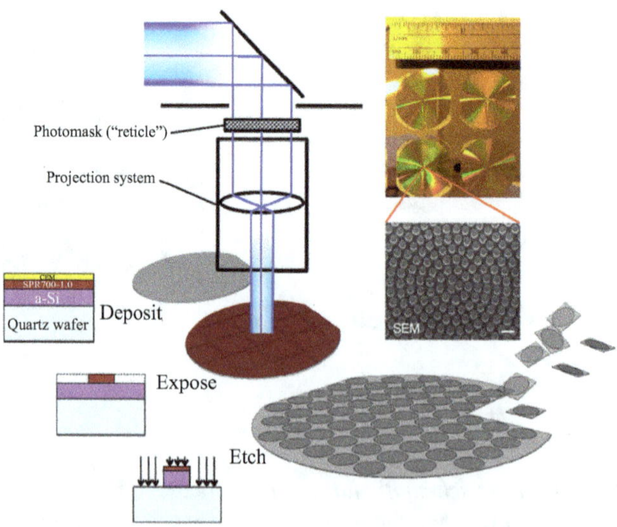

Figure 11.10 Schematic diagram of using photolithographic stepper technology for metalens fabrication over a 4-inch silicon wafer [15].

As the authors discussed in [15], one of the major challenges for scaled-up manufacturing of metasurfaces is the handling of very large metasurface layout files. To fulfill the subwavelength requirement, billions of meta-atoms with nanometer-scale precision are needed to fill the entire device aperture of a few centimeters in size, resulting in file sizes of hundreds of gigabytes. A multi-level compression algorithm taking advantage of rotational symmetry of metalens designs was thus developed to dramatically reduce the file size by several orders of magnitude [15]. Such an approach can be applied to other large-area fabrication methods for metasurfaces, such as nanoimprint and interference lithography.

Other challenges associated with fabricating large-area metasurfaces include local and global variance of the substrate over a large area, such as surface roughness, flatness, parallelism, etc. It should be noted that the impacts of these non-idealities are highly dependent on meta-atom design and meta-optic device architecture. For metalenses, variance of such parameters of commercial off-the-shelf wafers still guarantees good focusing and imaging performance at the expense of reduced optical efficiency, according to the analysis in [15]. For other types of devices (e.g., beam deflectors), however, such variance will enhance unwanted diffraction orders and result in cross-talks [76]. Thus, performance sensitivity to wafer parameter variances must be taken into account in the design phase with a thorough tolerance analysis.

11.3 Innovative solutions to nonconventional subwavelength optics designs

In the following sub-sections, we present the challenges and available solutions to the fabrication of three types of subwavelength optical structures: HAR structures, 3D structures, and devices on unconventional, non-planar substrates. Each sub-section begins by elaborating the significance of the type of structures and the unique optical functions they furnish. Various fabrication approaches are subsequently reviewed and compared.

11.3.1 HAR nanostructures

Advances in theoretical studies and numerical simulations of nanophotonic structures inspire scientists to further push the boundaries of fabrication capabilities. One of such avenues is to attain HAR nanostructures, which can enable significant improvements in imaging, sensing, energy harvesting, and secure communication. For instance, dielectric metalenses with HAR elements can achieve significantly higher focusing efficiencies and achromatic performance [64,122,123]; metamaterials and photonic crystals with narrow and deep trenches can support novel surface and bulk modes [124,125]; diamond nanowire-antenna arrays with color centers can unlock efficient on-chip single-photon generation [126]; ultra-narrow plasmonic gaps of <1 nm can lead to much stronger field enhancement, enabling quantum phenomena such as room-temperature strong coupling [127].

The scope of the developed techniques for achieving HAR structures is extremely broad, and the reader can refer to several reviews on the topic, for instance, the work by Lee *et al.* [128]. In this section, we would like to focus on several pattern transferring methods, which, in our opinion, are the most suitable for constructing high-quality subwavelength optical structures over large areas (mm-scale and larger) using commonly available cleanroom facilities. The quality of the HAR nano-optical devices is quantified by sidewall roughness ($<\lambda/100$), sidewall angle, and surface chemical contamination. Each of the following HAR techniques is presented below with a brief description of the technique and its principle of operation, general guidelines for achieving desired fabrication metrics, its advantages and limitations, as well as the state-of-the-art of fabricated devices.

11.3.1.1 Reactive-ion etching

Dry etching is an essential part of todays integrated circuit manufacturing line. Back in the early days of the semiconductor industry, wet etching was a dominant process used in pattern transfer processes. However, over the years wet etching was replaced with dry etching methods due to the necessity of reducing chemical waste and need to fabricate vertically standing elements with small feature sizes. One of the fascinating demonstrations of state-of-the-art etching technology are the 50:1 extremely HAR channels in a 3D flash memory stacks [129], known as the "Burj Khalifa towers" at the nanoscale.

RIE is the most common dry etching technique employed nowadays. In a RIE process, the material is selectively removed by a combination of chemical reaction and physical ion bombardment. The synergy between the two processes produces substantial increase in etching rates and selectivity [130]. RIE plasma is typically formed by collisions between accelerated electrons with the entering gas molecules, such as Ar, O_2, Cl_2, BCl_3, CF_4, CHF_3, SF_6, etc. These collisions lead to formation of ions and free radicals via ionization and dissociation processes, respectively. For example, electron collision with a tetrafluoromethane (CF_4) molecule can result in the following reaction:

$$CF_4 + e^- \rightarrow {CF_3}^+ + F + 2e^-.$$

This is only one example of the many other reactions occurring to CF_4. Therefore, in a RIE process, one has to consider a variety of ion groups, including ${CF_x}^+$, ${C_2F_x}^+$, F^+, as well as their negatively charged counterparts. Both positive and negative ions can participate in the etching processes, since the substrate is placed on a RF-biased electrode, which helps avoid charging of the sample.

The two main mechanisms to generate plasma are based on accelerating electrons with a parallel plate capacitor (capacitive coupled plasma or CCP) and with an induction coil (inductive coupled plasma or ICP). Historically, ICP RIE was invented later and hence this approach is more advanced than CCP. The development of ICP was mainly fueled by the industrial needs to achieve high-throughput etcher. ICP systems can generate plasma with extremely high densities, up to 10^{12} ions/cm^3, which is several orders of magnitude larger than the plasma densities in a CCP system. ICP RIE architecture was also designed to have a plasma source, ICP torch, outside of the main reactor. Hence, ICP machine design provides a separate control over ion energy and ion density.

Developing an etching recipe from scratch is a challenging task, however, here we would like to offer general strategies for achieving HAR structures with low sidewall roughness. HAR structures require a more anisotropic etch, which can be achieved by increasing ion kinetic energy and reducing their scattering by gas molecules. Therefore, increasing bias voltage and reducing gas pressure (~10 mTorr) generally lead to more vertical and smoother etch profiles. Additionally, sidewall passivation can be instrumental in further straightening of the walls by impeding lateral etching. The sidewall passivation is implemented by adsorption of carbon- and/or oxygen-containing molecules, which emanate either from photoresist mask erosion or as a constituent of the feeding gases. Passivation comes with the price of sidewall contamination, although there are existing approaches to remove the passivation layer, for instance, with a hydrogen plasma treatment [131]. Since Cl does not attack SiO_2, the oxide sidewall passivation layers may also form due to Si contaminants in the etching chamber during RIE based on Cl chemistries [70,132]. For the etching chemistries with Cl or F compounds, isotropic etching can also be suppressed by substrate backside cooling [133].

Numerous studies have demonstrated that conventional ICP RIE techniques can be used to fabricate HAR subwavelength optical structures from various materials with an AR up to 15 [62,122,126,134]. Here, we take RIE fabrication of broadband dielectric metalenses as an example [122]. The fabrication protocol mainly includes deposition of amorphous silicon film on top of quartz substrates; patterning of Al_2O_3 mask on top of the film; and dry etching through the mask followed by mask removal (Figure 11.11(a)). The etching process was performed inside an ICP RIE chamber with feed gases of SF_6 (40 standard cubic centimeters per minute or sccm) and O_2 (16 sccm). This is a standard chemistry used for etching deep trenches in Si, where sulfur fluoride ions act as etching reagents. Al_2O_3 is known to be extremely resistant to SF_6-based etching, and its selectivity to Si reaches 1:70,000 [135]. In addition, during the etching process, the substrate was kept at –100°C to retard the fluoride isotropic etching reaction. The fabricated

Figure 11.11 (a) Fabrication sequence for constructing Si-based metasurface. (b) SEM image depicting meta-atoms comprising broadband achromatic metalens, AR ~ 14, smallest feature size 100 nm [122].

metasurface is composed of 1.4-μm-tall meta-atoms having various shapes (Figure 11.11(b)). The smallest feature size is about 100 nm and it manifests a maximum AR of approximately 14. To the best of our knowledge, this is the largest AR value realized in metasurfaces.

Due to its wide availability, RIE is an attractive technique for implementing HAR nanophotonic structures. It is a relatively simple fabrication procedure typically consisting of only a few steps including mask patterning, etching, and mask removal. RIE is also a fast process with typical etching rates of about 100 nm/min. On the other hand, development and optimization of dry etching recipes, in particular for nonstandard materials, can be a time-consuming process. Often, the process is also sensitive to chamber contamination, which can be a major hurdle to maintaining process reproducibility especially in university foundries. Additionally, RIE is not suitable for etching metals such as Ag, Fe, Au, Ni, Co, and Pt, as their halides are nonvolatile compounds. The other common issue with RIE is the dependence of etching rate on feature size and pattern density due to the microloading effect. For instance, larger size openings and isolated standing structures etch faster than smaller and densely packed features.

RIE anisotropy can be further enhanced by increasing mask thickness. For instance, a thick hard mask can filter out most of the ions with a lateral momentum component. The inset of Figure 11.12(a) depicts two ions, where the first ion, incident at an angle, gets stuck by initially hitting the mask walls whereas the second ion, moving vertically, penetrates through the mask and impinges on the film surface. High *et al.* proposed and utilized this approach to etch epitaxially grown silver films and fabricated hyperbolic metasurfaces [136]. This visible-frequency metasurface was realized as a subwavelength grating with a width and height of 90 and 80 nm (AR ~ 1), respectively (Figure 11.12(b)). As previously mentioned, it is difficult to chemically etch metals since their halides (etching reaction products) are typically nonvolatile components. Thus, physical ion bombardment, for example, with Ar^+ ions is a viable solution to patterning silver. Physical etching with inert gases eliminates structure contamination. However, the etching process lacks selectivity and, therefore, thick masks made of hard materials, like aluminum oxide, are required.

11.3.1.2 Conformal atomic layer deposition on a template

As an alternative to the traditional approach of etching through a mask, HAR structures can be formed by conformally filling mask openings with the material of interest. For instance, conformal filling can be achieved by ALD, a vapor-phase technique capable of coating the substrate surface with layers whose thickness is of atomic precision. In 2016, Devlin *et al.* used the ALD-assisted patterning approach to produce dielectric metalenses in the visible spectrum [6]. Figure 11.13(a) illustrates the process flow, which includes the steps of patterning photoresist with EBL, ALD filling of titanium dioxide, and removing the capping layer with RIE.

More specifically, electron beam resist was spun onto a fused silica substrate, where the resist thickness defined the height of the final nanostructures. Afterwards, the resist was patterned with EBL to form a template. During the ALD

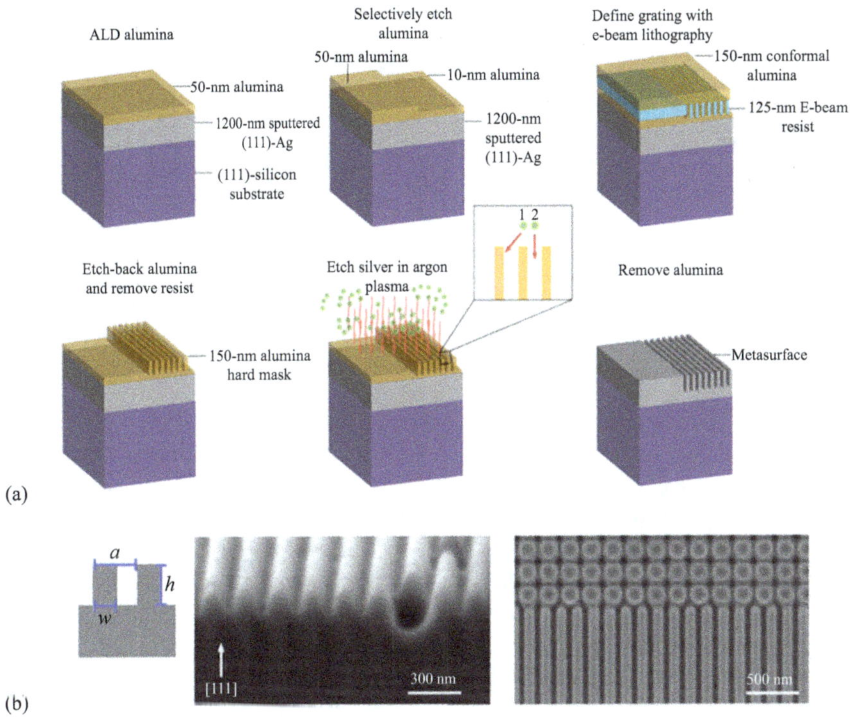

Figure 11.12 (a) Process cycle for transferring HAR anisotropic patterns into a metal film. (b) SEM images of fabricated hyperbolic metasurface [136].

process, the template was covered conformally with a layer of TiO_2. The ALD was terminated when the TiO_2 layer reached a thickness equal to half of the nanofin width. In a more general scenario, the ALD layer should be thicker than half of the maximum mask-opening width to ensure complete template filling. In the end, RIE was utilized to remove the residual TiO_2 capping film formed on top of the template surface and the resist was later dissolved to release the final metasurface.

The fabricated nanofins are 40 nm wide and 600 nm tall with an AR of 15, an almost impeccable sidewall angle of 89°, and miniscule RMS roughness of less than 1 nm (Figure 11.13(b) and (c)). Such smooth nanostructures were achieved due to the accurate monolayer-scale deposition control inherent to the ALD procedure, as well as that TiO_2 forms an amorphous-phase free-of-grain boundaries. Overall, this ALD template-filling technique can deliver spectacular fabrication results. However, on the downside, ALD is a slow process (~0.5 nm per cycle) and therefore cannot be used to generate large structures.

The ALD technique can also be applied to the construction of ultra-small gaps in metallic structures, which can produce significant electromagnetic field enhancement [137]. The process flow is illustrated in Figure 11.14(a). The first

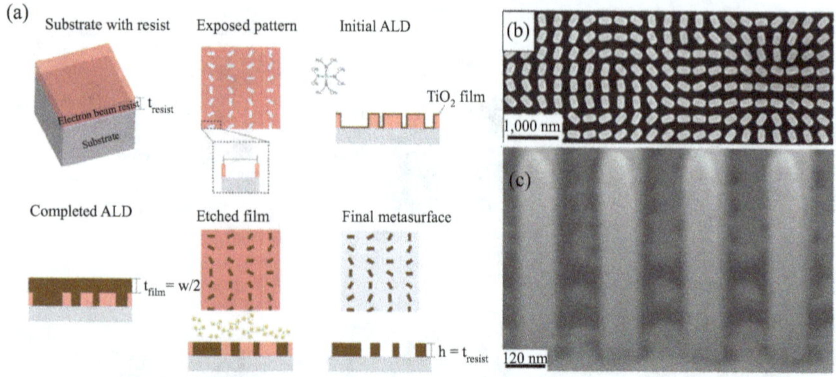

Figure 11.13 Fabrication of dielectric metasurfaces by conformally filling the resist-defined template with titanium oxide. (a) Process flow including the steps of EBL, ALD filling resist openings with TiO_2, and RIE of the extra material on top. (b) Top and (c) side views of the implemented metalens segment containing nanofins with various rotation angles. Nanofins are 600-nm tall with the smallest feature size of 40 nm (AR of 15) and smallest gap of 6 nm [6].

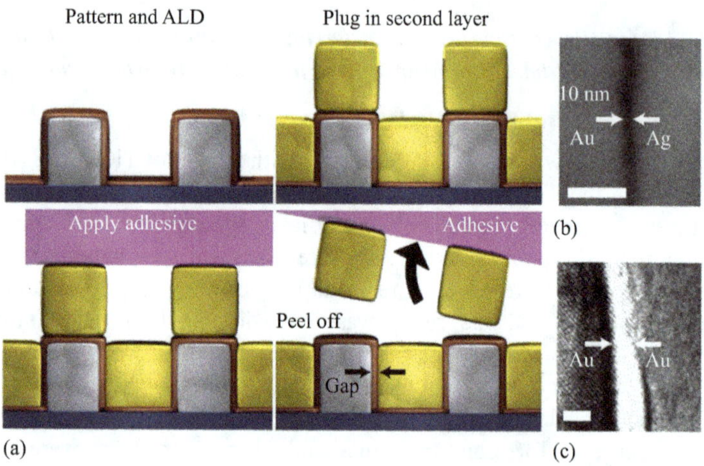

Figure 11.14 (a) Fabrication steps for constructing ultra-narrow plasmonic gaps. The procedure consists of two metal depositions, ALD conformal coating, and applying an adhesive tape to remove residual material. Cross-sectional SEM and TEM images of a (b) 10-nm and (c) 1-nm Al_2O_3 barrier inside the metal film [137].

metallic layer (colored in silver) was patterned by EBL, for example, into rectangular trenches. ALD was then utilized to conformally cover the whole pattern with a layer of dielectric (Al_2O_3), which in principle can be atomically thin. The gap size of the final structure is dictated by the ALD-layer thickness. Afterwards, the secondary metallic layer (shown with gold color) was directionally deposited. An adhesive tape was then attached on top of the structure. Since the tape only contacts the upper metallic layer, peeling off the tape selectively removes the upper layer and planarizes the whole structure. The resulting surface can be further smoothed by chemical-mechanical polishing.

As a result, a metallic film with ultra-narrow gaps of arbitrary forms and thicknesses can be constructed. Figure 11.14(b) and (c) displays SEM and TEM images of the fabricated 10 and 1 nm gaps. It is worth mentioning that the 1 nm gap was engraved in a 150-nm-thick planarized Au film, and thus the structure boasts a record large AR of ~150. The ALD coating further guarantees uniformity of the nanogaps, even when the first metallic layer comes with residual roughness after the patterning step. Additionally, this method enables the definition of gaps sandwiched between dissimilar materials, for instance, Au and Ag (Figure 11.14(c)). On the flip side, the sidewalls of the patterned metallic film must be vertical in order to physically separate the upper layer patterns and the structures located inside the initial openings. This condition is necessary for the peeling off step to work properly.

11.3.1.3 Metal-assisted chemical etching

Wet etching processes can also be employed to fabricate deep anisotropic patterns. One of such methods for achieving extremely HAR structures is metal-assisted chemical etching (MACE). In this technique, etching anisotropy originates from the metal-induced oxidation of a semiconductor material underneath and subsequent etching of the formed oxide. Typically, in MACE process, Au and Ag serve as catalyzers, H_2O_2 acts as an oxidant, and HF is an acidic solution to etch the oxide [138,139]. The typical MACE flow is depicted in Figure 11.15(a). Au is first patterned on top of a Si substrate. The etching process underneath the pattern produces vertical or porous pillars depending on consumption of generated holes at the Au-Si

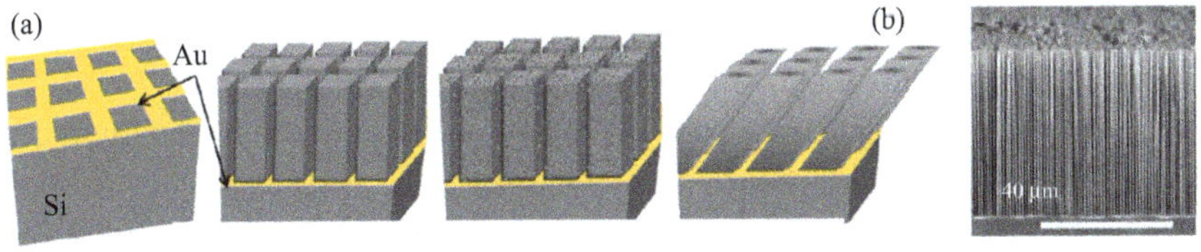

Figure 11.15 (a) Illustration of MACE technique principle, from left to right: metal pattern on top of the semiconductor; possible outcomes: vertical, porous, and slanted pillars [141]. (b) SEM image of 550-nm diameter Si nanowires with height of 51 μm, showing AR of ~93 [140].

interface. Additionally, at certain HF/H_2O_2 ratios the etching can lead to the formation of slanted pillars due to anisotropic etching rates along different crystallographic planes of Si. Figure 11.15(b) [140] shows an SEM image of the fabricated Si pillar array with an AR of 93.

In several studies, it was observed that vertical etching rate can be increased by reducing pattern period while horizontal etching rate and sidewall roughness depend on HF/H_2O_2 concentration ratio. High concentration H_2O_2 generates an excess of holes that cannot be consumed by HF, which increases roughness, while low concentrations of H_2O_2 limit the etching rate. Additionally, a low temperature environment can reduce roughness on sidewalls at the expense of much lower etching rates [142]. Typically, MACE minimally attaches the structure sidewalls and thus it can produce nanostructures with low-surface roughnesses of ~1 nm. Compared with other wet etching methods, the strongly anisotropic characteristic of MACE underpins its ability to achieve HAR. Nevertheless, MACE can only be applied to a narrow range of semiconductor materials, such as Si, Ge, GaAs, GaN, and InP, which limits its applications. Its etching direction depends on the material crystallinity and therefore only produces random structures when applied to etching polycrystalline materials.

Several variants of the standard MACE process were devised to overcome the drawbacks. For example, vertical directionality controlled MACE (V-MACE) [143] uses additional electron-hole-concentration balancing structures located in regions with large electron-hole concentration differences to restrict movement of metal catalysts, resulting in uniform etching rates and HAR structures. Inverse MACE (I-MACE) [144] uses a thick and insoluble oxide layer formed at the metal/semiconductor interface to serve as a barrier and limit hole injection in the vertical direction. In this way, semiconductor regions not interfaced with the catalyst layer can be etched preferentially. The resulting HAR structures can be narrower than the metal patterns with smooth sidewalls independent of the metal pattern edge roughness. Self-anchored-catalyst MACE (SAC-MACE) [145], which uses a nanoporous catalyst film to produce nanowires through pinholes, can confine the etching direction and enhance etch rate by increasing liquid access paths. This method can also etch through poly-Si/oxide/poly-Si stacks. Magnetic field-guided MACE (h-MACE) [146] enables the formation of periodic arrays of holes using isolated metal catalysts. HAR structures produced using the approach show applications over a broad range. For example, Si microwire with 2 μm diameter and an AR of 11.6:1 was fabricated by reducing the diffusion distance [147]. Divan *et al.* introduced a Ti adhesion layer to stabilize the migration of Au structures, a process used to fabricate linear Fresnel zone plates [148]. We foresee that further improvements of the versatile MACE process will facilitate its application in a broad range of areas.

11.3.2 3D structures

Conventional photonic devices are structured from thin films on flat substrates and are therefore 2D in nature. 3D structures make use of the underutilized third dimension and can offer unprecedented degrees of freedom for optical designers. Apparently, 3D structure fabrication demands novel fabrication schemes. Such 3D fabrication can, in some cases, leverage planar fabrication methods, while in some other scenarios unconventional routes are pursued. In this section, we sample a few examples of 3D fabrication technologies relevant to subwavelength optics processing.

11.3.2.1 Two-photon polymerization

Two-photon polymerization (TPP) takes advantage of highly localized nonlinear absorption in optical polymers to achieve high-resolution layer-by-layer object construction and serves as a versatile platform for fabricating 3D micro- and nano-structures [149]. In this approach, an ultrashort laser is focused into a photo-sensitive monomer and scanned along a design profile to cure the material. Nanophotonic fabrication poses stringent requirements on the TPP process in terms of resolution, dimensional accuracy, geometry fidelity, surface roughness, and position accuracy. In order to improve the fabrication quality, optimization with main control parameters of the TPP process (i.e., numerical aperture of the objective lens, laser power, and exposure time) have been performed to enhance resolution [150–153]. Geometry deformation is addressed by pre-compensating for structure shrinkage [154] and utilizing multi-path scanning and 2D slicing methods [155], [156]. Additionally, surface roughness can be improved by increasing overlapping between adjacent voxels [157], dividing a 3D microstructure into several sub-regions with different slicing thicknesses [158], and angular and equal-arc scanning methods [159,160]. The main limitations of TPP are associated with the relatively low-refractive indices of polymer materials as well as manufacturing throughput and scalability.

Such a maskless lithographic approach allows free shaping of complex 3D structures with high resolution (below 100 nm) for precise definition of sub-wavelength devices [153]. An array of photonic applications have been enabled with additional degrees of freedoms not available from conventional micro-fabrication methods, such as micro-optics [161–165], diffractive optics [166], 3D photonic crystals [166–168], light cages [169], phase masks [170], AFM probes [171], etc. Functional optical systems typically require more than one optical component while conventional assembly techniques developed for discrete optical components cannot provide sufficient precision and cost effectiveness for integration of subwavelength optics. TPP can potentially address this issue by monolithic fabrication of multiple elements without additional assembly steps. As demonstrated in [163] (Figure 11.16(a)), objectives consisting of stacked micro-lenses were developed, in which the multiple stages of aspheric lenses were written along with mechanical supporting structures using TPP, enabling monolithic integration of micro-scale high-performance optical systems immune from assembly misalignment.

3D photonic crystal structures with a total of 40 layers have been fabricated using TPP [167] (Figure 11.16(b)–(d)). Again, mechanical considerations were incorporated in the structure design to minimize distortions due to polymer

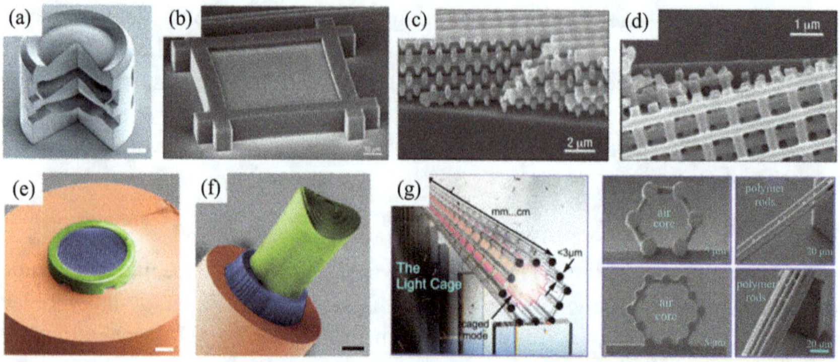

Figure 11.16 Implementation of TPP in micro- and nano-photonic device fabrication: (a) nano-printed triplet lens system (scale bar, 20 μm) [163]; (b)–(d) 3D photonic crystals with a total of 40 layers and surrounding walls written to prevent bending and reduce distortions due to polymer shrinkage during polymerization [167]; (e) a chiral photonic crystal structure for polarization control directly on an optical fiber tip (scale bar, 10 μm) [166]; (f) a free-form lens with patterned Fresnel zone plate for beam shaping on a fiber (scale bar, 25 μm) [166]; (g) schematic and fabricated light cage structures [169].

shrinkage during polymerization. In [166], chiral photonic crystals with an inter-layer distance of 400 nm were fabricated directly on an optical fiber tip for polarization control (Figure 11.16(e)). A free-form lens structure with Fresnel zone plate was subsequently fabricated on top of the photonic crystal structures for beam shaping (Figure 11.16(f)). This multi-scale photonic system is an excellent example showing the fabrication and integration versatility offered by TPP. Hollow core light cages fabricated by TPP were demonstrated [169], which consist of 6 or 12 free-standing polymer strands of 3 μm diameter around a central hollow core (Figure 11.16(g)). The light cage was written over 1 cm distance with diffraction-less propagation in the UV, visible, and near-infrared spectral regimes and high confinement of the fundamental mode. Such a platform is attractive to spectroscopic sensing and bioanalytics when integrated with optofluidic devices.

11.3.2.2 Interference lithography

Interference lithography (IL) (also known as holographic lithography) exploits the interference of two or more coherent laser beams to pattern fine features (below 50 nm) over large areas [172–177], which is of particular interest to large-scale fabrication of nanophotonic devices. Assuming a laser wavelength of λ and a refractive index of n for the immersion material, the minimum achievable feature size with a single exposure has a half-pitch of $\lambda/4n$.

The basic method of generating interference patterns for subwavelength optics fabrication involves multiple exposures [172,178] or multiple beams [179]. 3D photonic crystals have been realized using holographic lithography based on multibeam interference in a photoresist. Large-area, sub-micron structures were demonstrated [176,180]. With four non-coplanar beams from holographic lithography, diamond-like structures were fabricated with beams from all spaces [181]. Using five beams, the diamond-like structure can be created with only beams from half spaces [182,183].

In addition to bulk optical components, interference patterns can also be generated by passing the laser beam through a diffraction element mask [175,184–190]. For instance, multi-layer phase masks with orthogonally oriented gratings produced in a liquid crystal and photoresist mixture were developed to generate five-beam interference patterns [191] (Figure 11.17(a) and (b)). Diamond-like photonic crystal structures were created by exposing photoresist mixtures to the interference patterns formed by the phase mask (Figure 11.17(c) and (d)). Phase-engineered interference lithography can also be achieved using programmable spatial light modulators [192–198].

11.3.2.3 Membrane projection lithography

Membrane projection lithography (MPL) is an emerging method for fabrication of metamaterials and metasurfaces consisting of 3D unit cells [199–203]. As

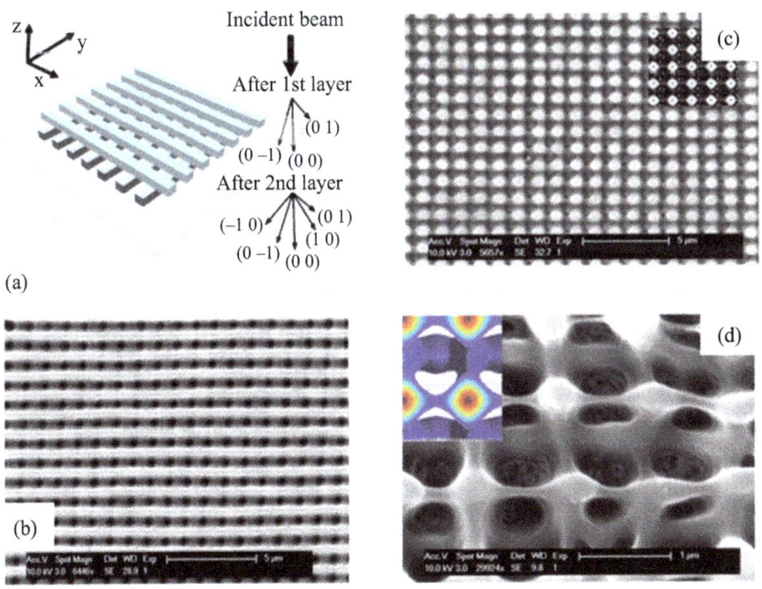

Figure 11.17 (a) Schematic of multi-layer phase mask IL setup [191]. (b) SEM of fabricated multiple-layer phase mask. (c) SEM of fabricated 3D photonic crystal structure using multi-layer phase mask IL method. (d) Enlarged view of the photonic crystal structure. Inset is the simulated five-beam interference pattern [191].

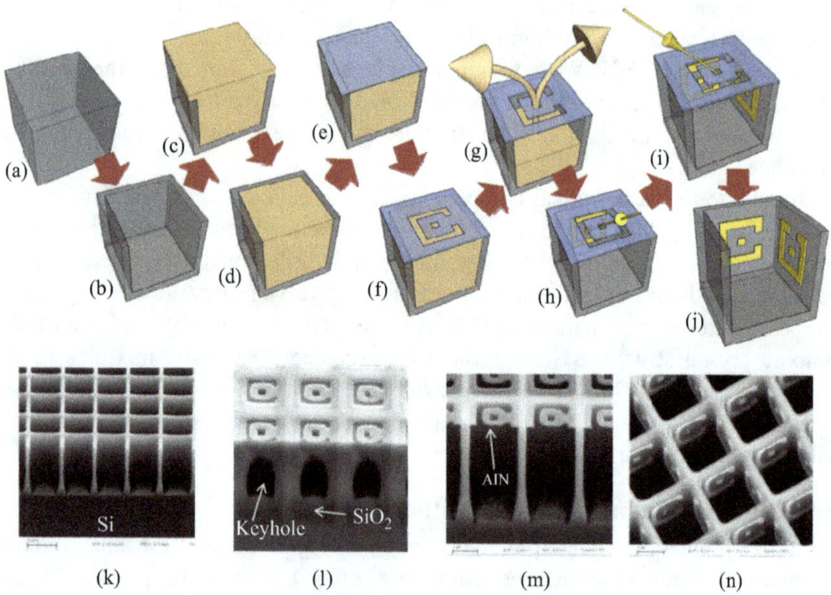

Figure 11.18 (a)–(j) Fabrication process flow of membrane projection lithography. (k) SEM image of cubic silicon cavities; (l) SEM image of backfilled cavities and patterned membrane; (m) SEM image of unit cells with patterned membrane suspended over cavity; (n) SEM image of final unit cells decorated with gold SRRs on sidewalls [199].

schematically illustrated in Figure 11.18 [199], a unit cell cavity was first etched into a Si substrate and backfilled with a sacrificial material. After the top surface was planarized using chemical-mechanical polishing (CMP), a membrane layer is deposited and patterned with through holes of target geometries. After the sacrificial layer is removed through wet etching, the membrane becomes a suspended shadow mask layer. Directional material depositions such as metal evaporation are then conducted to "decorate" the unit cell by projecting the membrane patterns onto the unit cell sidewalls. The "decoration" step can be repeated with different deposition flux directions to create patterns on multiple sidewalls. The membrane is finally removed to complete the fabrication process.

The CMOS compatible MPL process can be applied to generate a suite of unit cell geometries and sidewall pattern shapes such as vertically oriented split ring resonators (SRR), in metallic or dielectric materials. Structures with a linewidth down to ~100 nm have been demonstrated using MPL. The technique can also be adapted to fabricate 3D stacked photonic structures through layer-by-layer processing.

11.3.2.4 DNA-assisted nanoparticle assembly

Deoxyribonucleic acids (DNAs) are widely known as the essential components of all living creatures. These macromolecules serve as long-term storage of genetic information necessary for organism functioning, development, and reproduction. Back in 1996, it was proposed that DNAs can also be employed in programmable materials synthesis [204–206]. More specifically, rigid inorganic nanoparticles (NP) functionalized with DNAs can rationally conjugate to form self-organized structures ranging from microscopic molecule-like assemblies to macroscopic crystallographic arrangements [207,208]. Principle of binding gold colloids via DNAs is illustrated in Figure 11.19(a). The process typically consists of two steps. During the first phase, anchor-DNAs are attached to NP surface (illustrated as fully

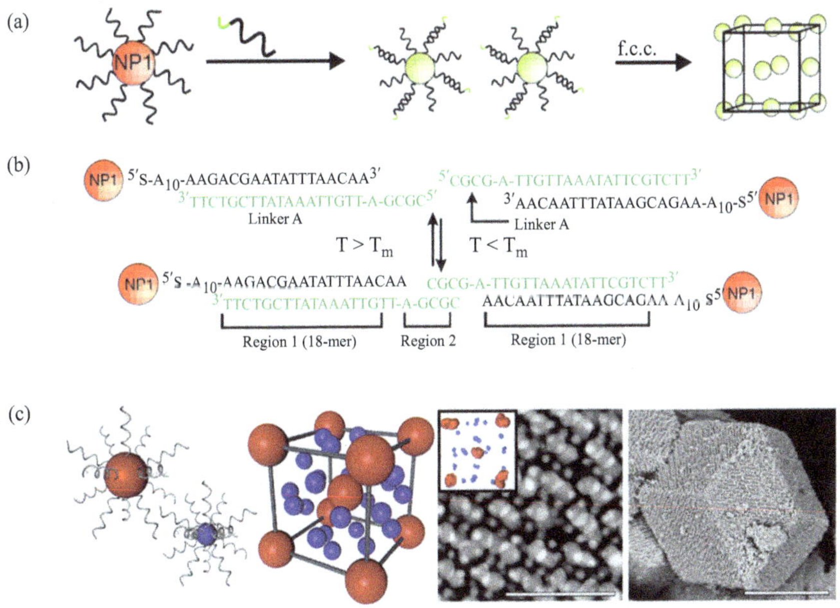

Figure 11.19 *(a) Gold colloids functionalized with DNA chains can be controllably connected with each other to form molecule- and crystal-like structures. (b) Detailed depiction of DNA binding procedure to form a binary NP molecule. Two molecules, each consisting of 18 nucleobases (black font color), directly anchor to the NPs. Then two linkers, each composed of 23 nucleobases (green font color), attach to the latter DNA strands (region 1) and connect with each other (region 2) via DNA base-pairing rules [207]. (c) Schematic of binary-component NP superlattice, TEM image of the Cs_6C_{60} isostructural lattice and SEM images of fabricated CsCl-lattice microcrystals, respective scale bars are 50 nm and 1 μm [208,209].*

black strands sticking out from an NP). In the second step, linker-DNAs are introduced (indicated as black strands with a green subsequences) in order to form a complete binding between the NPs. Figure 11.19(b) provides a detailed example of such a binding where four nucleotide sequences intertwine with each other and form three paired regions. Interestingly that DNA crosslinks can be reversibly disassembled if the substance is heated above the link (region 2) dissociation temperature [204]. The DNA-based conjugation approach can be extended further to bind together particles of different sizes to form various superlattices. Examples of the NP crystallographic arrangements mimicking lattices of Cs_6C_{60} and CsCl are demonstrated in Figure 11.19(c) [208,209]. DNA-assisted self-assembly is a very attractive technique due to the highly advanced capabilities of DNA synthesis, which allows automated generation of DNAs chains of arbitrary lengths and surface binding functionalities [210].

11.3.3 Fabrication on unconventional substrates

To date, most subwavelength optical structures are fabricated on flat, rigid substrates exemplified by silicon wafers. There is, however, growing interest in deploying these devices on unconventional, non-planar substrate platforms for many emerging applications. For example, integration of metasurface optics on arbitrary free-form substrates can decouple the optical capabilities of the component from its geometric configurations, the latter of which being dictated in many cases by considerations other than optical functions (e.g., hydrodynamics) [13,211–213]. For example, flexible devices built on mechanically compliant substrates with elastic properties matching those of biological tissues potentially enable minimally invasive biomedical sensors and actuators suitable, for instance, for conformal integration on human skin [214–217]; stretchable photonic and plasmonic structures provide a facile way for strain-mediated optical tuning and reconfiguration [218–223]; and nanostructures on the end facet of optical fibers can impart functions including antireflection, beam steering and shaping, optical coupling and sensing [224].

These tantalizing new capabilities nevertheless come with the price of added processing complexity. Tailoring standard optical and electron beam lithographic tools to patterning on these unconventional substrates is an applicable path in some cases, although the benefit is apparent only when extensive process customization is not mandated. Alternatively, the subwavelength structures can be fabricated on a flat handler substrate using classical methods and then transferred onto the unconventional substrates. Maintaining pattern fidelity, yield, and throughput constitutes the key to implementing a successful transfer process. Another popular alternative is to use direct laser beam or ion beam writing methods, although throughput and accuracy represent inherent trade-offs for these beam-based techniques. Simply put, there is no "universal" solution that applies to all substrate platforms, and in the following we will review the technology choices for different substrate classes.

11.3.3.1 Integration on flexible and stretchable substrates

Flexible and stretchable substrates are usually made of organic polymers and elastomers. Manufacturing electronic devices on these substrates has matured over the past several decades and is now a fairly well-established process [225]. However, printing or standard photolithography is routinely used for metal interconnect fabrication. When it comes to inorganic semiconductors, transfer printing becomes the preferred integration route [226,227].

These techniques can be adapted to subwavelength optics fabrication although process modifications are often mandated given the unique characteristics of optical devices [66,228–231]. For instance, unlike most electronic devices, optical devices are extremely sensitive to misalignment between the structural units, which readily modifies optical coupling. The sensitivity is further amplified in subwavelength optics due to strong field concentration common in these structures such that even displacements of a few nanometers can drastically change the device response. While the phenomenon has been harnessed for efficient optical tuning [232,233], it poses stringent requirements on pattern fidelity in particular to transfer printing methods, which are more prone to misalignment.

The technical challenge associated with misalignment can be obviated using anchored or interconnected subwavelength structures. As an example, Zhu *et al.* fabricated flexible high-contrast gratings (HCGs) exhibiting strain-tunable colors using the transfer printing method (Figure 11.20) [234]. The HCGs were patterned from an SOI wafer using SiO_2 as a etch mask. The structures were then released from the substrate through an HF vapor etching step, picked up from the substrate using a flat PDMS slab, and capped using a second PDMS layer. To ensure a high transfer yield, the HCGs assume the form of 100 μm × 100 μm pixel arrays, and the pixels are connected to a supporting frame with folded beams to minimize pattern distortion during structural release. The folded beams readily rupture during the PDMS delamination step without affecting the HCGs structures. Similarly, partially etched pedestals can be used to tether isolated structures (e.g., individual waveguides, meta-atoms, or nanoantennas) prior to their complete release from the handler substrate, thereby preventing undesired displacement or deformation

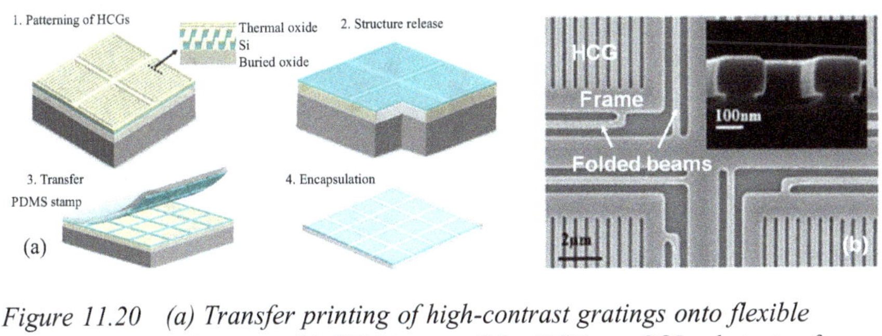

Figure 11.20 (a) Transfer printing of high-contrast gratings onto flexible substrates; (b) SEM image of the HCGs on SOI substrate after etching: inset shows the HCG cross section [234].

during transfer [235]. It has been demonstrated that the transfer printing process can be applied to create large-area (10 cm × 10 cm) structures on virtually arbitrary substrates [236].

Another commonly adopted method for subwavelength optics fabrication on flexible substrates involves first patterning the optical structure on top of a sacrificial layer on a rigid handler substrate, followed by encapsulation of the structure in PDMS elastomer, and finally delamination from the handler after removing the sacrificial layer [212,221,237]. Differing from the transfer printing approach, here PDMS is cast on the patterned structure and cured to encapsulate the entire structure. Figure 11.21(a) illustrates one instance of the process [212]. Here, a metasurface made of amorphous silicon (a-Si) nanopillars was first patterned using EBL and subsequent dry etching with an Al_2O_3 hard mask (Figure 11.21(b)). A two-layer PDMS film was then coated in two spin steps, the first with a diluted solution to enhance infiltration between the a-Si structures and the second with regular PDMS monomer. Once the PDMS layers were cured, the nanopillars were released from the handler wafer by wet etching of the Ge layer to form a free-standing, flexible metasurface structure. It should be noted that curing can lead to volume shrinkage of PDMS, and the resulting spacing change needs to be compensated for precise pattern pitch control [238].

An alternative to the transfer approach is direct patterning on flexible substrates. Metallic nanoantenna arrays can be readily deposited onto polymeric substrates via evaporation or sputtering and subsequent patterning using photolithography or EBL coupled with etching or liftoff [239–242]. The deposition process can also be masked to generate target patterns in a single step. For the masking purpose, nanosphere lithography [243–246] and block copolymer self-assembly [247] afford fine resolutions commensurate with subwavelength optical structure processing, although they only present limited options of possible unit cell shapes and lattice configurations. On the other hand, fine features of varying geometries can be formed using nanostencil lithography, an advanced shadow-masking

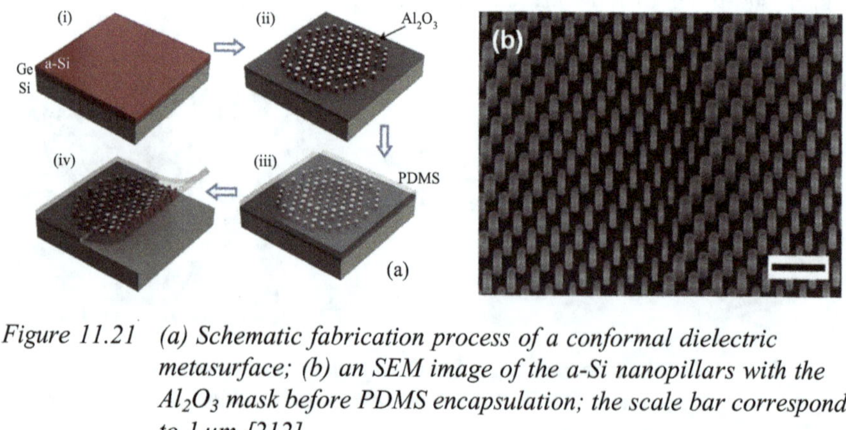

Figure 11.21 (a) Schematic fabrication process of a conformal dielectric metasurface; (b) an SEM image of the a-Si nanopillars with the Al_2O_3 mask before PDMS encapsulation; the scale bar corresponds to 1 μm [212].

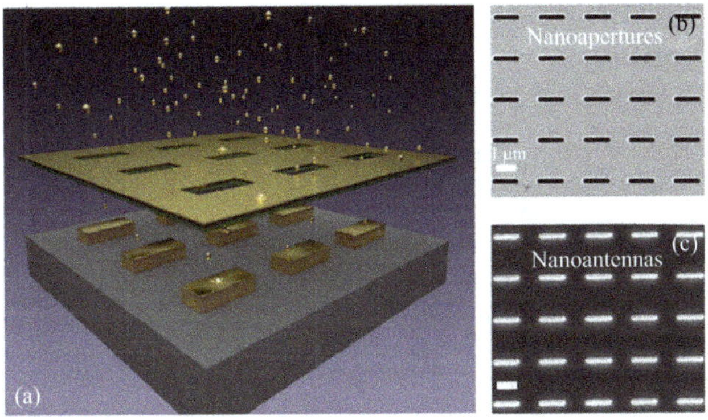

Figure 11.22 (a) Schematic illustration of the nanostencil lithography process; (b, c) SEM images of (b) the nanostencil and (c) fabricated nanoantennas [250].

technique (Figure 11.22) [248,249]. Plasmonic nanoantennas with a minimal feature size down to 25 nm have been experimentally realized using this method [248].

It is also possible to use a pre-patterned flexible substrate to template the subwavelength structures. Hokari *et al.* recently demonstrated embedded printing for sub-micron metallic structure fabrication [251]. The concept is depicted in Figure 11.23(a) [252]. The process began with imprinting a PMMA plastic substrate to form a groove pattern. The grooves were then filled with conductive ink containing metal nanoparticles assisted by capillary force. Once the residual ink outside the grooves was removed, the ink was sintered, transforming into solid metallic structures embedded in the plastic substrate (Figure 11.23(b)). It is worth noting that sintering results in dimensional changes of the metallic structures (Figure 11.23(d)), which must be properly taken into account to achieve desired optical functions (Figure 11.9(c)). Nanoimprint can be similarly combined with angled deposition to generate subwavelength plasmonic architectures [253].

Transfer-free fabrication of dielectric subwavelength structures on flexible substrates potentially offers the benefits of simplified processing as well as improved throughput and yield, albeit while imposing additional constraints on material selcction. In addition to possessing high refractive indices and low optical loss, the dielectric material should also be amorphous or polycrystalline to facilitate epitaxy-free deposition on polymers. Moreover, the deposition process must be thermally and chemically compatible with the flexible substrate: oxygen-plasma-assisted sputtering of oxide thin films, for instance, causes damage to most polymers. Possible material candidates that meet the standards include a-Si and silicon nitride grown with special low-temperature processes [254–256], sol-gel oxides such as TiO_2 [257,258], and vacuum deposited [100,259–265] or solution derived chalcogenides [68,266–268]. Figure 11.24 illustrates an exemplary process flow for monolithic photonic device fabrication on polymer substrates. In this embodiment, the flexible substrate comprises a polymer film initially spin coated on a handler wafer, and the target patterns can be defined using standard lithographic methods,

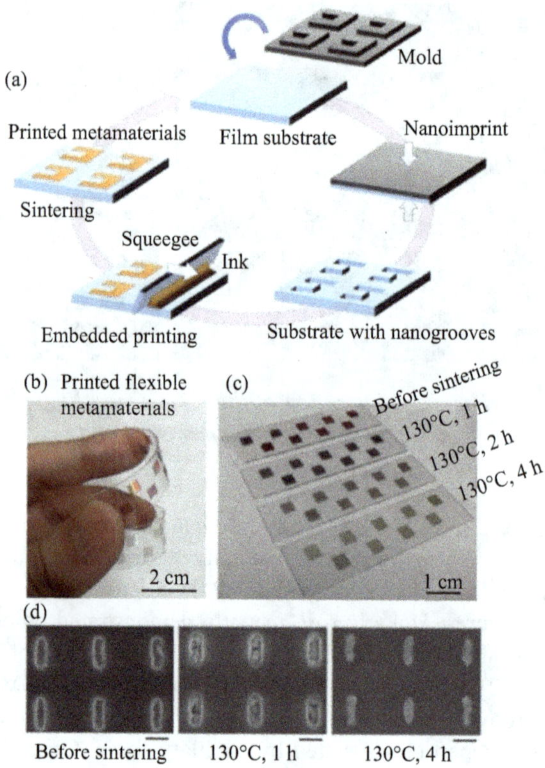

Figure 11.23 *(a) The embedded printing process; (b) printed flexible metamaterials; (c) photograph of printed flexible metamaterials before sintering and after sintering at 130°C for 1, 2, and 4 h; (d) SEM images of printed metamaterials for different sintering conditions. The black scale bars indicate 200-nm length [252].*

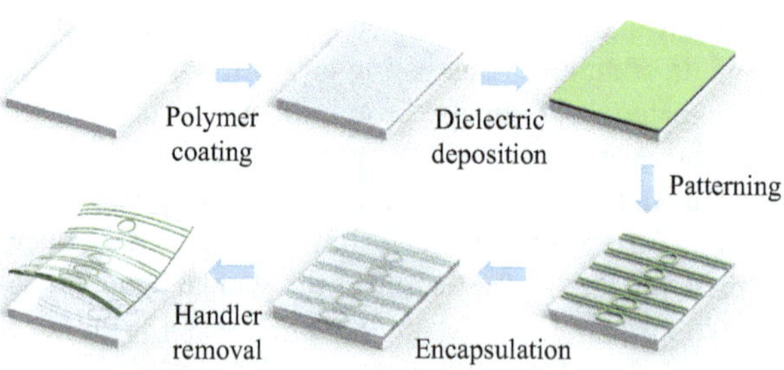

Figure 11.24 *A generic process flow for dielectric material deposition and device fabrication on polymer substrates [257].*

nanoimprint [269], or soft lithography. The optical structures are subsequently encapsulated and delaminated from the handler substrate to form a free-standing flexible device. A main advantage of the process is that it leverages standard nanofabrication infrastructures with minimal customization needed.

Additional challenges arise when it comes to dielectric material integration on elastomers. Elastomers typically exhibit gigantic coefficients of thermal expansion (CTE) that are one to two orders of magnitude larger than most dielectric and semiconductor materials. Take the most commonly used PDMS as an example; it has a tremendous linear CTE of 3.1×10^{-4}/°C [270], compared to those of Si (2.6×10^{-6}/°C), SiO_2 (0.55×10^{-6}/°C), and Si_3N_4 (3.2×10^{-6}/°C). To mitigate film delamination and cracking due to the large CTE mismatch, several measures can be taken. One option is to limit the thermal budget of the entire fabrication process, including that of the resist baking steps [271]. Other viable solutions include inserting a strain-releasing layer between the elastomer and the dielectric material [65,69,272] or arranging the dielectric structures into small isolated islands [273]. To accommodate the large elastic strain during stretching of elastomers, configurational designs matured in the stretchable electronics domain (e.g., serpentine line shapes and local substrate stiffening [274,275]) may be employed although optics-specific optimizations are often essential for photonic devices [69].

11.3.3.2 Subwavelength optics fabrication on non-planar and irregular substrates

Technologies for nanostructure fabrication on non-flat substrates can be generally classified into three categories: flexible membrane conformal integration, soft-lithography-based methods, and beam writing. In the first scheme, structures fabricated on a thin stretchable membrane are conformally attached onto a non-planar receiving substrate to impart the target functions, and the techniques described in Section 11.3.3.1 can be implemented for this purpose. In this section, we will therefore focus on the latter two approaches.

Soft lithography

Soft lithography is broadly defined as "a collection of techniques based on printing, molding and embossing with an elastomeric stamp" [276]. The elastomeric stamp can be readily deformed to conformally contact curved surfaces, enabling facile pattern transfer. A number of soft lithographic techniques have been applied to patterning on tightly curved surfaces with feature size down to 15 nm [277–282]. Figure 11.25(a) depicts the fabrication procedure of soft imprint molds capable of resolving such deep sub-micron features. The composite mold consists of three laminates: a PDMS/polyethylene terephthalate (PET) backplane and an epoxysiloxane imprint layer [283]. The soft PDMS/PET backplane can readily deform to conform to the shape of the substrate, whereas the epoxysiloxane film contains the replica molded features for imprint pattern transfer. Epoxysiloxane is a superior candidate for the imprint layer than the classically formulated PDMS given its larger stiffness, which minimizes pattern distortion during embossing. The mold

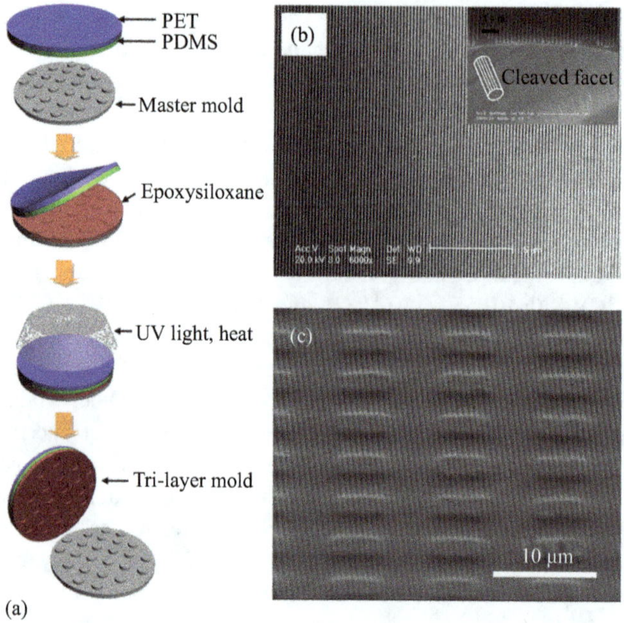

Figure 11.25 *(a) Composite mold fabrication for imprint on non-flat surfaces [283]; (b) SEM image of 200-nm pitch gratings imprinted on the surface of a 125-μm diameter fiber; inset shows the cleaved facet of the patterned fiber [277]; (c) SEM image of imprinted 500-nm pitch gratings over a 500-nm-thick micro-post array [283].*

can then be used to conformally imprint fine structures on curved (Figure 11.25(b)) or uneven surfaces (Figure 11.25(c)).

A variant of soft lithography (sometimes termed "optical soft lithography") entails creating an elastomeric photomask, through which a photoresist layer coated on a non-planar surface can be exposed to produce the target pattern [284–287]. So far, the minimum feature size of the method that has been experimentally validated is limited to 1 μm [284], which constrains the operation wavelength of viable optical structures. Notably, the method can also be utilized to generate gray-scale elastomeric masks [288].

Beam writing techniques

Focused energetic beams—laser, ion, or electron beams—can be used to either expose resist layers or directly inscribe patterns on non-planar or even irregular surfaces. For beam writing on a non-flat (or tilted flat) surface, the depth of field (DOF) that dictates the vertical distance over which the resolution of written patterns is not compromised is a critical parameter. For a focused laser beam, its field profile can be closely approximated with a Gaussian beam. Ideally, the focal spot (beam waist) should coincide with the substrate surface (or resist layer). Deviations from ideal focusing results in diffraction and blurring of the written pattern. The DOF is given by the Rayleigh range z_R of the beam [289]:

$$z_R = \frac{\pi w_0^2}{\lambda} \sim \frac{\lambda}{\pi \cdot NA^2}, \tag{11.1}$$

where w_0 gives the beam waist size, λ is the wavelength, and NA denotes numerical aperture of the lens used to produce the focused beam. Equation (11.1) indicates a trade-off between resolution (w_0) and DOF: a tightly focused laser beam with large NA contributes to enhanced resolution at the expense of DOF. For typical laser writing experiments, the DOF is generally less than 1 μm. Given this small DOF, the laser must be dynamically refocused via precision motion stages to follow the surface morphology. Figure 11.26(a) schematically illustrates one embodiment of such a laser writing system customized for patterning on curved surfaces [290]. In addition to autofocusing, the system is also capable of dynamically adjusting the substrate tilt such that the laser beam is always perpendicular to the local substrate surface to minimize pattern distortion. Figure 11.26(b) and (c) shows images of diffractive optical elements fabricated on a lens surface using the equipment. The system's resolution, dictated by the focal spot size, is approximately 1 μm, although further resolution enhancement is possible with improved optics or use of two photon lithography. It is also possible to use quasi-non-diffracting beams such as

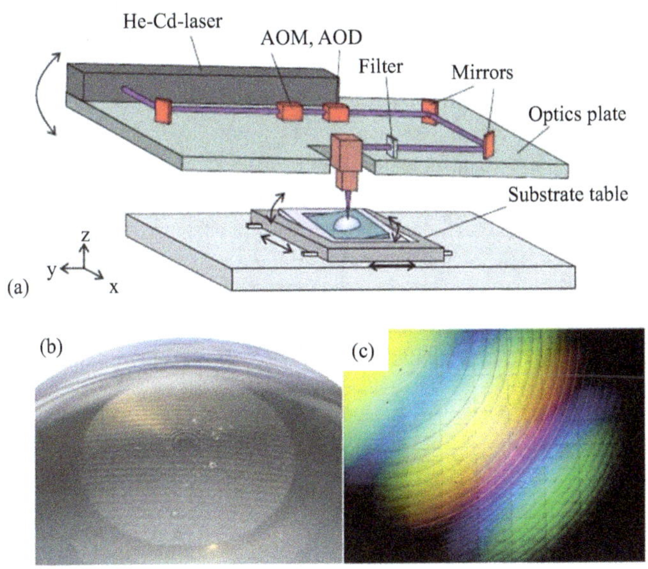

Figure 11.26 (a) A laser lithography system engineered for patterning on curved surfaces; (b) photograph of a diffractive element on a biconvex lens fabricated using the system; (c) microscope picture of the hybrid element taken in differential interference contrast mode [290].

Bessel beams or Airy beams to increase DOF with minimal loss in resolution [291], although their non-ideal beam profiles can be problematic for certain applications.

Electron beam and ion beam writing follow similar principles albeit with some important differences. Electron and ion beams have much smaller wavelengths than visible light: for electrons with 100 keV kinetic energy, their de Broglie wavelength is 0.039 Å, far smaller than that of lasers. These small wavelengths underlie the superior resolution of charged particle lithography. On the other hand, the NA of electron optics in EBL tools is small, typically on the order of 10^{-3}. The corresponding DOF calculated using the equation (11.1) is approximately a few microns. While the DOF is considerably larger than that of laser writing, the beam still needs to be refocused for samples with macroscopically non-flat features. Electron beam focus can be adjusted by dividing up the entire writing area into a number of smaller writing fields within which the height difference is smaller than the DOF [292], or constantly adjusted making use of the dynamic focusing function [293]. Compared to EBL, helium ion beam lithography has an even smaller NA, which contributes to its large DOF that can exceed 100 μm and facilitates patterning on uneven substrates [294].

Besides DOF considerations, one additional challenge for beam writing on non-flat substrates is how to obtain a uniform resist coating should a pattern transfer step be necessary. For a macroscopically curved substrate such as the surface of a lens with moderate NA, standard spin coating may work provided that the edge beads are properly taken care of. For irregular surfaces where spin coating fails, techniques including resist transfer [295], spray coating [296], evaporation [297,298], dry film resist application [299,300], plasma polymerization [301], the Langmuir–Blodgett method [302], and even the use of ice as a resist [303] have been implemented. As an example, Figure 11.27 shows structures written in thermally evaporated polystyrene resist on an AFM cantilever, highlighting versatility of the resist coating and patterning method [297].

Apart from focused particle beams, collimated ion beams can also be applied to patterning on non-planar surfaces. The technique is named ion beam proximity lithography (IBL). In IBL, a stencil mask is placed above the substrate and flooded with a collimated beam of energetic ions. Unlike focused beams, well-collimated ion beams offer an enormous DOF exceeding 1 cm [304,305]. Patterning of deep-sub-micron-sized features on non-planar substrates has been realized using IBL [306].

11.3.3.3 The tip of a fiber—a new dancing stage for photonic engineers

Having a size and form similar to that of an ordinary needle's point, the tip of a fiber is tiny from a macroscopic perspective. However, a fiber tip—a unique unconventional substrate with direct access to light—also presents big opportunities for photonic engineers. For instance, nanophotonic structures mounted on the fiber facet can reshape or redirect the output optical mode to modulate its spectral and/or spatial distributions, or facilitate coupling with other optical devices. Integration of photonic components on a fiber tip can also impart new

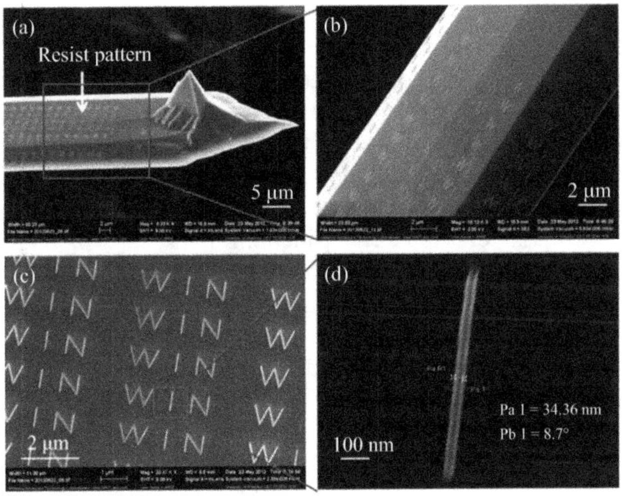

Figure 11.27 SEM images of patterned polystyrene resist on an atomic force microscope cantilever by EBL, taken at increasing magnifications from (a) to (d); the resist is deposited using thermal evaporation [297].

functionalities such as sensing and optomechanical manipulation, paving the path toward a versatile "lab-on-fiber" platform [307].

With its small cross-sectional area and a needle-like form factor, the fiber tip also presents unique challenges for subwavelength optical structure integration. It is possible to adapt most planar nanofabrication technologies for this purpose albeit with significant process customization. Alternatively, the structures can also be first fabricated on a standard planar substrate followed by transfer onto the fiber tip. Both approaches are elaborated further in this section.

Hybrid nanotransfer onto a fiber tip

By avoiding direct fabrication on a fiber tip, hybrid nanotransfer sidesteps the challenges linked to handling of fiber-associated widths during lithographic processing. In an exemplary process demonstrated by Smythe *et al.* [308], gold nanoantenna arrays were initially patterned on Si using EBL. The array was then stripped off the wafer using a thin polymer film, and attached onto a fiber facet along with the carrier film. In the last step, the sacrificial carrier is etched away, leaving the antenna array adhering to the fiber tip via van der Waals attraction. The fiber tip decorated with the nanoantenna array was used as an easy-to-use surface-enhanced Raman spectroscopy (SERS) probe operating in a reflective mode. In addition to SERS [309], metallic nanostructure arrays transferred onto fiber tips have also been applied to plasmonic refractive index sensing [310,311].

Other nanotransfer alternatives offer the added capability of aligning photonic structures to the fiber core. These methods are exemplified by the transfer of photonic crystal cavities onto fiber facets [312–315]. The processes in general involve release of a photonic crystal membrane from a handler substrate via sacrificial etching followed by transfer of the membrane via a micromanipulator [313,314] or direct gluing of the membrane to the fiber end [312]. Li *et al.* further extended the direct transfer approach by employing the transferred Si nanomembrane as a hard mask for subsequent patterning transfer [316]. Figure 11.28(a) depicts two variants of the process where the patterned Si membrane is first picked up from the starting substrate using a tungsten tip with PDMS adhesive, and then transferred to the final substrate to mask (i) etching or (ii) liftoff. Once the pattern transfer is complete, the mask is removed and can be reused multiple times. As a proof-of-concept, the authors implement the technique to pattern gold dot arrays aligned to a fiber core (Figure 11.28).

Here, we conclude this sub-section on nanotransfer by highlighting an impressive piece of work by Arce *et al.*, where they transferred an entire SOI photonic circuit to realize a fiber-tip-integrated refractive index sensor [317]. As shown in Figure 11.29, the compact photonic circuit encompasses a focusing grating coupler optically interfacing with the fiber core, a multimode

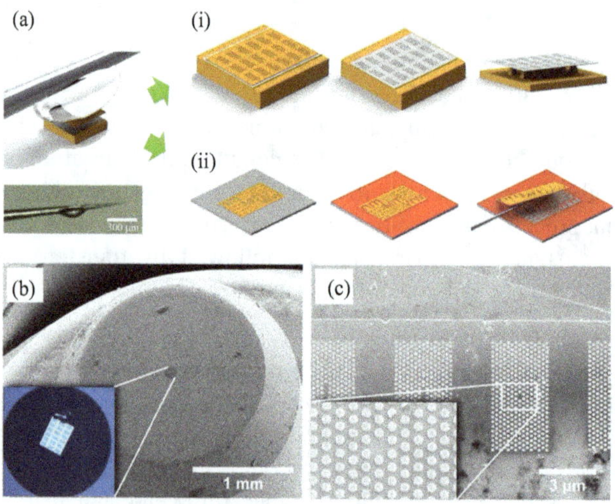

Figure 11.28 (a) Hard mask transfer process for (i) etching and (ii) liftoff; inset shows the image of a tungsten tip with attached PDMS adhesive used to pick up and position the nanomembrane mask; (b) a silicon membrane with patterned gold dot arrays on a fiber facet; inset shows an expanded view of the membrane on the fiber core; (c) patterned gold dot arrays on the fiber facet after removal of the nanomembrane mask using a tungsten tip; inset gives an expanded view of the gold dots [316].

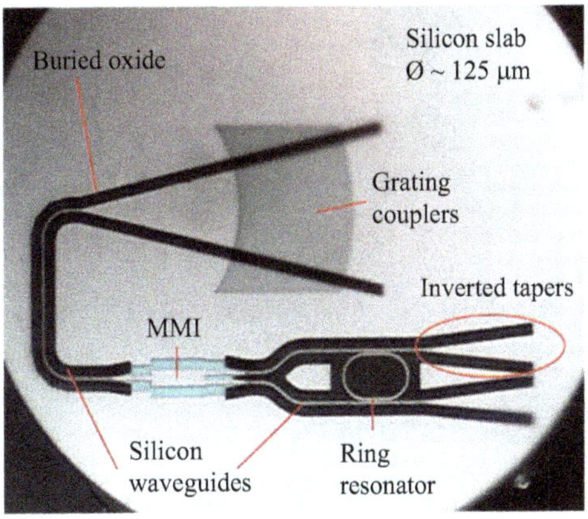

Figure 11.29 Top-view micrograph of a silicon photonic circuit integrated on a fiber tip [317].

interferometer, a racetrack resonator in an add-drop configuration, as well as inverted tapers to prevent back reflected light from coupling into the circuit. Both clockwise and counterclockwise propagating modes are excited by the bus waveguides such that the light is coupled back into the fiber to be analyzed. The demonstration reveals the possibility of integrating complex photonic functions onto the fiber tip to enable versatile fiber probe sensing systems.

Adapting planar nanofabrication techniques to direct patterning on a fiber tip

Over the past two decades, most standard wafer-scale manufacturing technologies have been adapted to photonic fabrication on a fiber tip. These methods include optical, electron beam, and ion beam lithography, nanoimprint, and laser direct writing. Unlike self-assembly based bottom-up methods [318–320] or photopolymerization using light guided in the fiber [321], top-down fabrication schemes are capable of reproducibly generating structures with well-defined geometry essential for many nanophotonic device applications.

For optical lithography and EBL patterning on a fiber tip, preparing a uniform resist coating on the fiber end sets the first technical challenge. Methods described in Section 11.3.3.3 for resist coating on irregular substrates such evaporation are equally applicable to fiber tips [322]. Resist can also be applied using classical spin coating while the fiber being supported on special holders or fiber ferrules [323–326]. Another technique involves dip and vibration coating, where the fiber end is first dipped into liquid resist and the excess resist is then removed by mechanical vibration [327]. Subsequent exposure of the resist is performed using standard UV lithography [328], interference lithography [329,330], digital micromirror device (DMD) projection [326], or EBL [323]. The structured resist layer can either act as the optical element itself or serve as a mask for etch pattern transfer into the fiber material.

Figure 11.30 (a) Microscope image of a fiber bundle wafer polished and ready for lithographic processing; (b) diffractive lenses fabricated on the wafer; (c) an element with multiplexed collimation and vortex phase functions fabricated on the end facet of an individual fiber rod [331].

The lithographic methods described above generally suffer from severely limited throughput as the process is carried out sequentially on individual fibers. Attempts have been made to parallelize the fabrication. Johnson *et al.* stacked fibers into a bundle using epoxy adhesive, sliced the bundle, and polished the end facets to form thin wafers comprising short segments of fiber cores arrayed in a close-packed hexagonal lattice (Figure 11.30(a)) [331]. The wafers can then be processed following standard handling protocols of classical semiconductor wafers. Figure 11.30 illustrates optical diffractive lenses and an element combining collimation and vortex phase generation functions fabricated using this approach. It should be noted that the photomask used along the wafer has to be custom-made following the fabricated fiber array geometry to accommodate random defects in the fiber lattice.

Nanoimprint and its variants are also popular choices for subwavelength optics integration on fiber tips [332–336]. Batch processing of up to 40 fiber facets with feature resolution down to 15 nm was also demonstrated [337]. Figure 11.31(a) illustrates an embodiment of the process, which combines UV nanoimprint and nanotransfer to produce grating couplers attached to a fiber facet [338]. A mold containing the grating pattern was first etched in an SOI wafer. A gold film was subsequently deposited on the mold, followed by selective removal of gold on top of the mold using microcontact printing, leaving gold only in the trench lines. The fiber was then dipped into a UV-curable resist and actively aligned to the SOI grating before cross-linking the resist. As the gold pattern adheres to the resist much better than to SOI, separation of the fiber and the SOI grating resulted in transfer of the metallic grating onto the fiber facet. The grating-coupler-integrated fiber probe can deliver (or extract) light into (from) a planar waveguide once directly contacting the waveguide surface, potentially enabling nondestructive wafer-scale testing of photonic circuits.

Direct write techniques such as focused ion beam (FIB) milling and direct laser writing present another solution to circumvent the technical complications associated with handling of fiber facets in lithographic patterning. FIB milling has been applied to fiber-tip integration of nanoapertures [339], plasmonics nanostructures

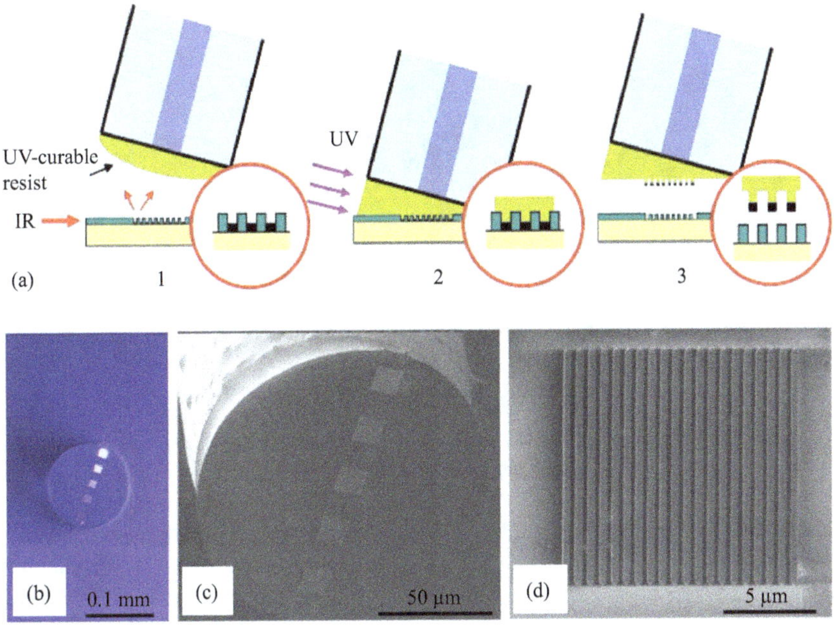

Figure 11.31 *(a) Schematic fabrication process of the grating-coupler-integrated fiber probe; (b) optical and (c) SEM micrographs of the fiber probe facet, showing the attached gratings; (d) top-view SEM image of the on-fiber grating coupler [338].*

[340,341], diffractive optical elements [342], photoacoustic sensors [343], as well as micro-channels for fluid filling in photonic crystal fibers [344]. FIB-etched surface corrugations can also be used as a template to define structures subsequently deposited on the fiber end facet [345]. It is worth pointing out that FIB milling inevitably causes contamination of the fiber by gallium ions, and thus the resulting fiber material optical property modification should be properly accounted for in the design phase [346,347].

Laser writing, in particular two-photo lithography or TPP, is uniquely poised for 3D optical structure fabrication on fiber facets with sub-micron alignment accuracy and resolution down to below 100 nm [348]. Single and compound microlenses [163,164], micro-prisms [165], phase masks [170], AFM probes [171], and photonic crystals (Figure 11.32) [168] are some examples of optical structures that have been integrated on optical fiber tips using TPP. Section 11.3.2.1 of this chapter provides further discussions about the technology for interested readers.

11.4 Summary and outlook

As the field of subwavelength optics continues to boom and novel nanophotonic device designs emerge (as other chapters in this book highlight), existing

Figure 11.32 False-color SEM image of a woodpile photonic crystal fabricated on a fiber facet using TPP [168].

nanofabrication methods will continue to evolve and new technologies will surface to keep pace with the increasingly sophisticated demands. We foresee that the crossroads of ingenious nanophotonic designs and nanofabrication techniques will thrive as a leading source of inspiration for researchers aspiring to explore the field. In this spirit, it is the authors' hope that this chapter can serve as a convenient solution guide to both theorists and experimentalists who are exploring unconventional subwavelength structures. Finally, we hope that the discussions here can further inspire design and process innovations in the dynamic field of subwavelength optics, where 'there is plenty of room at the bottom' for both dancing angels and optical engineers alike.

References

[1] C. Zheng, *Nanofabrication: principles, capabilities and limits*. 2008.

[2] A. A. Tseng, *Nanofabrication: fundamentals and applications*. World Scientific, 2008.

[3] C. Papadopoulos, *Nanofabrication: principles and applications*. Springer, 2016.

[4] J. A. Liddle and G. M. Gallatin, "Nanomanufacturing: a perspective," *ACS Nano*, vol. 10, no. 3, pp. 2995–3014, 2016.

[5] D. G. Baranov, D. A. Zuev, S. I. Lepeshov, *et al.*, "All-dielectric nanophotonics: the quest for better materials and fabrication techniques," *Optica*, vol. 4, no. 7, 2017.

[6] R. C. Devlin, M. Khorasaninejad, W. T. Chen, J. Oh, and F. Capasso, "Broadband high-efficiency dielectric metasurfaces for the visible spectrum," *Proc. Natl. Acad. Sci.*, vol. 113, no. 38, pp. 10473–10478, 2016.

[7] A. Y. Zhu, A. I. Kuznetsov, B. Luk'Yanchuk, N. Engheta, and P. Genevet, "Traditional and emerging materials for optical metasurfaces," *Nanophotonics*, vol. 6, no. 2, pp. 452–471, 2017.

[8] N. I. Zheludev and Y. S. Kivshar, "From metamaterials to metadevices," *Nat. Mater.*, vol. 11, no. 11, pp. 917–924, 2012.
[9] N. Meinzer, W. L. Barnes, and I. R. Hooper, "Plasmonic meta-atoms and metasurfaces," *Nat. Photonics*, vol. 8, no. 12, pp. 889–898, 2014.
[10] H.-T. Chen, A. J. Taylor, and N. Yu, "A review of metasurfaces: physics and applications," *Reports Prog. Phys.*, vol. 79, no. 7, p. 076401, 2016.
[11] A. M. Shaltout, A. V. Kildishev, and V. M. Shalaev, "Evolution of photonic metasurfaces: from static to dynamic," *J. Opt. Soc. Am. B*, vol. 33, no. 3, p. 501, 2016.
[12] M. Decker and I. Staude, "Resonant dielectric nanostructures: a low-loss platform for functional nanophotonics," *J. Opt.*, vol. 18, no. 103001, 2016.
[13] S. Walia, C. M. Shah, P. Gutruf, *et al.*, "Flexible metasurfaces and metamaterials: a review of materials and fabrication processes at micro-and nanoscales," *Appl. Phys. Rev.*, vol. 2, no. 1, p. 11303, 2015.
[14] V.-C. Su, C. H. Chu, G. Sun, and D. P. Tsai, "Advances in optical metasurfaces: fabrication and applications [Invited]," *Opt. Express*, vol. 26, no. 10, p. 13148, 2018.
[15] A. She, S. Zhang, S. Shian, D. R. Clarke, and F. Capasso, "Large area metalenses: design, characterization, and mass manufacturing," *Opt. Express*, vol. 26, no. 2, p. 1573, 2018.
[16] P. Genevet, F. Capasso, F. Aieta, M. Khorasaninejad, and R. Devlin, "Recent advances in planar optics: from plasmonic to dielectric metasurfaces," *Optica*, vol. 4, no. 1, p. 139, 2017.
[17] P. Cheben, R. Halir, J. H. Schmid, H. A. Atwater, and D. R. Smith, "Subwavelength integrated photonics," *Nature*, vol. 560, no. 7720, pp. 565–572, 2018.
[18] L. Zhang, S. Mei, K. Huang, and C. W. Qiu, "Advances in full control of electromagnetic waves with metasurfaces," *Adv. Opt. Mater.*, vol. 4, no. 6, pp. 818–833, 2016.
[19] H.-H. Hsiao, C. H. Chu, and D. P. Tsai, "Fundamentals and applications of metasurfaces," *Small Methods*, vol. 1, no. 4, p. 1600064, 2017.
[20] N. Yu and F. Capasso, "Flat optics with designer metasurfaces," *Nat. Mater.*, vol. 13, no. 2, pp. 139–150, 2014.
[21] N. Yu, P. Genevet, M. A. Kats, *et al.*, "Light propagation with phase discontinuities reflection and refraction," *Science*, vol. 334, no. 6054, pp. 333–337, 2011.
[22] E. Karimi, S. A. Schulz, I. De Leon, H. Qassim, J. Upham, and R. W. Boyd, "Generating optical orbital angular momentum at visible wavelengths using a plasmonic metasurface," *Light Sci. Appl.*, vol. 3, pp. 1–4, 2014.
[23] W. Cai, C. Jun, and J. White, "Plasmonics for extreme light concentration and manipulation," *Nat. Mater.*, vol. 9, no. 3, pp. 193–204, 2010.
[24] M. Pelton, J. Aizpurua, and G. Bryant, "Metal-nanoparticle plasmonics," *Laser Photon. Rev.*, vol. 2, no. 3, pp. 136–159, 2008.
[25] Y. Zhao and A. Alu, "Tailoring the dispersion of plasmonic nanorods to realize broadband optical meta-waveplates," *Nano Lett.*, vol. 13, 2013.

[26] K. Sarabandi and N. Behdad, "A frequency selective surface with miniaturized elements," *IEEE Trans. Antennas Propag.*, vol. 55, no. 5, pp. 1239–1245, 2007.

[27] P. Stute, S. Uniiwrsitj, and S. Collejie, "Flat lens antenna concept using aperture coupled microstrip patches," *Electron. Lett.*, vol. 32, no. 23, pp. 2109–2111, 1996.

[28] M. A. Al-Joumayly and N. Behdad, "A generalized method for synthesizing low-profile, band-pass frequency selective surfaces with non-resonant constituting elements," *IEEE Trans. Antennas Propag.*, vol. 58, no. 12, pp. 4033–4041, 2010.

[29] A. Boltasseva and H. A. Atwater, "Low-loss plasmonic metamaterials," *Science*, vol. 331, no. 6015, pp. 290–291, 2011.

[30] P. R. West, S. Ishii, G. V. Naik, N. K. Emani, V. M. Shalaev, and A. Boltasseva, "Searching for better plasmonic materials," *Laser Photon. Rev.*, vol. 4, no. 6, pp. 795–808, 2010.

[31] M. G. Blaber, M. D. Arnold, and M. J. Ford, "A review of the optical properties of alloys and intermetallics for plasmonics," *J. Phys. Condens. Matter*, vol. 22, no. 14, 2010.

[32] D. A. Bobb, G. Zhu, M. Mayy, *et al.*, "Engineering of low-loss metal for nanoplasmonic and metamaterials applications," *Appl. Phys. Lett.*, vol. 95, no. 15, p. 151102, 2009.

[33] M. Jablan, H. Buljan, and M. Soljačić, "Plasmonics in graphene at infrared frequencies," *Phys. Rev. B*, vol. 80, no. 24, p. 245435, 2009.

[34] S. A. Gregory, Y. Wang, C. H. De Groot, and O. L. Muskens, "Extreme subwavelength metal oxide direct and complementary metamaterials," *ACS Photonics*, vol. 2, pp. 606–614, 2015.

[35] S. Q. Li, P. Guo, L. Zhang, W. Zhou, T. W. Odom, and T. Seideman, "Infrared plasmonics with indium-tin-oxide nanorod arrays," *ACS Nano*, vol. 5, no. 11, pp. 9161–9170, 2011.

[36] M. Abb, Y. Wang, N. Papasimakis, C. H. De Groot, and O. L. Muskens, "Surface-enhanced infrared spectroscopy using metal oxide plasmonic antenna arrays," *Nano Lett.*, vol. 14, pp. 346–352, 2014.

[37] J. Kim, A. Dutta, B. Memarzadeh, A. V Kildishev, H. Mosallaei, and A. Boltasseva, "Zinc oxide based plasmonic multilayer resonator: localized and gap surface plasmon in the infrared," *ACS Photonics*, vol. 2, pp. 1224–1230, 2015.

[38] J. Park, J. Kang, S. J. Kim, X. Liu, and M. L. Brongersma, "Dynamic reflection phase and polarization control in metasurfaces," *Nano Lett.*, vol. 17, pp. 407–413, 2017.

[39] C. Rhodes, S. Franzen, J. P. Maria, *et al.*, "Surface plasmon resonance in conducting metal oxides," *J. Appl. Phys.*, vol. 100, no. 5, 2006.

[40] C. Rhodes, M. Cerruti, A. Efremenko, *et al.*, "Dependence of plasmon polaritons on the thickness of indium tin oxide thin films," *J. Appl. Phys.*, vol. 103, no. 9, 2008.

[41] K. Ellmer and R. Mientus, "Carrier transport in polycrystalline ITO and ZnO:Al II: the influence of grain barriers and boundaries," *Thin Solid Films*, vol. 516, no. 17, pp. 5829–5835, 2008.
[42] F. Lai, L. Lin, R. Gai, Y. Lin, and Z. Huang, "Determination of optical constants and thicknesses of In_2O_3:Sn films from transmittance data," *Thin Solid Films*, vol. 515, no. 18, pp. 7387–7392, 2007.
[43] M. Lv, X. Xiu, Z. Peng, *et al.*, "Structural, electrical and optical properties of zirconium-doped zinc oxide films prepared by radio frequency magnetron sputtering," *Thin Solid Films*, vol. 516, no. 8, pp. 2017–2021, 2008.
[44] J. G. Lu, Z. Z. Ye, Y. J. Zheng, *et al.*, "Structural, optical, and electrical properties of (Zn,Al)O films over a wide range of compositions," *J. Appl. Phys.*, vol. 100, no. 7, p. 073714, 2006.
[45] J. Kim, S. Choudhury, C. Devault, *et al.*, "Controlling the polarization state of light with plasmonic metal oxide metasurface," *ACS Nano*, vol. 10, pp. 9326−9333, 2016.
[46] S. Jahani and Z. Jacob, "All-dielectric metamaterials," *Nat. Nanotechnol.*, vol. 11, no. 1, pp. 23–36, 2016.
[47] A. I. Kuznetsov, A. E. Miroshnichenko, M. L. Brongersma, Y. S. Kivshar, and B. Luk'yanchuk, "Optically resonant dielectric nanostructures," *Science*, vol. 354, no. 6314, p. aag2472, 2016.
[48] A. Arbabi, Y. Horie, A. J. Ball, M. Bagheri, and A. Faraon, "Subwavelength-thick lenses with high numerical apertures and large efficiency based on high contrast transmitarrays," *Nat. Commun.*, vol. 6, no. pp. 1 10, 2015.
[49] L. Zhang, J. Ding, H. Zheng, *et al.*, "Huygens meta-optics," *Nat. Commun.*, no. 2018, pp. 1–9.
[50] I. Staude and J. Schilling, "Metamaterial-inspired silicon nanophotonics," *Nat. Photonics*, vol. 11, no. 5, pp. 274–284, 2017.
[51] M. Decker, I. Staude, M. Falkner, *et al.*, "High-efficiency dielectric Huygens' surfaces," *Adv. Opt. Mater.*, vol. 3, no. 6, pp. 813–820, 2015.
[52] A. B. Evlyukhin, C. Reinhardt, and B. N. Chichkov, "Multipole light scattering by nonspherical nanoparticles in the discrete dipole approximation," *Phys. Rev. B - Condens. Matter Mater. Phys.*, vol. 84, no. 23, pp. 1–8, 2011.
[53] E. Arbabi, A. Arbabi, S. M. Kamali, Y. Horie, M. S. Faraji-Dana, and A. Faraon, "MEMS-tunable dielectric metasurface lens," *Nat. Commun.*, vol. 9, no. 1, 2018.
[54] A. Arbabi, E. Arbabi, S. M. Kamali, Y. Horie, S. Han, and A. Faraon, "Miniature optical planar camera based on a wide-angle metasurface doublet corrected for monochromatic aberrations," *Nat. Commun.*, vol. 7, pp. 1–31, 2016.
[55] P. Spinelli, M. A. Verschuuren, and A. Polman, "Broadband omnidirectional antireflection coating based on subwavelength surface Mie resonators," *Nat. Commun.*, vol. 3, pp. 692–695, 2012.
[56] C. Shane, Z. Alan, B. Elyas, *et al.*, "Broadband transparent and CMOS-compatible flat optics with silicon nitride metasurfaces [Invited]," *Opt. Mater. Express*, vol. 8, no. 8, p. 2330, 2018.

[57] S. Colburn, A. Zhan, and A. Majumdar, "Metasurface optics for full-color computational imaging," *Sci. Adv.*, vol. 4, no. 2114, pp. 1–7, 2018.

[58] A. Zhan, S. Colburn, R. Trivedi, T. K. Fryett, C. M. Dodson, and A. Majumdar, "Low-contrast dielectric metasurface optics," *ACS Photonics*, vol. 3, pp. 1–6, 2016.

[59] G. E. Jettison, "Optical functions of GaAs, GaP, and Ge determined by two-channel polarization modulation ellipsometry," *Opt. Mater. (Amst).*, pp. 151–160, 1992.

[60] S. Liu, M. B. Sinclair, S. Saravi, *et al.*, "Resonantly enhanced second-harmonic generation using III–V semiconductor all-dielectric metasurfaces sheng," *Nano Lett.*, vol. 16, pp. 5426–5432, 2016.

[61] M. R. Shcherbakov, S. Liu, V. V. Zubyuk, *et al.*, "Ultrafast all-optical tuning of direct-gap semiconductor metasurfaces maxim," *Nat. Commun.*, vol. 8, pp. 1–5, 2017.

[62] S. Liu, G. A. Keeler, J. L. Reno, M. B. Sinclair, and I. Brener, "III–V semiconductor nanoresonators—a new strategy for passive, active, and nonlinear all-dielectric metamaterials," *Adv. Opt. Mater.*, vol. 4, no. 10, pp. 1457–1462, 2016.

[63] S. Person, M. Jain, Z. Lapin, J. Jose, G. Wicks, and L. Novotny, "Demonstration of zero optical backscattering from single nanoparticles," *Nano Lett.*, vol. 13, pp. 1806–1809, 2013.

[64] W. T. Chen, A. Y. Zhu, V. Sanjeev, *et al.*, "A broadband achromatic meta-lens for focusing and imaging in the visible," *Nat. Nanotechnol.*, vol. 13, no. 3, pp. 220–226, 2018.

[65] L. Li, H. Lin, Y. Huang, *et al.*, "High-performance flexible waveguide-integrated photodetectors," *Optica*, vol. 5, no. 1, pp. 44–51, 2018.

[66] J. Hu, L. Li, H. Lin, P. Zhang, W. Zhou, and Z. Ma, "Flexible integrated photonics: where materials, mechanics and optics meet [Invited]," *Opt. Mater. Express*, vol. 3, no. 9, pp. 1313–1331, 2013.

[67] L. Li, H. Lin, S. Qiao, *et al.*, "Integrated flexible chalcogenide glass photonic devices," *Nat. Photonics*, vol. 8, no. 8, pp. 643–649, 2014.

[68] J. Hu, L. Li, H. Lin, *et al.*, "Chalcogenide glass microphotonics: stepping into the spotlight," *Am. Ceram. Soc. Bull.*, vol. 94, no. 4, pp. 24–29, 2015.

[69] L. Li, H. Lin, S. Qiao, *et al.*, "Monolithically integrated stretchable photonics," *Light Sci. Appl.*, vol. 7, no. 2, p. 17138, 2018.

[70] Q. Du, Y. Huang, J. Li, *et al.*, "Low-loss photonic device in Ge – Sb – S chalcogenide glass," *Opt. Lett.*, vol. 41, no. 13, pp. 3090–3093, 2016.

[71] M. Wuttig, H. Bhaskaran, and T. Taubner, "Phase-change materials for non-volatile photonic applications," *Nat. Publ. Gr.*, vol. 11, no. 8, pp. 465–476, 2017.

[72] J. C. Ginn, I. Brener, D. W. Peters, *et al.*, "Realizing optical magnetism from dielectric metamaterials," *Phys. Rev. Lett.*, vol. 108, no. 097402, pp. 1–5, 2012.

[73] S. Liu, M. B. Sinchair, T. S. Mahony, *et al.*, "Optical magnetic mirrors without metals," *Optica*, vol. 1, no. 4, p. 250, 2014.

[74] S. Liu, J. F. Ihlefeld, J. Dominguez, *et al.*, "Realization of tellurium-based all dielectric optical metamaterials using a multi-cycle deposition-etch process," *Appl. Phys. Lett.*, vol. 102, no. 16, p. 161905, 2013.

[75] T. Lewi, H. A. Evans, N. A. Butakov, and J. A. Schuller, "Ultrawide thermo-optic tuning of PbTe meta-atoms," *Nano Lett.*, vol. 17, no. 6, p. 3940, 2017.

[76] L. Zhang, J. Ding, H. Zheng, *et al.*, "Ultra-thin high-efficiency mid-infrared transmissive Huygens meta-optics," *Nat. Commun.*, vol. 9, no. 1, p. 1481, 2018.

[77] J. Wang, J. Hu, X. Sun, *et al.*, "Structural, electrical, and optical properties of thermally evaporated nanocrystalline PbTe films," *J. Appl. Phys.*, vol. 104, no. 5, p. 053707, 2008.

[78] J. Wang, J. Hu, P. Becla, A. M. Agarwal, and L. C. Kimerling, "Resonant-cavity-enhanced mid-infrared photodetector on a silicon platform," *Opt. Express*, vol. 18, no. 12, pp. 12890–12896, 2010.

[79] J. Wang, T. Zens, J. Hu, P. Becla, L. C. Kimerling, and A. M. Agarwal, "Monolithically integrated, resonant-cavity-enhanced dual-band mid-infrared photodetector on silicon," *Appl. Phys. Lett.*, vol. 100, no. 21, p. 211106, 2012.

[80] Z. Han, V. Singh, D. Kita, *et al.*, "On-chip chalcogenide glass waveguide-integrated mid-infrared PbTe detectors," *Appl. Phys. Lett.*, vol. 109, no. 7, p. 071111, 2016.

[81] Q. Wang, E. T. F. Rogers, B. Gholipour, *et al.*, "Optically reconfigurable metasurfaces and photonic devices based on phase change materials," *Nat. Photonics*, vol. 10, no. 1, pp. 60–65, 2015.

[82] A. Tittl, A.-K. U. Michel, M. Schäferling, *et al.*, "A switchable mid-infrared plasmonic perfect absorber with multispectral thermal imaging capability," *Adv. Mater.*, vol. 27, pp. 4597–4603, 2015.

[83] A. Karvounis, B. Gholipour, K. F. MacDonald, and N. I. Zheludev, "All-dielectric phase-change reconfigurable metasurface," *Appl. Phys. Lett.*, vol. 109, no. 5, 2016.

[84] C. H. Chu, M. L. Tseng, J. Chen, *et al.*, "Active dielectric metasurface based on phase-change medium," *Laser Photonics Rev*, vol. 10, no. 6, 2016.

[85] X. Luo, M. Hong, and S. A. Maier, "Engineering the phase front of light with phase-change material based planar lenses," *Sci. Rep.*, vol. 5, no. 8660, pp. 1–7, 2015.

[86] X. Yin, T. Steinle, L. Huang, *et al.*, "Beam switching and bifocal zoom lensing using active plasmonic metasurfaces," *Light Sci. Appl.*, vol. 6, no. 7, p. e17016, 2017.

[87] Y. G. Chen, T. S. Kao, B. Ng, *et al.*, "Hybrid phase-change plasmonic crystals for active tuning of lattice resonances," *Opt. Express*, vol. 21, no. 11, pp. 1566–1569, 2013.

[88] Y. Zhang, J. B. Chou, J. Li, *et al.*, "Broadband transparent optical phase change materials for high-performance nonvolatile photonics," *Nat. Commun.*, vol. 10, no. 4279, pp. 1–9, 2019.

[89] K. Liu, S. Lee, S. Yang, O. Delaire, and J. Wu, "Recent progresses on physics and applications of vanadium dioxide," *Mater. Today*, vol. 21, no. 8, pp. 875 896, 2018.

[90] H. Kocer, S. Butun, B. Banar, *et al.*, "Thermal tuning of infrared resonant absorbers based on hybrid gold-VO_2 nanostructures," *Appl. Phys. Lett.*, vol. 106, 2015.
[91] K. Dong, S. Hong, Y. Deng, *et al.*, "A lithography-free and field-programmable photonic metacanvas," *Adv. Mater.*, vol. 30, 2018.
[92] Y. Ke, X. Wen, D. Zhao, R. Che, Q. Xiong, and Y. Long, "Controllable fabrication of two-dimensional patterned VO_2 nanoparticle, nanodome, and nanonet arrays with tunable temperature-dependent localized surface plasmon resonance," *ACS Nano*, vol. 11, pp. 7542–7551, 2017.
[93] J. Rensberg, S. Zhang, Y. Zhou, *et al.*, "Active optical metasurfaces based on defect-engineered phase-transition materials," *Nano Lett.*, vol. 16, pp. 1050–1055, 2016.
[94] O. L. Muskens, L. Bergamini, Y. Wang, *et al.*, "Antenna-assisted picosecond control of nanoscale phase transition in vanadium dioxide," *Light Sci. Appl.*, vol. 5, no. e16173-9, 2016.
[95] G. Kaplan, K. Aydin, and J. Scheuer, "Dynamically controlled plasmonic nano-antenna phased array utilizing vanadium dioxide [Invited]," *Opt. Mater. Express*, vol. 5, no. 11, pp. 2457–2463, 2015.
[96] S. K. Earl, T. D. James, T. J. Davis, *et al.*, "Tunable optical antennas enabled by the phase transition in vanadium dioxide," *Opt. Express*, vol. 21, no. 22, pp. 27503–27508, 2013.
[97] A. U. Michel, W. W. Tobias, T. Taubner, *et al.*, "Using low-loss phase-change materials for mid-infrared antenna resonance tuning," *Nano Lett.*, vol. 13, pp. 3470–3475, 2013.
[98] S. K. Earl, T. D. James, D. E. Gómez, R. E. Marvel, and R. F. Haglund, "Switchable polarization rotation of visible light using a plasmonic metasurface," *APL Photonics*, vol. 2, no. 016103, 2017.
[99] F. Ligmajer, L. Kejik, U. Tiwari, *et al.*, "Epitaxial VO_2 nanostructures: a route to large-scale, switchable dielectric metasurfaces," *ACS Photonics*, vol. 5, pp. 2561–2567, 2018.
[100] H. Lin, Y. Song, Y. Huang, *et al.*, "Chalcogenide glass-on-graphene photonics," *Nat. Photonics*, vol. 11, pp. 798–805, 2017.
[101] L. Ju, B. Geng, J. Horng, *et al.*, "Graphene plasmonics for tunable terahertz metamaterials," *Nat. Nanotechnol.*, vol. 6, no. 10, pp. 630–634, 2011.
[102] A. N. Grigorenko, M. Polini, and K. S. Novoselov, "Graphene plasmonics," *Nat. Photonics*, vol. 6, 2012.
[103] F. Withers, O. D. Pozo-Zamudio, A. Mishchenko, *et al.*, "Light-emitting diodes by band-structure engineering in van der Waals heterostructures," *Nat. Mater.*, vol. 14, no. 3, pp. 301–306, 2015.
[104] Z. Sun, A. Martinez, and F. Wang, "Optical modulators with 2D layered materials," *Nat. Photonics*, vol. 10, no. 4, pp. 227–238, 2016.
[105] M. Liu, X. Ying, E. Ulin-Avila, *et al.*, "A graphene-based broadband optical modulator," *Nature*, vol. 474, no. 7349, pp. 64–67, 2011.

[106] Y. Yao, R. Shankar, M. A. Kats, *et al.*, "Electrically tunable metasurface perfect absorbers for ultrathin mid-infrared optical modulators," *Nano Lett.*, vol. 14, pp. 6526–6532, 2014.
[107] V. W. Brar, M. S. Jang, M. Sherrott, J. J. Lopez, and H. A. Atwater, "Highly confined tunable mid-infrared plasmonics in graphene nanoresonators," *Nano Lett.*, vol. 13, no. 6, pp. 2541–2547, 2013.
[108] S. Thongrattanasiri, F. H. L. Koppens, and F. J. García De Abajo, "Complete optical absorption in periodically patterned graphene," *Phys. Rev. Lett.*, vol. 108, no. 4, pp. 1–5, 2012.
[109] Z. Fei, A. S. Rodin, G. O. Andreev, *et al.*, "Gate-tuning of graphene plasmons revealed by infrared nano-imaging," *Nature*, vol. 486, no. 7405, pp. 82–85, 2012.
[110] F. Bonaccorso, Z. Sun, T. Hasan, and A. C. Ferrari, "Graphene photonics and optoelectronics," *Nat. Photonics*, vol. 4, pp. 611–622, 2010.
[111] X. Gan, R. J. Shiue, Y. Gao, *et al.*, "Chip-integrated ultrafast graphene photodetector with high responsivity," *Nat. Photonics*, vol. 7, no. 11, pp. 883–887, 2013.
[112] Y. Hu, M. Pantouvaki, J. V. Campenhout, *et al.*, "Broadband 10 Gb/s operation of graphene electro-absorption modulator on silicon," *Laser Photonics Rev.*, vol. 10, no. 2, pp. 307–316, 2016.
[113] L. O. Nyakiti, V. D. Wheeler, N. Y. Garces, R. L. Myers-Ward, C. R. Eddy, and D. K. Gaskill, "Enabling graphene-based technologies: toward wafer-scale production of epitaxial graphene," *MRS Bull.*, vol. 37, no. 12, pp. 1149–1157, 2012.
[114] C. C. Huang, F. Al-Saab, Y. Wang, *et al.*, "Scalable high-mobility MoS_2 thin films fabricated by an atmospheric pressure chemical vapor deposition process at ambient temperature," *Nanoscale*, vol. 6, no. 21, pp. 12792–12797, 2014.
[115] L. Zhou, K. Xu, A. Zubair, *et al.*, "Large-area synthesis of high-quality uniform few-layer $MoTe_2$," *J. Am. Chem. Soc.*, vol. 137, no. 37, pp. 11892–11895, 2015.
[116] L. Colombo, R. M. Wallace, and R. S. Ruoff, "Graphene growth and device integration," *Proc. IEEE*, vol. 101, no. 7, pp. 1536–1556, 2013.
[117] B. Lee, G. Mordi, M. J. Kim, *et al.*, "Characteristics of high-k Al_2O_3 dielectric using ozone-based atomic layer deposition for dual-gated graphene devices," *Appl. Phys. Lett.*, vol. 97, no. 4, pp. 2008–2011, 2010.
[118] J. R. Williams, L. DiCarlo, and C. M. Marcus, "Quantum Hall effect in a gate-controlled p-n junction of graphene," *Science*, vol. 317, no. 5838, pp. 638–641, 2007.
[119] X. Wang, S. M. Tabakman, and H. Dai, "Atomic layer deposition of metal oxides on pristine and functionalized graphene," *J. Am. Chem. Soc.*, vol. 130, no. 26, pp. 8152–8153, 2008.
[120] W. Zhu, D. Neumayer, V. Perebeinos, and P. Avouris, "Silicon nitride gate dielectrics and band gap engineering in graphene layers," *Nano Lett.*, vol. 10, no. 9, pp. 3572–3576, 2010.

[121] J. Proust, F. Bedu, S. Frédéric, *et al.*, "Chemical alkaline etching of silicon Mie particles," *Adv. Opt. Mater.*, vol. 3, no. 9, pp. 1280–1286, 2015.

[122] S. Shrestha, A. C. Overvig, M. Lu, A. Stein, and N. Yu, "Broadband achromatic dielectric metalenses," *Light Sci. Appl.*, vol. 7, no. 1, 2018.

[123] S. Wang, P. C. Wu, V. C. Su, *et al.*, "Broadband achromatic optical metasurface devices," *Nat. Commun.*, vol. 8, no. 1, pp. 1–9, 2017.

[124] E. Shkondin, O. Takayama, J. M. Lindhard, *et al.*, "Fabrication of high aspect ratio TiO_2 and Al_2O_3 nanogratings by atomic layer deposition," *J. Vac. Sci. Technol. A Vacuum, Surfaces, Film.*, vol. 34, no. 3, p. 031605, 2016.

[125] O. Takayama, E. Shkondin, A. Bodganov, *et al.*, "Midinfrared surface waves on a high aspect ratio nanotrench platform," *ACS Photonics*, vol. 4, no. 11, pp. 2899–2907, 2017.

[126] T. M. Babinec, B. J. M. Hausmann, M. Khan, *et al.*, "A diamond nanowire single-photon source," *Nat. Nanotechnol.*, vol. 5, no. 3, pp. 195–199, 2010.

[127] R. Chikkaraddy, B. de Nijs, F. Benz, *et al.*, "Single-molecule strong coupling at room temperature in plasmonic nanocavities," *Nature*, vol. 535, no. 7610, pp. 127–130, 2016.

[128] S. Lee, B. Park, J. S. Kim, and T. Il Kim, "Designs and processes toward high-aspect-ratio nanostructures at the deep nanoscale: unconventional nanolithography and its applications," *Nanotechnology*, vol. 27, no. 47, pp. 1–15, 2016.

[129] R. Micheloni, L. Crippa, C. Zambelli, and P. Olivo, "Architectural and integration options for 3D NAND flash memories," *Computers*, vol. 6, no. 3, p. 27, 2017.

[130] J. W. Coburn and H. F. Winters, "Ion- and electron-assisted gas-surface chemistry—an important effect in plasma etching," *J. Appl. Phys.*, vol. 50, no. 5, pp. 3189–3196, 1979.

[131] J. P. Simko, "Removal of fluorocarbon residues on CF_4/H_2 reactive-ion-etched silicon surfaces using a hydrogen plasma," *J. Electrochem. Soc.*, vol. 138, no. 1, p. 277, 1991.

[132] S. Bouchoule, R. Chanson, A. Pageau, *et al.*, "Surface chemistry of InP ridge structures etched in Cl_2-based plasma analyzed with angular XPS," *J. Vac. Sci. Technol. A Vacuum, Surfaces, Film.*, vol. 33, no. 5, p. 05E124, 2015.

[133] V. M. Donnelly and A. Kornblit, "Plasma etching: yesterday, today, and tomorrow," *J. Vac. Sci. Technol. A Vacuum, Surfaces, Film.*, vol. 31, no. 5, p. 050825, 2013.

[134] A. Arbabi, Y. Horie, M. Bagheri, and A. Faraon, "Dielectric metasurfaces for complete control of phase and polarization with subwavelength spatial resolution and high transmission," *Nat. Nanotechnol.*, vol. 10, no. 11, pp. 937–943, 2015.

[135] K. Grigoras, L. Sainiemi, J. Tiilikainen, A. Säynätjoki, V. M. Airaksinen, and S. Franssila, "Application of ultra-thin aluminum oxide etch mask made by atomic layer deposition technique," *J. Phys. Conf. Ser.*, vol. 61, no. 1, pp. 369–373, 2007.

[136] A. A. High, R. C. Devlin, A. Dibos, *et al.*, "Visible-frequency hyperbolic metasurface," *Nature*, vol. 522, no. 7555, pp. 192–196, 2015.

[137] X. Chen, H. R. Park, M. Pelton, *et al.*, "Atomic layer lithography of wafer-scale nanogap arrays for extreme confinement of electromagnetic waves," *Nat. Commun.*, vol. 4, pp. 1–7, 2013.

[138] K. Peng, J. Hu, Y. Yan, *et al.*, "Fabrication of single-crystalline silicon nanowires by scratching a silicon surface with catalytic metal particles," *Adv. Funct. Mater.*, vol. 16, no. 3, pp. 387–394, 2006.

[139] K. Peng, H. Fang, J. Hu, *et al.*, "Metal-particle-induced, highly localized site-specific etching of Si and formation of single-crystalline Si nanowires in aqueous fluoride solution," *Chem. — A Eur. J.*, vol. 12, no. 30, pp. 7942–7947, 2006.

[140] K. Balasundaram, J. S. Sadhu, J. C. Shin, *et al.*, "Porosity control in metal-assisted chemical etching of degenerately doped silicon nanowires," *Nanotechnology*, vol. 23, no. 30, p. 305304, 2012.

[141] X. Li, "Metal assisted chemical etching for high aspect ratio nanostructures: a review of characteristics and applications in photovoltaics," *Curr. Opin. Solid State Mater. Sci.*, vol. 16, no. 2, pp. 71–81, 2012.

[142] H. Li, T. Ye, L. Shi, and C. Xie, "Fabrication of ultra-high aspect ratio (>160:1) silicon nanostructures by using Au metal assisted chemical etching," *J. Micromechanics Microengineering*, vol. 27, no. 12, p. 124002, 2017.

[143] C. Chang and A. Sakdinawat, "Ultra-high aspect ratio high-resolution nanofabrication for hard X-ray diffractive optics," *Nat. Commun.*, vol. 5, pp. 1–7, 2014.

[144] S. H. Kim, P. K. Mohseni, Y. Song, T. Ishihara, and X. Li, "Inverse metal-assisted chemical etching produces smooth high aspect ratio InP nanostructures," *Nano Lett.*, vol. 15, no. 1, pp. 641–648, 2015.

[145] J. D. Kim, M. Kim, L. Kong, *et al.*, "Self-anchored catalyst interface enables ordered via array formation from submicrometer to millimeter scale for polycrystalline and single-crystalline silicon," *ACS Appl. Mater. Interfaces*, vol. 10, no. 10, pp. 9116–9122, 2018.

[146] K. Balasundaram, P. K. Mohseni, Y.-C. Shuai, D. Zhao, W. Zhou, and X. Li, "Photonic crystal membrane reflectors by magnetic field-guided metal-assisted chemical etching," *Appl. Phys. Lett.*, vol. 103, no. 21, p. 214103, 2013.

[147] H. D. Um, N. Kim, K. Lee, *et al.*, "Versatile control of metal-assisted chemical etching for vertical silicon microwire arrays and their photovoltaic applications," *Sci. Rep.*, vol. 5, no. pp. 1–11, 2015.

[148] R. Divan, D. Rosenthal, K. Ogando, L. E. Ocola, D. Rosenmann, and N. Moldovan, "Metal-assisted etching of silicon molds for electroforming," *J. Vac. Sci. Technol. B, Nanotechnol. Microelectron. Mater. Process. Meas. Phenom.*, vol. 31, no. 6, p. 06FF03, 2013.

[149] K. Lee, D. Yang, S. H. Park, and R. H. Kim, "Review recent developments in the use of two-photon polymerization in precise 2D and 3D microfabrications," *Polym. Adv. Technol.*, vol. 17, no. 2, pp. 72–82, 2006.

[150] K. Satoshi, S. Hong-Bo, T. Tomokazu, and T. Kenji, "Finer features for functional microdevices," *Nature*, vol. 412, pp. 697–698, 2001.
[151] X. Z. Dong, Z. S. Zhao, and X. M. Duan, "Improving spatial resolution and reducing aspect ratio in multiphoton polymerization nanofabrication," *Appl. Phys. Lett.*, vol. 92, no. 9, pp. 90–93, 2008.
[152] D. Tan, Y. Li, F. Qi, *et al.*, "Reduction in feature size of two-photon polymerization using SCR500," *Appl. Phys. Lett.*, vol. 90, no. 7, pp. 1–4, 2007.
[153] X. Zhou, Y. Hou, and J. Lin, "A review on the processing accuracy of two-photon polymerization," *AIP Adv.*, vol. 5, no. 3, p. 030701, 2015.
[154] H. Sun, T. Suwa, K. Takada, *et al.*, "Shape precompensation in two-photon laser nanowriting of photonic lattices," *Appl. Phys. Lett.*, vol. 85, no. 17, pp. 3708–3710, 2004.
[155] C. Y. Liao, M. Bouriauand, P. L. Baldeck, J. C. Ĺon, Ć. Masclet, and T. T. Chung, "Two-dimensional slicing method to speed up the fabrication of micro-objects based on two-photon polymerization," *Appl. Phys. Lett.*, vol. 91, no. 3, pp. 1–4, 2007.
[156] D. Y. Yang, S. H. Park, T. W. Lim, *et al.*, "Ultraprecise microreproduction of a three-dimensional artistic sculpture by multipath scanning method in two-photon photopolymerization," *Appl. Phys. Lett.*, vol. 90, no. 1, pp. 2005–2008, 2007.
[157] K. Takada, H. B. Sun, and S. Kawata, "Improved spatial resolution and surface roughness in photopolymerization- based laser nanowriting," *Appl. Phys. Lett.*, vol. 86, no. 7, pp. 1–3, 2005.
[158] S. H. Park, S. H. Lee, D. Y. Yang, H. J. Kong, and K. S. Lee, "Subregional slicing method to increase three-dimensional nanofabrication efficiency in two-photon polymerization," *Appl. Phys. Lett.*, vol. 87, no. 15, pp. 1–3, 2005.
[159] R. Guo, S. Xiao, X. Zhai, J. Li, A. Xia, and W. Huang, "Micro lens fabrication by means of femtosecond two photon photopolymerization," *Opt. Express*, vol. 14, no. 2, p. 810, 2006.
[160] D. Wu, S. Z. Wu, L. G. Niu, *et al.*, "High numerical aperture microlens arrays of close packing," *Appl. Phys. Lett.*, vol. 97, no. 3, 2010.
[161] P. I. Dietrich, M. Blaicher, I. Reuter, *et al.*, "In situ 3D nanoprinting of free-form coupling elements for hybrid photonic integration," *Nat. Photonics*, vol. 12, no. 4, pp. 241–247, 2018.
[162] J. Serbin, A. Egbert, A. Ostendorf, and B. N. Chichkov, "Femtosecond laser-induced two-photon polymerization of inorganic—organic hybrid materials for applications in photonics," *Opt. Lett.*, vol. 28, no. 5, pp. 301–303, 2003.
[163] T. Gissibl, S. Thiele, A. Herkommer, and H. Giessen, "Two-photon direct laser writing of ultracompact multi-lens objectives," *Nat. Photonics*, vol. 10, no. 8, p. 554, 2016.
[164] C. Liberale, G. Cojoc, P. Candeloro, *et al.*, "Micro-optics fabrication on top of optical fibers using two-photon lithography," *IEEE Photonics Technol. Lett.*, vol. 22, no. 7, pp. 474–476, 2010.

[165] C. Liberale, G. Cojoc, F. Bragheri, *et al.*, "Integrated microfluidic device for single-cell trapping and spectroscopy," *Sci. Rep.*, vol. 3, p. 1258, 2013.

[166] S. Thiele, A. Herkommer, H. Giessen, and T. Gissibl, "Sub-micrometre accurate free-form optics by three-dimensional printing on single-mode fibre," *Nat. Commun.*, vol. 7, pp. 1–9, 2016.

[167] M. Deubel, G. von Freymann, M. Wegener, S. Pereira, K. Busch, and C. M. Soukoulis, "Direct laser writing of three-dimensional photonic-crystal templates for telecommunications," *Nat. Mater.*, vol. 3, no. 7, pp. 444–447, 2004.

[168] H. E. Williams, D. J. Freppon, S. M. Kuebler, R. C. Rumpf, and M. A. Melino, "Fabrication of three-dimensional micro-photonic structures on the tip of optical fibers using SU-8," *Opt. Express*, vol. 19, no. 23, pp. 22910–22922, 2011.

[169] C. Jain, A. Braun, J. Gargiulo, B. Jang, G. Li, and H. Lehmann, "Hollow core light cage: Trapping light behind bars," *ACS Photonics*, 2018.

[170] T. Gissibl, M. Schmid, and H. Giessen, "Spatial beam intensity shaping using phase masks on single-mode optical fibers fabricated by femtosecond direct laser writing," *Optica*, vol. 3, no. 4, pp. 448–451, 2016.

[171] B. J. Jung, H. J. Kong, Y. H. Cho, *et al.*, "Fabrication of sharp-needled conical polymer tip on the cross-section of optical fiber via two-photon polymerization for tuning-fork-based atomic force microscopy," *Opt. Commun.*, vol. 286, pp. 197–203, 2013.

[172] C. Lu and R. H. Lipson, "Interference lithography: a powerful tool for fabricating periodic structures," *Laser Photon. Rev.*, vol. 4, no. 4, pp. 568–580, 2009.

[173] S. R. J. Brueck, "Optical and interferometric lithography - nanotechnology enablers," *Proc. IEEE*, vol. 93, no. 10, pp. 1704–1721, 2005.

[174] C. Moormann, J. Bolten, and H. Kurz, "Spatial phase-locked combination lithography for photonic crystal devices," *Microelectron. Eng.*, vol. 74, pp. 417–422, 2004.

[175] I. Divliansky, T. S. Mayer, K. S. Holliday, and V. H. Crespi, "Fabrication of three-dimensional polymer photonic crystal structures using single diffraction element interference lithography," *Appl. Phys. Lett.*, vol. 82, no. 11, pp. 1667–1669, 2003.

[176] M. Campbell, D. N. Sharp, M. T. Harrison, and R. G. Denning, "Fabrication of photonic crystals for the visible spectrum by holographic lithography," *Nature*, vol. 404, no. 2, pp. 53–56, 2000.

[177] S. Yang, M. Megens, J. Aizenberg, P. Wiltzius, P. M. Chaikin, and W. B. Russel, "Creating periodic three-dimensional structures by multibeam interference of visible laser," *Chem. Mater.*, vol. 14, no. 7, pp. 2001–2003, 2002.

[178] F. Quiñónez, J. W. Menezes, and L. Cescato, "Band gap of hexagonal 2D photonic crystals with elliptical holes recorded by interference lithography," *Opt. Express*, vol. 14, no. 11, pp. 4873–4879, 2006.

[179] H. H. Solak, C. David, J. Gobrecht, *et al.*, "Multiple-beam interference lithography with electron beam written gratings," *J. Vac. Sci. Technol. B*, vol. 20, no. 6, 2002.

[180] S. Yang, M. Megens, J. Aizenberg, P. Wiltzius, P. M. Chaikin, and W. B. Russel, "Creating periodic three-dimensional structures by multi-beam interference of visible laser," *Chem. Mater.*, vol. 14, no. 7, pp. 2831–2833, 2002.

[181] O. Toader, T. Y. M. Chan, and S. John, "Photonic band gap architectures for holographic lithography," *Phys. Rev. Lett.*, vol. 92, no. 4, p. 4, 2004.

[182] Y. Lin, D. Rivera, Z. Poole, and K. P. Chen, "Five-beam interference pattern controlled through phases and wave vectors for diamondlike photonic crystals," *Appl. Opt.*, vol. 45, no. pp. 7971–7976, 2006.

[183] T. Y. M. Chan, O. Toader, and S. John, "Photonic band-gap formation by optical-phase-mask lithography," *Phys. Rev. E - Stat. Nonlinear, Soft Matter Phys.*, vol. 73, no. 4, pp. 1–11, 2006.

[184] C. Lu, X. K. Hu, I. V. Mitchell, and R. H. Lipson, "Diffraction element assisted lithography: pattern control for photonic crystal fabrication," *Appl. Phys. Lett.*, vol. 86, no. 19, p. 193110, 2005.

[185] Y. V. Miklyaev, D. C. Meisel, A. Blanco, *et al.*, "Three-dimensional face-centered-cubic photonic crystal templates by laser holography: fabrication, optical characterization, and band-structure calculations," *Appl. Phys. Lett.*, vol. 82, no. 8, pp. 1284–1286, 2003.

[186] Z. Poole, D. Xu, K. P. Chen, I. Olvera, K. Ohlinger, and Y. Lin, "Holographic fabrication of three-dimensional orthorhombic and tetragonal photonic crystal templates using a diffractive optical element," *Appl. Phys. Lett.*, vol. 91, no. 25, pp. 1–4, 2007.

[187] Y. Lin, P. R. Herman, and K. Darmawikarta, "Design and holographic fabrication of tetragonal and cubic photonic crystals with phase mask: toward the mass-production of three-dimensional photonic crystals," *Appl. Phys. Lett.*, vol. 86, no. 7, pp. 1–3, 2005.

[188] A. Shishido, I. B. Diviliansky, I. C. Khoo, *et al.*, "Direct fabrication of two-dimensional titania arrays using interference photolithography," *Appl. Phys. Lett.*, vol. 79, no. 3332, pp. 10–13, 2001.

[189] I. B. Divliansky, A. Shishido, I. C. Khoo, *et al.*, "Fabrication of two-dimensional photonic crystals using interference lithography and electro-deposition of CdSe," *Appl. Phys. Lett.*, vol. 79, no. 3392, pp. 1–4, 2005.

[190] Y. Lin, P. R. Herman, and E. L. Abolghasemi, "Proposed single-exposure holographic fabrication of microsphere-type photonic crystals through phase-mask techniques," *J. Appl. Phys.*, vol. 97, no. 9, pp. 4–7, 2005.

[191] Y. Lin, A. Harb, D. Rodriguez, K. Lozano, D. Xu, and K. P. Chen, "Holographic fabrication of photonic crystals using multidimensional phase masks masks," *J. Appl. Phys.*, vol. 104, no. 113111, 2008.

[192] J. Xavier and J. Joseph, "Complex photonic lattices embedded with tailored intrinsic defects by a dynamically reconfigurable single step interferometric approach," *Appl. Phys. Lett.*, vol. 104, no. 8, 2014.

[193] M. Kumar and J. Joseph, "Generating a hexagonal lattice wave field with a gradient basis structure," *Opt. Lett.*, vol. 39, no. 8, p. 2459, 2014.

[194] S. Behera, M. Kumar, and J. Joseph, "Submicrometer photonic structure fabrication by phase spatial-light-modulator-based interference lithography," *Opt. Lett.*, vol. 41, no. 8, p. 1893, 2016.
[195] J. Li, Y. Liu, X. Xie, *et al.*, "Fabrication of photonic crystals with functional defects by one-step holographic lithography," *Opt. Express*, vol. 16, no. 17, p. 12899, 2008.
[196] G. Zito, B. Piccirillo, E. Santamato, A. Marino, V. Tkachenko, and G. Abbate, "Two-dimensional photonic quasicrystals by single beam computer-generated holography," *Opt. Express*, vol. 16, no. 8, p. 5164, 2008.
[197] M. Kumar and J. Joseph, "Optical generation of a spatially variant two-dimensional lattice structure by using a phase only spatial light modulator," *Appl. Phys. Lett.*, vol. 105, no. 5, 2014.
[198] Y. J. Liu and X. W. Sun, "Electrically tunable two-dimensional holographic photonic crystal fabricated by a single diffractive element," *Appl. Phys. Lett.*, vol. 89, no. 17, pp. 1–4, 2006.
[199] D. B. Burckel, P. J. Resnick, P. S. Finnegan, M. B. Sinclair, and P. S. Davids, "Micrometer-scale fabrication of complex three dimensional lattice + basis structures in silicon," *Opt. Mater. Express*, vol. 5, no. 10, p. 2231, 2015.
[200] D. B. Burckel, J. R. Wendt, G. A. Ten Eyck, A. R. Ellis, I. Brener, and M. B. Sinclair, "Fabrication of 3D metamaterial resonators using self-aligned membrane projection lithography," *Adv. Mater.*, vol. 22, no. 29, pp. 3171–3175, 2010.
[201] D. B. Burckel, J. R. Wendt, G. A. T. Eyck, *et al.*, "Micrometer-scale cubic unit cell 3D metamaterial layers," *Adv. Mater.*, vol. 22, no. 44, pp. 5053–5057, 2010.
[202] D. B. Burckel, S. Campione, P. S. Davids, and M. B. Sinclair, "Three dimensional metafilms with dual channel unit cells," *Appl. Phys. Lett.*, vol. 110, no. 14, p. 143107, 2017.
[203] D. B. Burckel, "Device-level and module-level three-dimensional integrated circuits created using oblique processing," *J. Micro/Nanolithography, MEMS, MOEMS*, vol. 15, no. 3, p. 34504, 2016.
[204] C. A. Mirkin, R. L. Letsinger, R. C. Mucic, and J. J. Storhoff, "A DNA-based method for rationally assembling nanoparticles into macroscopic materials," *Nature*, vol. 382, no. 6592, pp. 607–609, 1996.
[205] A. P. Alivisatos, K. P. Johnsson, X. Peng, *et al.*, "Organization of nanocrystal molecules using DNA," *Nature*, vol. 382, no. 6592, pp. 609–611, 1996.
[206] X. Li, X. Yang, J. Qi, and N. C. Seeman, "Antiparallel DNA double crossover molecules as components for nanoconstruction," *J. Am. Chem. Soc.*, vol. 118, no. 26, pp. 6131–6140, 1996.
[207] S. Y. Park, A. K. R. Lytton-Jean, B. Lee, S. Weigand, G. C. Schatz, and C. A. Mirkin, "DNA-programmable nanoparticle crystallization," *Nature*, vol. 451, no. 7178, pp. 553–556, 2008.

[208] E. Auyeung, T. I. N. G. Li, A. J. Senesi, *et al.*, "DNA-mediated nanoparticle crystallization into Wulff polyhedra," *Nature*, vol. 505, no. 7481, pp. 73–77, 2014.

[209] R. J. Macfarlane, B. Lee, M. R. Jones, N. Harris, G. C. Schatz and C. A. Mirkin "Supplementary info: nanoparticle superlattice engineering with DNA," *Science*, vol. 334, no. 6053, pp. 204–8, 2011.

[210] M. R. Jones, K. D. Osberg, R. J. MacFarlane, M. R. Langille, and C. A. Mirkin, "Templated techniques for the synthesis and assembly of plasmonic nanostructures," *Chem. Rev.*, vol. 111, no. 6, pp. 3736–3827, 2011.

[211] Z. H. Jiang, L. Kang, and D. H. Werner, "Conformal metasurface-coated dielectric waveguides for highly confined broadband optical activity with simultaneous low-visibility and reduced crosstalk," *Nat. Commun.*, vol. 8, no. 1, p. 356, 2017.

[212] S. M. Kamali, A. Arbabi, E. Arbabi, Y. Horie, and A. Faraon, "Decoupling optical function and geometrical form using conformal flexible dielectric metasurfaces," *Nat. Commun.*, vol. 7, p. 11618, 2016.

[213] J. Cheng, S. Jafar-Zanjani, and H. Mosallaei, "All-dielectric ultrathin conformal metasurfaces: lensing and cloaking applications at 532 nm wavelength," *Sci. Rep.*, vol. 6, p. 38440, 2016.

[214] D.-H. Kim, N. Lu, R. Ma, *et al.*, "Epidermal electronics," *Science*, vol. 333, no. 6044, pp. 838–843, 2011.

[215] H. Zhao, K. O'Brien, S. Li, and R. F. Shepherd, "Optoelectronically innervated soft prosthetic hand via stretchable optical waveguides," *Sci. Robot.*, vol. 1, no. 1, p. eaai7529, 2016.

[216] S. Aikio, J. Hiltunen, J. Hiitola-Keinänen, *et al.*, "Disposable photonic integrated circuits for evanescent wave sensors by ultra-high volume roll-to-roll method," *Opt. Express*, vol. 24, no. 3, pp. 2527–2541, 2016.

[217] S. Aikio, J. Hiltunen, Johanna Hiitola-Keinänen, *et al.*, "Disposable (bio) chemical integrated optical waveguide sensors implemented on roll-to-roll produced platforms," *RSC Adv.*, vol. 6, no. 56, pp. 50414–50422, 2016.

[218] C. L. Yu, H. Kim, N. De Leon, *et al.*, "Stretchable photonic crystal cavity with wide frequency tunability," *Nano Lett.*, vol. 13, no. 1, pp. 248–252, 2012.

[219] M. L. Tseng, J. Yang, M. Semmlinger, C. Zhang, P. Nordlander, and N. J. Halas, "Two-dimensional active tuning of an aluminum plasmonic array for full-spectrum response," *Nano Lett.*, vol. 17, no. 10, pp. 6034–6039, 2017.

[220] W. Jinxiu, Z. Hongbo, C. Huanjun, Z. Weihong, and C. Jian, "Stretchable plasmonic substrate with tunable resonances for surface-enhanced Raman spectroscopy," *J. Opt.*, vol. 17, no. 11, p. 114015, 2015.

[221] H.-S. Ee and R. Agarwal, "Tunable metasurface and flat optical zoom lens on a stretchable substrate," *Nano Lett.*, vol. 16, no. 4, pp. 2818–2823, 2016.

[222] Y. Zou, L. Moreel, H. Lin, *et al.*, "Solution processing and resist-free nanoimprint fabrication of thin film chalcogenide glass devices: inorganic-

organic hybrid photonic integration," *Adv. Opt. Mater.*, vol. 2, no. 8, pp. 759–764, 2014.
[223] A. She, S. Zhang, S. Shian, D. R. Clarke, and F. Capasso, "Adaptive metalenses with simultaneous electrical control of focal length, astigmatism, and shift," *Sci. Adv.*, vol. 4, no. 2, 2018.
[224] G. Kostovski, P. R. Stoddart, and A. Mitchell, "The optical fiber tip: an inherently light-coupled microscopic platform for micro- and nanotechnologies," *Adv. Mater.*, vol. 26, no. 23, pp. 3798–3820, 2014.
[225] W. S. Wong and A. Salleo, *Flexible electronics: materials and applications*, vol. 11. Springer Science & Business Media, 2009.
[226] A. Carlson, A. M. Bowen, Y. Huang, R. G. Nuzzo, and J. A. Rogers, "Transfer printing techniques for materials assembly and micro/nanodevice fabrication," *Adv. Mater.*, vol. 24, no. 39, pp. 5284–5318, 2012.
[227] K. J. Yu, Z. Yan, M. Han, and J. A. Rogers, "Inorganic semiconducting materials for flexible and stretchable electronics," *NPJ Flex. Electron.*, vol. 1, no. 1, p. 4, 2017.
[228] H. Yang, D. Zhao, S. Chuwongin, *et al.*, "Transfer-printed stacked nanomembrane lasers on silicon," *Nat. Photonics*, vol. 6, no. 9, pp. 615–620, 2012.
[229] W. Zhou, Z. Ma, H. Yang, *et al.*, "Flexible photonic-crystal Fano filters based on transferred semiconductor nanomembranes," *J. Phys. D. Appl. Phys.*, vol. 42, no. 23, p. 234007, 2009.
[230] Y. Chen, H. Lin, J. Hu, and M. Li, "Heterogeneously integrated silicon photonics for the mid-infrared and spectroscopic sensing," *ACS Nano*, vol. 8, no. 7, pp. 6955–6961, 2014.
[231] Y. Chen, H. Li, and M. Li, "Flexible and tunable silicon photonic circuits on plastic substrates," *Sci. Rep.*, vol. 2, p. 622, 2012.
[232] I. M. Pryce, K. Aydin, Y. A. Kelaita, R. M. Briggs, and H. A. Atwater, "Highly strained compliant optical metamaterials with large frequency tunability," *Nano Lett.*, vol. 10, no. 10, pp. 4222–4227, 2010.
[233] F. Laible, D. A. Gollmer, S. Dickreuter, D. P. Kern, and M. Fleischer, "Continuous reversible tuning of the gap size and plasmonic coupling of bow tie nanoantennas on flexible substrates," *Nanoscale*, vol. 10, no. 31, pp. 14915–14922, 2018.
[234] L. Zhu, J. Kapraun, J. Ferrara, and C. J. Chang-Hasnain, "Flexible photonic metastructures for tunable coloration," *Optica*, vol. 2, no. 3, pp. 255–258, 2015.
[235] X. Xu, H. Subbaraman, A. Hosseini, C.-Y. Lin, D. Kwong, and R. T. Chen, "Stamp printing of silicon-nanomembrane-based photonic devices onto flexible substrates with a suspended configuration," *Opt. Lett.*, vol. 37, no. 6, pp. 1020–1022, 2012.
[236] D. Chanda, K. Shigeta, S. Gupta, *et al.*, "Large-area flexible 3D optical negative index metamaterial formed by nanotransfer printing," *Nat. Nanotechnol.*, vol. 6, p. 402, 2011.
[237] S. C. Malek, H.-S. Ee, and R. Agarwal, "Strain multiplexed metasurface holograms on a stretchable substrate," *Nano Lett.*, vol. 17, no. 6, pp. 3641–3645, 2017.

[238] P. Gutruf, C. Zou, W. Withayachumnankul, M. Bhaskaran, S. Sriram, and C. Fumeaux, "Mechanically tunable dielectric resonator metasurfaces at visible frequencies," *ACS Nano*, vol. 10, no. 1, pp. 133–141, 2016.

[239] X. Xu, B. Peng, D. Li, *et al.*, "Flexible visible–infrared metamaterials and their applications in highly sensitive chemical and biological sensing," *Nano Lett.*, vol. 11, no. 8, pp. 3232–3238, 2011.

[240] A. Di Falco, Y. Zhao, and A. Alú, "Optical metasurfaces with robust angular response on flexible substrates," *Appl. Phys. Lett.*, vol. 99, no. 16, p. 163110, 2011.

[241] J. Burch, D. Wen, X. Chen, and A. Di Falco, "Conformable holographic metasurfaces," *Sci. Rep.*, vol. 7, no. 1, p. 4520, 2017.

[242] F. Andrea Di, P. Martin, and F. K. Thomas, "Flexible metamaterials at visible wavelengths," *New J. Phys.*, vol. 12, no. 11, p. 113006, 2010.

[243] P.-C. Li and E. T. Yu, "Flexible, low-loss, large-area, wide-angle, wavelength-selective plasmonic multilayer metasurface," *J. Appl. Phys.*, vol. 114, no. 13, p. 133104, 2013.

[244] N. Sharac, H. Sharma, M. Veysi, *et al.*, "Tunable optical response of bowtie nanoantenna arrays on thermoplastic substrates," *Nanotechnology*, vol. 27, no. 10, p. 105302, 2016.

[245] G. Zhang, C. Lan, H. Bian, R. Gao, and J. Zhou, "Flexible, all-dielectric metasurface fabricated via nanosphere lithography and its applications in sensing," *Opt. Express*, vol. 25, no. 18, pp. 22038–22045, 2017.

[246] M. Kahraman, P. Daggumati, O. Kurtulus, E. Seker, and S. Wachsmann-Hogiu, "Fabrication and characterization of flexible and tunable plasmonic nanostructures," *Sci. Rep.*, vol. 3, p. 3396, 2013.

[247] J. Y. Kim, H. Kim, B. H. Kim, *et al.*, "Highly tunable refractive index visible-light metasurface from block copolymer self-assembly," *Nat. Commun.*, vol. 7, p. 12911, 2016.

[248] S. Aksu, M. Huang, A. Artar, *et al.*, "Flexible plasmonics: flexible plasmonics on unconventional and nonplanar substrates," *Adv. Mater.*, vol. 23, no. 38, p. 4421, 2011.

[249] O. Vazquez-Mena, T. Sannomiya, M. Tosun, *et al.*, "High-resolution resistless nanopatterning on polymer and flexible substrates for plasmonic biosensing using stencil masks," *ACS Nano*, vol. 6, no. 6, pp. 5474–5481, 2012.

[250] S. Aksu, A. A. Yanik, R. Adato, A. Artar, M. Huang, and H. Altug, "Nanostencil lithography for high-throughput fabrication of infrared plasmonic sensors," in *SPIE Defense, Security, and Sensing*, 2011, vol. 8031, p. 5.

[251] R. Hokari, K. Kurihara, N. Takada, and H. Hiroshima, "Development of simple high-resolution embedded printing for transparent metal grid conductors," *Appl. Phys. Lett.*, vol. 111, no. 6, p. 63107, 2017.

[252] R. Hokari, K. Kurihara, N. Takada, and H. Hiroshima, "Printed optical metamaterials composed of embedded silver nanoparticles for flexible applications," *Opt. Express*, vol. 26, no. 8, pp. 10326–10338, 2018.

[253] L. Duempelmann, J. A. Müller, F. Lütolf, B. Gallinet, R. Ferrini, and L. Novotny, "Controlling the color of plasmonic substrates with inkjet printing," *Adv. Opt. Mater.*, vol. 5, no. 17, p. 1700153, 2017.

[254] L. Fan, L. T. Varghese, Y. Xuan, J. Wang, B. Niu, and M. Qi, "Direct fabrication of silicon photonic devices on a flexible platform and its application for strain sensing," *Opt. Express*, vol. 20, no. 18, pp. 20564–20575, 2012.

[255] R. Madaka, V. Kanneboina, and P. Agarwal, "Low-temperature growth of amorphous silicon films and direct fabrication of solar cells on flexible polyimide and photo-paper substrates," *J. Electron. Mater.*, vol. 47, no. 8, pp. 4710–4720, 2018.

[256] Z. Shao, Y. Chen, H. Chen, *et al.*, "Ultra-low temperature silicon nitride photonic integration platform," *Opt. Express*, vol. 24, no. 3, pp. 1865–1872, 2016.

[257] L. Li, P. Zhang, W.-M. Wang, *et al.*, "Foldable and cytocompatible sol-gel TiO_2 photonics," *Sci. Rep.*, vol. 5, 2015.

[258] L. Li, H. Lin, S. Geiger, *et al.*, "Amorphous thin films for mechanically flexible, multimaterial integrated photonics," *Am. Ceram. Soc. Bull.*, vol. 95, no. 4, pp. 34–36, 2016.

[259] L. Li, Y. Zou, H. Lin, *et al.*, "A fully-integrated flexible photonic platform for chip-to-chip optical interconnects," *J. Light. Technol.*, vol. 31, no. 24, pp. 4080–4086, 2013.

[260] R. G. DeCorby, N. Ponnampalam, H. T. Nguyen, and T. J. Clement, "Robust and flexible free-standing all-dielectric omnidirectional reflectors," *Adv. Mater.*, vol. 19, no. 2, pp. 193–196, 2007.

[261] J. D. Musgraves, N. Carlie, J. Hu, *et al.*, "Comparison of the optical, thermal and structural properties of Ge-Sb-S thin films deposited using thermal evaporation and pulsed laser deposition techniques," *Acta Mater.*, vol. 59, no. 12, pp. 5032–5039, 2011.

[262] N. Carlie, J. D. Musgraves, B. Zdyrko, *et al.*, "Integrated chalcogenide waveguide resonators for mid-IR sensing: leveraging material properties to meet fabrication challenges," *Opt. Express*, vol. 18, no. 25, pp. 26728–26743, 2010.

[263] J. Hu, V. Tarasov, N. Carlie, *et al.*, "Exploration of waveguide fabrication from thermally evaporated Ge–Sb–S glass films," *Opt. Mater. (Amst).*, vol. 30, no. 10, pp. 1560–1566, 2008.

[264] J. Hu, N. N. Feng, N. Carlie, *et al.*, "Optical loss reduction in high-index-contrast chalcogenide glass waveguides via thermal reflow," *Opt. Express*, vol. 18, no. 2, p. 1469, 2010.

[265] L. Petit, N. Carlie, B. Zdyrko, *et al.*, "Development of novel integrated bio/chemical sensor systems using chalcogenide glass materials," in *2nd International Symposium on Transparent Conducting Oxides*, vol. 6, no. 9, pp. 799–815, 2009.

[266] Y. Zou, H. Lin, O. Ogbuu, *et al.*, "Effect of annealing conditions on the physio-chemical properties of spin-coated As_2Se_3 chalcogenide glass films," *Opt. Mater. Express*, vol. 2, no. 12, pp. 1723–1732, 2012.

[267] S. Novak, P. T. Lin, C. Li, *et al.*, "Electrospray deposition of uniform thickness Ge23Sb7S70 and As40S60 chalcogenide glass films," *Jove-Journal Vis. Exp.*, no. 114, 2016.

[268] Y. L. Zha, M. Waldmann, and C. B. Arnold, "A review on solution processing of chalcogenide glasses for optical components," *Opt. Mater. Express*, vol. 3, no. 9, pp. 1259–1272, 2013.

[269] Y. Zou, D. Zhang, H. Lin, *et al.*, "High-performance, high-index-contrast chalcogenide glass photonics on silicon and unconventional non-planar substrates," *Adv. Opt. Mater.*, vol. 2, no. 5, pp. 478–486, 2014.

[270] M. V Kunnavakkam, F. M. Houlihan, M. Schlax, *et al.*, "Low-cost, low-loss microlens arrays fabricated by soft-lithography replication process," *Appl. Phys. Lett.*, vol. 82, no. 8, pp. 1152–1154, 2003.

[271] W. Chen, R. H. W. Lam, and J. Fu, "Photolithographic surface micromachining of polydimethylsiloxane (PDMS)," *Lab Chip*, vol. 12, no. 2, pp. 391–395, 2012.

[272] L. Li, H. Lin, J. Michon, *et al.*, "A new twist on glass: a brittle material enabling flexible integrated photonics," *Int. J. Appl. Glas. Sci.*, vol. 8, no. 1, pp. 61–68, 2017.

[273] J.-Y. Sun, N. Lu, J. Yoon, K.-H. Oh, Z. Suo, and J. J. Vlassak, "Inorganic islands on a highly stretchable polyimide substrate," *J. Mater. Res.*, vol. 24, no. 11, pp. 3338–3342, 2009.

[274] J. A. Rogers, T. Someya, and Y. Huang, "Materials and mechanics for stretchable electronics," *Science*, vol. 327, no. 5973, pp. 1603–1607, 2010.

[275] D.-H. Kim, R. Ghaffari, N. Lu, and J. A. Rogers, "Flexible and stretchable electronics for biointegrated devices," *Annu. Rev. Biomed. Eng.*, vol. 14, pp. 113–128, 2012.

[276] D. Qin, Y. Xia, and G. M. Whitesides, "Soft lithography for micro- and nanoscale patterning," *Nat. Protoc.*, vol. 5, p. 491, 2010.

[277] W. Li, H. Ge, W. Wu, *et al.*, "Hybrid nanoimprint−soft lithography with sub-15 nm resolution," *Nano Lett.*, vol. 9, no. 6, pp. 2306–2310, 2009.

[278] K. E. Paul, M. Prentiss, and G. M. Whitesides, "Patterning spherical surfaces at the two-hundred-nanometer scale using soft lithography," *Adv. Funct. Mater.*, vol. 13, no. 4, pp. 259–263, 2003.

[279] R. J. Jackman, J. L. Wilbur, and G. M. Whitesides, "Fabrication of submicrometer features on curved substrates by microcontact printing," *Science*, vol. 269, no. 5224, pp. 664–666, 1995.

[280] D. Zhang, W. Yu, T. Wang, Z. Lu, and Q. Sun, "Fabrication of diffractive optical elements on 3-D curved surfaces by capillary force lithography," *Opt. Express*, vol. 18, no. 14, pp. 15009–15016, 2010.

[281] K. Thomas, H. Lukas, B. Iris, and M. Michael, "Nanoimprint lithography on curved surfaces prepared by fused deposition modelling," *Surf. Topogr. Metrol. Prop.*, vol. 3, no. 2, p. 24003, 2015.

[282] B. Farshchian, A. Amirsadeghi, S. M. Hurst, J. Wu, J. Lee, and S. Park, "Soft UV-nanoimprint lithography on non-planar surfaces," *Microelectron. Eng.*, vol. 88, no. 11, pp. 3287–3292, 2011.

[283] B. Li, J. Zhang, and H. Ge, "A sandwiched flexible polymer mold for control of particle-induced defects in nanoimprint lithography," *Appl. Phys. A*, vol. 110, no. 1, pp. 123–128, 2013.

[284] A. M. Bowen and R. G. Nuzzo, "Fabrication of flexible binary amplitude masks for patterning on highly curved surfaces," *Adv. Funct. Mater.*, vol. 19, no. 20, pp. 3243–3253, 2009.

[285] H. Huan, Y. Junghoon, M. Glennys, C. Yaofeng, A. S. Mark, and P. K. William, "Nano-fabrication with a flexible array of nano-apertures," *Nanotechnology*, vol. 23, no. 17, p. 175303, 2012.

[286] J. G. Kim, N. Takama, B. J. Kim, and H. Fujita, "Optical-softlithographic technology for patterning on curved surfaces," *J. Micromechanics Microengineering*, vol. 19, no. 5, p. 55017, 2009.

[287] J. Park, H. Fujita, and B. Kim, "Fabrication of metallic microstructure on curved substrate by optical soft lithography and copper electroplating," *Sensors Actuators A Phys.*, vol. 168, no. 1, pp. 105–111, 2011.

[288] A. M. Bowen, J. A. Ritchey, J. S. Moore, and R. G. Nuzzo, "Programmable chemical gradient patterns by soft grayscale lithography," *Small*, vol. 7, no. 23, pp. 3350–3362, 2011.

[289] M. Duocastella and C. B. Arnold, "Enhanced depth of field laser processing using an ultra-high-speed axial scanner," *Appl. Phys. Lett.*, vol. 102, no. 6, p. 061113, 2013.

[290] D. Radtke and U. D. Zeitner, "Laser-lithography on non-planar surfaces," *Opt. Express*, vol. 15, no. 3, pp. 1167–1174, 2007.

[291] M. Duocastella and C. B. Arnold, "Bessel and annular beams for materials processing," *Laser Photon. Rev.*, vol. 6, no. 5, pp. 607–621, 2012.

[292] P. Mouroulis, D. W. Wilson, P. D. Maker, and R. E. Muller, "Convex grating types for concentric imaging spectrometers," *Appl. Opt.*, vol. 37, no. 31, pp. 7200–7208, 1998.

[293] J. Zhang, B. Shokouhi, and B. Cui, "Tilted nanostructure fabrication by electron beam lithography," *J. Vac. Sci. Technol. B*, vol. 30, no. 6, p. 06F302, 2012.

[294] F. Ranveig, A. Akshay, H. Richard, M. G. Martin, H. Bodil, and K. B. Karl, "Exploring proximity effects and large depth of field in helium ion beam lithography: large-area dense patterns and tilted surface exposure," *Nanotechnology*, vol. 29, no. 27, p. 275301, 2018.

[295] J. Chang, Q. Zhou, and A. Zettl, "Facile electron-beam lithography technique for irregular and fragile substrates," *Appl. Phys. Lett.*, vol. 105, no. 17, p. 173109, 2014.

[296] J. Linden, C. Thanner, B. Schaaf, S. Wolff, B. Lägel, and E. Oesterschulze, "Spray coating of PMMA for pattern transfer via electron beam lithography on surfaces with high topography," *Microelectron. Eng.*, vol. 88, no. 8, pp. 2030–2032, 2011.

[297] J. Zhang, C. Con, and B. Cui, "Electron beam lithography on irregular surfaces using an evaporated resist," *ACS Nano*, vol. 8, no. 4, pp. 3483–3489, 2014.

[298] C. Con, J. Zhang, and B. Cui, "Nanofabrication of high aspect ratio structures using an evaporated resist containing metal," *Nanotechnology*, vol. 25, no. 17, 2014.

[299] P. W. Leech, N. Wu, and Y. Zhu, "Application of dry film resist in the fabrication of microfluidic chips for droplet generation," *J. Micromechanics Microengineering*, vol. 19, no. 6, p. 65019, 2009.

[300] L. Duanhui, L. Lan, J. Bradley, *et al.*, "Wafer integrated micro-scale concentrating photovoltaics," *Prog. Photovoltaics Res. Appl.*, vol. 26, no. 8, pp. 651–658, 2018.

[301] R. H. Pedersen, M. Hamzah, S. Thoms, P. Roach, M. R. Alexander, and N. Gadegaard, "Electron beam lithography using plasma polymerized hexane as resist," *Microelectron. Eng.*, vol. 87, no. 5, pp. 1112–1114, 2010.

[302] I. R. Peterson, "Langmuir–Blodgett electron-beam resists," *IEE Proc. I - Solid-State Electron Devices*, vol. 130, no. 5, pp. 252–255, 1983.

[303] G. M. King, G. Schürmann, D. Branton, and J. A. Golovchenko, "Nanometer patterning with ice," *Nano Lett.*, vol. 5, no. 6, pp. 1157–1160, 2005.

[304] P. Ruchhoeft and J. C. Wolfe, "Optimal strategy for controlling linewidth on spherical focal surface arrays," *J. Vac. Sci. Technol. B*, vol. 18, no. 6, pp. 3185–3189, 2000.

[305] P. Ruchhoeft, M. Colburn, B. J. Choi, *et al.*, "Patterning curved surfaces: template generation by ion beam proximity lithography and relief transfer by step and flash imprint lithography," *J. Vac. Sci. Technol. B*, vol. 17, no. 6, pp. 2965–2969, 1999.

[306] D. Parikh, B. Craver, H. N. Nounu, F. O. Fong, and J. C. Wolfe, "Nanoscale pattern definition on nonplanar surfaces using ion beam proximity lithography and conformal plasma-deposited resist," *J. Microelectromechanical Syst.*, vol. 17, no. 3, pp. 735–740, 2008.

[307] A. Ricciardi, A. Crescitelli, P. Vaiano, *et al.*, "Lab-on-fiber technology: a new vision for chemical and biological sensing," *Analyst*, vol. 140, no. 24, pp. 8068–8079, 2015.

[308] E. J. Smythe, M. D. Dickey, J. Bao, G. M. Whitesides, and F. Capasso, "Optical antenna arrays on a fiber facet for in situ surface-enhanced Raman scattering detection," *Nano Lett.*, vol. 9, no. 3, pp. 1132–1138, 2009.

[309] D. J. Lipomi, R. V. Martinez, M. A. Kats, *et al.*, "Patterning the tips of optical fibers with metallic nanostructures using nanoskiving," *Nano Lett.*, vol. 11, no. 2, pp. 632–636, 2010.

[310] P. Jia and J. Yang, "Integration of large-area metallic nanohole arrays with multimode optical fibers for surface plasmon resonance sensing," *Appl. Phys. Lett.*, vol. 102, no. 24, p. 243107, 2013.

[311] X. Zhang, F. Liu, and Y. Lin, "Direct transfer of metallic photonic structures onto end facets of optical fibers," *Front. Phys.*, vol. 4, p. 31, 2016.

[312] G. Shambat, J. Provine, K. Rivoire, T. Sarmiento, J. Harris, and J. Vučković, "Optical fiber tips functionalized with semiconductor photonic crystal cavities," *Appl. Phys. Lett.*, vol. 99, no. 19, p. 191102, 2011.

[313] I. W. Jung, B. Park, J. Provine, R. T. Howe, and O. Solgaard, "Highly sensitive monolithic silicon photonic crystal fiber tip sensor for simultaneous measurement of refractive index and temperature," *J. Light. Technol.*, vol. 29, no. 9, pp. 1367–1374, 2011.

[314] B. Wang, T. Siahaan, Dündar, *et al.*, "Photonic crystal cavity on optical fiber facet for refractive index sensing," *Opt. Lett.*, vol. 37, no. 5, pp. 833–835, 2012.

[315] M. Boerkamp, Y. Lu, J. Mink, Ž. Zobenica, and R. W. van der Heijden, "Multiple modes of a photonic crystal cavity on a fiber tip for multiple parameter sensing," *J. Light. Technol.*, vol. 33, no. 18, pp. 3901–3906, 2015.

[316] L. Li, I. Bayn, M. Lu, *et al.*, "Nanofabrication on unconventional substrates using transferred hard masks," *Sci. Rep.*, vol. 5, p. 7802, 2015.

[317] C. L. Arce, K. De Vos, T. Claes, K. Komorowska, D. Van Thourhout, and P. Bienstman, "Silicon-on-insulator microring resonator sensor integrated on an optical fiber facet," *IEEE Photonics Technol. Lett.*, vol. 23, no. 13, pp. 890–892, 2011.

[318] M. Pisco, F. Galeotti, G. Quero, *et al.*, "Nanosphere lithography for optical fiber tip nanoprobes," *Light. Appl.*, vol. 6, 2017.

[319] P. Di Palma, C. Taddei, A. Borriello, *et al.*, "Self-assembled colloidal photonic crystal on the fiber optic tip as a sensing probe," *IEEE Photonics J.*, vol. 9, no. 2, 2017.

[320] Q. Giuseppe, Z. Gianluigi, Managò Stefano, *et al.*, "Nanosphere lithography on fiber: towards engineered lab-on-fiber SERS optrodes," *Sensors*, vol. 18, no. 3, 2018.

[321] R. Bachelot, C. Ecoffet, D. Deloeil, P. Royer, and D. J. Lougnot, "Integration of micrometer-sized polymer elements at the end of optical fibers by free-radical photopolymerization," *Appl. Opt.*, vol. 40, no. 32, pp. 5860–5871, 2001.

[322] P. S. Kelkar and J. Beauvais, "Nano patterning on optical fiber and laser diode facet with dry resist," *J. Vac. Sci. Technol. A*, vol. 22, no. 3, pp. 743–746, 2004.

[323] M. Prasciolu, D. Cojoc, S. Cabrini, *et al.*, "Design and fabrication of on-fiber diffractive elements for fiber-waveguide coupling by means of e-beam lithography," *Microelectron. Eng.*, vol. 67–8, pp. 169–174, 2003.

[324] N. Wang, M. Zeisberger, U. Hubner, and M. A. Schmidt, "Nanotrimer enhanced optical fiber tips implemented by electron beam lithography," *Opt. Mater. Express*, vol. 8, no. 8, pp. 2246–2255, 2018.

[325] V. M. Sundaram and S. B. Wen, "Fabrication of micro-optical devices at the end of a multimode optical fiber with negative tone lift-off EBL," *J. Micromechanics Microengineering*, vol. 22, no. 12, 2012.

[326] J. B. Kim and K. H. Jeong, "Batch fabrication of functional optical elements on a fiber facet using DMD based maskless lithography," *Opt. Express*, vol. 25, no. 14, pp. 16854–16859, 2017.

[327] Y. B. Lin, J. P. Guo, and R. G. Lindquist, "Demonstration of an ultra-wideband optical fiber inline polarizer with metal nano-grid on the fiber tip," *Opt. Express*, vol. 17, no. 20, pp. 17849–17854, 2009.

[328] J. J. Kaufman, E. H. Banaei, A. F. Abouraddy, and Ieee, "Inscription of photonic devices on the tip of a chalcogenide glass fiber," in *Conference on Lasers and Electro-Optics (CLEO)*, 2012.

[329] S. F. Feng, X. P. Zhang, H. Wang, M. D. Xin, and Z. Z. Lu, "Fiber coupled waveguide grating structures," *Appl. Phys. Lett.*, vol. 96, no. 13, 2010.

[330] H. Hemmati, Y. H. Ko, and R. Magnusson, "Fiber-facet-integrated guided-mode resonance filters and sensors: experimental realization," *Opt. Lett.*, vol. 43, no. 3, pp. 358–361, 2018.

[331] E. G. Johnson, J. Stack, T. J. Suleski, C. Koehler, and W. Delaney, "Fabrication of micro optics on coreless fiber segments," *Appl. Opt.*, vol. 42, no. 5, pp. 785–791, 2003.

[332] G. Kostovski, D. J. White, A. Mitchell, M. W. Austin, and P. R. Stoddart, "Nanoimprinted optical fibres: biotemplated nanostructures for SERS sensing," *Biosens. Bioelectron.*, vol. 24, no. 5, pp. 1531–1535, 2009.

[333] G. Calafiore, A. Koshelev, F. I. Allen, *et al.*, "Nanoimprint of a 3D structure on an optical fiber for light wavefront manipulation," *Nanotechnology*, vol. 27, no. 37, p. 375301, 2016.

[334] S. Tabassum, Y. Wang, J. Qu, *et al.*, "Patterning of nanophotonic structures at optical fiber tip for refractive index sensing," in *SENSORS, 2016 IEEE*, 2016, pp. 1–3.

[335] J. Viheriala, T. Niemi, J. Kontio, T. Rytkonen, and M. Pessa, "Fabrication of surface reliefs on facets of singlemode optical fibres using nanoimprint lithography," *Electron. Lett.*, vol. 43, no. 3, pp. 150–151, 2007.

[336] G. Calafiore, A. Koshelev, T. P. Darlington, *et al.*, "Campanile near-field probes fabricated by nanoimprint lithography on the facet of an optical fiber," *Sci. Rep.*, vol. 7, no. 1, p. 1651, 2017.

[337] G. Kostovski, U. Chinnasamy, S. Jayawardhana, P. R. Stoddart, and A. Mitchell, "Sub-15nm optical fiber nanoimprint lithography: a parallel, self-aligned and portable approach," *Adv. Mater.*, vol. 23, no. 4, pp. 531–535, 2011.

[338] S. Scheerlinck, D. Taillaert, D. Van Thourhout, and R. Baets, "Flexible metal grating based optical fiber probe for photonic integrated circuits," *Appl. Phys. Lett.*, vol. 92, no. 3, p. 31104, 2008.

[339] S. Pilevar, K. Edinger, W. Atia, I. Smolyaninov, and C. Davis, "Focused ion-beam fabrication of fiber probes with well-defined apertures for use in near-field scanning optical microscopy," *Appl. Phys. Lett.*, vol. 72, no. 24, pp. 3133–3135, 1998.

[340] A. Dhawan, J. F. Muth, D. N. Leonard, *et al.*, "Focused ion beam fabrication of metallic nanostructures on end faces of optical fibers for chemical sensing applications," *J. Vac. Sci. Technol. B Microelectron. Nanom. Struct. Process. Meas. Phenom.*, vol. 26, no. 6, pp. 2168–2173, 2008.
[341] M. Principe, M. Consales, A. Micco, *et al.*, "Optical fiber meta-tips," *Light Sci. Appl.*, vol. 6, no. 3, p. e16226, 2017.
[342] F. Schiappelli, R. Kumar, M. Prasciolu, *et al.*, "Efficient fiber-to-waveguide coupling by a lens on the end of the optical fiber fabricated by focused ion beam milling," *Microelectron. Eng.*, vol. 73, pp. 397–404, 2004.
[343] C. Ma and A. Wang, "Optical fiber tip acoustic resonator for hydrogen sensing," *Opt. Lett.*, vol. 35, no. 12, pp. 2043–2045, 2010.
[344] F. Wang, W. Yuan, O. Hansen, and O. Bang, "Selective filling of photonic crystal fibers using focused ion beam milled microchannels," *Opt. Express*, vol. 19, no. 18, pp. 17585–17590, 2011.
[345] A. Micco, A. Ricciardi, M. Pisco, V. La Ferrara, and A. Cusano, "Optical fiber tip templating using direct focused ion beam milling," *Sci. Rep.*, vol. 5, p. 15935, 2015.
[346] Y. Fu and N. K. A. Bryan, "Investigation of physical properties of quartz after focused ion beam bombardment," *Appl. Phys. B*, vol. 80, no. 4–5, pp. 581–585, 2005.
[347] H. Lin, L. Li, F. Deng, *et al.*, "Demonstration of mid-infrared waveguide photonic crystal cavities," *Opt. Lett.*, vol. 38, no. 15, pp. 2779–2782, 2013.
[348] J. F. Xing, X. Z. Dong, W. Q. Chen, *et al.*, "Improving spatial resolution of two-photon microfabrication by using photoinitiator with high initiating efficiency," *Appl. Phys. Lett.*, vol. 90, no. 13, p. 131106, 2007.

Index

www.ingramcontent.com/pod-product-compliance
Lightning Source LLC
LaVergne TN
LVHW022313150926
848178LV00003B/56